MANUFACTURING a basic text for industrial arts

James Fales
Everett Sheets
Gregg Mervich
John Dinan

MANUFACTURING: a basic text for industrial arts

James Fales
Everett Sheets
Gregg Mervich
John Dinan

McKNIGHT Publishing Company
Bloomington, Illinois

FIRST EDITION

**Copyright 1980
by McKnight Publishing Company, Bloomington, Illinois**

All rights reserved. No part of this book may be reproduced or utilized in any form or by any means, electronic or mechanical, including photocopying, recording, or by any information storage and retrieval system, without permission in writing from the publisher.

Lithographed in U.S.A.

Library of Congress Catalog Card Number: 79-91205

SBN: 87345-586-X

Editors : Robert W. Todd
Carole Fletcher

Design : Jim Coventry

Art : Stan Crum
Howard Davis
Jack Kershner
Roger Herberts

Copy Editing : Carole Fletcher

Layout : Elizabeth Purcell

Production : Sandy Baker
Sandy Savage

To the Students —

Are you ready to develop skills in working with tools and materials? Can you accept the challenge of planning and making a product yourself? Would you like to work with others and put together several products for your family and friends? Or how about making some products to sell to others? Studying this book and using the plans and suggestions given in it will help you to do any or all of these things. It will help you to make products. That's what **manufacturing** is — making products.

While you are learning to make products, you will also be learning about the industry in the United States that makes products. That is manufacturing, too — the **manufacturing industry**. Many of the procedures and processes used in the industry are the same as those used in industrial arts. They may be done by more people using larger, more complicated machines, but still they are basically the same, as you will see.

Many of you will be reading, studying, and using all parts of this book in one school term. Others will use parts of it in several school terms.

The book is divided into five units. Unit I is the **Introduction to Manufacturing** — both the industry and the practices. You will learn the basics of the manufacturing industry and the tools, materials, processes, and safety procedures that are basic to making products. Would you like to make a skateboard or a belt buckle? Choose these or any of eighteen other products suggested for this unit.

Unit II is **Research and Development**. This unit tells how product ideas are made into products. You'll read about interesting research done in the industry. You'll learn how product decisions are made there, and you will learn to make them for yourself. You may design and build your own product or choose from five product suggestions for this unit.

Unit III is **Production**. This is producing products. How is it done in the industry? How will you do it? What product will you choose? A rocket? A toolbox? Sixteen more products are also suggested for you.

Plans for products are given in Unit IV, **Product Plans**. Forty-three plans are included in this unit. They offer you the opportunity to learn the processes used to make products. You will gain experience in working with different materials and tools. Some of the products are for fun, such as the Pet Rooster and the Target Game. Others are useful, such as the Salt and Pepper Shakers and the Screwdriver. There's even a plan for a Gizmo! All of the products can be interesting to make.

Instructions for using the product plans are given in **Let's Go to Work** following Chapter 4, **Workers and Safety**. Read both the instructions and the chapter **before** you try to make a product. You **must** understand what to do and how to work safely in order to be successful in making products.

Unit V will help you to **know** what to do. This unit is **Tools, Materials, Processes**. Do you need to drill a hole? Read Topic 6, **Drilling and Boring.** You will learn about drilling tools and how to use them. Would a bending jig make your work easier? Read Topic 19, **Metal Forging and Bending.** You will learn about metal bending processes. Should a race car be made from wood or plastic? Read Topic 26, **Woods,** and Topic 28, **Plastics,** to help you decide. A total of 29 topics are presented in this unit.

This, then, is what our textbook contains. And yet it's only a glimpse of all that's presented. Besides the basic information, you'll read about real happenings in manufacturing — the inventors and the inventions, the designers and the designs, the manufacturers and the products. For this book is about manufacturing — as it was yesterday, as it might be tomorrow, but mostly what it is today, in industry and in industrial arts.

— The Authors

TABLE of CONTENTS

PHOTO ESSAY Manufacturing and American Industries . . . 1

UNIT I Introduction to Manufacturing . . . 13

CHAPTER 1 What is Manufacturing . . . 14

☐ **The Manufacturing Industry 14** ☐ **Tools, Materials, and Processes 15** ☐ **Types of Production 15** ▫ Custom Production ▫ Job-Lot Production ▫ Line Production ☐ **Manufacturing Must Be Planned 18** ▫ Identifying Consumer Demand ▫ Designing and Engineering Products ▫ Planning Production ▫ Tooling-Up for Production ▫ Planning Quality Control ☐ **Production Procedure 22** ▫ Converting Raw Materials ▫ Making Standard Stock ▫ Making Components ▫ Assembling Components into Subassemblies ▫ Making Finished Products ▫ Preparing for Distribution ☐ **Looking Ahead** ☐ **New Terms** ☐ **Study Guide**

CHAPTER 2 The Development of Manufacturing . . . 26

☐ **How Manufacturing Developed 26** ▫ Bartering and the Money System ▫ The Industrial Revolution ▫ Working Conditions in the Factory ☐ **Manufacturing in America 29** ▫ A New Process ▫ Interchangeable Parts ▫ The American System ▫ The Assembly Line ▫ Other Important Developments ▫ Improved Planning ☐ **Recent Developments 33** ▫ Automation ▫ Computers ▫ Recycling and Energy Conservation ☐ **Manufacturing in the Future 35** ▫ Use of Outer Space ▫ Better Use of Resources ▫ Importance of Future Changes ☐ **Looking Ahead** ☐ **New Terms** ☐ **Study Guide**

CHAPTER 3 How is Manufacturing Done . . . 38

☐ **What Is Needed for Manufacturing? 38** ☐ **Tools 38** ▫ Basic Machines ▫ Hand Tools ▫ Portable Power Tools ▫ Power Machines and Equipment ▫ Industrial Machines and Equipment ☐ **Materials 42** ▫ Basic Types of Materials ▫ Wood ▫ Metal ▫ Plastics ▫ Other Materials ☐ **Processes 46** ☐ **Forming Processes 47** ▫ Casting or Molding ▫ Compressing or Stretching ▫ Conditioning ☐ **Separating Processes 49** ▫ Shearing ▫ Chip Removing ▫ Other Separating Processes ☐ **Combining Processes 49** ▫ Mixing ▫ Coating ▫ Bonding ▫ Mechanical Fastening ☐ **Looking Ahead** ☐ **New Terms** ☐ **Study Guide**

CHAPTER 4 Workers and Safety ... 52

☐ **People – The Most Important Input 52** ☐ **Worker Attitudes 52** ☐ Promptness ☐ Regular Attendance ☐ Cooperation ☐ Reliability ☐ Persistence ☐ Patience ☐ Advantages of Good Attitudes ☐ **People and Safety 54** ☐ Safety in Manufacturing ☐ Safety in the Shop ☐ **Personal Safety Practices 56** ☐ **General Safety Practices 57** ☐ **Hand Tool Safety Practices 58** ☐ **Machine Safety Practices 58** ☐ **Looking Ahead** ☐ **New Terms** ☐ **Study Guide** ☐ **Let's Go To Work: Understanding Product Plans 60** ☐ **Unit I Product Previews 65**

UNIT II Research and Development ... 69

CHAPTER 5 Ideas Are Powerful ... 70

☐ **Where Do New Ideas Come From? 70** ☐ **Never Ending Research and Development 70** ☐ Results of Research and Development ☐ **What are Research and Development? 72** ☐ Stating the Design Problem ☐ Research ☐ Development ☐ **Why Are Research and Development Important? 73** ☐ **Who Does Research and Development? 74** ☐ What Can Be Accomplished in Research and Development? ☐ **Research and Development in Action 76** ☐ The Most Amazing Car Never Built ☐ The Designer and the Design Process ☐ Designing Air Transportation ☐ **Looking Ahead** ☐ **New Terms** ☐ **Study Guide**

CHAPTER 6 Selecting Products ... 80

☐ **Finding a Product 80** ☐ Where are New Product Ideas Found? ☐ Big Ideas from Common People ☐ The Company Idea Teams ☐ Considering Product Ideas ☐ **Consumer Demand 83** ☐ Finding Consumer Demand ☐ **Company Limitations 84** ☐ Knowledge and Skill Limitations ☐ Machine and Tool Limitations ☐ Material Limitations ☐ Space Limitations ☐ Time Limitations ☐ Money Limitations ☐ **Final Selection 86** ☐ **Looking Ahead** ☐ **New Terms** ☐ **Study Guide**

CHAPTER 7 Designing and Engineering Products ... 87

☐ **Designing 87** ☐ The Design Process ☐ **Presenting Design Ideas 88** ☐ Sketches ☐ Mock-Ups ☐ **Making Design Decisions 89** ☐ **Engineering 91** ☐ Drafting the Plans ☐ **The Prototype 93** ☐ Trying Out the Plans ☐ Testing the Prototype ☐ **Final Approval 94** ☐ **Looking Ahead** ☐ **New Terms** ☐ **Study Guide** ☐ **Let's Go To Work: Research and Development 96** ☐ **Unit II Product Preview 102**

UNIT III Production...103

CHAPTER 8 Starting and Organizing Manufacturing Companies...104

☐ **What Is a Company? 104** ☐ Where Do Companies Come From? ☐ Building Better Milk Shakes ☐ Building Better Hamburgers ☐ **Types of Ownership 106** ☐ **Proprietorship 106** ☐ A Successful Proprietorship ☐ **Partnership 107** ☐ A Successful Partnership ☐ **Corporations 107** ☐ A Successful Corporation ☐ The Corporation — A Separate "Person" ☐ Forming a Corporation ☐ **How Does a Company Operate? 110** ☐ Management ☐ Administration ☐ Personnel Department ☐ Research and Development Department ☐ Production Department ☐ Marketing or Sales Department ☐ Everyone Must Cooperate ☐ **Looking Ahead** ☐ **New Terms** ☐ **Study Guide** ☐ **Let's Go To Work: Starting a Manufacturing Company 117**

CHAPTER 9 Planning a Production System...119

☐ **What Is Production Planning? 119** ☐ Types of Production ☐ Production Planning and Product Planning ☐ **Why Is Production Planning Important? 122** ☐ Importance of Costs ☐ **Production Planning 124** ☐ Product Analysis ☐ The Production Flow Chart ☐ Cost Considerations ☐ Operation Sheets ☐ The Plant Layout ☐ **Planning a Materials-Handling System 128** ☐ Conveyors ☐ Other Transporting Methods ☐ **Looking Ahead** ☐ **New Terms** ☐ **Study Guide** ☐ **Let's Go To Work: Production Planning 130**

CHAPTER 10 Getting Ready to Manufacture...132

☐ **Ordering Materials and Supplies 132** ☐ Sources of Supply ☐ Make It or Buy It? ☐ Preparing to Order ☐ Inventory ☐ **Estimating Production Time 133** ☐ Timing Activities ☐ Work Measurement ☐ **Estimating Costs 134** ☐ Material Costs ☐ Labor Cost ☐ Overhead ☐ The Total Cost ☐ Lowering Production Costs ☐ **Tooling-Up 137** ☐ Jigs and Fixtures ☐ Molds and Dies ☐ Setting Up ☐ **Hiring and Training Workers 140** ☐ Job Openings ☐ Applying for Jobs ☐ The Job Application ☐ The Interview ☐ Hiring and Training ☐ Unions ☐ Successful Workers ☐ **Looking Ahead** ☐ **New Terms** ☐ **Study Guide** ☐ **Let's Go To Work: Getting Ready for Production 143**

CHAPTER 11 Producing the Products...145

☐ **Where Does Production Begin? 145** ☐ **Quality Control 146** ☐ Purpose of Quality Control ☐ **Inspecting 146** ☐ Material Inspection ☐ Inspection During Production ☐ Use of Measuring Devices ☐ Visual Inspection ☐ Role of the Inspector ☐ Final Inspection ☐ Correcting Problems ☐ **The Trial Run 149** ☐ **The Production Run 150** ☐ Producing Components ☐

Preparing for Assembly □ Assembling Products □ **Controlling Production 153** □ The Orders □ The Schedule □ The Assembly Lines □ Corrections During Assembly □ Working Together □ **Inventory Control 154** □ Categories of Inventory □ Inventory Records □ **Packaging and Storing 155** □ Why Package a Product? □ Storage □ **Looking Ahead** □ **New Terms** □ **Study Guide** □ **Let's Go To Work: Producing the Product 158**

CHAPTER 12 Marketing: Profit or Loss . . . 159

□ **Marketing 159** □ **Market Research 159** □ **Advertising 161** □ Newspapers: Immediacy □ Magazines: Permanency □ Radio: Drama □ Television: Action □ Billboards: Impact □ Other Means of Advertising □ A Special Way to Advertise □ **Sales 164** □ **Distribution 165** □ **Figuring Profit or Loss 165** □ Keeping Accurate Records □ Costs □ Income □ Loss □ Profit □ The Importance of Profit in the Manufacturing Industry □ **The American Economic System 169** □ Supply and Demand □ Free Enterprise □ Laws and Free Enterprise □ **Looking Back** □ **New Terms** □ **Study Guide** □ **Let's Go To Work: Marketing 172** □ **Unit III Product Preview 173**

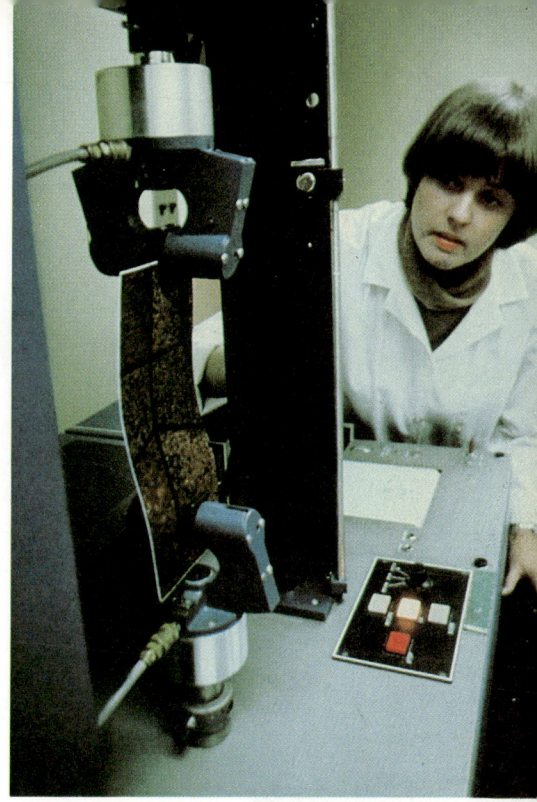

UNIT IV Product Plans . . . 177

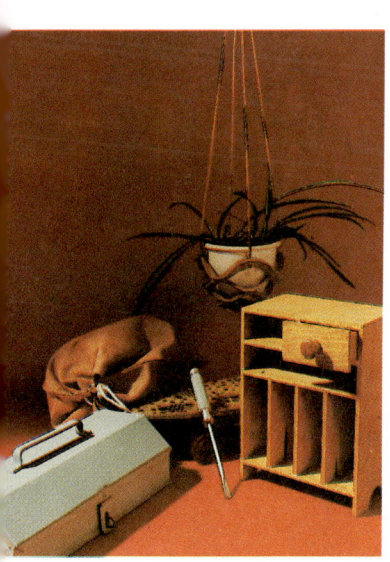

Wood Products

Whistle . . . 178
Xylobox . . . 180
Jump-A-Peg . . . 182
Salt and Pepper Shakers . . . 184
Super Q . . . 186
Note Holder . . . 188
Marble Drop . . . 190
Pet Rooster . . . 192
Handy Dandy Tennis . . . 196
Paper Towel Stand . . . 198
Candle Holder . . . 202
Cutting Board . . . 204
Marble Shoot . . . 206
Plant Stand . . . 210
Coaster Set . . . 212
Target Game . . . 216
Desk Caddy . . . 220
Mirror Shelf . . . 224
Gumball Machine . . . 228

Metal Products

Name Badge . . . 232
Bike Beverage Holder . . . 234
Belt Buckle . . . 236
Charcoal Tongs . . . 240
Planter . . . 244
Screwdriver . . . 246
Toolbox . . . 248
Desk Lamp . . . 252

Plastic Products

Calculator Stand . . . 256
Letter Opener . . . 258
Gizmo . . . 260
Funnel . . . 262
Trivet . . . 266
Skateboard . . . 268

Products Made of Other Materials

Tote Bag . . . 272
Shop Apron . . . 274
Plant Holder . . . 276
Camera Case . . . 278
Rocket . . . 282

Products to Research and Develop

Race Car . . . 286
Pinball . . . 288
Sports Equipment Rack . . . 290
Penny Sports . . . 292
Locker Organizer . . . 294

UNIT V Tools, Materials, Processes . . . 295

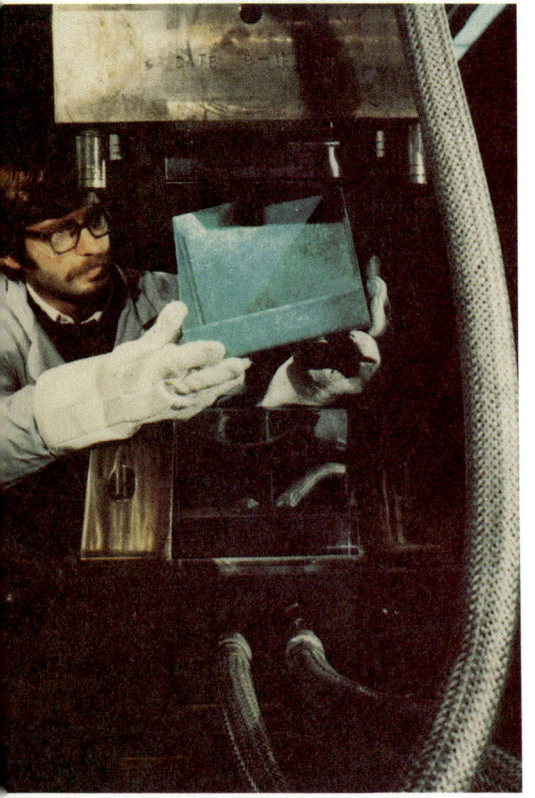

TOPIC

1 Metrics . . . 296
2 Layout . . . 300
3 Bench Tools . . . 305
4 Shearing . . . 308
5 Sawing . . . 311
6 Drilling and Boring . . . 318
7 Wood Planing, Shaping, and Turning . . . 322
8 Metal Turning . . . 330
9 Threaded Fasteners and Taps & Dies . . . 334
10 Non-Threaded Fasteners and Fastening Tools . . . 338
11 Glues, Cements, and other Adhesives . . . 343
12 Soldering . . . 347
13 Oxyacetylene Brazing, Welding, and Cutting . . . 350
14 Electric Welding . . . 355
15 Filing and Sanding . . . 358
16 Grinding . . . 365
17 Induced Fracture and Etching . . . 367
18 Metal Casting . . . 369
19 Metal Forging and Bending . . . 375
20 Plastic Casting, Molding, and Forming . . . 380
21 Stamping Tools . . . 385
22 Heat Treating . . . 388
23 Finishing . . . 390
24 Clamping . . . 398
25 Production Tooling . . . 403
26 Woods . . . 410
27 Metals . . . 414
28 Plastics . . . 420
29 Other Materials . . . 424

Photo Notes . . . 430

Index . . . 432

It's early morning. Like drowsy giants, cities awaken. Millions of cars full of workers begin streaming onto the roads of America...

MANUFACTURING and AMERICAN INDUSTRIES

Charles Moore, Black Star

MANUFACTURING and AMERICAN INDUSTRIES

AMERICA GOES TO WORK Union Carbide Corporation
In a shower of sparks like a fireworks display, this worker in a manufacturing plant is helping to shape one of the millions of products produced every day in the United States. Individuals and industries depend on manufactured products.

. . . Truck drivers pull out of warehouses in Ohio and Michigan. Their trucks are loaded with spools of wire and stacks of wallboard. Typists and telephone crews in Tennessee begin work. Plumbers and carpenters in Colorado take up their tools and head for construction sites. Pilots dock ore boats in harbors along the Great Lakes to begin unloading. In factories and mills from Seattle to New York, workers start up machines that will cut, drill, sew, or form industrial materials. In millions of places, Americans roll up their sleeves and go to work.

The endless activities of a busy nation center around industries. **The purpose of industries is to provide for people's needs and wants.** Millions of people must have food, shelter, and clothing. They need the right tools and supplies to do their work. They need homes that are heated and lighted. They also want such items as cars, television sets, and golf clubs. These things make life easier and more fun. To fill these needs and others, industries expand. They search. As a result, many marvels are produced. Satellites are produced that foretell the weather, and computers are made that do the work of many people. Living longer becomes a promise with miracle medicines. Living better becomes the promise of industries. Food is ever more plentiful. Replacements for body parts are made. Industries provide products and services to meet all needs.

There are several kinds of industries. Generally, they may be referred to as manufacturing, mining,

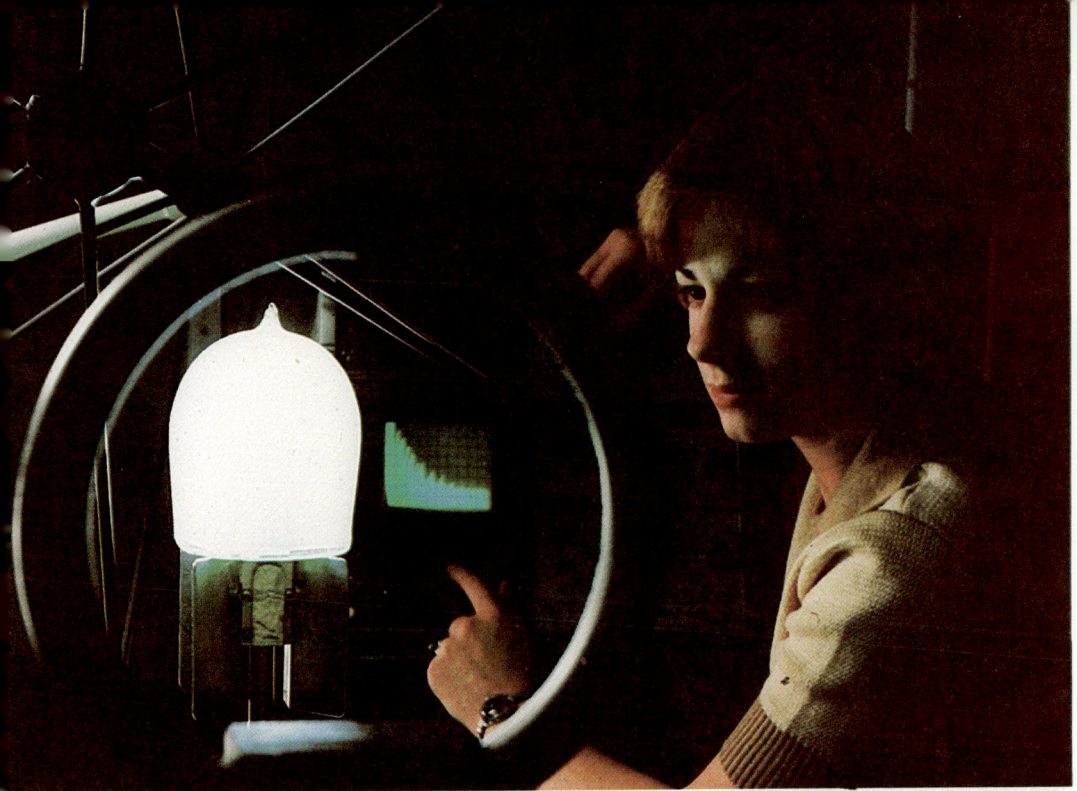

General Electric

EXPLORING NEW IDEAS

Products begin with ideas. People in manufacturing try to find new, improved ways of doing things. Here, in an Ohio research center, the search is on for better, less costly ways of producing light.

Dan McCoy, Rainbow

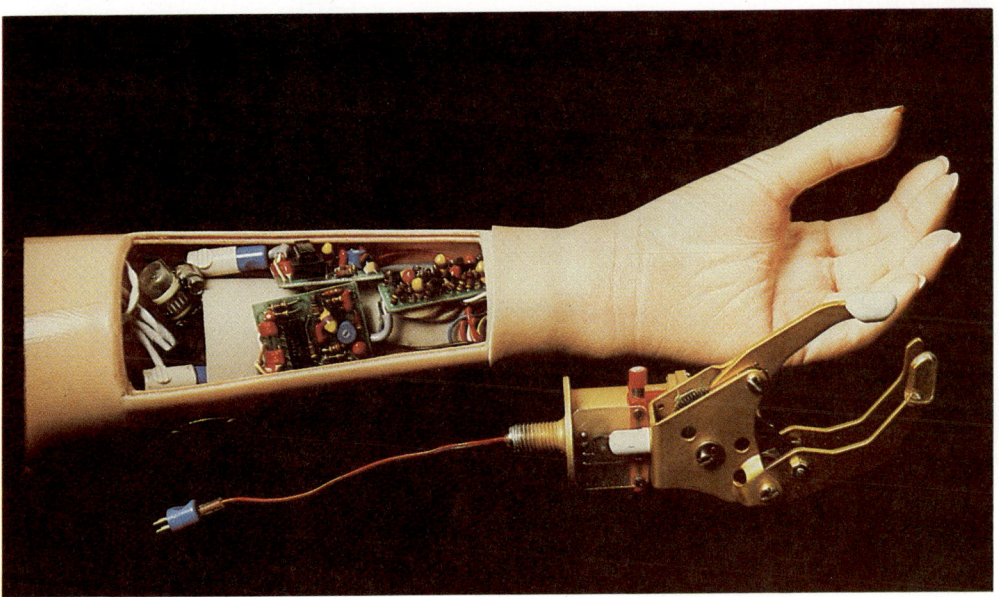

HELPING THOSE WHO NEED HELP

Amazing progress is being made in replacing lost or damaged body parts. Shown here is a replacement forearm. It was developed at the New York Institute of Rehabilitation Medicine. Other artificial parts are being manufactured. One man can actually see through replacement eye parts!

MINING FOR EARTH'S TREASURES

Allied Chemical Corporation

The earth does not give up its riches easily. Minerals must be dug, drilled, or blasted loose. Deep in the tunnels of this Wyoming mine, a machine operator loads mineral ore that is the raw material for many products.

PRODUCING THE PRODUCTS

Steel wheels for the nation's trains are made stronger by being heated. They must take a pounding as they speed products and passengers from coast to coast.

United States Steel Corporation

agriculture, lumbering, service industries, such as communications and transportation, and construction. **Manufacturing is central to all industries**. Through manufacturing, all of them have the machinery and tools they need. Machinery gives workers thousands of extra hands and arms. It gives them the power of millions of horses. It gives them the ability to perform magic with materials.

Industries are very different. Yet they are bound together because they must depend on each other. Each industry does its own work. But each needs the products and services of other industries. In the manufacturing industry, for example, steel companies produce huge amounts of metal. Other manufacturers use it to make many different products. It may be made into machinery, nuts and bolts, wire, pipes, safety pins, or other items. The construction industry uses steel beams to build bridges. The transportation industry uses vehicles made of steel, such as trucks and trains. Towers made of steel help the communications industry to pass along messages. Steel equipment used by power companies sends electricity across the land.

To fill all needs, steel companies turn to other manufacturers for products. They must have fireproof bricks to line blast furnaces. Cranes are used to unload ore from ships. Storage bins hold raw materials. Workers must be supplied with safety goggles and clothing that will protect them. As you can see, steel companies need the products of other manufacturers just as other manufacturers need steel.

Steel companies depend on other industries. Raw materials are transported to the companies. Finished products are shipped out. Steel plants must have power and light. Companies must communicate with other companies and with customers.

Each industry must serve and be served by other industries. Thus, through manufacturing and through this vast dependence on each other, industries meet the needs of the nation.

The earth holds great treasures. Coal, oil, and minerals are raw materials taken from the earth. Industries turn raw materials into power and products.

Coal has many uses. It is a source of energy. It is burned to change water into steam. This steam drives the giant engines that produce electrical power. In turn, electricity serves all industries. Chemicals made from coal are used by manufacturing companies. Products such as perfumes, explosives, and paints are also made from coal.

From oil are manufactured lubricants, chemicals, and fuels. Gasoline and other liquid fuels are energy sources. They power airplanes,

International Paper Company

SEARCHING FOR RAW MATERIALS
Oil and natural gas wait far below. Determined workers force drilling equipment deep into the earth in search of these raw materials. Homes and industries everywhere need heat and power.

cars, and ships. They drive millions of machines in factories, in mills, and on farms.

From the storehouse of oil, chemical companies have unlocked thousands of useful products. Their chemical stews burbling in giant steel tanks help produce such products as synthetic rubber and plastic coatings. Some chemicals are used in making nylon fibers for carpets. Others are used to produce plastics for aircraft windows, face shields, and eyeglass lenses. ==The mining industry provides raw materials, and manufacturing makes products.== Their combined efforts can build giant rockets. Chemical products help the rockets to withstand the heat, cold, and strange forces beyond the earth's atmosphere.

Chemical products are also used on farms. Without them the world would suffer an instant shortage of food. But chemical fertilizers are tilled into farmlands. Harvests are abundant.

REAPING THE HARVEST

Deere and Company

Giant combines rumble through America's fields, harvesting golden grain. About two billion bushels of wheat are produced each year in the United States. Agriculture depends on manufactured machines and equipment to harvest, transport, and process the bountiful crops.

Farm machinery is manufactured. American agriculture relies on machines. In this way, farms are much like modern factories. There are tractor-drawn plows, harrows, and cultivators. There are power-driven combines. Attachments and rigs dig furrows and plant seeds. They fertilize the soil. They cover the seeds and press them into the earth. Chemicals are sprayed to keep the weeds from sprouting. Some machines perform all of these steps in a single operation. They are truly marvels of manufacturing.

The agricultural industry is typical of all industries. It depends on the products, power, and machines of other industries to get its own job done. Take a bottle of ketchup, for instance. A California farmer harvests tons of tomatoes. Trucks carry them to a processing plant. Machines reduce them to a tomato paste. The new product is then loaded into a special railroad tank car. It is shipped to the ketchup manufacturer's plant in Ohio. Garlic bulbs, onions, and red peppers are also harvested, processed, and sent to Ohio. In the meantime, a cargo ship from the Orient docks in Brooklyn. It is carrying black pepper and spices. In Louisiana, a sugar mill crushes sugar cane. The sugar is refined. It is then funneled into heavy sacks for a trip by barge up the Mississippi River.

Getting tomatoes into a ketchup bottle includes nonfarm activities, too. A logging crew in West Virginia cuts pulpwood for a Georgia paper mill. This mill manufactures cases. The ketchup company will use the cases to ship its product. In New York state, a glass factory is making millions of bottles. At the same time in Pennsylvania, a tin-plate rolling mill stamps out metal caps. In Illinois, big presses are printing ketchup labels.

When ingredients and supplies arrive at the Ohio plant, processing begins. When it is finished, the ketchup flows to the automatic filling machines.

FROM THE FIELDS...   Ogden Corporation

Only the best tomatoes will be used in canned or bottled products. Good quality is ever a vital concern in manufacturing.

...TO THE FACTORIES   Tillie Lewis Foods Division

The ketchup is in the bottles, and the bottles are being capped. Soon this product will be on its way to grocery stores, where it will be sold to individuals.

As fast as the bottles are filled, capped, and labeled, they are put in cases. They then travel on a conveyor line into the company's storeroom. In one day, more than 100,000 bottles of ketchup are produced.

The agricultural industry has always depended on energy from the sun to make plants grow. Recently, more interest has been shown in this energy source. People want to use it in other ways. For example, it may be used to heat buildings. People are also working on ways to change sunlight into electricity. Solar energy is "clean." Using it will not put waste materials into the air. **The sun is a source of energy that cannot be used up.**

Trees use energy from the sun to grow. A little fir tree growing in a forest is hardly noticed. Yet, when the tree is harvested at full growth, a

ENERGY FROM THE SKY

The sun that heats the earth can also provide electricity. People have invented and manufactured devices, such as this solar module, that turn sunlight into power. They are especially useful in places far away from other sources of power.

Exxon Corporation

FROM RAW MATERIAL...

Waste material from the manufacture of one product may become the raw material for another. Here a massive "mountain" of sawdust and shavings is waiting to be processed into paper or other products.

Southwest Forest Industries, Snowflake, Arizona

variety of products can be made from it. These products serve many industries. Lumbering companies change wood into planks and plywood. The construction industry uses these products to build houses, offices, and farm buildings. From walnut and cherry trees come choice woods for the manufacture of beautiful tables and chests. Parts of trees go into the making of the see-through wrap used to package many products. Other products made from trees are rayon, chemicals, and dyes. Indeed, few parts of a tree go to waste. Even chips and shavings are made into paper.

Vast amounts of paper are used by communications companies. Publishers use it to produce books, magazines, and newspapers. These feed the hungry minds of millions of people seeking information or entertainment. Huge quantities, too, are made into notebooks, memo pads, and bank checks. Maps, catalogs, games, and greeting cards — there seems no end to the list of items made of paper. The manufacture of these printed products requires the skills of many workers. People operate typesetting machines. They reproduce colorful art and photographs. They run printing presses and binders. Other products are required. Film, colored ink, and glue are supplied by the manufacturing industry.

...TO A FINISHED PRODUCT

Many manufacturers are making wise use of resources. In this Arizona paper mill, 155,000 tons of newsprint paper are produced each year. Sawmill wastes, such as sawdust, and newsprint wastes make up 83 percent of the raw materials used to make the paper. Only 17 percent comes directly from the forests.

Weyerhauser Company

KEEPING AN EAR TO THE WORLD
Looming over the Eagle River valley, this dish antenna has helped to improve communications for the people in Alaska.

The communications industry needs manufactured products for sending and receiving messages. To manufacture them, steel, aluminum, glass, plastic, and many other industrial materials must be available. In return, the communications industry becomes the eyes and ears of the people.

Products such as telephones and radios have made the world seem smaller. Instant voice-to-voice contact is possible. For every thousand people in the U.S., there are 718 telephones and more than 1800 radios.

Radios installed in cars offer news and entertainment. Ship-to-shore radios improve safety at sea. Workers on opposite ends of huge construction projects, factories, or warehouses can talk by two-way radio. Astronauts on the moon talked by radio-telephone with people on earth.

Voices are recorded on manufactured phonograph records and tapes. Using cameras and film, pictures are made. Television and movies combine both sound and pictures. All of these products provide entertainment. And they are sources of information. Business firms, schools, and public agencies often use them in learning programs. These products may be used to store information.

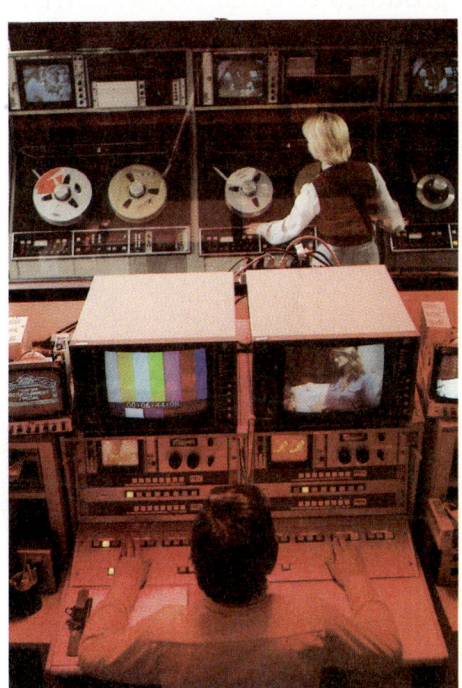

INFORMATION SAVED AND SHARED
History-in-the-making, information, and entertainment can be saved on videotape. These machines are used to tape and play back TV programs for showing on local stations.

SUPPLYING THE POWER

At a heart-fluttering height above the Niagara River, a worker, with safety belt locked to a steel cable, fearlessly checks lines on a transmission tower. These towers were built to carry the power lines that supply electricity to millions of homes and businesses.

Niagara Mohawk Power Corporation

Bill Osmun, Air Transport Association

TRANSPORTING THE PRODUCTS

Like a gigantic whale with open jaws, the nose cone of a cargo plane lifts up to swallow freight for a long-distance flight. Around the clock, all kinds of products are moved on over 13,000 flights made daily by the airplanes.

MINDS AND MACHINES

Tomorrow's products may begin as designs on the TV-like screen of a computer. This designer is using the computer to design a part for a product. When the design is complete, the computer will print drawings that show others how to make the part.

Niagara Mohawk Power Corporation

BUILDING FOR TODAY AND TOMORROW
Nearly 2000 workers, from dozens of trades, are building this nuclear power station on the shore of Lake Ontario. So huge and so complex is this power plant that its construction is taking nine years to complete. It will cost over a billion dollars.

Information can also be stored by computers. Workers in all industries use computers. Computers can give fast, accurate answers to questions. Some solve math problems. Some translate languages. Some design products or give directions. For example, they may control machines that cut logs or put parts together. In time, computers will be regular household equipment. On tiny chips will be stored the information that will cause washing machines and robot vacuum cleaners to clean a home. In every industry, computers will perform wonders as part of the daily work.

Day after day, raw materials and products pour from America's mines, factories, farms, and processing plants. Some of them are whisked away in trucks. Some are bundled into boxcars or poured into tank cars. Or they may be loaded into truck trailers to ride piggyback on the nation's railroads. Still others are lifted into huge cargo planes and flown within hours to Florida, Hawaii, or Alaska.

Were it not for all these ways to move products, grocery shelves would be bare. Homes would be cold. The sick would die for need of medicines. Without transportation, the manufacturing industry could never produce millions of products. These two industries work together to make the American way of life comfortable, enjoyable, and productive.

America has an amazing ability to manufacture products. It also has vast numbers of ships and planes to carry cargo. Combined, these things make America the world's greatest trading nation. In 1977, the United States exported more than $100 billion worth of goods. It helps supply food to hungry people in other countries. It sends machines, tools, and many other products to every part of the globe. Shipping products from farms and factories fills needs in other lands. It also helps American industries to grow and prosper.

The construction industry is important in America. Many homes and other buildings have been built. Whole cities have been raised to meet the needs for homes, businesses, and industries. Plants have been built that provide heat, light, and water. Everything that is fastened to the earth is the work of construction. Workers build big dams to hold back rivers. They build bridges across the waters and ports and piers for ships.

FLYING WORKHORSE
What an eye-popping sight! Industries combine their know-how to do things that seem impossible. Here the shell of a house built in a factory is speedily carried by helicopter to its homesite.

Sikorsky Aircraft

RCA Corporation

AMERICA TRADES WITH THE WORLD
Products made in the fifty states are bought and sold inside the World Trade Center in New York City. Is manufacturing there? Yes — Manufacturing is anyplace that there are products, a part of all industries and a part of our lives.

They criss-cross the country with concrete highways. Rocket-launching sites and radar stations are constructed. Workers build runways and airports for planes. Cities are linked with railroad tracks. Pipelines are laid for gas and oil.

To do so many kinds of work, the construction industry needs manufactured items. Huge earthmoving machinery, such as power shovels, are often used. Cranes are needed to lift concrete slabs into place.

The construction industry depends on transportation. Trucks bring steel beams, lumber, piping, glass, and wiring to the building site. Electrical power is needed to run the machines that cut, saw, and drill. Construction is like all industries. It could not get its jobs done without the products and services of other industries.

Machines, workers, and manufacturing know-how help America produce more products and services than any other country in the world. Whether workers are building skyscrapers high in the sky or drilling for earth's treasures far below, manufacturing is there. It is a major industry in America. Manufacturing makes products for America's people and for others in the world. It makes products for you. This is the purpose of manufacturing. This is manufacturing.

MANUFACTURING and AMERICAN INDUSTRIES

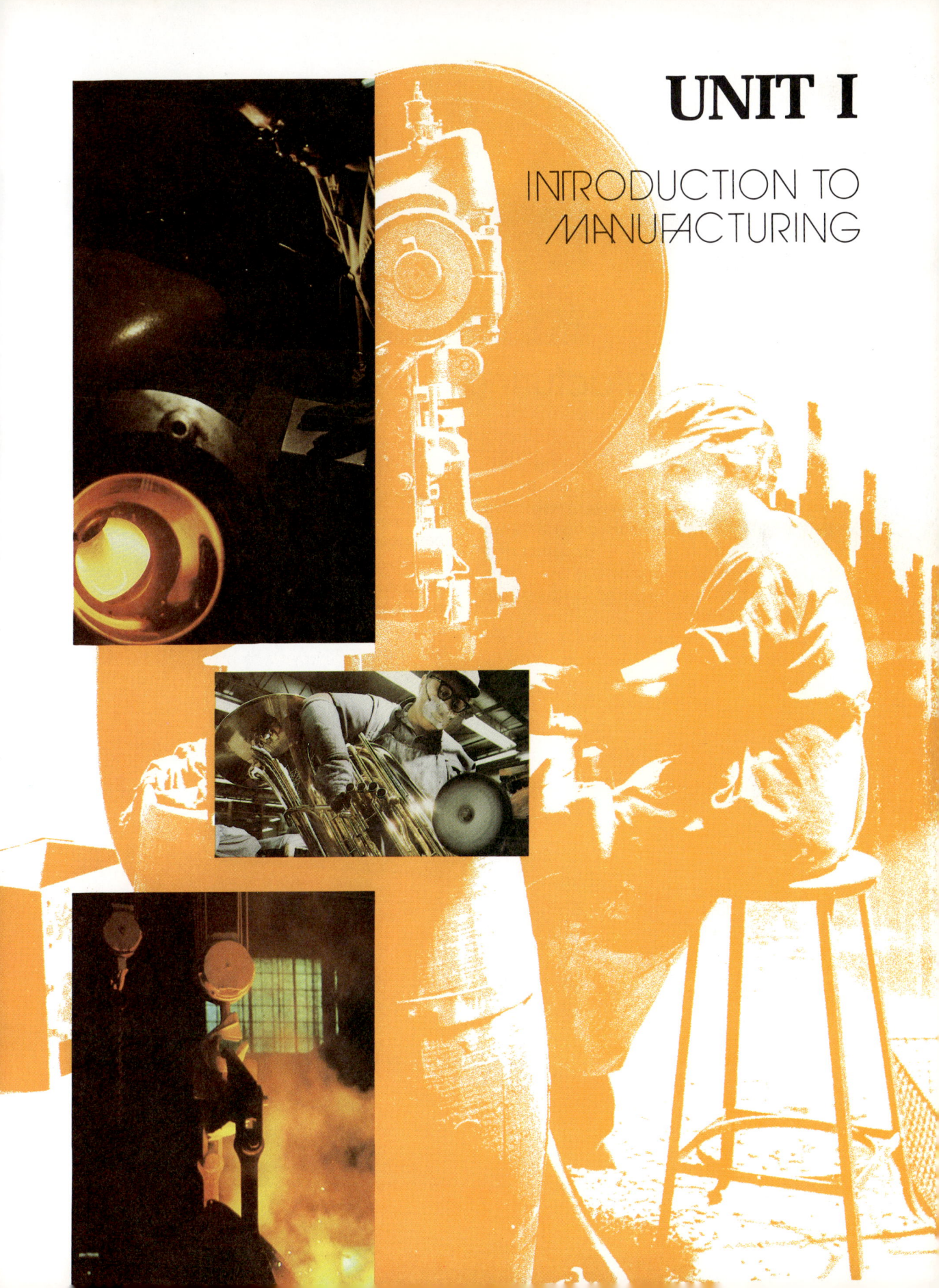

UNIT I

INTRODUCTION TO MANUFACTURING

chapter 1

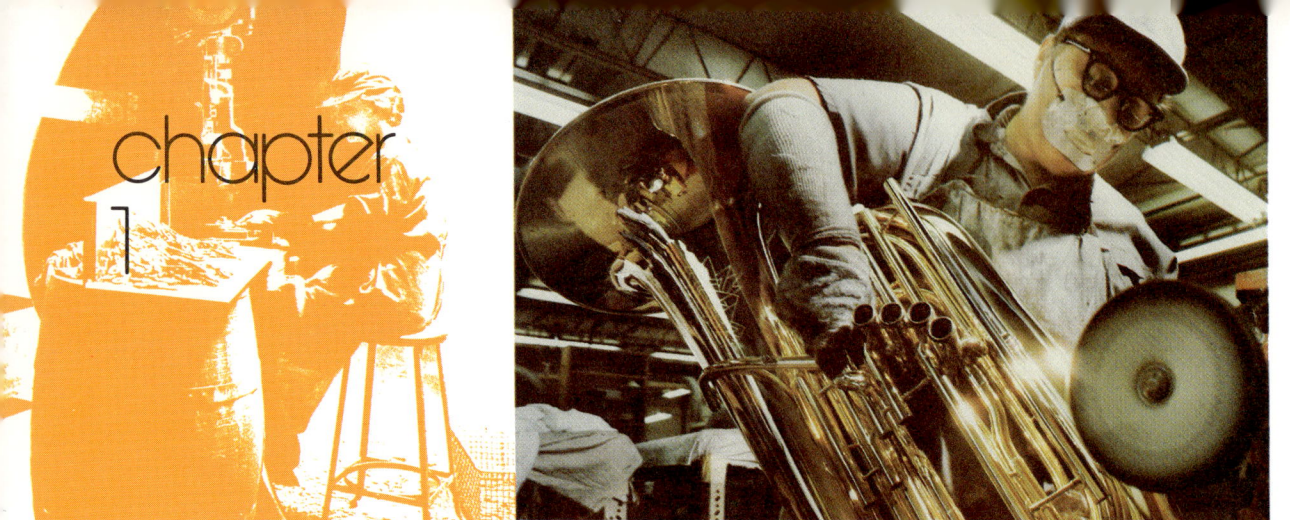

What Is Manufacturing

Look around your classroom. As you look, don't look at the people, look at the objects. Name four things you see around you. You probably just identified four manufactured products. You are sitting in a manufactured chair. You are reading a manufactured book and are wearing manufactured clothes. You use a manufactured pen or pencil to write on manufactured paper. Most of what you see around you has been manufactured. What is manufacturing?

The Manufacturing Industry

Stop for a moment and consider (think about) how manufacturing affects your life. Where did all the manufactured items around you come from? How much do you know and really understand about the production system — the **manufacturing industry** — that makes the products you use and depend on every day? In **Manufacturing: A Basic Text for Industrial Arts,** you will be exploring how the manufac-

Figure 1-1
Imagine how different your life would be without manufactured products.

What Is Manufacturing   15

Figure 1-2

The space shuttle, "Enterprise," was largely researched, developed, and produced by the manufacturing industry. The Enterprise is launched into space from an airplane, but it is able to fly back to earth and land on its own.

The Boeing Company

turing industry makes products for you and for everyone.

In **manufacturing,** products are usually made in special buildings. When the products are finished, they are shipped (moved) to **consumers** (people who will use them). Turning small pieces of plastic into a phonograph record is manufacturing. Assembling (putting together) bicycles and designing and making new cars are manufacturing. Even producing a space vehicle such as the space shuttle is manufacturing. See Figure 1-2.

The manufacturing industry not only supplies many of the products we use, but it is also one of the biggest areas of employment. Over 21 million people in the United States work in jobs related to manufacturing. That is 23% (almost one-fourth) of the **labor force** (people who work). Every year in the United States over one trillion dollars ($1,000,000,000,000) is spent on manufactured products. This is over $4500 for every person.

Tools, Materials, and Processes

In manufacturing, a combination of tools, materials, and processes are used to make products. See Figure 1-3. As you learn about manufacturing, you will see how **materials** are changed. A material such as wood or metal may be changed by sawing, casting, bending, assembling, or other **processes.** During these processes, **tools** are used. When all three things are used according to a carefully prepared plan, the result is a finished product.

This is only a glance at how manufacturing changes materials into products. You will take a closer look in a later chapter.

Types of Production

Almost everywhere you go manufacturing is being done. Manufacturing can be done by a small business with only one person running

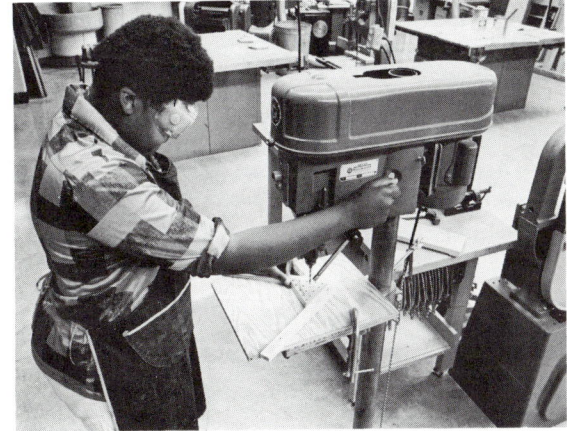

Figure 1-3

In manufacturing, a material (wood) is changed by a person using a tool (drill) to do a process (drilling).

16   Introduction to Manufacturing

Figure 1-6
Wagons were built a few at a time using the job-lot production system. Sometimes different shops produced the various wagon parts. Then, one shop would assemble the parts into wagons.

Smithsonian Institution

Figure 1-5
Primitive people custom-built sleds by tying sticks together with leather.

Figure 1-4
Small as well as large manufacturing plants use planned production systems.

the **company** and making products. It can also be done by a large corporation that has thousands of workers, many factories, and acres of land. Every company, whether large or small, uses a planned system of production. See Figure 1-4.

There are three types of production systems: custom, job-lot, and line production. Let's see how these types of production have come into being. The following example explains how people have produced ways to carry loads.

One of the first products manufactured to help carry things was a sled that was pulled along the ground. See Figure 1-5. These sleds were made one at a time, and each one was different. This type of production is called **custom production.**

Eventually, wagons were invented. Many wagons were made by the custom-production system. No two were the same. However, as more and more wagons were needed, people began to use the **job-lot production** system. One person or shop would specialize in making one kind of wagon. Several of each wagon part, such as wagon wheels, could be made. This is called a **job lot.** The parts were then assembled with the other parts to make the wagons. Using the job-lot production system, limited numbers of the same kinds of wagons were produced.

When automobiles were invented and the demand for them grew, there was a need for new and improved ways to produce them in large numbers. **Line production** was developed to meet the demand for this more efficient (faster and better) way to carry people and products. See Figure 1-7. In line production, the car is moved slowly past the workers. Each worker adds a new part. Workers become very good at doing their jobs and can accomplish their tasks quickly. Many products can be produced in a short time.

Each type of production is still used today and each has its place. Let's see how custom, job-lot, and line production are used for modern manufacturing.

Custom Production

In custom production, one person or several people work on one product at a time. They do all the processes on the one product until it is finished. This type of manufacturing is generally used when only one product of a certain type is wanted. For example, someone may want a custom-made boat. The buyer would go to a boat manufacturer who does custom work. The boat would be designed, plans drawn up, and one boat would be made. This type of production is usually done by smaller manufacturing companies. Generally, each worker must have a variety of skills for custom production.

Today, custom production is used in small craft industries. Leather products and jewelry are custom-produced. It is also used on a large scale for big products, such as a ship or an airplane, when only one product at a time is built. See Figure 1-8. Custom production is very expensive and can take a long time because each product is different.

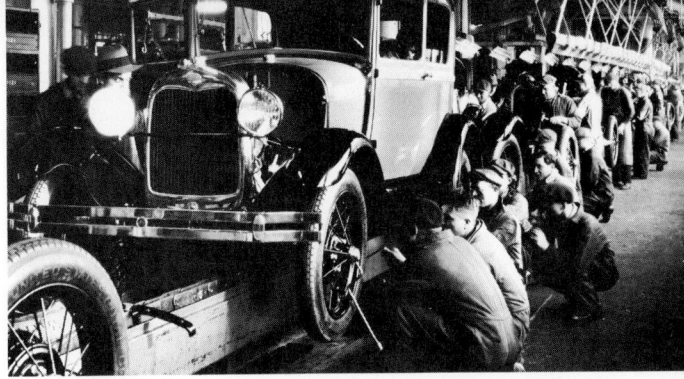
Ford Motor Company

Figure 1-7
Line production allowed more products to be made in less time. The cost of producing each one was less. The price of cars went down when they were made on an assembly line.

Job-Lot Production

Job-lot production is used when only a limited number of the same product is wanted. See Figure 1-9. The company takes orders from individuals or from another company for a specific number of products. The machines are **set up** (prepared for production) and the products are made. Perhaps 10, perhaps 1000 products will be made. A common job-lot size is 500 products. After the products are finished, a new order for a different product is taken. The machines are set up for the new product, and production starts.

Suppose a company makes only small electric motors. Another company that produces electric trains may send in an order for 500 special-size electric motors. Workers in the **job**

Figure 1-8
Most airplanes are built by custom production. Here an engine is being attached to a plane.

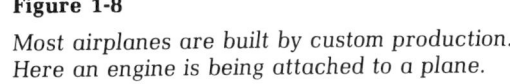

Sheller-Globe Corporation

Figure 1-9
These instrument panel pads for cars are made in job lots according to the design, color, and quality ordered by automobile manufacturers.

Have a Heart!

Worcester Historical Museum

The various types of production methods are used as they are needed. An example of this can be seen in the experience of **Esther Howland** of Worcester, Massachusetts.

In 1848, young Howland, who had seen English valentines, decided to design her own and sell them. She made a few valentines and asked her brothers to show them to people and take orders for them. To her surprise, they received orders for 5000 valentines! She had to have the help of friends and neighbors to make enough to fill the orders. She organized the work into a production line and accomplished the task efficiently. Esther Howland had discovered a product that people wanted and a way to supply it.

The business grew from there. Eventually, Howland had **dies** (special cutting tools) made to produce paper lace. She bought pictures and large amounts of other materials needed to make valentines. Work which Esther Howland began as job-lot production quickly became a line-production process. This successful business lasted many years. She sold it in 1876.

---

shop (place that does job-lot production) would set up equipment and produce 500 motors. Then they would begin working on another order for a different electric motor.

Workers in job-lot production must be able to learn new tasks easily. Jobs are likely to change when different products are made.

Chevrolet Motor Division, General Motors

Figure 1-10

Automobile manufacturers use the line-production system. Workers stay in one area while cars are moved past them by a conveyor system.

Line Production

When large quantities of a product are to be made, **line production** is used. In this type of production, all the jobs are broken down into single operations. People are assigned to jobs. Machines and usually a **conveyor system** (to move parts around the factory) are set up in positions where they will remain for long periods of time. Once the production starts, each worker does one job continually (over and over). By doing the job this way, the worker becomes specialized (very good at one job). Line-production workers generally do not need a variety of skills.

Line production moves quickly and smoothly. Many products can be made in a short period of time. Line production is the most efficient and least expensive way to manufacture large numbers of products. See Figure 1-10.

Manufacturing Must Be Planned

No matter which type of production a company uses, manufacturing must be planned.

One of the biggest challenges in manufacturing is to plan a system to produce as many good products as are needed in the shortest amount of time. Following are descriptions of the five basic steps needed in planning a manufacturing system.

Identifying Consumer Demand

What products do people want? How many of them? The manufacturing company must find out what people want in order to be sure that the product will sell when it is finished.

Most products come into being when someone thinks of something people need or want. A product is then designed to meet that need. The toothbrush was invented by William Addis in this way. Addis was put in an English prison around 1770 for provoking (starting) a riot. Since he didn't have much to do, he spent his time trying to think of ways to make a living once he got out. One morning while he was cleaning his teeth (by rubbing a rag against them as most people did at that time), he realized that this method did not work well for cleaning out food particles. The following day he had an idea. He saved a bone and persuaded a guard to give him some bristles (stiff hairs). Addis made the world's first toothbrush when he drilled holes in the bone and glued the bristles in them.

When Addis was released from prison, he went into the business of manufacturing toothbrushes. He was an instant success. Why? Because he had found a product for which there was a **consumer demand** (want).

Designing and Engineering Products

Once it is known what people want, it is time to design the product. Each detail is considered. How the product will look (shape, size, color) as well as how it will work is decided carefully. All the information is put together into a set of plans for the product. This process can take many years before a final, acceptable product is developed.

Sometimes new products are the results of searching and experimenting to solve other problems. A type of chewing gum was made in this way. **Thomas Adams** developed it in 1872. Long before that time, Mexican Indians had begun to gather and use gum from the chicle tree. Adams obtained a lump of the gum and experimented for two years trying to create a substitute for rubber. One day he put a piece of gum into his mouth and began to chew it. He liked it so well that he tried to sell his idea to a big

Edison National Historic Site

Figure 1-11

Thomas A. Edison understood consumer demand. In his laboratory in Menlo Park, New Jersey, he invented products to fill the needs and wants of people. One of the products he invented was the phonograph. Can you name another? What is he holding in his hand?

company. When the company turned him down, he went into production himself. By 1890, Adams had a six-story factory with 250 workers producing chewing gum, not rubber.

Even though some products are designed "accidentally," they are nearly always the result of creative thinking. Designing and engineering products takes time and thought.

Planning Production

A system must be planned for producing the product. The jobs that are needed and how

Identifying Consumer Demand

A manufacturing company must find out what consumers want before producing a new product.

Designing and Engineering Products

Designers and engineers work closely together to decide how a product will look and operate.

Figure 1-12
In manufacturing, every detail is planned.

Planning

A system must be planned for manufacturing the product. All

Production	Tooling-Up	Planning and Quality Control

details of production are planned by manufacturing engineers.	Special jigs and machine setups are made, and all equipment is moved into position so that production can begin.	Checking and testing products are very important. Successful manufacturers make certain that high-quality products are made.

they will be done must be decided. The decisions made will depend greatly on what type of production system is used (custom, job-lot, or line production).

Planning can take a long time. Every detail is determined before the production is started. For example, if a company is going to produce transistor radios, **manufacturing engineers** must make plans for producing the radio. Which electronic parts will be bought, and which will be made? How will the radio case be made? What machines will be used to make the radios? How many people will be needed, and what will each person do? How long will each job take? Not only do all these things need to be decided, but the manufacturing engineers must find the most efficient way to do each job.

Tooling-Up for Production

All tools and machines must be made ready to do specific jobs. This is called **tooling-up.** Tools and machines must be placed in their right locations. Special holding devices (jigs) may be needed for use with the machines.

If specific machines must be built, these must be designed and ordered from companies that manufacture them. It is not uncommon for one specially built machine to cost $200,000 or more. Just the **punch** and **die** (part of a machine that cuts a shape, as a cookie cutter) for cutting the steel for the cab of a Tonka Toy truck can cost over $80,000!

Sometimes companies actually set up their equipment in a separate building. Then they try it out in a **trial run** for several weeks before moving it to the final production line. This way many problems are discovered before the actual production begins.

Planning Quality Control

The quality of a product is determined by how well it is designed and made. Planning must be done and a method set up to make sure each job is done properly and well. Then the finished product will be "good quality."

What would happen if a company had no plan for checking how well its products were being made? In a chair manufacturing plant, for example, just one person who failed to do the job right could produce chairs with faulty legs. If no one checked the quality, then hundreds of these chairs could be shipped to customers. There would be many unhappy customers when the mistake was finally found.

Some examples of planning for manufacturing are shown in Figure 1-12.

Production Procedure

The methods used to produce finished products are different because products are different. Yet the procedure followed to make the products and prepare them for **distribution** (shipping) is generally the same. There are six basic steps in this procedure. However, not all of the six steps may be necessary. This depends on the product being made.

Converting Raw Materials

Raw materials are provided by nature. They may be grown such as trees, cotton, and grain, or raised such as sheep and cattle. Others are **extracted** (taken) from the ground, water, or air such as iron ore, petroleum, and oxygen.

The first step in production is to change the raw material into a form useful for making a product. Iron ore is a raw material, but it is useless until it is **refined** (has impurities removed). Then it can be combined with other refined minerals to make steel. In the same way, a tree cannot be used to make a product until it is changed into lumber.

Making Standard Stock

A raw materal is more useful when it is converted (changed) into **standard stock.** This means that the material is formed into a standard (widely used) size and shape. Think how hard it would be to work with and transport (move) a thick piece of steel the size of your classroom. Standard stock sizes are smaller and easier to handle. Also, planning products is easier when the sizes and shapes of materials are known. Lumber and sheet metal, for example, can be purchased in standard sizes.

Making Components

Standard stock is made into the **components** (individual parts) of a product. Some products have one component, such as a comb, a key, or a coin. Many products are made up of many individual components. The handle grip is one component of a motorcycle. The lead and the eraser are components of a pencil.

Assembling Components into Subassemblies

The components are **assembled** (put together in a planned way) with other components. This can be done by a number of methods. Components may be put together with glue, by welding, or with screws, bolts, or nails. A combination of methods may be used. The motorcycle handle grip is assembled with the handlebar to make a **subassembly.** The lead and eraser are assembled with other components to make a pencil.

Making Finished Products

When all the components are assembled and inspected, the result is a finished product. Motorcycles, pencils, nuclear submarines, space shuttles, and safety pins are all finished products. Often, finished products are components of **other** finished products. For example, the light bulb is a finished component of a desk lamp. Tires are finished components of the motorcycle.

Preparing for Distribution

The finished product is prepared for distribution to protect it during shipping. Many products need to be placed in packages to protect them or to make them easy to handle.

Products are usually stored in a **warehouse** (storage building) before shipment to consumers. They are then loaded onto trucks, trains, boats, or planes, or pumped into pipelines to be transported to the consumer.

Several examples of production procedure are shown in Figure 1-13.

Looking Ahead

This chapter has given you an overview of manufacturing — making products. In the next chapter, we will begin taking a closer look.

Do you know
- if "Industrial Revolution" refers to a war between industries?
- how much iron a worker in the late 1800's would have to load in one day to earn $1.85?
- why medicine may someday be produced in space stations?

These topics and others are mentioned or discussed in the next chapter. Look for them as you learn how manufacturing has developed into what it is today and what it might become in the future.

New Terms

assembled	labor force
company	line production
components	manufacturing
consumer	engineer
consumer demand	punch and die
conveyor system	raw materials
custom production	refined
distribution	set up
extracted	standard stock
industry	subassembly
job-lot production	tooling-up
job lots	trial run
job shop	warehouse

Study Guide

1. Give some reasons why manufacturing is so important to our country.
2. How are custom production and job-lot production different?
3. Why is line production used to make automobiles?
4. Why does a company identify consumer demand before a product is made?
5. How do people get ideas for new products?
6. Why are manufacturing engineers important to the manufacturing industry?
7. What is meant by "tooling-up for manufacturing"?
8. Name the six basic steps of production, from raw materials to the finished product.
9. What is standard stock? Give three examples.
10. What are components? Name the components of two manufactured products in your classroom.

	CONVERTING RAW MATERIALS	MAKING STANDARD STOCK	MAKING COMPONENTS
AUTOMOBILES	 Iron ore is made into steel.	 Rolls of sheet metal are made.	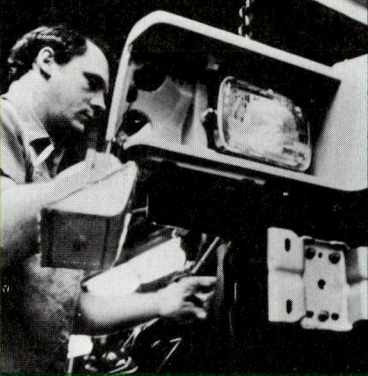 Headlights are finished components.
FURNITURE	 Logs are made from trees.	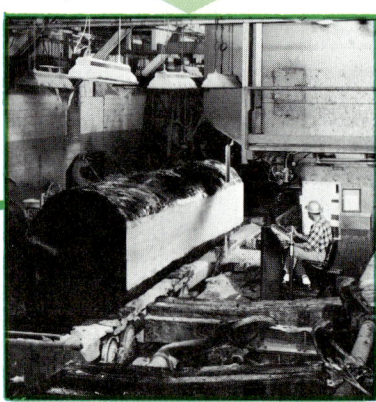 Logs are cut into lumber.	 Furniture parts are made.
CLOTHING	 Separating usable cotton from seeds and waste.	 Cotton thread is woven into cloth.	 Parts of clothing are cut out.

Figure 1-13
Products follow the same general production procedure.

ASSEMBLING COMPONENTS INTO SUBASSEMBLIES	MAKING FINISHED PRODUCTS	PREPARING FOR DISTRIBUTION
 The body is assembled.	 The finished car awaits distribution.	 Cars are often shipped by rail.
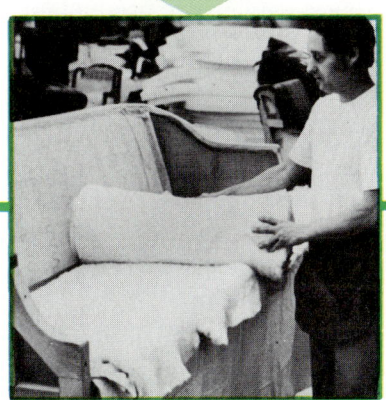 The parts of a couch are assembled.	 The finished couch must be inspected.	 The couch is packaged for shipping.
 The parts are sewn together.	 Finished clothing may be stored briefly before shipping.	 Clothing is packaged for shipping.

chapter 2

The Development of Manufacturing

The products we buy today are most often manufactured in large numbers. Efficient methods are used. This was not the case in the past. As you shall see, manufacturing has changed from crude beginnings to become what it is today. It is still changing. The manufacturing industry is growing. New materials and new uses for materials are being discovered or invented. New processes and better ways of making products are being found. The development of manufacturing continues. What will it be tomorrow?

How Manufacturing Developed

Modern manufacturing is very different from manufacturing of long ago. In earlier times, each person or family had to make everything that was needed in order to live. Families had to grow their own food and build their own houses. They had to make their own clothes, tools, weapons, and utensils (household tools such as spoons). It took a long time to make each item, and the necessary materials were not always available. This method of making things was not efficient. See Figure 2-1.

At a later time, people began to specialize in doing one kind of work. Some people did only farming. Others baked bread, made shoes, or produced tools. They even developed names for their crafts — baker, shoemaker, and blacksmith, for example. These work activities became known as **cottage crafts,** because the work was usually done in the family's home or cottage. See Figure 2-2.

Bartering and the Money System

In order for each family to have the other items it needed, a way of trading called **bartering** developed. Bartering is trading one product for another. People who made shoes could trade them for food. People who made tools could trade them for shoes. The same type of trading was done with all products.

Bartering was an inefficient system. For example, a person who raised chickens probably had to carry a big pen of them to the marketplace. The chickens were then traded for necessities (things that were needed).

Later, people began to use money. Products were no longer bartered, but were sold for money. Then the money was used to buy other needed products. This money system is still used today.

The Industrial Revolution

During the late 1700's, cottage crafts became less and less important. Changes began to take place in manufacturing:

- **Machines** instead of only hand tools were used to make products.
- **Engines** were used to power the machines.
- Products were made in **factories** instead of homes.

The process of making these changes is called the **Industrial Revolution.** It began in Great Britain and the countries of western Europe.

Making products by hand at home took a long time. Machines could do the work much faster. Since machines were usually powered by a

The Development of Manufacturing

waterwheel, shops were located next to a stream or river. See Figure 2-3.

Machines were invented to do many kinds of work. The **textile** or cloth-making industry was the first to change because of machines. A spinning jenny was invented to spin yarn. Later, a loom was invented that was run by an engine. It could weave cloth much faster than a hand-powered loom.

In 1769, an Englishman, **James Watt,** invented a steam engine that could be used to provide power. Machines powered by a steam engine could be used almost anyplace. No longer was it necessary to build a shop by a river. Shops could be built close to the source of raw materials. For easy shipment, products could be manufactured near seaports and cities. At first the workers did not like steam-powered machines. They were afraid that they would lose their jobs because of these machines. Some workers resisted the change by smashing the machinery.

Figure 2-2
Eventually, families began to specialize in making one kind of product. Tinsmiths made pans and other utensils out of tin.

Figure 2-1
There was a time when each family made or grew everything it needed for daily living.

Figure 2-3
Moving water caused the waterwheel to turn. This action powered the machines inside early textile mills.

Figure 2-4

James Watt's steam engine helped to change the way America manufactured products. Engines were an important part of the Industrial Revolution.

The new machines were large and expensive. Individually, most skilled workers could not afford to buy machines. Even if they could, the machines wouldn't fit inside their houses or shops. People with money decided to work together. They formed manufacturing companies. Then they built special buildings called **factories** and put the large machines inside them. They hired workers and paid the workers' wages. Usually the people hired were not skilled workers. They were taught to run machines and make parts. A person might run a machine that did just one operation. The worker who did not change jobs never learned the other steps in making a product.

Skilled workers also had to find jobs in factories. Goods could be made faster and more cheaply by machines than by even the most skilled hand labor.

Working Conditions in the Factory

In the early days, working in a factory was usually not pleasant. Because many people needed jobs, factory owners could pay low wages. They also made the people work long

Figure 2-5

In some early factories, children worked long hours under difficult conditions. Not until the 1930's were laws passed that finally began to put an end to child labor in America.

Speeding Up the Flow of Progress

In the 1700's, the steam engine was operated by controlling two valves.* Someone had to open and close each valve continually to keep the engine running. A boy named Humphrey Potter grew bored while opening and closing the valves. He devised a method by which the action of the engine would control the opening and closing of the valves in proper order. Putting his idea to work not only doubled the speed of the engine, but was the invention of the **automatic valve**.

* A **valve** is a mechanical device used to control the flow of a liquid or a gas in a passageway such as a pipe. An example of a valve is a faucet.

---

hours. A worker sat or stood at a machine all day with only a few minutes of rest.

Most factories were gloomy. They seldom had heat in the winter or fresh air in the summer. Factory work was often dangerous. Machines had no guards or safety devices.

Frequently, children as young as seven years old worked in factories side by side with adults. See Figure 2-5. They often worked from 5 a.m. to 8 p.m., six or seven days a week. Beginning in 1833, effective laws about child labor were passed in England. First the working hours of young children were reduced. But finally, children were not allowed to work in factories at all.

Other kinds of laws about work were also passed. Factory owners were made to shorten working hours and raise wages. They had to put in safety devices. As the factory system grew, workers banded together and formed **labor unions**. The unions helped to get the shorter hours, higher wages, and better working conditions.

Manufacturing in America

The Industrial Revolution spread to this country when America was a young nation. **Samuel Slater,** an English immigrant (settler from England), helped to start it. He and two partners established the first successful textile factory in America. The factory was located in Pawtucket, Rhode Island. Slater used new management and marketing (selling) methods. And he used his knowledge of textile machines.

In 1790, he, with others, built textile machines of English design. These machines changed textile manufacturing. The idea behind them changed all manufacturing. Much of the handwork could now be done by machines. Previously, highly skilled people were needed to do it. Now the work was divided into simple steps. Less skilled people could work at each step. The idea of **division of labor** began to grow in America.

New products were invented. The telephone, typewriter, light bulb, and airplane each caused new manufacturing companies to be formed. New materials were discovered and developed. For example, iron replaced wood in many products. Then steel replaced iron. More recently, plastics are often used instead of glass and metal. Lightweight aluminum frequently replaces heavier iron and steel.

A New Process

In 1886, a 22-year-old American inventor, **Charles Hall,** worked in a laboratory set up in his family's woodshed. Here he made a most important discovery. Using a crude apparatus (simple equipment), a secondhand gasoline stove, and a borrowed battery, he carefully mixed and melted some chemicals. When the mixture was ready, he poured the molten contents into one of his mother's old kitchen

Figure 2-6
Aluminum, the most plentiful metal on earth, must be processed into a usable form. Charles Hall developed a practical method for processing aluminum.

Aluminum Company of America

skillets. He allowed the molten mixture to become solid. Then he smashed it with a hammer. The little blobs of silver metal were just what he was looking for, **aluminum.** Charles Hall had invented a practical way to make aluminum.

Aluminum was first manufactured in France around 1855. At that time, it cost about $500 to make one pound of aluminum. Hall's method was much cheaper. By 1893 the price dropped below $1 per pound. In the 1930's, the price was lowered to 20¢ a pound. Charles Hall, using his new process, helped to form a business that eventually became the Aluminum Company of America, ALCOA.

Interchangeable Parts

The discovery of new materials encouraged the growth of manufacturing in America. Several other factors contributed as well. **Mass production** was one factor. This meant that large quantities of parts and products could be produced within a short time. Mass production is impossible unless all the same parts or components are the same size and shape. They must be very nearly identical (alike). They must be **interchangeable.** Any one of several thousand pieces of one part must fit with any one of several thousand pieces of a joining part. Up to this time, each part had to be handmade and carefully fit by hand to the part next to it. Parts must be accurately made to be interchangeable.

When parts are put together using the line-production system, interchangeable parts are picked at random (by chance). For example, electric fans are made in this manner. There may be hundreds of interchangeable fan blades. A worker assembling the fans may choose any one of the blades. Any blade chosen can be attached to any fan. The fan will operate in the same manner with any blade.

The American System

Mass-production methods were not always used in manufacturing. For example, the first guns and muskets produced were all custom-made. Each gun was unique (different). Even though two guns appeared to be identical, their parts were not interchangeable. Each part was "hand-fit" to a particular gun by a skilled worker. This took a great deal of time and was very expensive.

In the early 1800's, the system of manufacturing began to change. New machines were invented. New processes were developed. Manufacturers such as **Simeon North** of Connecticut and **John Hall,** a fellow Yankee, were able to mass-produce guns with interchangeable parts. See Figure 2-7. This revolutionized (changed greatly) the manufacturing process. Guns could now be made faster, better, and less expensively. This system of using mass production and interchangeable parts proved to be so efficient that it is still the basis for manufacturing today. British visitors to American factories at the time of North and Hall were impressed with the system. They called it the "American System" of manufacturing.

These principles were used a short time later by **Eli Terry** to make clocks. Terry had a shop in Plymouth, Connecticut. He made clocks by hand, one at a time. Each part was carefully and slowly hand-fit to the next part. Because of this, clocks were expensive. Few people could afford to own one. Terry began to apply the idea of interchangeable parts in his clock business. In just a few years, he was manufacturing large numbers of wooden shelf clocks that sold for only about $15 apiece.

Terry's neighbor and assistant, **Jerome Chauncey,** took over the clock business after Terry. Chauncey built on Terry's ideas. Chauncey's company made a small one-day

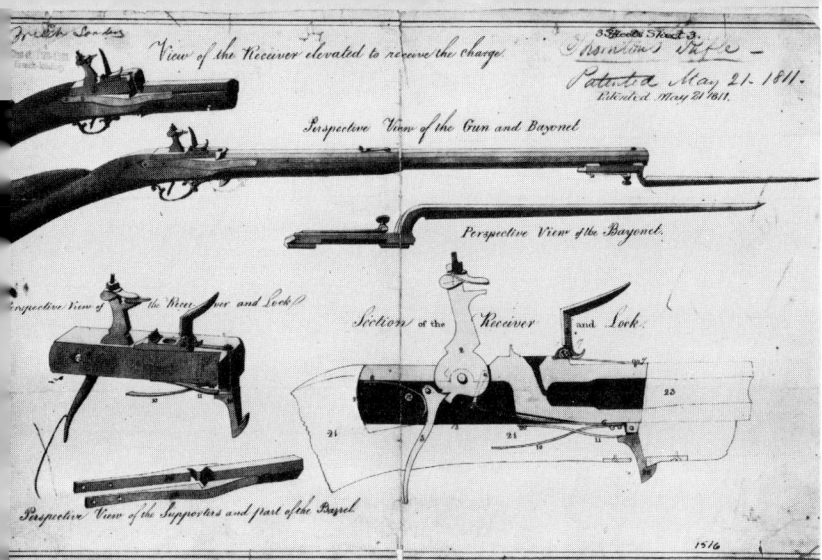

Figure 2-7

This rifle was designed by John Hall. Its simple design made it well suited for mass production. John Hall's production system and the standards he set for interchangeable parts improved early manufacturing.

Figure 2-8

The Ford Motor Company's first moving assembly line was set up in 1913. The car body was lowered onto the frame outside the factory.

clock out of brass. Before this time, brass clocks were very expensive. Chauncey made them popular by mass-producing them. They could then be sold for about $2 each.

The brass clock opened the **export market.** (markets in foreign countries). Clocks made of wood could not be shipped overseas because moisture would cause them to swell and stick. In 1842, Chauncey shipped some of his one-day brass clocks to England.

At that time, taxes had to be paid on imported goods. The amount of tax was based upon the value of the product as stated by the owner. Some owners tried to pay less in taxes by saying their products were worth less than they really were. To prevent this, a British law allowed custom officials (people who set the amount of the taxes) to buy the shipment at the **declared value** plus 10%. Because of using interchangeable parts, Chauncey's prices were low. The British did not believe him. They seized his clocks and "punished him" by paying him 10% above his declared value. Chauncey was excited. He not only sold his clocks right at the dock, but he received 10% **extra** profit. He sent a second boatload of clocks to England. Again he was punished in the same manner. When he sent the third boatload, however, they gave up and let it go through at the declared value.

The Assembly Line

There were other important developments in American manufacturing. The use of an **assembly line** to make products was one of them. On the assembly line, a worker stays in one place. The parts and work are moved to the worker.

Ford Motor Company

In the early 1900's, **Henry Ford** made popular the use of this new development in manufacturing. See Figure 2-8. By using and developing the assembly line and other techniques, he was able to lower the cost of making cars. Before he found ways to make them more efficiently, many cars were priced as high as $5000 to $10,000 each. Ford's Model T was first priced at $850. When it was made on an assembly line, the price was lowered to $360. Many people were then able to afford a car. The assembly line saved time. Before this

Figure 2-9

Special tools speed up the work. This worker can fasten all the wheel lugs in one operation by using this tool.

method was used, a total of 12½ hours of labor was required to build a car. The same processes required only 1 hour and 33 minutes of labor on an assembly line. Even though the prices of Ford's cars went down, the wages of his workers went up. More cars were made and sold.

Other Important Developments

Closely related to the assembly line are some other ideas that helped manufacturing to develop. The use of **special tools** is an important contribution to the American system. Designing special tools for each worker or task allows the work to be done faster. See Figure 2-9.

Job specialization is another important development. It is the practice of assigning a limited number of tasks to a worker. This allows the worker to specialize in one thing. **One worker doing one job** can be very efficient. Also, with job specialization less knowledge is needed for some jobs. Unskilled workers can perform simple tasks over and over again.

Improved Planning

Job planning is figuring out the most efficient way to do a job. It is a part of American manufacturing that has developed over the years. In the late 1800's, a man named **Frederick Taylor** worked for the Bethlehem Steel Company. He began to study various jobs to see if they could be improved. In one experiment, he studied some men who were shoveling iron. The men tired easily. Taylor began planning ways to improve the job. He thought that the men were working too hard and too long. They needed rest periods to recover from the hard work. He reasoned that shortened work periods and a different way of shoveling would help. Each man loaded about 12½ tons (25,000 pounds) of iron per day and earned $1.15 per day. One man, named Schmidt, was chosen for an experiment. He earned $1.85 per day by following Taylor's directions.

Schmidt was told how to pick up, carry, and put down the iron. He was also told to take frequent rest periods. Schmidt followed directions and loaded 47 tons (94,000 pounds) in one day. Soon the company had many people wanting to earn $1.85 a day. Taylor's planning certainly improved production.

As Frederick Taylor thought about the workers he observed, so **Lillian and Frank Gilbreth** thought about all workers. Lillian and Frank were both engineers. They worked together for many years on **time-and-motion studies,** finding the most efficient ways of performing tasks. They shared their management ideas with others and became known as **efficiency experts.** After Frank died in 1924, Lillian continued the work they had begun. Lillian believed, as had Frank, that the management in business and industry should consider the human side of workers. For example, she realized that people work faster and do better jobs when given tasks they like to do or that add variety to their workdays. She also believed that workers take more of an interest when encouraged to think of better or different ways to perform their tasks. Lillian reasoned that good worker attitudes improve production. She taught her ideas in classes on management and even gave on-the-spot advice in factories where problems existed.

The work that Lillian and Frank Gilbreth began in the early 1900's has had great and lasting influence. Now the interests and skills of people are matched with job requirements. Tests have also been developed to help determine the jobs or careers for which people are best suited. The idea of applying skills from one job to another job has become widely accepted.

From the Gilbreths we learn that when plans are made, the people who will do the work must be considered. When workers are happy in their jobs, both they and the company benefit.

Recent Developments

Manufacturing has developed over the years into a highly efficient system for making products. Changes and improvements have slowly but steadily come about. Today, improvements are being made at an extremely fast rate. Research has increased the knowledge of almost every human activity. This is especially true in manufacturing.

ACF Industries, AMCAR Division

Figure 2-10

Planning everything, even the smallest detail, can improve production. Here, a supervisor times each movement as a worker performs one task.

Figure 2-11

Lillian Gilbreth was a pioneer in time-motion studies. She combined an active engineering career with raising a family of twelve children.

Brown Brothers

Figure 2-12

Chevrolet Motor Division, General Motors

Automation is important in automobile manufacturing. Here, spot welders automatically extend, weld, and pull back as the floor pan of a car is moved past.

Automation

Automation is a way of making a machine, process, or system operate automatically (without constant human control). For example, most buildings today are heated automatically. A thermostat is the control. All that a person must do is set the thermostat at the temperature desired. When the temperature falls below this set point, the thermostat automatically causes the furnace to start. The furnace continues to heat until the set temperature is reached. Then the thermostat causes the furnace to shut off. This system of heating is done by automation.

Only recently has automation become a factor in American manufacturing. As assembly line work increases, more and more machines are made to operate with less human control. Processes are planned to be completed automatically. See Figure 2-12.

Automation allows more work to be done. Machines can work 24 hours a day without rest breaks. One worker can often operate two or more machines. This increases efficiency and lowers costs.

Computers

The use of **computers** in manufacturing today has helped to make the system more efficient. Computers are useful in many ways. They can be used to keep records and store information. They can be made to write orders and make decisions. They are also used for calculating with numbers.

A common use of a computer is to keep an **inventory** (list of items on hand) of materials. When the supply of materials becomes low, the computer can write an order for more materials.

Another important use of these "electronic brains" is to control the machines that make parts. "Speed up," "slow down," "change tools" — these are commands that computers put into effect in controlling machines. Computers can also indicate to workers when machines need attention. For example, worn

Figure 2-13

Using this small computer, one worker can operate the large paper machine better than a team of workers.

Georgia-Pacific Corporation

Figure 2-14

Each year in the United States, 16.7 million tons of paper are gathered for recycling.

tools may need to be replaced. Many parts of the manufacturing process can be controlled by computers.

Recycling and Energy Conservation

Waste and pollution are problems today. **Recycling** is one way that many manufacturers are helping to cut down on the amount of both in our environment.

Recycling means processing again. For example, empty aluminum cans can be melted and made into new cans. This saves energy. Recycling saves natural resources. Recycling **twenty** cans requires the same amount of energy needed to produce **one** can from ore. Since the aluminum is reused, less raw material is required. Reuse of cans and bottles reduces the litter in our environment. And, less space is needed for dumping waste materials.

In 1978, over seven billion (7,600,000,000) aluminum beverage cans were repurchased for recycling. Fifty-four million dollars ($54,000,000) was spent in this effort. One company alone, the Aluminum Company of America, spent over eighteen million dollars ($18,500,000) to buy back more than two billion of the cans. This effort has provided jobs and income for many people. It shows, also, the determination of American industry to improve the lives of all of us.

Many factors have contributed to the growth of manufacturing in America. As it continues to expand, there will be many more new developments. Manufacturers will continue to improve their products and production methods.

Manufacturing in the Future

How will daily life benefit from manufacturing in the future? What will manufacturing be like in 50 years? No one knows for sure. Experts predict that in the future manufacturing will be faster, more automated, and more complex. It will also be more efficient and better able to make products at much less cost.

We can try to make forecasts by looking at changes that have recently taken place. Some of the latest technical and scientific developments have come from our national space program. Developments there give us a hint of what manufacturing might be like in the future.

Use of Outer Space

Experiments were done aboard a spacecraft. The results proved that certain types of manufacturing can be done in outer space. Outer space is clean. There is no effect of gravity (earth's pull). Certain types of parts or products will actually be easier to produce. Four areas seem especially suited for outer-space manufacturing:

- Making perfect spheres (ball-like objects),
- Mixing different liquids,
- Refining pure metals, and
- Growing crystals.

Spheres, such as the steel balls used in ball bearings, can be made better in outer space than on earth. Because of gravity, spheres made here on earth are slightly egg-shaped. They must be processed further to make them round. But spheres made in space will be perfectly round. There is no gravity to distort them (change their shapes).

Liquids that won't ordinarily mix here on earth will mix well in outer space. Oil and water, for example, won't mix. The oil is lighter than the water and floats on top. In space, everything is weightless. The difference will not matter. Thus, lubricating fluids might be made in a space station factory. Liquid medicines could be made in the same manner. Imagine, drinking cough syrup that was manufactured in outer space!

The processing of precious **metals,** such as gold and silver, is improved when there is no gravity and no air. Ordinarily, metal must be melted in a container. When the container is heated, the metal melts. The metal is usually contaminated (made impure) by dirt, dust, and exposure to air. Outer space is a vacuum. There is no air. There is no gravity. The metal is weightless and floats. A magnetic field (area of force) keeps the metal suspended (held floating) in one place. It can touch nothing. The metal can be processed without contamination.

Crystal chips for electronic products are easier to produce in outer space. These can be made more quickly in a vacuum. Crystal chips are used in calculators and wristwatches.

Figure 2-15

Successful manufacturing experiments were performed inside the Skylab space station. During its six years in space (1973-1979), the Skylab made use of energy from the sun. Notice the solar panel on the right side.

Figure 2-17

People are working to find efficient ways to recover methane gas from buried garbage. This worker is measuring the amount of methane produced in a New Jersey landfill area.

Figure 2-16

Wind may be used as a source of energy. This experimental windmill is expected to provide enough electricity to power about 30 homes.

Better Use of Resources

Manufacturing will, of course, continue here on earth. Machines and processes are being developed that will enable us to make better use of our natural resources. For example, new energy devices, such as **solar collectors** and **wind generators**, will provide heat and power. See Figure 2-16. Valuable petroleum and natural gas can then be used as raw materials for other products rather than burned as fuel.

Recycling will continue to be an important way to make wise use of materials and energy. Using garbage as a fuel to run boilers and generators will save money and space. Now much garbage is buried to dispose of it. As it rots, methane gas is given off. Methane can be used as a fuel. See Figure 2-17.

Recovering our resources will be helpful. Extracting (separating) the metal, glass, and other reusable material before burning or burying the garbage will cost money. However, it will save money in the long run. As an example, mixing scrap steel from car bodies with iron ore during processing will reduce the cost of producing new steel. The mining and shipping costs are less for the amount of steel produced.

Importance of Future Changes

What will these changes mean to us? The way we live will be changed by improvements made in manufacturing. For example, education, medicine, and transportation will be changing. Just as tape recorders and calculators have improved our **education** in the past, some new products will be developed that will help us learn better.

We will continue to receive better **medical treatment** because of improvements in manufacturing. Artificial parts such as hearts, knees, and hands are a reality now. Continuing research will improve these parts and lead to the manufacture of other replacement parts as well. See page 3. Medical equipment will be improved.

Even our ways of **traveling and shipping** goods will change. Can you imagine the kinds of vehicles we will have in the future?

We will probably enjoy a higher **standard of living**. There will be more products available for us to buy and use. Workers will receive higher incomes. However, the **cost of living** may also be higher. Because workers are paid more, products and services are likely to be more expensive.

There will be more leisure time (free time), because the amount of time spent working will be shorter. Instead of the 40-hour **work week** we have today, a 30-hour work week may be common. Less time will be needed to manufacture products. Many people will be working in jobs that don't even exist today.

We **think** these things will happen. It is impossible to know for sure. We can only look at the past and try to predict the future. One thing is sure, however. The manufacturing industry will continue to look for ways to efficiently meet our needs and wants.

Looking Ahead

This chapter has shown how manufacturing began. You saw the progress it has made. You even looked into its future. Next, you will begin to see some of the elements (basic parts) of manufacturing.

Do you know
- what toothbrushes and parachutes have in common?
- what kind of drawing requires no writing tool?

In the next chapter you will learn the "what and how" of manufacturing — tools, materials, and processes.

New Terms

assembly line	interchangeable
automation	job specialization
bartering	labor unions
cost of living	mass production
cottage crafts	recycling
declared value	solar collectors
division of labor	standard of living
efficiency experts	textiles
export market	time-and-motion
factories	studies
Industrial Revolution	work week

Study Guide

1. In early times, each family made everything it needed. Why was this system inefficient?
2. Why did bartering lead to the money system?
3. How did the Industrial Revolution affect the way products were manufactured?
4. What major change did the invention of the steam engine bring about? Why was this change important?
5. Describe the working conditions in early factories.
6. How did new inventions lead to the formation of companies?
7. What feature of manufactured products makes the assembly line possible?
8. Why was Henry Ford able to raise his worker's wages even though he charged less for each automobile?
9. How has improved planning helped manufacturing develop?
10. How are computers used in modern manufacturing?
11. What might manufacturing be like in the future?

chapter 3

How Is Manufacturing Done

Manufacturing means making products. You learned that earlier in this book. You also learned that there is a great variety of manufactured products, and you know how heavily we depend upon them. Now, you will see what is needed to make products.

What is Needed for Manufacturing?

All products are made up of parts or components. These parts are made from **industrial materials**. Nylon, steel, particle board, and glass are examples of industrial materials. Nearly all parts of all products are made by processing materials. **Processing** means changing materials into something useful. Processes, such as drilling, cutting, and welding, are used to change industrial materials into parts.

To perform these processes, **tools** are required. Hammers, drills, saws, and other tools are used.

Since all products are made by using tools to perform processes on materials, it is important to know **how** this happens.

Suppose you want to make the tool tray shown in Figure 3-1. What material can be used? Wood? Metal? Plastics? Other? How will you make the parts? What processes will you use? Cutting? Bending? Nailing? Finally, what tools are needed? Saws? Hammers? Screwdrivers? Using this book will help you to answer these questions. You will learn more about the various kinds of tools, the industrial materials, and the different processes that can be used. Then you can decide which is best to use for each product you want to make.

Tools

A tool is a device that helps a person to do work. Tools are extensions of (additions to) hands and arms. With tools, work can be done faster, better, and more safely. There are many kinds of tools. A lawn mower is a tool that helps you cut the grass. A fork is a tool with which you eat. Many tools are used in manufacturing. See Figure 3-2.

Tools used in manufacturing products come in all sizes, from very small to very large. They may be common tools, such as hammers and screwdrivers. Or they may be specially designed, one-of-a-kind tools.

Figure 3-1
How would you make this tool tray?

Figure 3-2

How many of these hand tools can you name? Tools may be small enough to hold in your hand or as large as this press which forms automobile body parts.

Sheller-Globe Corporation

Tools can be classified as
- Hand tools,
- Portable power tools,
- Power machines and equipment, and
- Industrial machines and equipment.

Some tools are simple, and some are complex (have many parts), but all are related to one or more of the basic machines.

Basic Machines

There are six basic machines: the wheel, the lever, the inclined plane, the pulley, the screw, and the wedge. These basic machines change the speed or force produced by a person. All tools are related to one or more of these six. See Figure 3-3.

The blade in a circular saw uses sharp teeth (wedges) around the edge of a **wheel** to cut wood. A pair of pliers is a form of **lever**. It increases the user's grip on an object. A twist drill is really an **inclined plane** in spiral form (winding from point to point). The blade in a band saw runs on **pulleys**. A **screw** thread allows a C-clamp to be tightened. A block plane

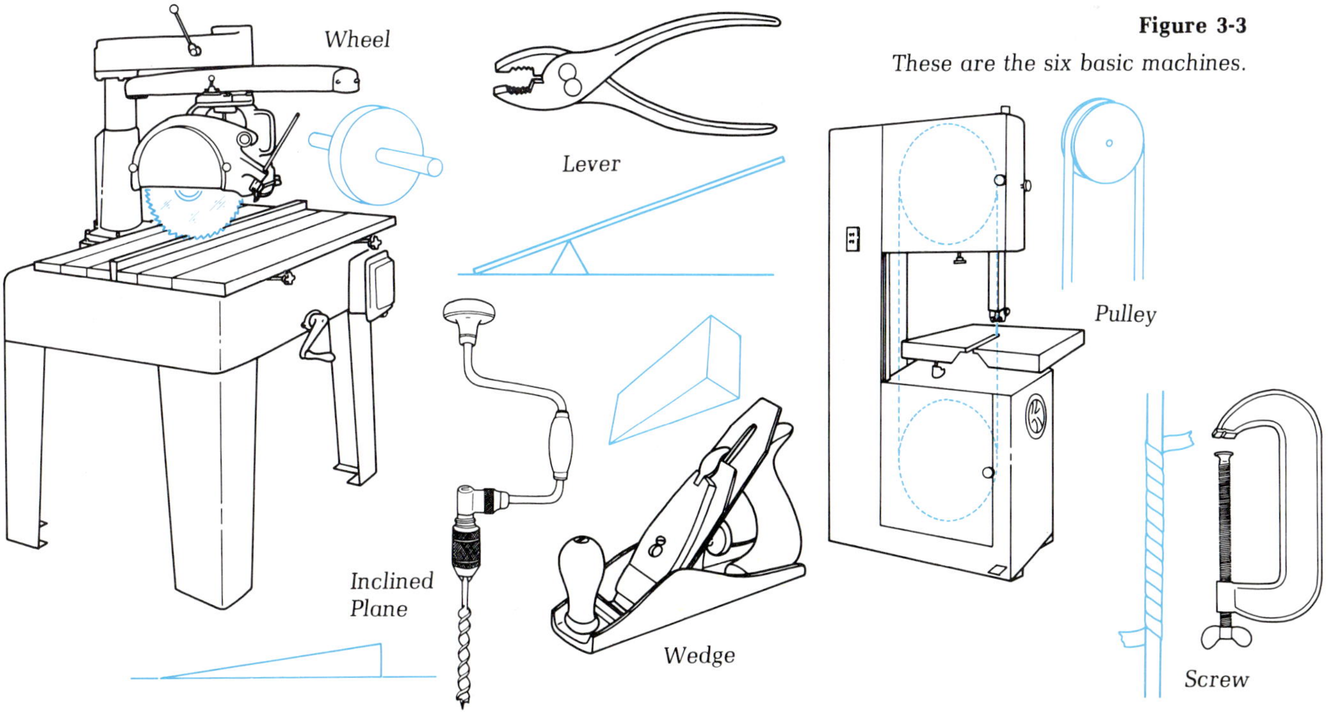

Figure 3-3

These are the six basic machines.

Wheel

Lever

Pulley

Inclined Plane

Wedge

Screw

uses a **wedge** to shave thin strips of wood. No matter how large or how small, no matter how it is used, every tool is related to one or more of the six basic machines.

Hand Tools

Hand tools are simple tools that are powered by human muscle. Common hand tools are available in many stores. Most families own some. (As you make the products presented in this book, you will need to use many hand tools to do the different processes.) See Figure 3-4.

Although many hand tools have been around for hundreds of years, new hand tools are still being developed. **William Petersen** was a blacksmith in Dewitt, Nebraska during the 1920's. He needed a tool that would firmly grip a piece of steel that was being hammered. An ordinary pliers worked fine as long as he kept his hand squeezed tight. But if he relaxed his hand, the steel would slip from the pliers. Petersen searched for a way to lock the pliers and hold them shut. Finally he solved the problem. He invented the locking pliers. Petersen called his new tool VISE-GRIP* locking pliers. In 1924, he formed the Petersen Manufacturing Company and hired his neighbors to help build his unique tools. Today many people own and use these handy tools. See Figure 3-5.

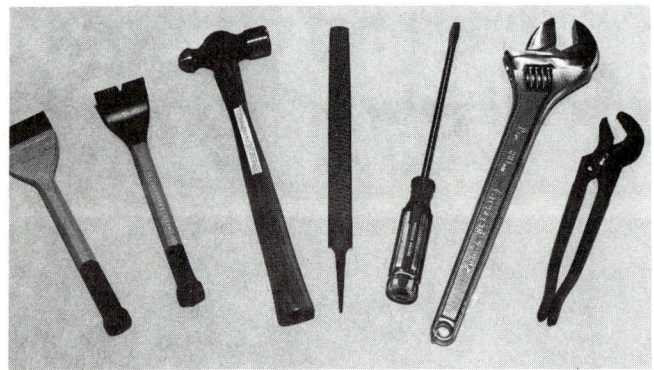

Figure 3-4

Hand tools require muscle power. Which of these tools would you use to drive a nail? to turn a screw? to smooth a rough metal surface?

Portable Power Tools

Portable power tools can be moved easily from place to place. Basically, they are the same as hand tools. There is one major difference. They have a motor that provides the power to do the work. The operator guides and directs the portable power tool. A portable electric drill, for example, does the same job as a hand drill. However, it does the job faster and with less effort from the operator. Portable power tools were developed by people who wanted to do more work in less time. See Figure 3-6.

Electric drills were first invented in Europe in the early 1900's, but they were bulky and hard to handle. In 1914, in Maryland, two young men, **Samuel Black** and **Alonzo Decker,** made improvements on the drill. They added a pistol grip and a trigger switch. These additions made the drill much easier to use. See Figure 3-7.

* VISE-GRIP is a trademark of Petersen Mfg. Co., Inc.

Figure 3-5

William Peterson was a blacksmith who needed a different kind of pliers. In 1924, he invented one. Now thousands of these locking pliers are manufactured daily.

How is Manufacturing Done   41

Power Machines and Equipment

Power machines and equipment are also described as "universal" machines and equipment. They can be used in a variety of operations.

Power machines are bigger and more powerful than portable power tools. However, the same kind of work is done with them. And they are generally more accurate. Power machines are stationary (not portable). They are mounted on **pedestals** or permanent (lasting) bases. Usually, they are not moved. These machines are run by electric motors.

Relatively large, permanently based tools that are not run by motors are called **equipment.** A squaring shears is a stationary piece of equipment. It is used for cutting sheet metal. The

Figure 3-6

These portable power tools use electricity or compressed air for power.

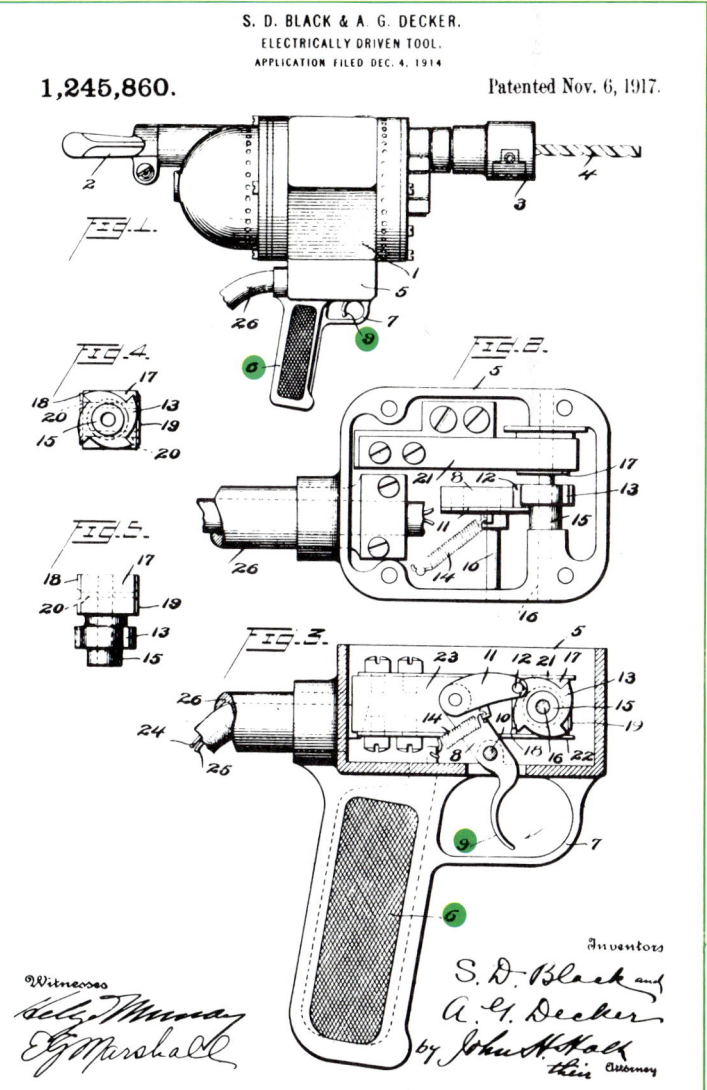

Figure 3-7

This sketch of the Black & Decker drill was drawn in 1917. Notice the pistol grip (6) and trigger switch (9). These new features made the portable electric drill easier to use.

Figure 3-8

"Universal" power machines and equipment can be used in a variety of operations.

A table saw is a power machine used to cut wood.

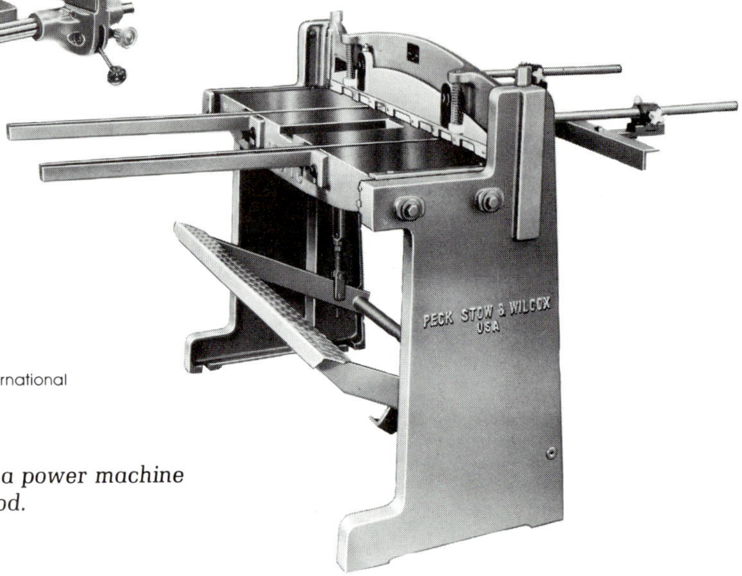

A squaring shears is a piece of equipment used to cut sheet metal.

operator supplies the power for cutting by pushing down on a foot lever. See Figure 3-8.

An oxyacetylene welding outfit is another example of equipment that is not motor-driven. Ovens, furnaces, and kilns are used to heat materials. They are also equipment.

Figure 3-9

The riveter is an industrial machine. Here it is used to fasten the wing assembly of an airplane.

Industrial Machines and Equipment

Machines and equipment used in the manufacturing industry are usually large and powerful. Most of these machines are designed for special purposes. See Figure 3-9.

Heavy work can be done faster on these machines. They are designed for frequent use. Sometimes they run without stopping. The operator may simply load the work into a machine and push the "start" button. The machine does the rest.

The machines and equipment used in the manufacturing industry are larger and more powerful than the tools in the school shop. However, they do the same basic work.

Materials

Our world is made up of a great variety of materials. Steel, glass, nylon, aluminum, and wood are just a few of the materials from which products are made. These materials are called industrial materials.

Years ago only a few materials were known and used to make products. Today there are many different materials that can be used. For example, in 1900, about 100 different materials

How is Manufacturing Done   43

were used in making a car. Today, each car contains over 4000 different materials.

Materials must be chosen wisely for each part of a product. See Figure 3-11. Before selecting materials, a number of questions must be answered. Will the material stand the "wear and tear" that it will receive when used in a product? Will it be light enough or heavy enough? Will it rust? or split? or warp? Can it be welded? bent? painted? These are just some of the questions that might need to be answered when selecting a material.

Basic Types of Materials

Several major kinds of industrial materials are commonly used in manufacturing. Wood, metal, and plastics are examples. Industrial materials can be classified as natural or synthetic.

Natural materials are those that can be found in nature. Woods and metals are examples of natural materials. They have been around for thousands of years. Materials such as plastics are not natural materials. People

Figure 3-11
Rubber, steel, aluminum, and vinyl were considered by the manufacturer to be the best materials to make a snowmobile.

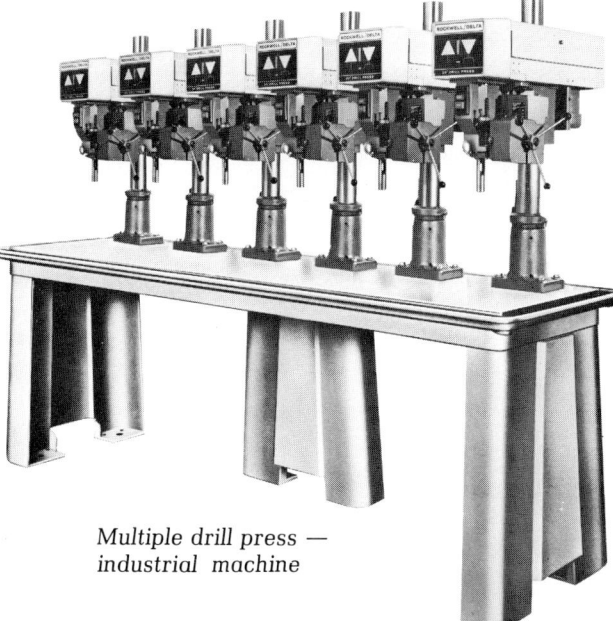

Drill press — power machine

Hand drill — hand tool

Portable electric drill — portable power tool

Multiple drill press — industrial machine

Figure 3-10
All of these tools are drills, but each is in a different tool classification.

make plastics. Materials that are made by people are called **synthetic.**

During World War II, the supply of Japanese silk was cut off. The military needed a material to replace the silk used in making parachutes. Military people decided to use a new synthetic fabric that could be made by combining chemicals. Nylon, a synthetic material developed by DuPont in 1934, was first used to make bristles for toothbrushes. It also proved to be the ideal fabric for military equipment. Nylon was lightweight and strong. Nearly four million parachutes were made of it. Lifeboats, combat clothing, tents, and hammocks were also produced from this new material.

Today, nylon is commonly used. Tires are reinforced (made stronger) with nylon. Clothing and carpets are made of nylon. These are just a few of the products that were once manufactured from natural materials, but are now made from nylon.

Wood

Wood is a natural material that is used to make many products. It is relatively easy to obtain. It is also easy to process. Trees are the source of wood. Wood has several qualities that make it useful for products. Some woods have a beautiful grain pattern. Some are pleasant smelling. Others are tough and rot-resistant. The kind of wood selected for a part or product depends upon how it will be used. The appearance of the product must also be considered. See Figure 3-12.

Not all wood is cut into boards. Some is sliced into thin layers that are glued together. This is called **plywood.** Some wood is used to make other kinds of wood products. **Particleboard** is made of many small chips compressed (squeezed together) into a sheet. **Fiberboard** (hardboard) is made of wood fibers that have been compressed in the same manner.

Metal

Metal is a natural material. It is strong and durable. Metal does not come from the earth in usable form. It is found combined with other substances. This combination is called ore. The ore is dug from mines in the earth. It must be processed to make usable metal.

Steel, for example, does not come from the earth ready to use. Iron ore must be made first into iron and then into steel. The steel is made into standard forms and sizes. It can be formed into sheets, rods, or bars.

Copper, aluminum, and other metals are processed in much the same way as steel. Many parts and products are made of metal. See Figure 3-13.

Plastics

Plastics are synthetic materials. Vinyl and acrylic are plastics. The raw materials used to make plastics are natural substances, such as petroleum or cellulose (material from plants such as cotton).

Plastics are popular materials for manufacturing. They are cheap and easily formed. Many parts that were once made of metal or wood are now made of plastics. Some plastics are unbreakable. Others are heat-resistant. Most plastics are good electrical **insulators.** They do not allow electricity to flow through them.

Plastics are made in different forms. They can be obtained in sheets, foam, pellets (pieces), liquid, and many other forms. If you look inside a car, you can see many plastic parts. The dashboard, steering wheel, knobs, and seat covers may all be made of plastics. See Figure 3-14.

Figure 3-12

The type of wood selected for a product depends on the wood's special qualities. Why do you think the manufacturer picked this type of wood for the chair? For its strength? Or for its beautiful grain pattern?

How is Manufacturing Done 45

Plastic Gets Behind the Eight Ball

In 1868, the production of a certain kind of item was threatened. Previously, ivory had been used to make the product. But now ivory was in short supply.

An acceptable substitute had to be found or production would cease. A totally new material was needed. As incentive (encouragement), a $10,000 prize was offered for the person who developed the new material.

John Wesley Hyatt of Albany, New York, won the prize. He experimented until he developed the right combination of cellulose and solid camphor to produce Celluloid®, the first **plastic**. It was used successfully to make the product, and Hyatt became known as "the father of the plastics industry." In 100 years this industry grew to include over 5700 companies.

But what was the product for which the prize money was offered? Camera film? Comb? Raincoat? Electrical insulator? None of these. Plastic was **first** used as a substitute for ivory in the manufacture of billiard balls.

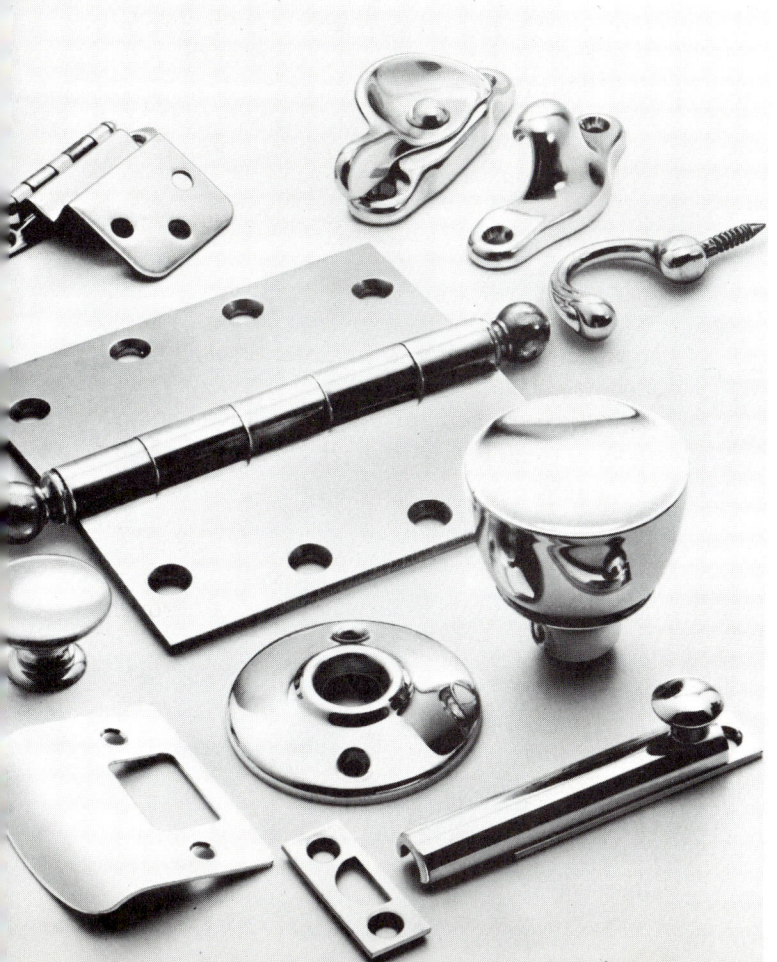

Figure 3-13

These door attachments are made of brass. This material is strong and has a bright, shiny appearance.

Figure 3-14

The body of this car is made of fiber glass reinforced plastic. This material is very strong. Is strength a good quality for a car body to have?

Bradley Automotive

Other Materials

Many other kinds of materials are used in manfacturing. Glass-like glazes and clay that is fired (baked in a furnace) are called **ceramics.** The sand (silica) and clay for making ceramics are natural materials that are taken from the earth. Bricks, dishes, and sinks are made of ceramic material. Some spacecraft have nose cones made of ceramic material.

Fabric materials are called **textiles.** Textiles are made from natural fibers (thread-like material) or from synthetic fibers. Cotton, wool, and silk are natural fibers. Nylon and polyester are synthetics. Some fabrics are a blend (mixture) of several materials, for example, 50% cotton and 50% polyester.

Leather is a natural material. It comes from animal hides. Some leathers are smooth and shiny. Others are soft and furry. A wide variety of products are made from leather. In addition to shoes, belts, and coats, leather is used to make suitcases and upholstery (seat coverings).

Rubber is still another important material used in manufacturing. Rubber can be hard or soft. It may be either natural or synthetic. Tires, of course, are made of rubber. Other products are hoses, V-belts for machinery, gaskets, and inner tubes.

Still other materials are used in manufacturing. Wood, metal, plastics, ceramics, textiles, leather, and rubber are some of the most common ones.

Processes

Processes are activities that change materials according to a plan. Sawing, drilling, sanding, grinding, bending, and painting are **methods** (specific ways) of changing industrial materials into products. Three major types of processes are used in manufacturing:

- Forming,
- Separating, and
- Combining.

Forming and separating are the two basic ways to change materials into parts or components. When components are put together into products, the process is called **combining** or assembling. The manufacture of most products requires all three processes. See Figure 3-16.

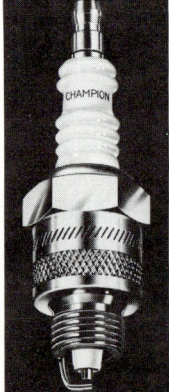

Ceramic-insulated spark plug

Leather purse

Figure 3-15

Materials are selected that best suit the uses of each product.

Rubber tire for an earthworking machine

Fabric-covered chair

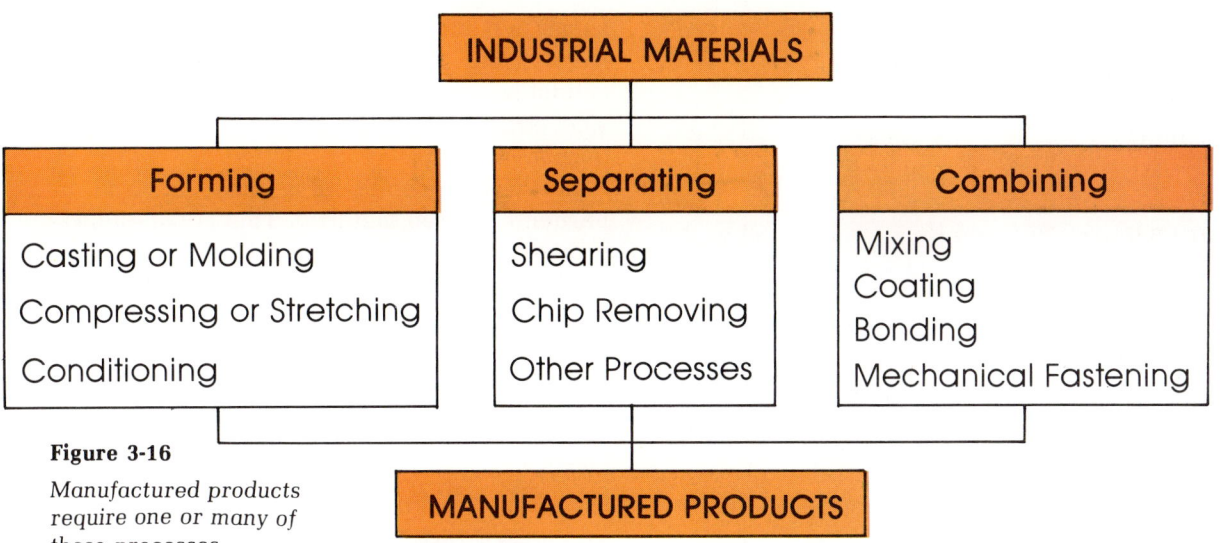

Figure 3-16
Manufactured products require one or many of these processes.

People are always trying to develop new and better ways to make and assemble parts. For example, **Vernon Krieble,** a college chemistry professor, had been searching for a new process for assembling metal parts. He invented the first practical glue for bonding metal.

When he retired from teaching, the 70-year-old man went into business for himself. He made some glue and gave samples to manufacturers for testing. He convinced the engineers at International Harvester Company to use his glue for assembling tractor parts. His new process was also approved for use at General Motors for bonding parts of cars. The glue, Loctite®, was a success. Soon other manufacturers began using the glue. With the introduction of these new "super glues," it was no longer necessary to weld or rivet metal parts together.

Today many products have parts held in place by glue. These adhesives are so strong that even parts of airplanes and spacecraft are assembled with glue. Vernon Krieble had developed a new method for assembling materials.

Forming Processes

In **forming,** the shape of a material is changed. No material is added or taken away. Forming processes are often used because little material is wasted. The forming processes are

- Casting or molding,
- Compressing or stretching, and
- Conditioning.

Casting or Molding

In **casting or molding,** a solid material is first changed to a liquid. This is usually done by heating the material until it melts. It is then poured or forced into a hollow form called a **mold.** See Figure 3-17. The space inside the form is called a **cavity.** This cavity must be the desired shape of the part. As the material cools, it hardens. It can then be removed from the mold. Making ice cubes is a very simple form of casting.

Figure 3-17
In casting or molding, liquid metal may be poured or forced into a cavity and allowed to harden.

same. In manufacturing, a common example is magnetizing a piece of steel. The difference cannot be seen. A change takes place in the inside structure of the steel that causes it to become magnetic.

In **thermal conditioning,** heat is used to harden or soften material. This is done to improve the material or make it easier to work. Metal, for example, can be made very hard by treating it with heat. See Figure 3-18.

Chemicals can be added to cause changes in some materials. A **catalyst** is a chemical that

Figure 3-18
This large gear is being thermally (heat) conditioned to make the teeth hard.

Figure 3-19
Separating processes.

Compressing or Stretching

In **compressing or stretching,** the material is shaped by squeezing or pulling. The specific methods are listed below.
1. **Forging** means hammering or squeezing material into shape.
2. **Bending** is a simple way to form materials such as metal or paper. The shape of a sheet of paper can be changed into a paper airplane by bending or folding it.
3. **Rolling** material will make it thinner by squeezing it. A material may be forced under a roller or between two rollers. The roller or rollers press against the material. In a kitchen, pie or cookie dough is made thinner by using a rolling pin to flatten it.
4. **Compression molding** is done by squeezing small solid particles of material into a closed mold. The particles are packed tightly together and come out as a solid.
5. **Drawing** or stamping is stretching a material over a pattern or shape called a **die.** Automobile wheel covers, pots and pans, and motorcycle fenders are formed by this method from flat sheets of metal.

Conditioning

Usually during **conditioning,** only the inside structure of a material is changed. The outside appearance of the material is seldom affected.

An example of **physical conditioning** is boiling an egg. The heat causes the inside to change. The outside appearance remains the

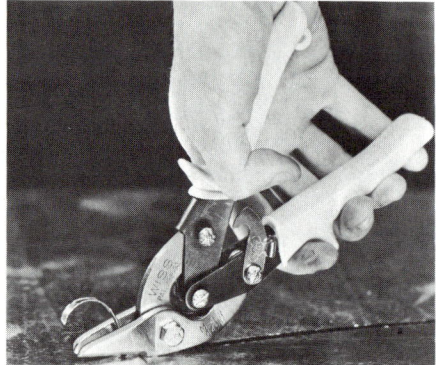
Shearing

Chip removing
Thermal erosion

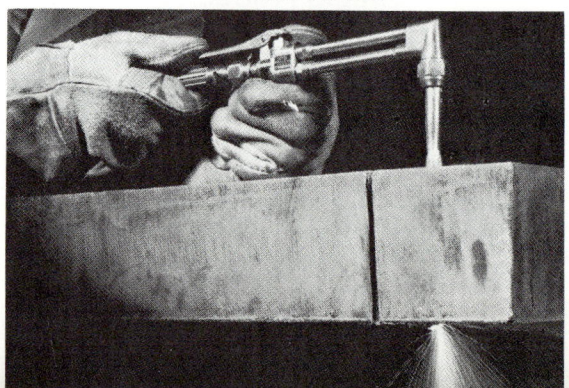

changes the makeup of a material. Epoxy glue, for example, will become hard when it is **chemically conditioned** by adding a catalyst.

Separating Processes

Separating is a widely used type of process. It is used to change the shape of material. Some of the material is removed. Separating can be done in a number of ways:

- Shearing,
- Chip removing, and
- Other separating processes.

Chemical separation

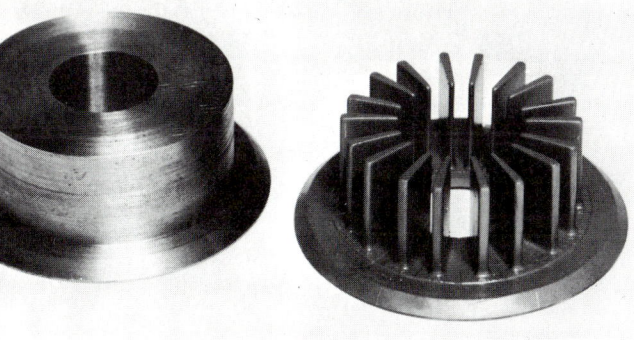

Electrochemical separation

Induced fracture

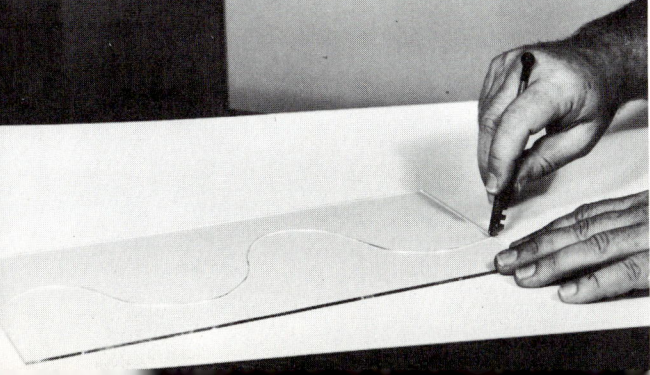

Shearing

In shearing, part of a solid piece of material is separated or cut apart from the rest of the material. For example, cutting a piece of paper with scissors is **shearing.** Sheet metal and plastics are often sheared. Punching holes is also a kind of shearing.

Chip Removing

In **chip removing,** material is removed in the form of small chips, such as sawdust or metal shavings. Chip removing can be done by many different methods. Sawing, drilling, sanding, grinding, planing, and turning are some examples. Anytime you separate by chip-removing processes, there will always be small pieces of material that are removed. These are called **waste.**

Other Separating Processes

Materials can be separated by other processes as well. **Heat** can be used to burn away material. Cutting a piece of steel with a cutting torch is an example.

Chemicals can also be used for separating. For instance, acid can be used to eat away metal in a pattern or design. This is called **etching.** Electrical current is sometimes combined with chemicals to separate material.

Glass cutting is a common process. Glass isn't really cut. It is broken or fractured. A line is scribed (scratched) into the surface of the glass. When pressure is applied, the glass breaks along the scratched line.

Combining Processes

When two or more materials or parts are put together, the process is called **combining.** Most products are combinations of many parts. Even a simple pencil has at least five parts: the lead (really graphite), two wooden halves, the eraser, and a metal band. Parts and materials may be combined by several processes. Four combining processes are widely used:

- Mixing,
- Coating,
- Bonding, and
- Mechanical fastening.

50   Introduction to Manufacturing

Figure 3-20
These dryer drums are being coated with a finish that will help to prevent rusting.

Mixing

The gases, liquids, or solids that go into a mixture are its **ingredients.** In the **mixing** process, the bits or particles of each ingredient are spread evenly throughout the mixture. Motor oil, cough syrup, and paint are liquids that have been manufactured by mixing. Solid ingredients that make up powdered soft drinks, aspirin tablets, and raisin cereal are mixed.

Coating

Coating is covering one material with another. This is usually done to protect or to decorate the product. A car is coated (painted) to give it a good appearance. The coating also protects the car from rusting. Painting is a common way of coating. See Figure 3-20.

Another method of coating is **plating.** The bumper on a car is shiny because the steel has a coat of chrome metal on the outside. It has been plated with chrome.

Bonding

A major combining process is **bonding.** In bonding, the place where the materials are joined is called the **joint.** Many types of glue are used to bond materials together. This method is called **adhesion.** Using an adhesive (usually glue), different kinds of material can be combined.

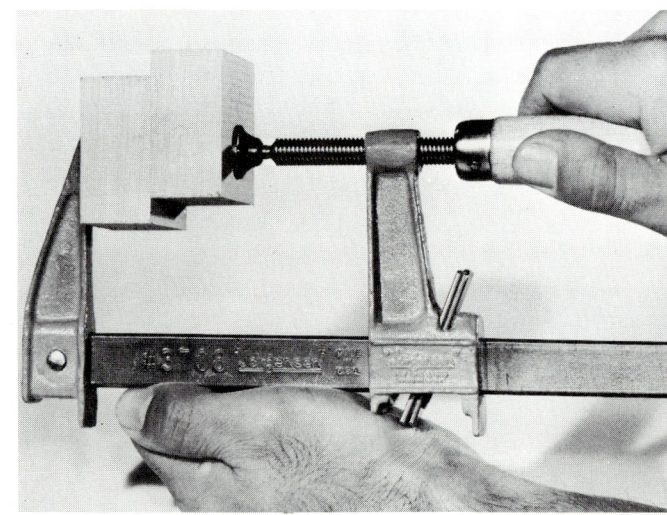

Figure 3-21
Bonding is a way to join parts permanently. These blocks of wood are joined with glue and clamped to seal the bond.

Two similar materials can be joined by **fusion.** Welding is a method of bonding by fusion. If the materials are melted at the joint, they fuse (blend together). Metals and plastics are often welded. Metals are welded by using heat. Plastics can be welded by using heat or a chemical to melt them. Usually, only materials that are alike can be combined by fusion bonding.

Mechanical Fastening

Many parts are combined by **mechanical fastening.** Examples of mechanical fasteners are screws and nails. These fasteners depend on friction (resistance) for holding power. Some hold parts together permanently. Others allow the parts to be disassembled (taken apart).

Parts that will need to be replaced later should be assembled with threaded fasteners. Bolts and screws are common examples of threaded fasteners.

Some mechanical fasteners do not have threads. Nails, staples, and rivets are non-threaded fasteners. They are used to combine parts permanently. Other non-threaded fasteners, such as cotter pins, retainer rings, and clips, are used in assemblies that may need to be taken apart later for repairs or adjustments. Many mechanical fasteners have been designed for special purposes.

Looking Ahead

The materials and tools necessary to produce products are the "what" of manufacturing. The processes used to change materials are the "how." One more element is needed — people, the "who" of manufacturing.

Do you know
- which is more dangerous: a sharp tool or a dull tool?
- where safety zones are located in your school shop?
- how to be a safe and successful worker?

In the next chapter, ways of being a good, safe worker are discussed. Learning these things can help you later in whatever work you do. It will help you now as you work in the school shop.

New Terms

adhesion
bonding
casting
catalyst
ceramics
chip removing
coating
combining
compression mold
equipment
etching
fiberboard
forging
forming
fusion
industrial materials
ingredients
insulator
joint
mechanical fastening
method
mixing
mold
particle board
pedestal
plating
plywood
separating
shearing
synthetic
thermal conditioning

Study Guide

1. How are the basic machines related to tools?
2. What is the main difference between power tools, power machines, and equipment?
3. What are some examples of industrial materials?
4. Explain each type of process used in manufacturing:
   a. Forming
   b. Separating
   c. Combining
5. Name a product that was made by using:
   a. One type of process.
   b. Two types of processes.
   c. All three types of processes.
6. Explain how pots and pans are formed.
7. Name two products in which nylon is used today. What qualities make nylon useful in products?
8. When selecting material for a product, why may wood be chosen over metal or plastic?
9. List at least three ceramic products that may be found in the home.
10. What major questions must be answered when selecting a material for a product?
11. What materials were used to make your desk? What process or processes do you think were used to put the parts together?

chapter 4

Workers and Safety

People are vital (necessary) to manufacturing. People with good **attitudes** who practice safety are good workers. You, as a student, are about to participate (take part) in the manufacturing process. Soon you will begin your first project. You need to know the attitudes of good workers and the safety practices to follow in the school shop. Using this knowledge will help you to produce quality products efficiently and safely.

People — The Most Important Input

What would manufacturing be without people? There would be no one to run machines or to design new products. Supplies would not be ordered, and products would not be sold. In short, manufacturing could not be done.

People are the most important part of manufacturing. People make the production system work. Manufacturing needs all kinds of people: short and tall, big and small, young and old, skilled and unskilled. There is a place in manufacturing for everybody. See Figure 4-1.

Worker Attitudes

Nothing is as important to a company as its workers. The attitudes of workers toward their jobs and their safety are very important to the success of a company. Let's look at some good attitudes that one must have or develop in order to be a safe and successful worker.

Promptness

Promptness is being on time. When work starts, the worker needs to be there. Money is wasted and time is lost when a worker is late. Also, the worker may rush the job while trying

Figure 4-2

Don't rush to work, but do get there on time.

Figure 4-1
No matter what your interests are, there's a job for you in manufacturing.

to catch up. Such hurrying often results in mistakes and sometimes accidents as well.

Regular Attendance

Millions of dollars are lost each year by the manufacturing industry because workers fail to show up for work. Workers who are ill should stay home. However, any other reason for missing work must be a good one. It is important for a worker to be on the job. Each person has tasks to perform every day. It is often a big problem to have someone else do them. Also, others may have to work extra hard to make up for a co-worker's absence.

Cooperation

Most people who lose their jobs do so because they cannot get along with fellow workers. A worker must cooperate with others. See Figure 4-3. No one should always demand his or her own way. It is also important to listen to the supervisor. A worker must follow the supervisor's directions carefully.

Reliability

A **reliable** worker can be trusted to do a job. Such a worker is valuable to the company. The worker accepts responsibility. The more responsibility a person is able to handle, the more

Figure 4-3

Learn to get along with your fellow workers. Cooperation makes everyone's job easier.

that person will be given. This usually results in a pay raise.

Persistence

A good worker is **persistent.** This means that the worker does not give up easily. A company needs people who will stick to difficult tasks and see their jobs through to completion. See Figure 4-4.

Patience

A good worker needs to be patient. It is hard to work with someone who "blows up" when things go wrong or when there is a delay. Such a person is also a danger to others.

Advantages of Good Attitudes

The more good attitudes a worker develops in life, the more valuable that person will be to the company or employer. This will help a worker to keep a job or to advance to a better one.

Good attitudes make a person a much safer worker. Working safely is a protection for the worker. A good safety record also increases the value of an employee.

A person is never too young to begin developing good worker attitudes. In fact, as time goes by, work will become much easier for a young person who has developed these attitudes. Such a person will be able to deal with more and more responsibility.

People and Safety

Why is it important to think about safety? Because every day over 167,000 people in the U.S. are injured. There are 360 accidents per minute or six every second! Some are very

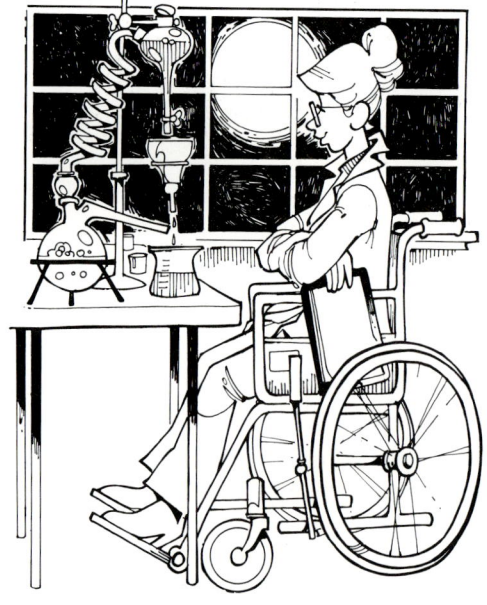

Figure 4-4

Finish each task, even though it may be difficult.

Workers and Safety 55

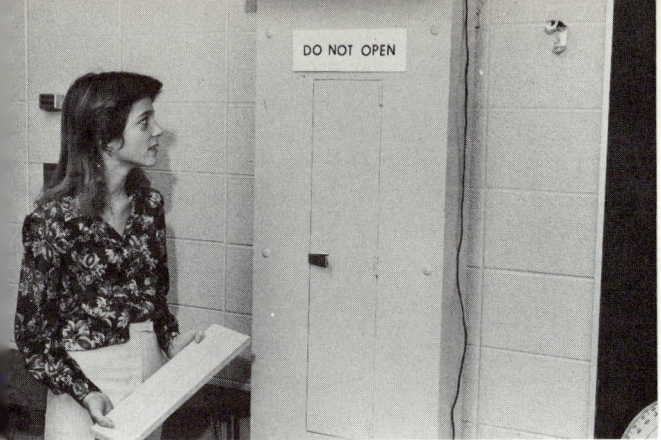

Figure 4-5

Caution! Obey the signs and protect yourself from injury.

or courses for teaching about safety or the improvement of job skills.

The United States Government is also concerned with safety. An agency called the Occupational Safety and Health Administration (OSHA) sets safety standards (acceptable conditions). This agency checks to see that places of work are safe.

serious accidents. Most of them happen because of carelessness. Accidents happen at home, at school, on the job, while playing, anywhere. Many of the 10,000 daily accidents could be avoided if people would simply think about safety. Accidents are not planned. They happen when we fail to take steps to avoid them. It does not matter if a person is in the shop, at home, or on the street riding a bike. Wherever one happens to be, that person should think ahead to avoid an accident. Safety should be an important concern in each person's life all of the time.

A Safety First

In 1850, an American, **Margaret E. Knight,** invented a safety device for a textile loom. It was designed to automatically shut down the loom if a steel-tipped shuttle fell out. The invention, however, was less remarkable than the inventor. For in 1850, Margaret E. Knight was only twelve years old. She went on to invent many other things.

Safety in Manufacturing

Manufacturers have responsibilities to the people who work for them, too. Companies should provide their workers with a safe place in which to work. Dangerous machines should have safety guards. Special safety devices, such as hard hats and safety glasses, should be provided to workers. There should be adequate (enough) lighting. Excess noise should be controlled.

Many companies have accident prevention programs. Some companies have special schools

Figure 4-6

Your personal safety depends on following these safety rules.

Safety in the Shop

In the shop, it is important to **work safely at all times.** Many tools and machines, if not used properly and safely, can cause serious accidents. Students need to know how to avoid **safety hazards** (dangerous situations) in the shop.

Learn **how** to use each tool and machine safely **before** you begin to use it. Be aware of the general safety practices for working in the shop. By learning how to work safely in the school shop, you will develop good safety habits. These will help you now and in the future when you get a job.

56 Introduction to Manufacturing

Safety glasses

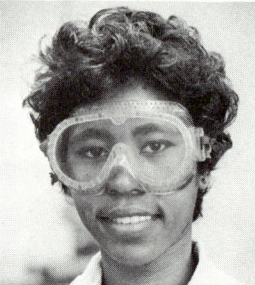

Goggles

Face shield

Figure 4-7
Proper eye protection will prevent flying chips and splinters from injuring your eyes.

Figure 4-8
Remember, allow two inches for safety. Your hands should never be closer to the blade than two inches when sawing.

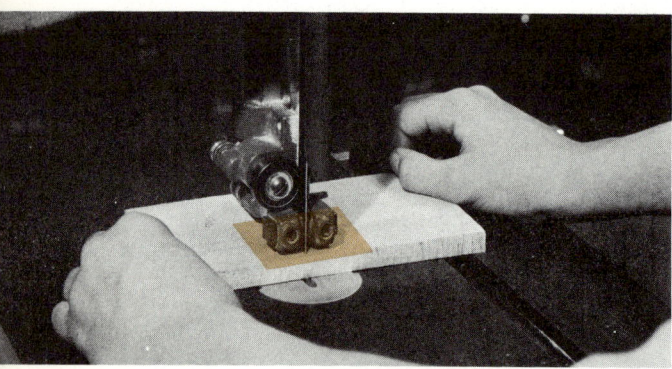

Learning the correct and safe way to use tools and machines will help you to be a more efficient worker. You will be less likely to make mistakes or get hurt.

Personal Safety Practices

It is important to protect yourself from possible accidents. This protection requires that you **plan ahead** to avoid hazardous (dangerous) situations. Special care must be taken for the following:

- **Eyes** — Wear safety eyeglasses or goggles at all times in the shop, but especially for all machine operations. Safety glasses and goggles prevent chips from striking the eye. Some goggles protect the eyes from bright, damaging light. See Figure 4-7.
- **Ears** — Loud, continuous noise may damage hearing. When working under those conditions, wear ear protectors. You will be able to hear voices, but not noise. Ear protectors can be seen in the photograph at the beginning of this chapter.
- **Hands and Fingers** — Keep your hands and fingers clear of all blades and cutting tools at **all** times. Do not wear gloves while operating machines in the shop. See Figure 4-8.
- **Hair** — Long hair can get caught in moving parts of machines. Tie the hair back or wear a protective net or cap. See Figure 4-9.

Figure 4-9
Tie it back! Long hair can get caught in a machine and cause a serious injury.

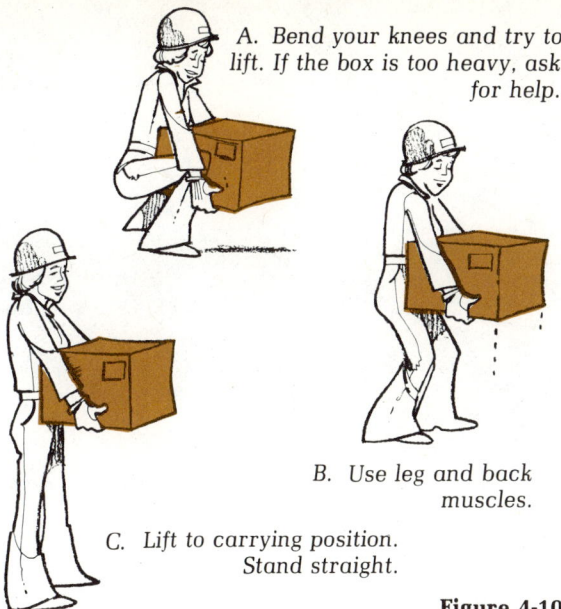

A. Bend your knees and try to lift. If the box is too heavy, ask for help.

B. Use leg and back muscles.

C. Lift to carrying position. Stand straight.

Figure 4-10

Here's the safe way to lift and carry a box.

- **Clothing** — Loose clothing, long sleeves, ties, and shirttails can get caught in machines. Take off your coat, roll up long sleeves tightly above the elbow, and tuck in your shirttails. Tie a shop apron behind your back to keep the strings away from the machine.
- **Jewelry** — Necklaces, bracelets, watches, and rings can get caught in machines. They should be removed before working with machines.
- **Mouth** — Placing objects, such as nails, in the mouth is very dangerous. They can be swallowed easily.
- **Back** — Use caution when handling large or heavy materials. When lifting, bend your knees, not your back. See Figure 4-10.
- **Feet** — Often there are small sharp objects and scraps on the floor. Wear shoes at all times. Sandals and open-toed shoes do not protect your feet in the shop. Hot materials could easily splatter or fall on your feet.

General Safety Practices

Pay attention to what you are doing at all times.

"Goofing around" and throwing objects should **never** be done. They are dangerous things to do. They can cause serious accidents.

Report all accidents (even though minor) to the instructor **immediately.**

Rags with oil or paint thinner on them can start a fire. Place them in a covered metal container.

In case of fire, know the locations of exits and fire extinguishers. Also be sure to know the proper type of extinguisher to use to put out a fire. When ordinary **combustibles** such as papers are on fire and no electricity is involved, use an extinguisher marked with an **A**. When liquids such as gasoline or oil are burning, use an extinguisher marked with a **B**. To put out electrical fires, use an extinguisher marked wih a **C**. See Figure 4-11.

 green

FOR ORDINARY COMBUSTIBLES
Put out a Class A fire by lowering its temperature using a water or water-based extinguisher. Wet fire to cool, and soak to stop smoldering.

 red

FOR ORDINARY COMBUSTIBLES
Put out a Class B fire by smothering it. Use extinguisher giving a blanketing, flame-interrupting effect. Cover the whole flaming liquid surface.

 blue

FOR ELECTRICAL EQUIPMENT
When live electrical equipment (Class C fire) is involved, always use a nonconducting extinguishing agent (not water) to avoid receiving an electric shock. Shut off power as quickly as possible.

Metal Manufacturing Technology by Jack W. Chaplin

Figure 4-11

These labels on fire extinguishers help you to know how to put out a fire. Each label has a different letter, shape, and color for the three kinds of fires. Learn them now. You may not have time to read when a fire occurs.

Use extreme caution when handling combustible liquids or chemicals that may burn the skin or eyes. Avoid breathing the vapor (fumes) from these chemicals. Wear rubber gloves when needed.

Handle hot metals carefully. Use tongs or pliers to pick them up. If you are not sure if a piece of metal is hot, treat it as hot. See Figure 4-12.

Report to the instructor any tools or machines that are broken, dull, or do not work properly.

Clean up material scraps and spills immediately. Otherwise, someone may trip or slip on them. See Figure 4-13.

As a protection for yourself and your classmates, report any violations of safety rules to the instructor.

To avoid making mistakes and having accidents, **listen** carefully to the instructor at all times.

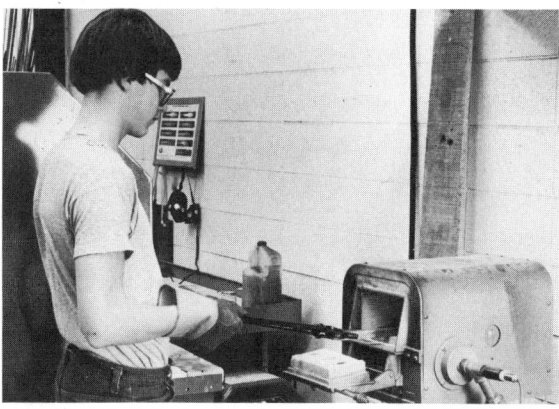

Figure 4-12
Use tongs or pliers to handle hot metal. Don't take a chance that the metal may have cooled.

Figure 4-13
Take the time to clean up spills or material scraps. Cleaning up will help prevent slipping or tripping by fellow workers.

Figure 4-14
Hold all tools securely. Close materials tightly in a vise or clamp if necessary.

Hand Tool Safety Practices

If the wrong tools are used, they can break or chip and cause an injury. **Use the right tool for the right job.**

A dull tool is a dangerous tool. Accidents are more likely to happen when you try to use a dull tool. You must push harder. The tool could slip. Keep tools sharp. Protect yourself and others from injury. Carry sharp tools carefully, with points down. **Never** carry tools in your pockets.

Hold all tools and materials securely. Use **vises** and clamps when needed. See Figure 4-14.

Machine Safety Practices

Be sure to have permission before using machines.

Only the person operating the machine is allowed inside the **safety zone** around the machine. Unless otherwise marked, this area is approximately an arm's distance away from the machine operator. See Figure 4-15.

Distracting or hurrying the people using machines can cause them to make mistakes. They may seriously injure themselves. Let them concentrate on what they are doing.

Remove small scraps of material from a machine only **after** the machine has come to a complete stop. Use a brush to clean the area.

Machine guards keep your fingers safe. They also protect you from material that might be thrown from the machine. **Keep guards in position.** See Figure 4-16.

Figure 4-15

The safety zone helps to prevent bumping or distracting the machine operator. Here, one worker waits outside the safety zone while the other worker operates the machine.

Figure 4-16

Always use machine guards. They can keep your fingers from slipping into the blade and the board from flying out of the machine.

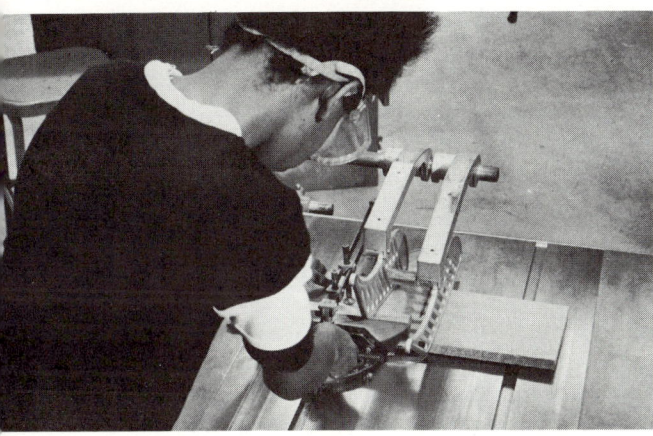

Before measuring work or adjusting the machine, make sure the machine is at a complete stop.

When in doubt about any machine operation, **ask the instructor.**

Looking Ahead

Good workers are important to manufacturing. You are important to manufacturing. You are about to produce your first product.

Workers and Safety 59

Remember, having good worker attitudes and following safety rules will help you to become a successful manufacturer.

Next, ideas and manufacturing will be discussed. You will discover some of the ways that people get ideas for products and processes. You will learn how ideas are developed into products, and you will begin to understand some of the reasons why products succeed or fail.

Do you know
- why paper bags have flat bottoms?
- what tiny development is changing giant industries?
- if a prototype is a special printing machine?
- an invention that is invisible?

In the next chapter, we will examine the research and development part of manufacturing, where **ideas** become realities.

New Terms

attitudes	reliability
combustibles	safety hazards
machine guards	safety zone
OSHA	vise
persistence	

Study Guide

1. Why are workers important to the manufacturing industry?
2. Name some attitudes that a worker must have (or develop) in order to be a successful worker. Explain each one.
3. How can a worker protect the following when working in the shop?
   a. Eyes
   b. Ears
   c. Hands
   d. Hair
   e. Mouth
   f. Back
   g. Feet
4. What safety hazards can cause fires and burns in the shop?
5. Why is it important to use the right tool for a job?
6. Name some safety rules for carrying tools.
7. List several safety rules that you should follow when operating machinery.

LET'S GO TO WORK

Understanding Product Plans

One of the most important things in making a product is knowing how to get started. If this is your first product, you should choose a simple one. As you learn more about tools, materials, and processes you can make products that are more difficult.

Product Decisions

In order to manufacture a product in your school shop, you will have to make several decisions. First, you must decide what **product** to make. Will it be one of the products in this book? Perhaps your teacher has another product idea. Maybe **you** have seen a product you would like to make.

Next, you must decide what type of **production system** to use: custom, job-lot or line (mass) production. Will you make many of the products or only one? What **materials** will you need?

If you are making a product from this book you will find a list of parts and materials for each product. Even if you use this list you will need to make additional decisions about materials. For example, the parts and materials list may specify (say to use) wood. You must decide what **kind** of wood to use.

You will need to know what **tools** to use. Selecting the proper tool for each job will help you to make a better product. You will also be able to work more efficiently. Consider all questions carefully. This experience will help you to develop skills and to understand the processes of manufacturing. You will learn to waste little material. It is also a chance to practice working **safely**.

Product Plans

Manufacturing a product requires planning. The **plans** for products shown in this book are made up of several items.

- The **photograph** shows the finished product.
- The **dimensions list** gives the measurements of the product and product parts.
- The **working drawings** show what the product looks like.
- The **parts and materials list** gives information about materials and other items needed.
- The **production flow chart** shows the overall plan for making the product.
- The **procedure chart** tells how to make and assemble the parts.

As you read further, refer to the information given on pages 62 and 63 for the first product, a football kicking tee, as an example.

Working Drawings

To make a product, you must first learn to read **working drawings.** The working drawings show all the **sizes** of the parts. They also show how the parts **fit** together.

Study the drawings of the football kicking tee. The drawings show the top, front, and side views. Notice the dashed lines in the front and side views. Like an X ray, they show parts that are hidden from your eyes. The dashed lines in these views indicate that the pegs or dowels go all of the way to the bottom of the base.

In this plan, a **pictorial view** is also given. A pictorial view shows the three sides of the ob-

Introduction to Manufacturing **61**

ject using only one view. Pictorial views may or may not be given in other plans.

Dimensions List

The dimensions, or sizes, of each part are given on the drawing. These sizes are indicated by letters instead of numbers. Locate the letters on the **dimensions list.** Then use either the metric or the customary measurements when making the parts. Check yourself according to the plan. What size is the base of the kicking tee in metric? in customary? Because the metric measurements may not be exactly the same as the customary measurements, you should not mix measurements. Use **either** the metric or the customary but **not both.** If you would like to use metric measurements but do not understand the metric system, refer to Topic 1, **Metrics.**

Parts and Materials List

The materials and other items needed to make the product are shown on the **parts and materials list.** The list for the kicking tee names all of the parts and materials needed to make **one** kicking tee. If you want to manufacture 10 kicking tees, you will probably need 10 times the amount of material.

You will need to make some decisions about materials to use. For example, the parts and materials list shows that the base of the kicking tee is made of wood. What kind of wood will you use? Can you use a different material? Refer to Topics 26 through 29 for information about materials.

Production Plans

The working drawings **show** the parts of the product and how they should fit together. Production plans **explain** how to make the parts and assemble them. First, look at the **production flow chart.** It is a diagram that outlines the steps in making and assembling the parts. Each operation has a code number. These give the order in which the operations are to be done and when the components should be assembled. The symbol **TS** in the final step stands for **Temporary Storage.**

The **procedure chart** tells how to make the product. Information given here is more detailed than that given on the production flow chart. Study the various columns of the chart. Again each operation has a code number. This is the same number that is on the flow chart. The information in the operation column tells what operation must be done. Operation B-1 is "Cut to size." You must look at the drawing and the dimensions chart to find the correct sizes or dimensions. For operation B-2, "Drill holes," refer again to the drawings and to the dimensions chart. Find the correct diameter, depth, and locations for drilling the holes. Measurements that are given in numerals on the drawings or on the parts and materials list are in inches unless another kind of measurement is given. For example, "3/8 drill" on the kicking tee plan means 3/8-**inch** drill.

Notice the column marked "Tools & Equipment." Listed here are the tools and equipment needed for the operations. You will need to choose the best tool or tools available to you for each operation. If you need to learn more about the tools, information is provided in the topics in Unit V.

You may also wish to learn about the materials and processes used when making the product. The **topic numbers** given on the chart refer to the topic or topics related to each operation. Be sure to read and review the topics, especially when you have questions about a certain operation. It is easier to avoid mistakes than to correct them. Always ask your teacher for help when you need it.

Special information is given in the "Notes" column. These suggestions may help you to perform an operation more efficiently. They may even include an alternate (another) method.

Abbreviations are sometimes used in the product plans. These abbreviations and what they mean are listed below.

DIA.	— diameter	CSK	— countersink
PH	— pan head (screws)	OD	— outside diameter
		GA.	— gage
FH	— flat head	#	— number
RH	— round head		

Layout

Transferring measurements from the drawings to the material and marking their locations

Football Kicking Tee

PARTS AND MATERIALS

Qty.	Part	Size	Material
1	Base	D x A x B	Wood
2	Long Pegs	3/8" dia. x F	Wood Dowel
2	Short Pegs	3/8" dia. x G	Wood Dowel
			Glue
			Finish

DIMENSIONS

Dimension Symbol	Metric mm	Cust. in.
A	114	4-1/2
B	108	4-1/4
C	20	3/4
D	25	1
E	16	5/8
F	70	2-3/4
G	45	1-3/4
H	3	1/8

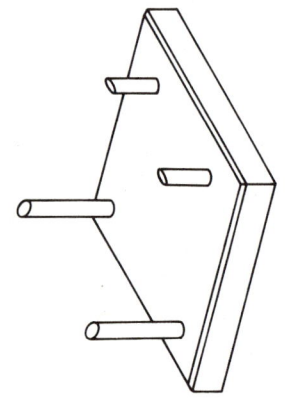

PICTORIAL VIEW

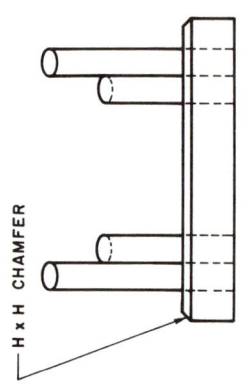

H x H CHAMFER

SIDE VIEW

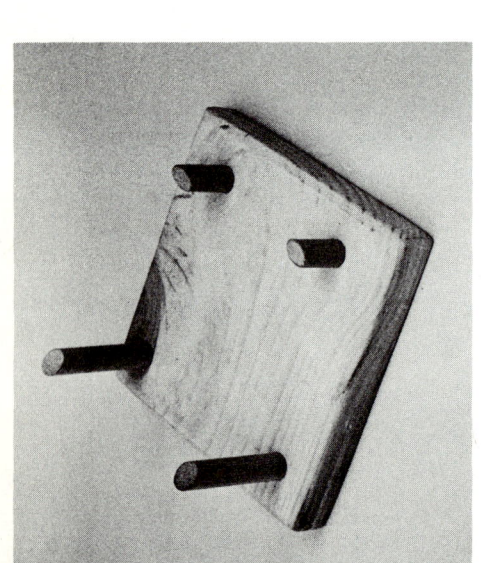

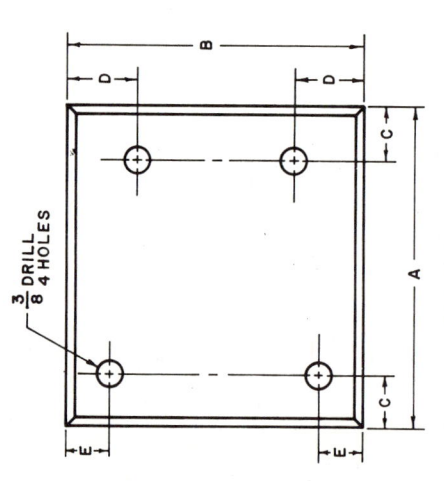

3/8 DRILL 4 HOLES

TOP VIEW

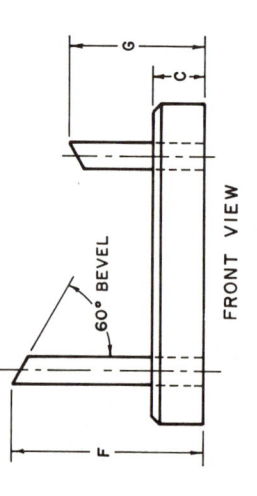

60° BEVEL

FRONT VIEW

Football Kicking Tee

PROCEDURE

Operation Number	Operation	Tools & Equipment	Topics	Notes
BASE				
B-1	Cut to size	Handsaw or power saw	5	
B-2	Drill holes	Portable electric drill or drill press, 3/8" twist drill	6, 25	A drilling fixture may be used.
B-3	Chamfer edges	Hand plane or Uniplane or disk sander	7 or 15	
B-4	Sand edges	Belt sander or disk sander or finishing sander, abrasive paper	15	
B-5	Sand faces	Belt sander or finishing sander, abrasive paper	15	
PEGS				
P-1	Cut to length	Miter saw or coping saw or backsaw	5	
P-2	Bevel ends	Disk sander	15	
ASSEMBLY				
FKT-1	Assemble pegs to base	Mallet or hammer	3, 11	Use glue. Be sure bevels are positioned as shown.
FKT-2	Finish	Spray equipment or brush	23	

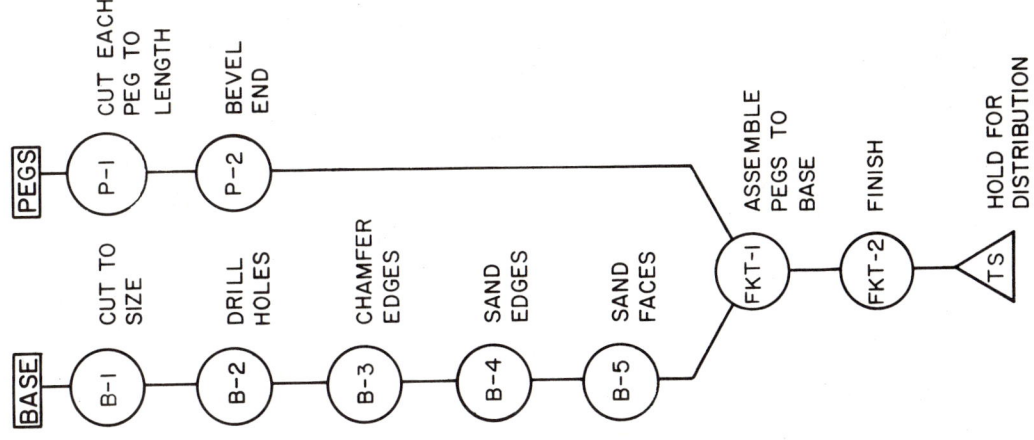

64 Introduction to Manufacturing

is called layout. Notice that the production flow chart does not show any layout operations. This is because the layout depends on the type of production system you will use.

Before starting your layout, refer to Topic 2, **Layout.** This topic discusses layout procedures and the kinds of tools used most often in laying out work.

If you are making a quantity of the same product, refer to Topic 25, **Production Tooling.** Here you will find information about jigs and fixtures and other special tools. These special tools make layout work much easier.

Product Previews

At the ends of Units I, II, and III are **Product Previews.** Products related to the units are listed in the previews. Photographs of the products and some information about them are given. These previews will help you to select a product to make. Each preview is a little different from the others. The Product Preview for Unit I follows this section.

The products suggested for Unit I help you to gain experience in using tools, materials, and processes. The stars indicate how easy or how difficult a product might be to make. Products with one star are the easiest to make. Products with two stars are of medium difficulty. Products with three stars are the most difficult. More information about making the products is given in Unit IV, **Product Plans.** Page numbers of the product plans for each product are listed in the last column of the Product Preview.

The products listed for Unit II, **Research and Development,** help you to learn to design and engineer products on your own. This preview tells you what information is given on the plans and what you will need to find out or to decide for each product.

The preview for Unit III, **Production,** suggests products that can be mass-produced. Again the level of difficulty is shown by stars. Time is also a factor in mass production. Products requiring the least amount of time to produce have one star in the time column. The more stars in this column, the greater the amount of time needed to produce the product.

Products

This book contains drawings and production plans for 44 products. Some products are made of wood, some of metal, some of plastics, and some are made of other materials. Different skills may be required to make each product. Some products are easy to make. Some are more difficult. As you make products, you will increase your knowledge of the processes to follow and the tools to use. You will improve your skills. You will also learn more about the kinds of materials from which products can be made. Knowledge of tools, materials, and processes and experience in using them will help you to become a skilled manufacturer.

Let's Go To Work!

Unit 1 Product Review

Note: Before making any of the products presented in Unit IV, read **Workers and Safety** and **Understanding Product Plans**, pages 52 through 64.

Product	Primary Material	Level	Special Notes	Page
XYLOBOX — Makes "beautiful" music.	Wood	★	Learn how to make simple wood joints.	180
PET ROOSTER — Definitely not for the birds.	Wood	★	Offers lots of creative opportunity for decorating.	192
CANDLE HOLDER — This could turn you on to wood turning.	Wood	★★	Design can easily be changed.	202
CUTTING BOARD — Bet you can't make just one!	Wood	★★	Requires careful cutting. Two different kinds of wood are needed.	204
DESK CADDY — Organize your clutter.	Wood	★★★	Requires careful assembly.	220

(continued next page)

Product		Primary Material	Level	Special Notes	Page
MIRROR SHELF Your work reflects on you.		Wood	★★★	Gain experience in working with a router.	224
NAME BADGE Join the name game.		Metal	★	Use when class members do not know each other.	232
BIKE BEVERAGE HOLDER Pedal with your "pop."		Metal	★	Requires careful layout.	234
CHARCOAL TONGS Smolder holder.		Metal	★★	Choose from two different designs.	240
SCREWDRIVER Not a boring product.		Metal	★★★	Requires careful attention to details.	246

CALCULATOR STAND You can count on it.	Plastic	★	Must be formed accurately.	256
LETTER OPENER A new twist in opening envelopes.	Plastic	★	You could make a bracelet instead.	258
FUNNEL Fill it up!	Plastic	★★	Choose either the straight or the offset version.	262
TRIVET This one can take the heat.	Plastic	★★	Surround an item with clear plastic. Requires a pattern and a mold.	266
SKATEBOARD You can really start rolling with this one.	Plastic	★★★	Requires a lot of time.	268

(continued next page)

Product		Primary Material	Level	Special Notes	Page
BELT BUCKLE Buckle up in style.		Metal Ceramic	★★	Each one is different.	236
TOTE BAG You can take it with you.		Fabric	★	No sewing required. Any fabric can be used.	272
SHOP APRON Keep your clothes clean.		Fabric	★	No sewing required.	274
PLANT HOLDER Support your local plant.		Leather	★	Requires careful cutting.	276
CAMERA CASE Picture your camera in this.		Leather	★★	Requires assembly. Case can be decorated. May be made into a calculator case instead.	278

UNIT II

RESEARCH AND DEVELOPMENT

chapter 5

Ideas Are Powerful

Ideas have changed the world. They are important in all areas of our lives. Products we use began as ideas. People in manufacturing work to make ideas into realities.

Where Do New Ideas Come From?

Products don't just fall out of the sky. They don't suddenly show up on the assembly line of a manufacturing company. Much work and time goes into finding new ideas for products. Every new idea does not necessarily become a new product. Every new product is not always the result of a good idea. In modern manufacturing, **research and development** are the processes used to find new ideas and to develop them into successful products.

Never Ending Research and Development

Research and development have been going on since earliest times. Beginning with the very first products (probably primitive tools), constant attempts have been made to improve products or to find new and better ones.

Many times a new product or an improvement on a product is the result of creative thinking by one person. The product may be as common as a paper bag. In the late 1860's, paper bags were already being produced. **Margaret E. Knight** looked at them, however, and decided they would be more useful and hold more if the bottoms of the bags were flat. She devised machinery to produce these bags in 1870. The creative thinking of one person long ago has made our everyday lives just a little easier.

Today it usually takes many good ideas to come up with one that will work as a product. All ideas do not become products, only the best ones. It is important to find out which new ideas will work and which ones won't.

Designers work with the ideas and try to make them into products. Many other people are also involved in this process. Skilled model makers build **mock-ups** (detailed models) to see

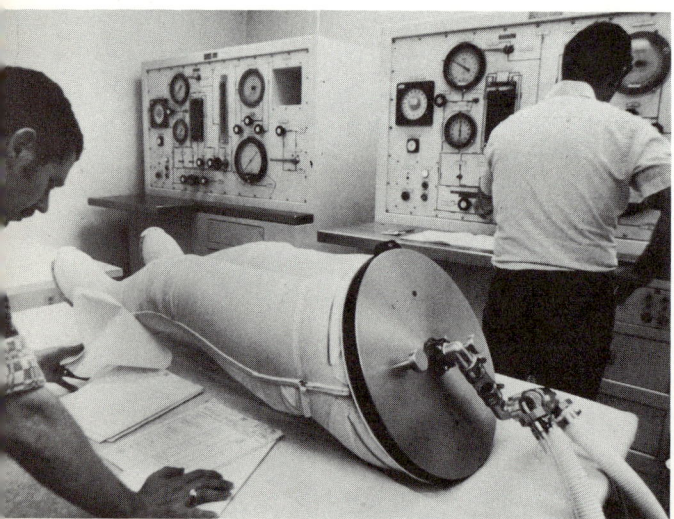

United Technologies

Figure 5-1
New products have been created for U.S. trips into space. Once this space suit is fully tested, it will be worn by a crew member of the U.S. Space Shuttle Orbiter.

Figure 5-2

Sheller-Globe Corporation

Even a school bus starts out as an idea on someone's sketch pad. Here a clay mock-up is being made to show the designers and others what the bus will look like.

Results of Research and Development

Most products we have and enjoy today are the result of research and development. For example, small electronic calculators are common products in many homes today. This was made possible by the development of a new component for the calculator. It is so tiny that you could hold it on the tip of your finger. It is called a **microcomputer.**

A microcomputer is a chip of silicon (a mirror-like material) less than 1/5 of an inch or 5 mm square. It is treated chemically to control the flow of electricity in very precise (exact) ways. One microcomputer can do the same work that once required 30,000 transistors and other electronic parts. This new development makes calculators easier to manufacture. They cost less to produce. Therefore, companies can charge less for each one, and more people can afford to buy them. See Figure 5-3.

Now microcomputers are also found in such items as televisions, telephones, microwave ovens, and electronic games. Microcomputers and the ways in which they are used are results of research and development.

Electronic digital clocks and watches were unknown several years ago. All watches had gears and dials with clock hands. Because of research and development in electronics these new watches and clocks are common products today. See Figure 5-4.

how ideas will shape a product. See Figure 5-2. Engineers make sure that the product will work and can be produced. People who do sales research can tell if consumers will want the product. And there is a very important need for **financial backing.** Someone must provide the money for expensive research and development processes. These are only a few of the people needed to develop a new product.

Figure 5-3

Making better products at a lower cost is the goal of research and development. The tiny microcomputer used in today's calculators can be made cheaply. Each one replaces many other parts. The cost of producing calculators is reduced — and so is the selling price.

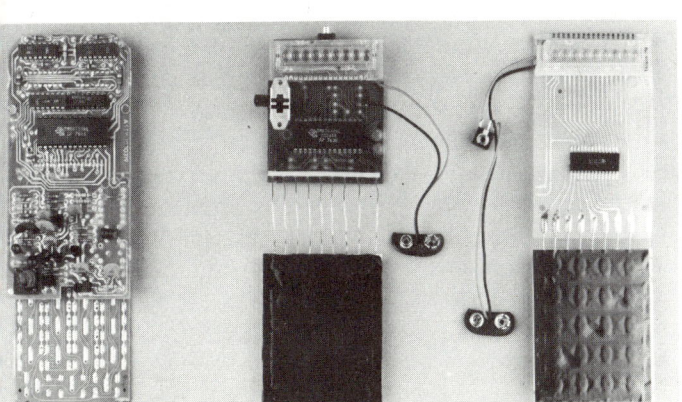

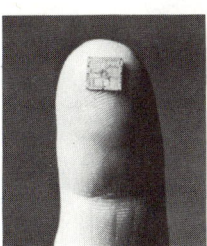

Texas Instruments

Figure 5-4

This liquid crystal watch is a "spin-off" of research and development done by NASA. NASA learned much while developing liquid crystal dials for spacecraft. This information helped others to develop the watches. Can you name any other spin-offs from space R & D?

What are Research and Development?

Research and development (R & D) are processes used to make ideas work. The new idea might be a totally new product. Or it might be an improvement of a product already being made. New ideas about production processes are also developed. These ideas may help to produce the product faster, easier, or more economically (less cost for the time spent). Research and development are done in basically the same way for any new product.

Stating the Design Problem

Making a clear, specific statement of the problem that must be solved is important in the R & D processes. It has been said that a problem well stated is 60% solved. A sample design problem might be, "Design and build a skateboard with brakes." This would be the statement of the problem. Using research and development, the designer and the engineer would look for a solution.

Figure 5-5

Peanuts were just a children's treat before George Washington Carver's research. His work produced new foods, dyes, a fertilizer for the soil, and many other products.

Brown Brothers

Research

Research is used to gather information. When done to solve a problem, it is known as **applied research.** Sometimes it is done just to learn new things. This is **basic research.**

Both terms may be used to describe **George Washington Carver's** work with peanuts. Carver was a teacher and an agricultural researcher at Tuskegee Institute in Alabama. In 1915, farmers were having a problem selling the peanuts that they raised. Carver began experimenting to see what could be made from peanuts. Before his death in 1943, he had developed over 300 products. These range from food products, such as cheese and milk, to materials such as plastics and synthetic marble — all made from peanuts. Other people built factories to produce these products. Thus Carver's research created new markets for the farmers and new products for people to use.

Researchers find information in three ways. They

- **Retrieve** information that has already been discovered by someone else. It is usually found in magazines or books.
- **Describe** what exists right now. What do people like or dislike? What products are available now? What is needed?
- **Experiment** to learn things that no one knew before.

Suppose you worked in the R & D department of a tire manufacturing company. You might be assigned the job of making a better dirt tire for a motorcycle. By looking in books or magazines you can find out what has already been done to make back tires. This would be **retrieving** information. Then you could talk with dirt bike riders. You could see what tires are used now. You could ask what problems they have with these tires. You could find out what type of back tire they want. This would be **describing**. See Figure 5-6. Finally, you could **experiment** with new types of materials, designs, and processes. You would do this until you found a better way to make a tire.

Development

In development, the new information provided by research is used to make a new or better

Figure 5-6
This interviewer's job is to describe the kind of tire the dirt bike riders want.

Figure 5-7
Researchers give the information to development. The result is an improved dirt bike tire.

product. Designing and engineering are development processes. In the motorcycle tire example, the information found during research would be used to make a new and better dirt bike tire. See Figure 5-7.

Research people and development people work closely together. While a new product is being developed, new questions may come up. More research may be needed to answer the new questions.

Why Are Research and Development Important?

Research and development have given us thousands of new products and materials. Many of these did not exist when your parents were born. When you get home, ask your parents to name some products you use today that were not available when they were your age. These new items are the results of research and development. Because of R & D, life for you is a little easier than it was for your parents.

A manufacturing company spends a lot of time and money on research and development. Producing a new product is expensive. To avoid mistakes, as much planning as possible is done before production is started. The company wants to be sure that the product it chooses will be successful. Many companies have gone out of business because their products would not sell. Research and development must make certain that the product produced is a good one. See Figure 5-8.

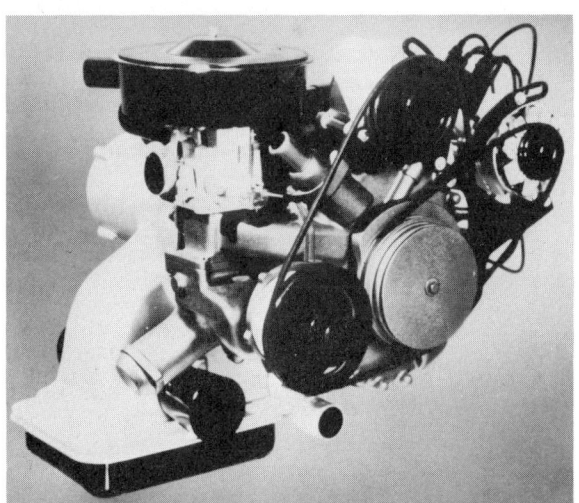

General Motors Technical Center

Figure 5-8
It started out as a good idea. General Motors spent seven years and millions of dollars on the Wankel rotary engine. But after all the R & D, General Motors decided not to produce it. It had poor gas mileage and high exhaust pollution.

Who Does Research and Development?

The people who work in R & D need a variety of skills. Experience in research is helpful. Knowledge of science and mathematics is very important. A high degree of curiosity is needed. Research and development people must be able to

- **Create new ideas,**
- Think about the **ideas of others,** and
- **Explain ideas** in order to share them with others.

Some people in research and development work in the R & D department of a large company. Others may work for a company that does **only** research and development. This type of company specializes in designing products or in solving problems. Smaller companies may not need R & D departments of their own. Sometimes they can't afford them. They go to private R & D companies. People there listen to the problems or needs. Then they do the research and development necessary to solve the problems or fill the needs.

NASA

Figure 5-10

Flight faster than sound was made possible by government-sponsored research at Dryden Flight Research Center.

The U.S. Government sponsors a great deal of research and development. Some of this R & D is done in colleges and universities. The National Aeronautics and Space Administration (NASA) and the Department of Energy are government agencies that provide money for research. Some of their problems may be researched by private industries. See Figure 5-10.

What Can Be Accomplished in Research and Development?

People in research and development have already accomplished many things. Often these things seemed impossible not many years ago. R & D people have invented many new items. They have also improved existing items.

Improvements are often made in ways that people seldom think about. For example, consider the glass used in picture frames. Marked on the labels of many frames are such words as "Made with non-glare (non-reflecting) glass." This means that when you look at a picture in this frame, you will not see the glass, or a scene that the glass is reflecting, or a glare of light. You will see the picture inside the frame clearly and distinctly. The glass itself is "invisible." A technical development such as glass that will not reflect light is most often accepted without

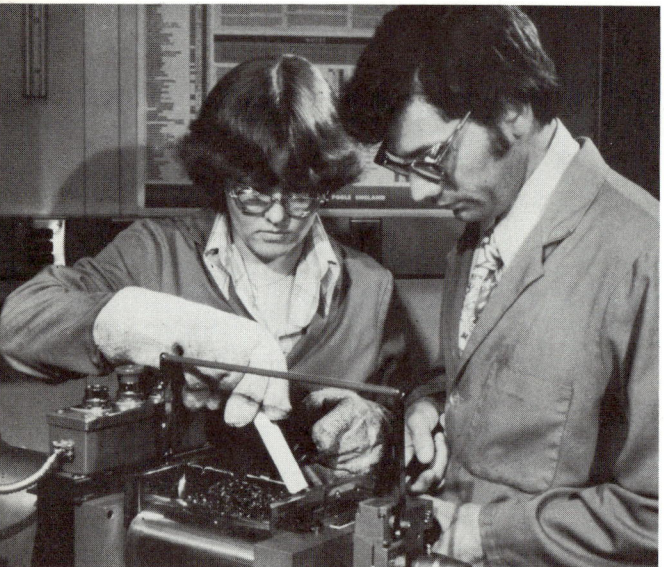

Figure 5-9 Western Electric

A curious mind helps a researcher to improve products. Soon old plastic will be recycled into a new shape. Engineers are working on a way to get the plastic into usable form again.

"Smile, Grandpa!"

In February 1979, **Barbara Askins** received a very special award. She received this honor for inventing an **autoradiographic process**. Her invention is described as a "practical photographic image enhancement process." Simply, it improves faded photographs or photographic film that was underexposed (not enough time allowed for film to receive a clear image).

In the autoradiographic process, underexposed photographic film is coated with a mildly radioactive chemical. It is then placed in contact with photographic paper or film. The result is a 90 percent improvement. The photograph would have nearly normal detail. This same process can restore an old, faded photograph to its original quality.

This process has many possible uses. One use could affect us directly. The process can be used on X-ray film. It would shorten the time required for us to be exposed to X rays by 80 to 90 percent.

This autoradiographic process will influence the manufacturing industry. It can be used to enhance (make better) photographs of earth made from aircraft. This will help in locating raw materials. It may be used with industrial X rays. Sometimes, welded parts are inspected in this manner.

Barbara Askins works in research. Her work will affect people and industries. She **studied** (retrieved and described) what others had done and were doing. She **experimented** and improved upon their work. She **invented** a new and beneficial (helpful) process, and **became** the 1978 National Inventor of the Year.

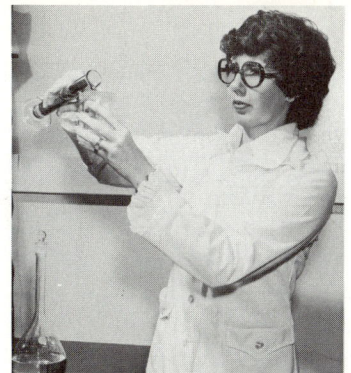

This historic photo of construction workers in the 1900's was restored by the process developed by Barbara Askins.

thought. Yet someone had to develop a technique (way) to create it. One someone in this case was **Katharine B. Blodgett**. She discovered a coating for glass that caused it to become non-reflecting. Blodgett developed a **prototype** (original model) of "invisible" glass in the Research Laboratory of the General Electric Company in 1938. This invention can be used in many ways. Applications are found not only in picture frame glass but windshields, eyeglasses, camera lenses, and many other items.

Figure 5-11

Katherine Blodgett's research has improved products we use today. One discovery in research can sometimes improve many products.

Katharine Blodgett worked in research for many years. She began in 1918 and retired in 1963. During that time, she applied her knowledge of physics and chemistry to a variety of projects. During the years of World War II (early 1940's), she had military projects. She worked on ways to keep ice off the wings of airplanes. She also developed a smoke screen for use in combat. Later, Blodgett helped develop an instrument that could be carried by a weather balloon. It measured humidity (moisture) in the upper atmosphere. She has also worked on projects that have helped others in research and development.

The results of Katharine B. Blodgett's work affect us today, even though some of her projects were developments needed for her time. What might research and development people accomplish in **your** time? What might **you** accomplish in research and development?

Research and Development in Action

The Most Amazing Car Never Built

During the 1930's, an interesting series of events took place. They had to do with a research and development problem. **R. Buckminster (Bucky) Fuller,** an engineering genius, set out to design the perfect car. The result was a car called the **Dymaxion.** See Figure 5-12. It could hold up to 11 people, and run at speeds up to 120 miles per hour. It could travel 40 miles on one gallon of gasoline. But the Dymaxion never went into production.

It all began when Bucky Fuller matched his ideas with Nannie Biddle's money and faith. Nannie Biddle came from a wealthy family but didn't like the social life. She was always willing to try something new and different. Bucky Fuller was a young engineer. He had new ideas but no money. Fuller convinced Biddle to **finance** the development of a totally new three-wheeled car he was designing. She seemed to be on her way to becoming a wealthier woman. Suddenly her plans were ruined by a freak accident and a political cover-up.

This is the way it happened. Fuller and an engineering partner hired 27 skilled workers to build their prototype Dymaxion. They drew **sketches** to share their ideas. They did research to find out which ideas would work best. Then they designed and built models to test their ideas. They engineered and tested to determine (figure out) the best engine and steering system. The **working drawings** were made for each part. Then the workers began producing the prototype. This original model was intended to be the standard for the future products.

Road and Track Magazine

Figure 5-12
This prototype automobile looks like a pickle on wheels, but it was designed in the 1930's to be safe, efficient, and save on gas mileage. A freak accident crushed the designer's dream of mass-producing this car.

The plans showed an 85 hp V-8 engine in the back. The Dymaxion had front wheel drive. Since steering was by a single back wheel, the car could turn around in one spot. The driver and passenger seats both had rearview periscopes (viewers much like those used in submarines). The car was 19 feet long. It was planned to seat four but had room for eleven. Most of the material was aluminum. The vehicle weighed only 2300 pounds (about the weight of a small compact car of today). The prototype was built and ready for testing in five months. This first car was a great success. Remember this was in the 1930's.

The Dymaxion was unveiled in a blaze of publicity. Fuller enjoyed driving it around New York City and creating traffic jams by the attention he drew. Finding someone to set up production and manufacture the car was the next problem. This is when a tragedy ended the Dymaxion.

The company's test driver was driving an interested buyer to the airport after a test drive. A politician came alongside and challenged them to a race. They raced through the streets at 70 miles per hour. The politician lost control and crashed into the Dymaxion. The driver was killed, and the passenger was seriously injured. The politician's car was quickly towed away before newspaper photographers swarmed to the scene. Newpaper headlines across the nation read, "Three-wheeled Car in Disaster," "Freak Car Kills Driver." Some reports even said it struck a bump and flipped over.

The truth came out when the passenger finally recovered and testified. It was too late. No one wanted to produce the car with a bad image. The unfavorable publicity finished the Dymaxion.

The Designer and the Design Process

Bucky Fuller is a designer who is willing to be different and to try something new. To design the Dymaxion, he used the same basic research and development steps used by industries today. The Dymaxion example shows how ideas are developed into useful products. Using sketches, mock-ups, working drawings, and finally prototypes, new products are developed.

Tom Munk, photographer, Buckminster Fuller Archives

Figure 5-13

R. Buckminster (Bucky) Fuller is a designer. He is shown here standing in front of a geodesic dome, a structure he designed. Bucky Fuller and other designers look at the things around them and think of ways to improve them. How might you improve the things around you?

The failure of the three-wheeled car to become a commercial success did not stop Fuller from creating other designs. He has invented a variety of things. Fuller has designed a jet-powered airplane as well as a one-piece bathroom (complete). He made a new projection* for a map of the earth. He has also designed a seven-room house that can be mass-produced. He is probably best known as the inventor of the "geodesic dome." See Figure 5-13.

There are hundreds of areas of product research and development. Clothing, tools, cosmetics, appliances, boats, and food are a few other examples. Let's look at research and development in the area of air transportation.

Designing Air Transportation

People have always wanted to fly. Through the years this has lead to many new and unusual inventions. See Figure 5-14. From the first kites and hot air balloons, to modern jets, men on the moon, and space shuttles, research and development have played an important part in getting people off the ground. The design of the airplane is a good example.

Two brothers, **Wilbur and Orville Wright,** ran a bicycle shop in Dayton, Ohio. In the early 1900's, they became interested in flying. They experimented with gliders to learn how to keep

---

* A projection is a way to show something round (ball-shaped) on a flat surface.

Figure 5-14

Inventors dangled from kites and airships before the Wright Brothers found a better way to fly at Kitty Hawk.

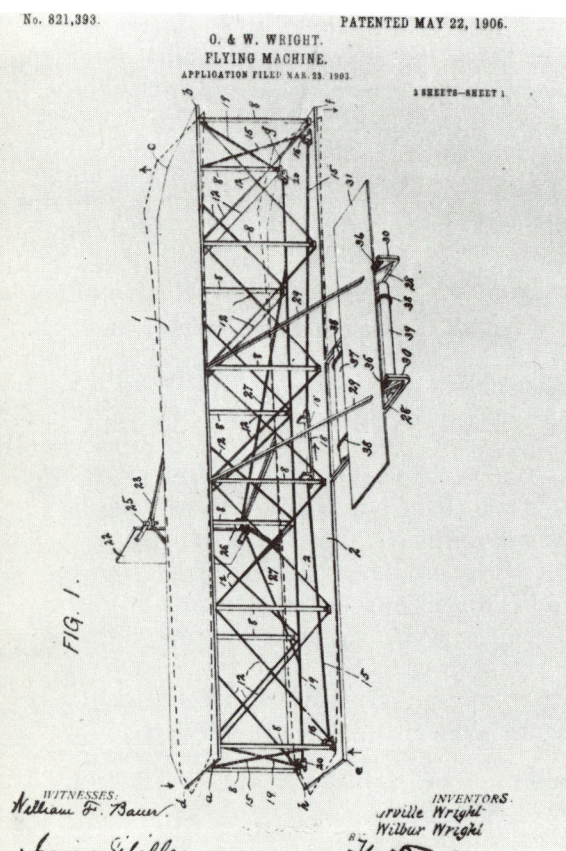

Figure 5-15

In 1903, the Wright Brothers overcame design and engineering problems to create the first successful "flying machine."

an airplane in the air. Their biggest problem was finding a way to keep the airplane balanced.

One day a man came into their bicycle shop to buy an inner tube. Wilbur held a cardboard box while the customer looked at various tubes. Suddenly, an idea came to Wilbur. He realized he was holding the box in a way that could be used to balance the airplane. Wilbur quickly made sketches. He discussed the idea with his brother. Off they went to build a model. When they tested their design, it worked! They immediately began making drawings for a full-size prototype. It would be powered by an engine. They planned that the airplane would carry one person into the air.

On December 17, 1903, Wilbur and Orville were ready to test their finished airplane. They were at their seaside laboratory in Kitty Hawk, North Carolina. It was Orville's turn to make the test. He climbed aboard and became the first person to pilot an airplane. The flight lasted 12 seconds and covered a distance of 120 feet. See Figure 5-15. Four flights were made that day. The longest flight was 59 seconds and covered 852 feet.

The Wright brothers used the same basic steps to design their airplane that today's designers use to design cars, furniture, sports equipment, and other products. The brothers had ideas that were sketched on paper. They could see them and discuss them. They built models to find out if their design would work. From these they drew detailed drawings of all the parts. The drawings showed how everything would fit together. Finally, the first product, the prototype, was built and tested. Unlike most designers of today, the Wright brothers did not design their first product to be mass-produced and sold. They were more interested in just making it work. Later they continued to improve their design. They wanted to be sure it was reliable before selling it to a company for production.

Research and development are very important parts of manufacturing. Products made by the manufacturing industry go through the research and development process. Research and development continue to provide new and improved products to meet the constant demand for less expensive, better, and more efficient products.

Looking Ahead

Research and development make product ideas into real items. But companies cannot make **every** possible product. Companies carefully select the products they will produce. Do you know

■ from an average list of 60 good product ideas, how many are likely to become successful products?
■ what single limiting factor has the most influence on product selection?

These items and others are discussed in the next chapter. You will learn how companies select their products.

New Terms

applied research	mock-ups
basic research	prototype
describe	research and
Dymaxion	development
experiment	retrieve
financial backing	sketches
microcomputer	working drawings

Study Guide

1. What are research and development?
2. Explain the difference between applied research and basic research. Give an example of each.
3. What are three ways in which we can find new information through research?
4. In research and development, of what value is "development"? Why is development very difficult without research?
5. If you wanted to see research and development going on, name some places you would go.
6. Why do manufacturing companies spend a great deal of money on research and development?
7. Explain the research and development of the Dymaxion. Why wasn't this car successful?
8. Of what importance was the prototype made by the Wright brothers?
9. If you were building a plant stand for your bedroom window, would research and development be needed?

chapter 6

Selecting Products

Product ideas can come from anybody. Sometimes companies are formed to produce a new product. Common people may make good suggestions to companies about new products or product improvements. Or ideas may come from the people who work in research and development. But all product ideas are not good ones. And sometimes an idea might be good, but the company cannot produce the product for another reason. Many factors must be considered as a company selects its product.

Finding a Product

Searching for new product ideas and picking the best ones are difficult tasks. Keeping in mind **consumer demand** (what people want), the designer looks for new product ideas. This is the beginning of most new products. The idea begins in the mind of the designer. See Figure 6-1.

Finding the right product for the company to manufacture is very important. See Figure 6-2. The more good ideas designers have, the more chances of developing a successful product. It takes about 60 good new product ideas for one to become a successful product. It the company chooses the wrong product, no one will buy it. The company will lose money. It may even go out of business.

Where are New Product Ideas Found?

People get ideas for new products or product improvements in many ways. Sometimes ideas come while looking at other products. For example, many years ago, while looking at canned goods, **Tillie Lewis** noticed that expensive canned tomato products were **imported.** They were shipped to the United States from Italy. This

Figure 6-1
Imagination is the key to product ideas.

Selecting Products   81

This is a mini-typewriter which will write what is said aloud.

Figure 6-2

Are these pens in your future? A pen company is researching new ways to write. The pen company did not say when, or if these pens would be produced.

This pen contains a small computer which stores information. When the pen is put into a special machine, the information is printed.

Wide World Photos

---

gave her an idea. She wondered why the special tomatoes used in these products could not be grown and canned in the United States. In 1934, she traveled to Italy. There she formed a company with an Italian canner. He provided money and equipment to start a cannery and gave her enough seed to begin raising these pear-shaped tomatoes commercially (to sell) in the United States. Tillie Lewis worked hard. She developed a simple observation of fact into a very successful business.

All designers of products stay alert to what is around them. This helps them to **formulate** (create) new ideas. The following list shows a few of the many places where ideas for new products might be found:

- Nature
- Hobby Stores
- Gift Shops
- Catalogs
- TV Ads
- Workshops
- Magazines
- Newspaper Ads
- Around the Home
- Traveling on Public Vehicles

Big Ideas from Common People

Many good ideas come from private **inventors**. They tinker with products to improve them. For example, in the early days of telephones, it was necessary to first call the operator. Then the operator connected the caller's telephone line with the line of the person being called. **Almon B. Strowger** became impatient with this system. In 1889, he invented an automatic switching system. This led eventually to dial telephones.

Some people **feel** the need for something new. Such was the case with **Murray Spangler**,

Figure 6-3

Ideas for new products can come from the things around us. Tillie Lewis built a multi-million dollar business from a simple observation.

82 Research and Development

a janitor. Spangler suffered from asthma (a condition that affects breathing). The dust raised by his broom affected the asthma. In 1908, he invented a type of vacuum cleaner. Then the dust went into the cleaner, not into the air.

The Company Idea Teams

In companies, people often work together as teams. They discuss ideas with others.

Brainstorming is a good way to stimulate thinking (get ideas). Brainstorming can be done by one person. The person lists all the product ideas that come to mind. Later, the ideas are shared with other members of the team.

Brainstorming can also be done in a group. A meeting of the team of company designers or managers might be held. Each person shares ideas with the group. Someone writes down all ideas. Even those that do not sound good are accepted. They might help someone else think of another idea. These meetings can be very short, but they often last several hours or all day.

Considering Product Ideas

To select a new product, the **management personnel** of the company (those who run the company) look over all the product ideas. They discuss each idea carefully. They must determine whether the company has the ability (equipment and knowledge) needed to produce the product.

Figure 6-4

This idea for a tire was taken from nature. A cat can extend its claws when necessary. This tire could be made to do much the same thing. It would have normal tread, but when needed, more air could be pumped into it and studs would appear.

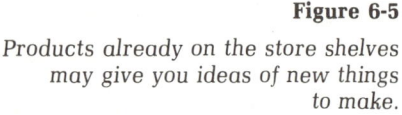
Creative Design Department, Trenton State College

Figure 6-5

Products already on the store shelves may give you ideas of new things to make.

A question is considered. "Will people buy the product?" It is not possible to fully know the answer ahead of time. Even so, as much information as possible about how the product will sell is found before the final product is chosen.

Consumer Demand

Consumers are the people who buy products. When a company is planning a new product, it is interested in finding out what people want. If people want or need a certain product, there is a demand for it. This is known as consumer demand. Is there a demand from consumers for black toothpaste? for a car that gets 70 miles per gallon of gasoline? for an electric fork? If there is a demand for these things, then someone will try to design and produce them.

Smithsonian Institution

Figure 6-6

Crank-type starters were dangerous. Consumers wanted a better way to start cars.

Can You Guess the Name of This Company?

Conditions were bad when this company was started. It was in 1946, shortly after World War II. The small city of Hamamatsu, Japan, still showed the effects of bombing. People had a hard time going from place to place. Trains were crowded. Private ownership of a car was only a dream. One person who lived in this city saw the need for cheap, convenient transportation. He had an idea.

The man hired 12 workers. In a small, crowded shack, they began producing motorized bicycles. At first, they mounted surplus (left over) engines from army equipment on regular bicycles. The fuel commonly used was poor in quality, and the motors were difficult to start. Yet the motorized bicycles sold as quickly as they were produced.

The production rate began at one per day. As this increased, the demand for the cycles also increased. The company grew and expanded. Eventually it became a corporation.

Product designs changed. Motorized bicycles were no longer made. The products were motorcycles. Soon factories were established in other countries. People in many parts of the world including the United States began to ride motorcycles made by this company. The one-per-day production rate of 1946 developed into as many as 10,000 per day in 1979. Now the company produces cars and other related products. The man who began this company had never foreseen that it would become so successful.

The company and the products carry the man's name. Do you know the name? See page 84.

84 Research and Development

Cars are products that have changed through the years because of consumer demand. At first the engines were started by turning a crank by hand. See Figure 6-6. It was a dangerous operation. Often the engine backfired. This reversed the direction of the crank. People had to move their arms out of the way quickly because the crank could break an arm as it spun around. People wanted a safer way to start cars. To meet this demand, an electric starter motor was developed in 1911.

Finding Consumer Demand

Consumer demand is found by taking **consumer surveys** (asking consumers questions about the products they use). Much time and money is spent each year trying to find out what people want and like. Consumer surveys are done in many ways. **Product samples** can be shown or given to people. Then their opinions are asked. In **telephone surveys**, people can be asked if they have used certain products. They can tell what they think of the products they have used. **Written surveys** are often mailed to people to get their opinions. Surveys are given to a **cross section** of people. This is a sampling of people of different ages, from a variety of areas, with different incomes, and other differences that may be important. These people are then interviewed (asked questions) for their opinion of the product.

After the information is collected from the consumers, the company managers study the results. They then have a much better idea of what people like and don't like. This will help them in choosing a new product to make. See Figure 6-7.

Company Limitations

It is not possible for a company to make anything it wants. Many factors must be considered before the product is finally chosen. These factors determine what a company can make. They are called **limitations.**

Knowledge and Skill Limitations

The product selection of a company is limited by the knowledge and skills of its employees. The company should not select a product that is too difficult for them to make. Suppose a company has never made airplanes. It would not be wise to simply begin producing airplanes. If a change in product is planned, employees must be **retrained.** Then they will have the knowledge and skills needed to produce the product. See Figure 6-8.

Machine and Tool Limitations

A major limitation on a company is the machines and tools it has. Sometimes the company cannot afford the new equipment needed to make a product. It must look for a different product.

Figure 6-7

Consumers want products that will use less energy. This air-conditioner can be run more efficiently by adjusting the energy-saving switch.   Fedders Corporation

> "Can You Guess the Name of This Company?"
> Answer: Honda Motor Company, Ltd.
> The man's name is Soichiro Honda.

Selecting Products 85

Figure 6-8
Product selection must be done carefully. Workers may need to be retrained.

others. A difficult product might take a long time to research and develop. The company may not be able to afford to wait.

Money Limitations

The biggest limitation on every company is money. If the company has the money, it can usually overcome all other limitations. A company can make a new product if it can afford to

- Train the workers,
- Buy new equipment,
- Buy the right materials,
- Expand its factory for more production, and
- Take the time needed to get everything done.

Material Limitations

The materials needed to make the product may not be available near the company. Shipping the material long distances may be too expensive.

Space Limitations

A company may not have room to make bigger products. Suppose a company has been manufacturing jewelry. It would be foolish to consider making mobile homes.

Time Limitations

A company must have products to sell. Some products take more time to produce than

Figure 6-9
A company can make a new product if it can afford to buy the new equipment.

Figure 6-10
When considering new product ideas, management must keep in mind the company's limitations.

Final Selection

The final selection of the product is made by the management personnel of the company. Many meetings are held. Each company limitation is discussed for each product being considered. The product ideas that are **feasible** (reasonable for the company to make) are discussed. Then the ideas are developed further to determine which one would be best.

At the final meeting, the product is selected. The top management people in the company make the final choice. The new product idea is then given to development. Designers and engineers determine the product's appearance and the way that it will work.

Looking Ahead

Selecting a product may be difficult. But developing the product into a useful item or one that people will want usually involves a great deal of time, thought, and work.
Do you know
- what looks like a product, may work like a product in many ways, but is not a product?
- why prototypes are sometimes made to be destroyed?
- how "bugs" affect product development?

In the next chapter, you will see in detail how products are designed and engineered.

New Terms

brainstorming
consumer surveys
cross section
feasible
formulate
imported
inventors
limitations
management personnel
product samples
retrained

Study Guide

1. Why does a company have to be so careful in choosing a new product to make?
2. How do designers find ideas for new products?
3. How is brainstorming used in the development of new products?
4. A designer has an idea for a new product. How will the company determine whether this product will be manufactured?
5. What is consumer demand? How is it determined?
6. Name six things that limit what a company can produce.
7. Look around your school shop. Think about the kinds of products that could be made.
   a. Name three products which you would like to make that are too large to make in your shop.
   b. Name one product which you would like to make, but doing so would cost too much.

chapter 7

Designing and Engineering Products

A new product idea has been chosen. The next task is to design and engineer the product. In **designing** and **engineering**, all details of the product are worked out. Designers and product engineers decide such things as

- How the product will look,
- How it will work,
- What shape and size it should be,
- How parts will fit together, and
- How the product will be powered.

To make a successful product, designers and engineers must have good ideas. They must be able to solve many problems. Even a good idea for a new product may fail. Designers and engineers must give the product a good appearance and make certain that it works properly, or the product may not be accepted by consumers.

Designing

Before starting the design process, it is important to state the problem properly. The statement describes what will be done. It must be simple and clear. "Design a flying motorcycle" could be a sample design problem. Every step of the design process should bring the designer closer to solving the problem.

The Design Process

To design a new product, the designer and the engineer follow the same basic steps for

Figure 7-1
Working together, the designer and engineer use their special talents to create a new product.

88 Research and Development

every product. Some steps are used more than others for some products. This depends on which steps or design "tools" will best solve the problem. The steps are usually followed in this order:

1. Collecting Ideas
2. Sketching Ideas
   - Thumbnail sketches
   - Rough sketches
   - Renderings
3. Making Mock-ups
   - Paste-up mock-ups
   - Appearance mock-ups
   - Hard mock-ups
4. Drawing Plans
   - Detail drawings
   - Assembly drawings
   - Schematic drawings
5. Building Prototypes

Presenting Design Ideas

A new design starts with ideas. The designer must discuss these ideas with others. Sketches and mock-ups help people to understand the product ideas.

Sketches

Designers sketch ideas on paper so they can be seen. Sketches help the designer to solve

Figure 7-2

The designer begins by sketching ideas.

RCA Consumer Electronics Division

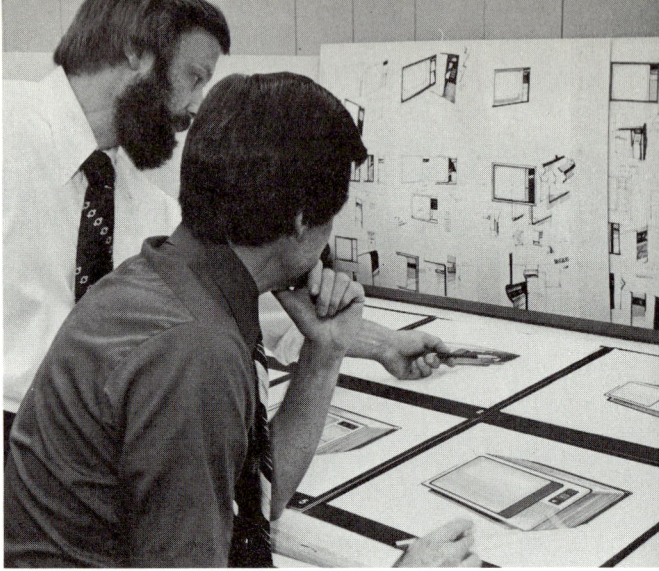

RCA Consumer Electronics Division **Figure 7-3**

Renderings are made to show the best design ideas. Neat and full of detail, the renderings are presented to management for discussion.

design problems. Different solutions can be sketched and studied. Sketches are cheap to make and can be done quickly. Three kinds of sketches are used by designers: thumbnail, rough, and rendering.

The easiest type of sketch to make is called a **thumbnail sketch.** These are usually small, quick sketches that capture the designer's ideas. Sometimes they lead to new ideas. The designer may make many thumbnail sketches. The best ideas are chosen from among them.

The next step is to make better, more detailed sketches of the best ideas. These are called **rough sketches.** Several "roughs" may be made. Sometimes color is used to show different components. The thumbnail sketches and rough sketches are done quickly. Both types of sketches would probably be shown only to other designers to get their suggestions. In this way, the designer gathers additional ideas.

The final step in sketching is to pick the most promising rough sketches. Each one of these is made into a sketch called a **rendering.** Renderings are very neatly done and show all the details of the product design. Color is added with felt tip markers or with colored pencils.

A product planning meeting is then called. Here the sketches are presented to the company managers, and the designer's ideas are discussed. The best design ideas are released for the next step in the design process.

Mock-ups

From sketches, the designer and management can get only a partial (not complete) idea of how the product will look. Sketches are only **two-dimensional** (height and width) presentations of the product idea. To show three dimensions (height, width, and depth), mock-ups are made. It is much easier to see and understand the design by looking at a mock-up. There are three basic types of mock-ups made by designers: paste-up, appearance, and hard mock-ups.

A **paste-up mock-up** is done in the same manner as a thumbnail sketch. It is made quickly to get a rough idea of the design. Materials are usually pasted together. Then the designer and management can see the general size and shape of the product. Cardboard and STYROFOAM® are often used to make paste-ups.

The **appearance mock-up** looks like the product. It can be made out of any suitable material. Cardboard, clay, plastic, or wood may be used. Even though it looks like the actual product, none of the parts work. For example, if a mock-up is made of a sewing machine, none of the knobs or levers would work. They would only look as if they do. See Figure 7-4.

A more accurate mock-up is called a **hard mock-up.** This mock-up is often made of the final material that will be used to make the product. Many of the parts, such as knobs, handles, and doors, will work. Engines or motors, or other mechanical parts that make the product operate, would **not** be included. A hard mock-up of a refrigerator would have a door that opens. The trays would probably slide out. However, it would not have a motor and a cooling system in it.

Most often, appearance mock-ups and hard mock-ups are used in two ways. Many **pictures** of products you see in catalogs are appearance mock-ups or hard mock-ups. They are used so that catalogs can be printed before production of the product begins. Both mock-ups are used when taking **consumer surveys.** This helps the designer find out what suggestions the consumers have.

Making Design Decisions

The sketches and mock-ups represent the designer's ideas of how to solve the design problem. Meetings are held quite often between management and the designer through all the steps of design. Ideas for solving the design

Figure 7-4
It looks real, but the appearance mock-up is just the shell of the product.
RCA Consumer Electronics Division

Figure 7-5
The mock-up is presented to management for final consideration. Management will accept the design, reject it, or suggest changes.

RCA Consumer Electronics Division

problem are discussed. Management makes suggestions, and the designer makes any necessary changes.

Many factors must be considered when deciding which design is best. Some of these are included in the list below.
1. Will it work?
   - **Appearance** — Does the design look good? Will people like the way it looks? Can the appearance be improved?
   - **Function** — Will the product do what it is supposed to do? If it is a shirt, for example, are the holes for the arms and head large enough? Is the shirt long enough?
   - **Safety** — Is the design safe? Does it have sharp edges? Can it break easily?
   - **Maintenance** — Can people take care of the product easily? Is the product easy to fix? Are changes needed that will make maintenance easier?
2. Can the company make it?
   - **Cost** — Would it be too expensive to make the design? Could it be changed to make it cost less?
   - **Producibility** — Can the product be made with the machines and the workers that the factory has? Are changes necessary to save time and money on production?
3. Will it sell?
   - **Market** — Is there a **market** (demand) for the product? Will it sell? This is the most important question that must be answered.

Each design is considered and discussed carefully. Necessary changes are made. Then, consumer opinions are taken. Sometimes more changes are required before the final design solution is made. This is usually presented to management as an appearance or hard mock-up. If the design is satisfactory, it is approved by management. Engineering can begin.

Figure 7-6
Questions must be asked and answered when deciding on a product.

Figure 7-7
Engineers can answer the question, "Will it work?" Using math and science, they can determine how a product should work.

Engineering

Product engineers work from the sketche and mock-ups made by the designer. They figure out all of the details. The engineer must make sure that the product works. A product may be attractive, but if it does not work well, few customers will buy it.

Engineers decide exactly what parts will be needed. They determine the shape and size of the parts. They decide where holes must be located and how the parts will fit together. If the product has working parts, engineers must make sure that they will work together properly.

Engineers use mathematics and science to solve problems. Suppose engineers are designing a bicycle. The sprockets on a bicycle work with the chain to make the bicycle move. Engineers must determine several things. How large should each sprocket be? How many teeth should each sprocket have? Engineers use a knowledge of mathematics and science to answer these and other questions. They know how parts work together. They can figure out how fast the bicycle will go, and how much force must be used to operate it. Other engineering problems can also be solved by using mathematics and science.

Drafting the Plans

The engineer must put ideas and plans in specific detail. These details are described in a set of working drawings. A person called a **drafter** makes these drawings. The drafter uses the designer's sketches and the engineer's calculations (figuring) to draw the product and each of its parts. Three main types of drawings are needed: **detail, assembly,** and **schematic drawings.** Sometimes the drawings are made by a computer instead of a person. See Figure 7-8. An example of each type of drawing is shown in Figure 7-9.

A set of working drawings includes all drawings needed to make the product. A drawing is made of each part of every product. Imagine how many drawings were made for all parts of the bicycle! A parts list is also included in a set of working drawings. See Figure 7-10. After drawings are finished, they must be approved. Then copies are made. These are given to the departments and workers who will make the product. The copies are usually called **prints.**

Figure 7-8

The computer has become a drafter. It uses the same information a human uses — designer's sketches and engineer's calculations — to draw a product and each of its parts. Here a computer produces a sketch which will help engineers decide how much weight the tower can handle.

Niagara Mohawk Power Corporation

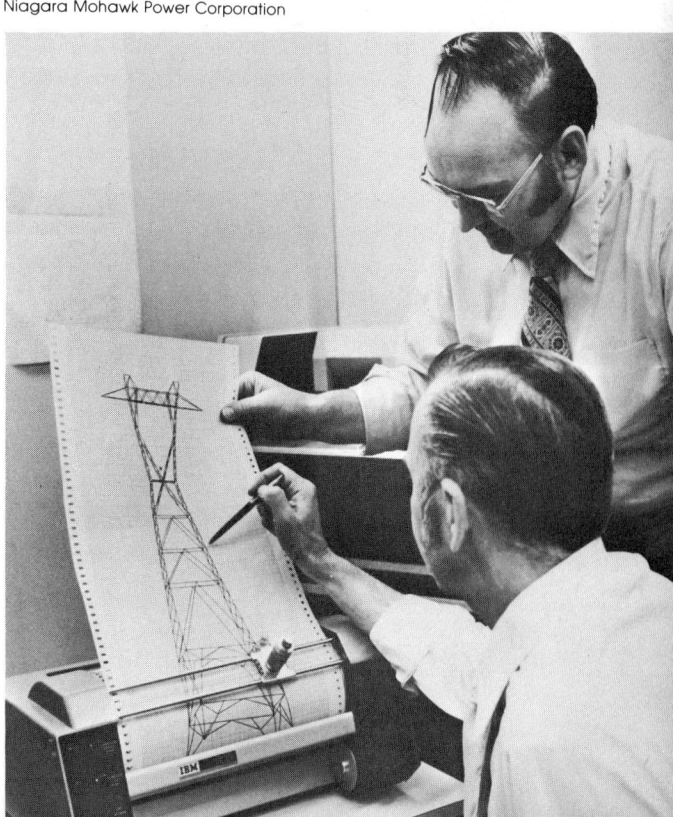

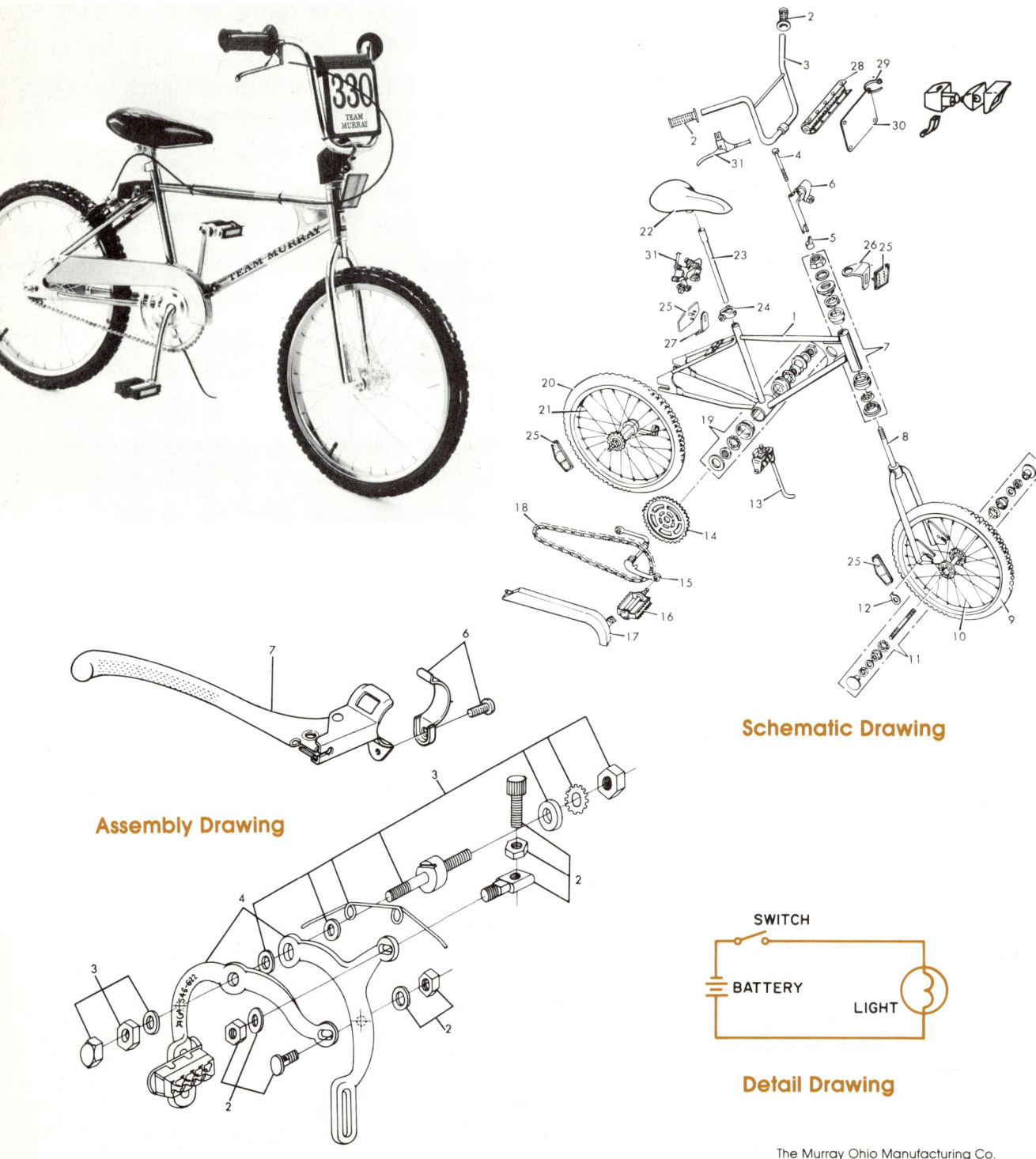

Figure 7-9
Working drawings show how to turn a product idea into a real product.
No detail is too small to include.

KEY NO.	PART NO.	DESCRIPTION
1	34750C AAC	Frame
2	34781	Grips
3	34771C	Handlebar Assembly
4	302108	Binder Bolt
5	302107	Wedge
4-6	32752	Handlebar Stem
7	303384	Head Bearing Set
8	34784C	Fork
9	303005 CCD	Front Wheel - Less Tire (Tire Size 20X1.75)†
10	303155	Front Spoke & Nipple Set (6 each)
11	303520	Axle Bearing Set
12	32375Z	Front Wheel Retainer
13	27679Z	Kickstand Assembly
14	14590C	Sprocket
15	12300C	Crank
16	32809	Pedals
17	34779C	Chain Guard
18	12834	Chain & Link
19	303333	Crank Hanger Bearing Set
20	303013 CCD	Rear Wheel - Less Tire (Tire Size 20X1.75)†
21	303155	Rear Spoke & Nipple Set (6 each)
22	98X250	Saddle
23	32942Z	Seat Post
24	303404Z	Seat Post Clamp Assembly
25	34350	Reflector Package
26	34792Z	Front Reflector Bracket
27	34800Z	Rear Reflector Bracket
28	34696	Handlebar Pad
29	32618	Tie Straps
30	34764	Number Plate
31	34783PA	Caliper Brake
*	64X286	Front Plate Decal
*	F-4661	Owner's Manual

* Not illustrated in manual.
† Wheel Size

Figure 7-10

All of these parts are needed for one bicycle.

The Prototype

Engineers use working drawings to have prototypes made. There are three main reasons for making a prototype:

■ To test the plans,
■ To test the product, and
■ To see how to produce the product.

Using the prototype to plan production will be discussed in Chapter 9.)

Trying Out the Plans

If the plans are followed to make a product, will all the parts fit together? To find out, a prototype is built according to the plans. Engineers can check to see that all components are in the right places. They can make sure that parts fit together and work. If not, a second prototype or even a third may have to be made.

A prototype is custom-made. Only one or just a few are produced. They are assembled mostly by hand to make sure that everything fits. If a part doesn't fit, changes are made on the plans. A prototype is very expensive. It can cost 100 times more than the final mass-produced product.

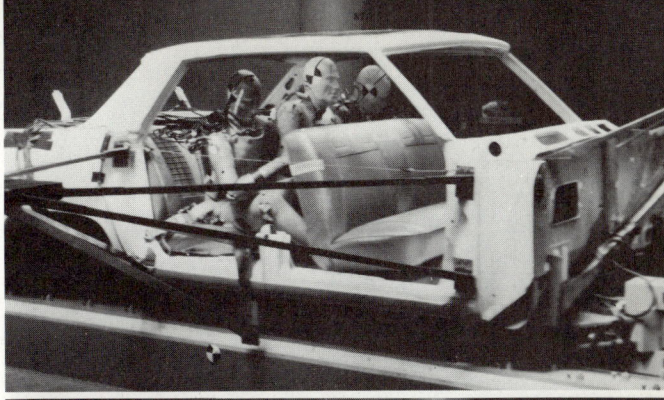
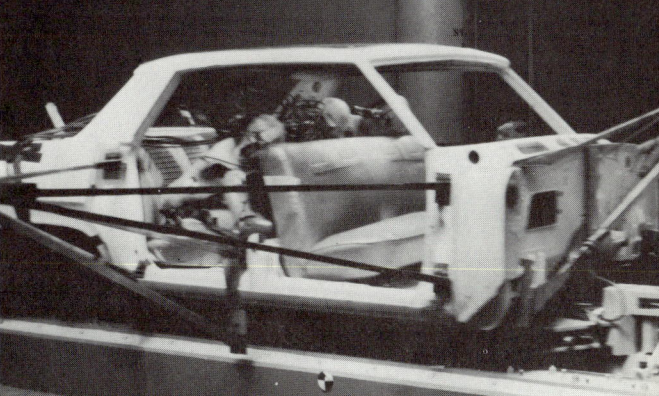

General Motors Corporation **Figure 7-11**

The passengers in this test are dummies. The dummies help car manufacturers determine that the cars are safe. Here, General Motors is testing to see how well the seat belts work.

Testing the Prototype

After the prototype is finished, it is tested. Since this is the first example of the product, it is important to find out how well it will work. Sometimes prototypes are completely destroyed to test them. Cars are often smashed into solid walls to test for passenger safety. See Figure 7-11. Engineers test the prototype carefully. This helps them answer questions about it. Does it do what it is supposed to do? Will it break? Will it wear out too fast? Is it safe? A prototype that fails any test is changed before production begins. This is done to avoid problems in the final product.

Before giving final approval for production, the company will probably want one more opinion from the consumers. Since the prototype looks and works just like a final product, it is a good example to use for consumer surveys. It

Figure 7-12
For each product, there's a test. In this case, a machine that is three stories tall presses on the world's heaviest tire.

The Goodyear Tire & Rubber Company

can be used to demonstrate the product. Consumers can try it out. The real question can be answered at this time. Will the consumer **want** the product once it is produced?

Final Approval

Surveys and testing continue until all "bugs" (problems) have been worked out of the prototype. Then management makes the final decision. Most often the product is released for **production planning**. Much time and money have been spent to design the product. Now the best way to produce many identical products must be decided. You will learn about production planning in Chapter 9.

Looking Ahead

Designers and engineers do the work of developing a product. However, decisions are made by people in management. Many people are involved in producing a product. They are all part of the company.
Do you know
- that games can be big business?
- in what way a company may be treated as a person?
- if bankruptcy is having too much money or not enough?
- if stockholders are people who place products on shelves to sell them?

All manufacturing companies do not produce the same products. Yet they are alike in many ways as you will see in the next chapter.

New Terms

appearance mock-up	prints
	producibility
assembly drawings	production planning
designing and engineering	rendering
detail drawings	rough sketches
drafter	schematic drawings
hard mock-up	
market	three-dimensional
paste-up mock-up	thumbnail sketch
	two-dimensional

Study Guide

1. Explain the role of the designer in the development of a new product.
2. Explain the role of the engineer in the development of a new product.
3. What kinds of sketches would a designer show to another designer in order to get more ideas for a new product?
4. What kind of sketch would a designer show to the management of the company?
5. Suppose a person wants to know how a design will look as a finished product. Can the person get a better idea from a sketch or from a mock-up? Why?
6. What is the difference between an appearance mock-up and a hard mock-up? How are these mock-ups used?
7. What factors must a company consider when deciding which design is best for a new product?
8. After management approves the design for a new product, what happens in the next step?
9. How is the drafter important in the creation of new products?
10. What are the main reasons for making a prototype?
11. A prototype has just been made for a new product. What steps will be taken before final approval is given?

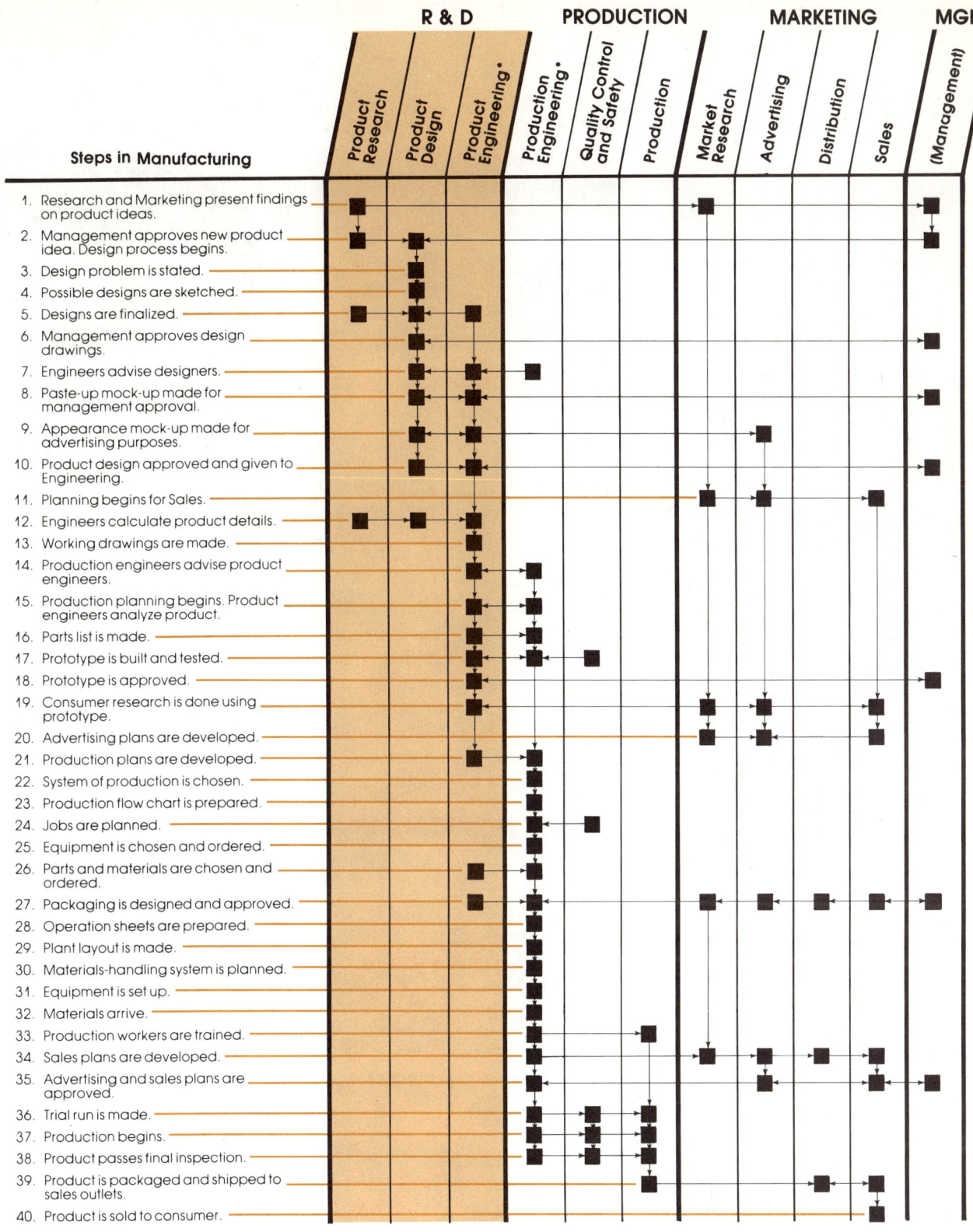

Figure 7-13

The R & D department must work with other departments to make sure the product can be made and will sell.

LET'S GO TO WORK

Research and Development

Activity: Sketching Ideas

Follow these steps to make sketches of design ideas.

Making Thumbnail Sketches

Using a pencil or felt tip marker, draw small, quick sketches of design ideas. Make a lot of thumbnail sketches. Don't worry about being neat. Just get the ideas down on paper.

Making Rough Sketches

Examine your thumbnail sketches. Pick several of the best ideas to redraw. Improve the drawings, and make them neater. Put in more detail to show the design more clearly. Do three or four rough sketches. Share the ideas with several friends and ask them for suggestions.

Making a Rendering

Draw a rendering of the design idea. Pick the best ideas from the rough sketches for one final sketch. On one sheet of paper, **neatly** draw your design. Make it look as much like the product as possible. Using colored pencils or felt tip markers will help to make the sketch more realistic.

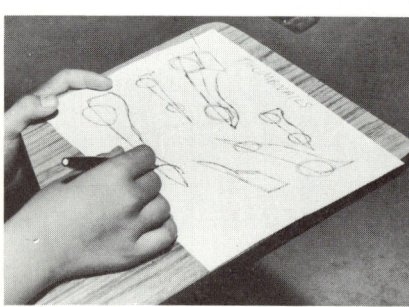

Figure 7-13

Let your ideas come tumbling out onto paper. That's the idea of thumbnail sketches.

Figure 7-14

Pick the best ideas for rough sketches. At this stage you'll want to add more detail to the design.

Figure 7-15

It's decision time. Select your best design, and draw a rendering that looks as much like the final product as possible.

Figure 7-16
This mock-up is the actual size and shape of the race car, but the body is Styrofoam®. A coat of paint can be added to make it look like the real product.

Activity: Making Mock-ups

Follow these instructions to make a mock-up.

Making the Basic Shape

Paper, cardboard, wood, and plastics are some of the materials that can be used to make a mock-up. Cut paper or cardboard with scissors or a utility knife. Paper or cardboard can be assembled using glue or staples.

Cut wood or Styrofoam®* with a saw. A coping saw is especially good for cutting Styrofoam®. Styrofoam® or wood can be smoothed into shape by using fine abrasive paper or a file. Assemble wood or Styrofoam® with white glue.

Look carefully at your mock-up. If the shape of the mock-up is not right, make the necessary changes to correct it.

Finishing the Mock-up

Make the mock-up look more like the actual product. Paint it or put shelf paper on it. The self-stick kinds work very well. Aluminum foil glued on with white glue will give the mock-up a "metal" look. Cut colored paper into shapes or designs. Glue them to the mock-up for trim.

*STYROFOAM® is a trademark for one kind of polystyrene foam.

Topics to Review

Topic 4, **Shearing**
Topic 5, **Sawing**
Topic 10, **Non-Threaded Fasteners and Fastening Tools**
Topic 11, **Glues, Cements, and Other Adhesives**
Topic 15, **Filing and Sanding**
Topic 23, **Finishing**
Topic 26, **Woods**
Topic 28, **Plastics**

Review any other topics that you feel will help you. Do this especially if the tools, materials, or processes that you plan to use are different from those suggested.

Activity: Drafting the Plans

Getting to Know the Equipment

Look over the basic drafting equipment shown in Figure 7-17. Learn to identify the different pieces of equipment.

Making Working Drawings

Study carefully the step-by-step pictures that show how to use drafting equipment. Follow these same steps to make working drawings of your product.

Finishing the Drawings

Erase any mistakes, projection lines, and other marks that you do not need. Using capital letters as shown, letter the name of the drawing, the name of each view (front, top, side) and your name and class.

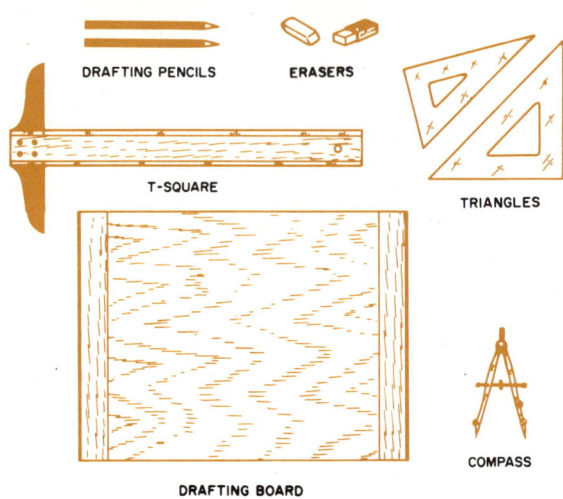

Figure 7-17
These tools and equipment are needed for drafting.

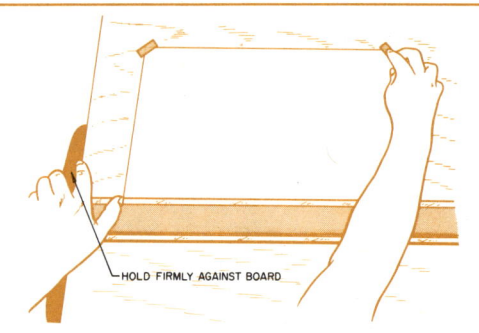

Figure 7-18   *Placing paper on the drafting board.*

1. Hold the T-square firmly against the drawing board.
2. Place the bottom of the paper against the T-square.
3. Put a small piece of tape on each corner to hold paper in position.
4. Make sure that the paper is straight.

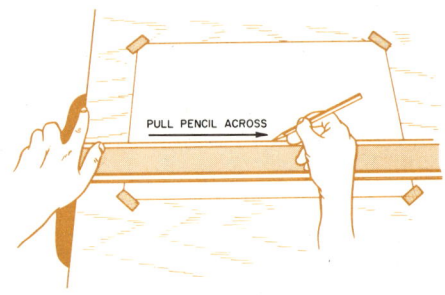

Figure 7-19   *Drawing horizontal lines (across the paper).*

1. Place the T-square where the horizontal line will be drawn.
2. Hold the T-square firmly against the side of the boards.
3. Draw the line. Pull the pencil, don't push it. Turn the pencil slowly between your fingers as you draw. Doing this will help you to make a smooth, even line.

Figure 7-20   *Drawing vertical lines (up and down).*

1. Place the T-square **below** the point where the vertical line will end.
2. Position the triangle where the line will be drawn. Hold the bottom of the triangle against the T-square.
3. Keep the T-square tight against the drawing board.
4. Draw the line. Pull the pencil from the top to the bottom of the paper. Again, turn it as you make your line.

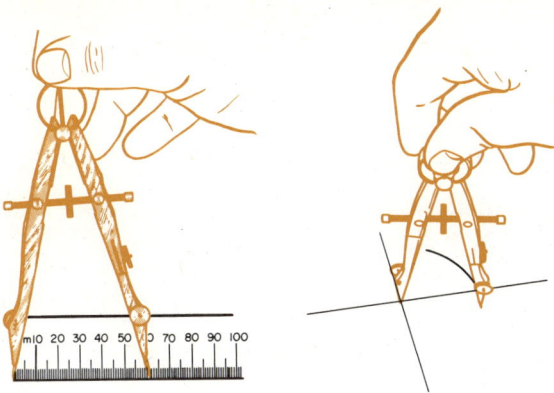

Figure 7-21 *Using a compass.*

Setting the compass.
1. Place the point end on the edge or zero mark of your rule.
2. Adjust the pencil end to the **radius** size of the circle you want.
3. Clamp the compass to keep this radius (if the compass has a clamp).

Drawing a circle.
1. Place the point at the center of the planned circle. Push the point slightly into the paper.
2. Hold the compass between your forefinger and thumb.
3. Twist it between your fingers to make it turn.

Figure 7-22 *Lettering and numbering on drawings.*

All lettering is done in **CAPITAL** letters. Lettering and numbering must be done **very neatly.** Other people must read drawings to make the products and set up machines. Make the letters and numbers as shown. Practice making them until you can make them neatly.

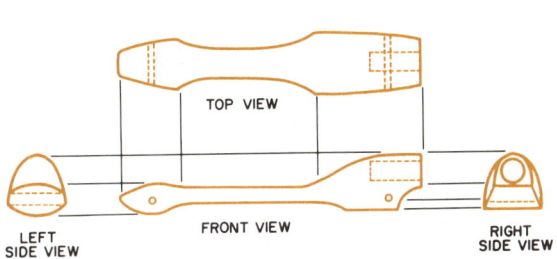

Figure 7-23 *Drawing the object.*

Most drawings are done with at least three views of the object. The front view shows the object as it will appear when you look directly at it. The side view shows what the right side will look like. The top view shows the top as it will appear when you look down at it. Notice that each view is directly in line with the front view.

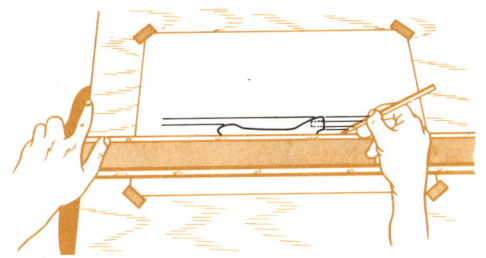

Figure 7-24 *Drawing the front view.*

1. Draw the front view.
2. Using the T-square, draw very light lines called transfer lines from each part of the front view to the side view.

Figure 7-25 *Drawing the side view.*

1. Draw the side view using the T-square and the triangle. Use the transfer lines to guide you. The side view must be directly in line with the front view.
2. Erase all transfer lines.

Figure 7-26 *Transferring lines for the top view.*

1. Using the T-square and the triangle, draw very light lines from each part of the front view to the top view.

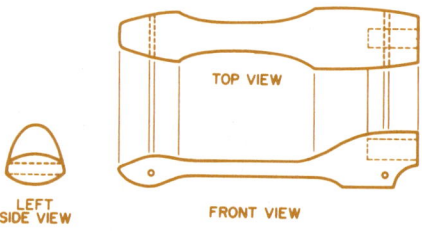

Figure 7-27 *Drawing the top view.*

1. Draw the top view using the T-square and the triangle. Use the transfer lines to guide you. The top view must be directly in line with the front view.
2. Erase all transfer lines.

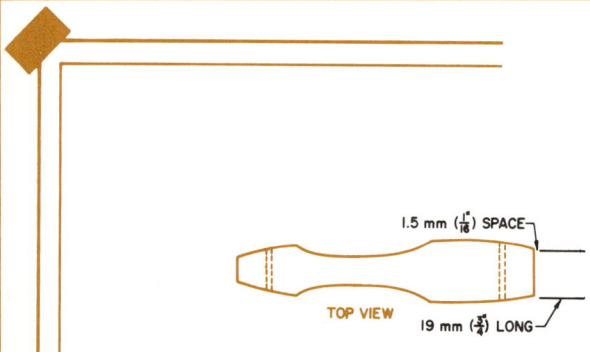

Figure 7-28 *Drawing extension lines.*

1. Using the T-square (or triangle for vertical extension lines), draw two **extension lines** straight out from the object.
2. **Extension lines** should not touch the object. Make them lighter than the lines of the drawing. They must not look like part of the object.
3. Make each **extension line** about 3/4" (19 mm) long.

Figure 7-29 *Drawing dimension lines.*

1. Using the T-square and triangle (or just the T-square for horizontal lines) draw **dimension lines.**
2. Each dimension line should be about 1/2" (13 mm) away from the object.
3. Leave a space in the middle of each line to put in the dimension number.

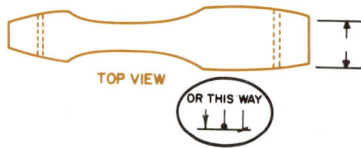

Figure 7-30 *Drawing arrowheads.*

1. Neatly draw arrowheads on both ends of each dimension line.
2. Arrowheads **must** touch the **extension lines.**
3. Note the alternate (different) ways to make arrowheads.

Figure 7-31 *Writing the dimension.*

1. Neatly write the dimension within the break of the dimension line. Put in a very neat number to show what size the object is. The dimension can be found by measuring the mock-up.
2. Put down an indication of what type of measurement is used. Inches (") and millimeters (mm) are used most often.

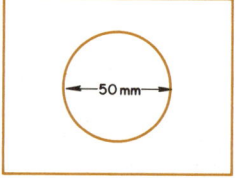 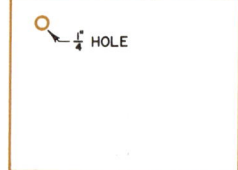

Figure 7-32 *Dimensioning a circle and a hole.*

If the dimension will fit inside a hole or circle, put it in as shown. The arrowheads should touch the circle. Use your T-square to draw the dimension line. Small holes can be dimensioned as in the other example. The arrow should point toward the center of the hole and touch the edge of the circle.

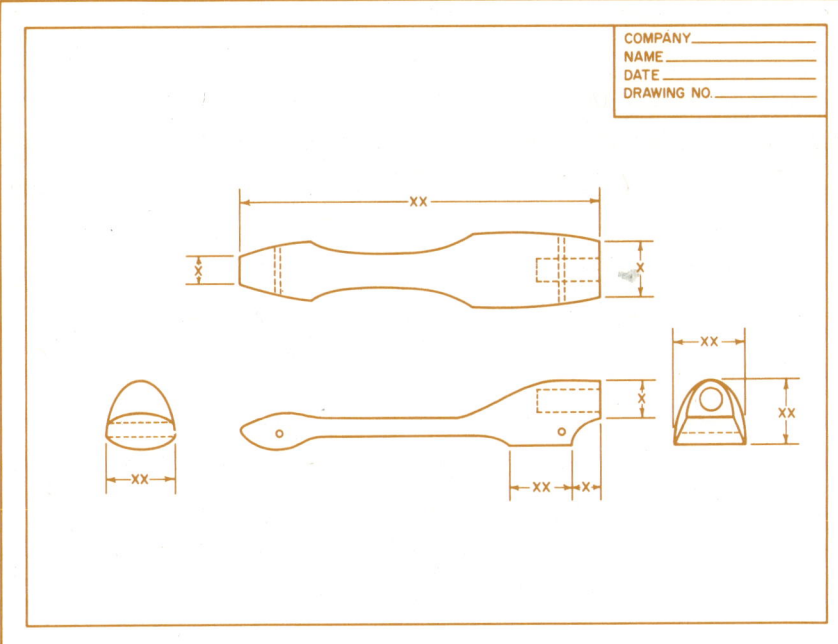

Figure 7-33

These are the **working drawings** of the race car. Notice that the major dimensions should be marked. Much more detail will be given on the **detail drawings** of each part.

Activity: Making a Prototype

Each prototype will be made differently. The exact steps to follow depend on the product being made and the material being used. However, the following steps will help you to get started no matter what product you are making.

Planning a Procedure

Use the working drawings to plan a procedure chart. In order, list each operation required. This chart should show step by step how to make the prototype. List also the machine or tool which will be used to do each operation.

Checking the Plans

Have the procedure chart checked by the teacher. Make any necessary changes.

Following the Plans

Do each operation listed on the procedure chart to produce the prototype. Follow all safety rules when doing each operation.

Topics to Review

All topics that will help you make your prototype.

Figure 7-34

The prototype starts to take shape. Follow your procedure chart exactly to build your prototype. And follow safety rules!

Unit 11 PRODUCT REVIEW

Note: Before making any of the products presented in Unit IV, read Workers and Safety and Understanding Product Plans, pages 52 through 64.

Product	Information Given	Research & Development Needed	Page
RACE CAR — This one really goes!	Material, Size, Wheels, Engine, Axles, Bearings	Body shape, Lengths of axles, Wheelbase, Finish	286
PINBALL — Just like the real thing.	Material, Size, Flippers, Shooter	Game layout, Accessories (Helpful additions), Scoring, Finish	288
SPORTS EQUIPMENT RACK — A place for everything and everything in its place.	Material, Sizes, General idea	Types of equipment to be held, Size, Method of attaching to wall, Finish	290
PENNY SPORTS — A penny for your sports!	Material, Size, Kicker or shooter, Idea	Type of game, Accessories, Finish, Game rules	292
LOCKER ORGANIZER — Find things fast!	General idea	Size, Shape, Materials, Spaces to hold things, Finish	294

UNIT III
PRODUCTION

chapter 8

Starting and Organizing Manufacturing Companies

In preceding chapters you learned that new ideas are behind all new products. You also learned that companies cannot use all new ideas for products. They are limited to making those products that are suited to their equipment, the skills of employees, and other resources.

After a company has made its decision to produce something needed by the customer, much more has to happen before the product appears on the store shelves. That is what this unit, **Production,** is all about. You will be learning about manufacturing companies, where they come from and how they produce the products we enjoy and depend on every day. Since you are making products in this course, you may even decide to form a company in your class.

Figure 8-1
Companies cannot be successful unless people buy their products.

What is a Company?

A company may be formed by one person. Or a group of people may organize to make products or perform a service. The goal of every company is to make a **profit.** Profit is the money left after the company pays all of its expenses.

A company can be small. One person producing handcrafted leather products in a garage may be a company. Or it can be large. For example, the General Motors Corporation produces thousands of cars each day. No matter what the size of a company, it is organized to do business. Some of the larger companies are well known. Can you name a few of these companies?

Every manufacturing company is basically an **input-process-output system** for making products. Every company needs **inputs.** Money, machines, people, and materials are inputs. They must be brought into the company in order for it to function (work). The company **processes** materials. This means that materials are changed into products. **Output** is what the company produces. These products are shipped out of the factory to consumers. See Figure 8-3.

Where Do Companies Come From?

Companies, like products, grow from ideas. One person or a group of people come up with an idea for a new product or a better product. To begin production they form a company.

Ideas and new inventions are the key to any new company. Perhaps the company can make

The Procter and Gamble Co.

Figure 8-3

The company's output, its product, is shipped from the factory to consumers in different ways. Note railway and truck shipping docks.

Figure 8-2

Companies may be large or small. Sometimes small companies grow into large ones. This plant was first established in 1886. It has been expanded many times since.

The Procter and Gamble Co.

something new that people want. Or maybe it can make a better or less expensive item than other companies make. Either way, there is a good chance of success for that company.

Building Better Milk Shakes

Ray Kroc is a man who built the McDonald's Corporation into a multi-billion dollar business. He first worked as a paper cup salesman in Chicago. One of Kroc's customers was a chain of ice cream parlors with a problem. They kept burning out their mixers. They couldn't make milk shakes fast enough to meet the demand. One of the ice cream parlor owners invented a new mixer that had a heavy duty motor. Five milk shakes at one time could be made. Ray Kroc saw an opportunity. In 1939 he quit selling paper cups and became a partner in Prince Castle Sales Company. This company produced milk shake machines called Multimixers.

Ray Kroc's company began to grow. In a short time, people from different parts of the country began requesting Multimixers. The McDonald brothers owned a small roadside restaurant in San Bernardino, California. In the restaurant they used not one but eight mixers. Forty shakes were made at a time. Kroc's curiosity about the McDonald business began to grow. Why did they need so many mixers? He caught a plane to California to check out the McDonald's restaurant.

Building Better Hamburgers

When Ray Kroc arrived in California, he watched swarms of people move into the restaurant at lunch time. They gave their orders and were gone in minutes. Ray Kroc was impressed. He'd never seen anything like this before.

Kroc tried to convince the brothers to begin a chain of McDonald's restaurants. He could see them all over the country with eight of his Multimixers in every one. But the McDonalds weren't interested. They liked the simple life of San Bernardino. And they didn't want the big responsibility.

Figure 8-4

Ray Kroc, founder of the McDonald's Corporation, saw new ideas as new opportunities.

Figure 8-5   McDonald's Corporation

A fast hamburger for 15 cents! That was the going price in 1955 when the first McDonald's restaurant opened in Des Plaines, Illinois.

Kroc didn't give up this new idea. He convinced the McDonald brothers to let him build fast-food restaurants and use their name. It wasn't easy getting started. It took a lot of hard work. Now the McDonald's Corporation has more than 5000 restaurants. These employ over 300,000 people. Yearly sales exceed (go over) four and a half billion ($4,500,000,000) dollars.

Anyone with an idea can form a company. But the success of the company depends on certain factors. First, is the idea that started the company a good one? If it is, then careful planning and organization are needed. Equally important is the amount of money available to the company. People may have good ideas and the know-how to make them work, but poor organization or a lack of money could cause their company to fail. Let's look at how a company can be organized. Let's see how money can be raised to run it.

Types of Ownership

Companies can be owned and operated under one of three different organizations:

- Proprietorship,
- Partnership, or
- Corporation.

Much of the planning and making of products is the same for all three types of manufacturing companies. Actual ownership and leadership is different for each type.

Proprietorship

A **proprietorship** is a company that one person owns and runs. The person comes up with a good idea and wants to start a business. A company is formed. The proprietor (owner) usually puts up the money needed to get the company started. The proprietor may hire people to do some of the work. All of the decisions and the running of the company are handled by the proprietor.

A Successful Proprietorship

Johnson Wax is a large company today. It started as a proprietorship. **Samuel C. Johnson** started his company in 1886. The company first manufactured a fancy wood flooring known as parquet flooring. It was installed in homes. People asked Johnson for help in protecting their wood floors. He saw this as a chance to provide an additional service to his customers. Perhaps he could even expand his business. Johnson experimented with paste waxes for wood floors. Soon he was sending a can of Johnson's Paste Wax with each new floor he sold.

By 1898, the income from the sale of floor waxes and finishes was more than what he made from the flooring. At the same time, people's tastes in flooring were changing. Johnson had problems selling his parquet flooring. People bought a simpler wood flooring. But this flooring also needed a wax protection. Johnson's wax still sold as fast as it was made. The company was changed from a floor company to a wax company.

Because his company was a proprietorship, Samuel Johnson was responsible for many things. As the single owner and operator of the company, he was the salesman, bookkeeper, and business manager. Five days a week he traveled. He sold his products to customers and stores. On Saturdays he returned home. There he filled new orders. He **kept the books** (records) and paid bills. He did everything else that was needed to run the company.

Samuel Johnson had some advantages in running his company as a proprietorship. He made all of the decisions and ran the company his way. He also kept all of the profits.

As proprietor he also had some disadvantages. There was no one in the company to help him make difficult decisions. And he took all the risks. Suppose the company had failed. He would have been responsible for paying all the bills. He could even have lost personal property, such as his home.

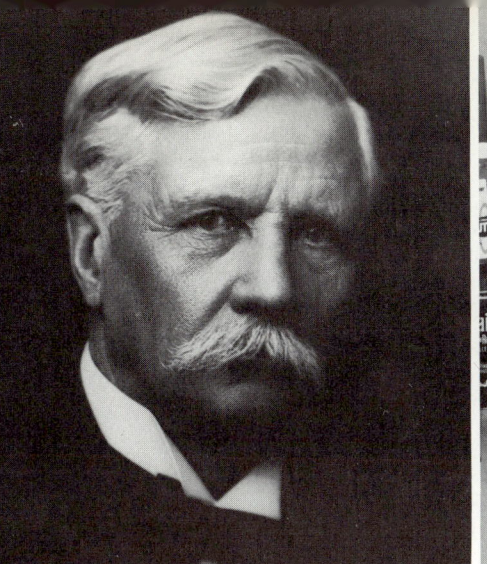

Figure 8-6

S.C. Johnson & Son, Inc.

Samuel Johnson started a flooring company in 1886 as a proprietorship with one product. Today, the company no longer produces flooring. It makes a variety of wax and other kinds of products.

Partnership

Sometimes two or more people share the responsibility for forming and running a company. This is called a **partnership.** Often a partnership is formed when a proprietorship grows and expands. The proprietor can no longer do all the work. Help is needed in making decisions. When this happens, the proprietor may take on a partner. The partner can share the responsibilities and help to finance (provide funds for) the company. This is exactly what Samuel Johnson did with his company.

Twenty years after Johnson started his business (1906), the company had grown to a point where he needed help. He brought his son, Herbert, into the business as a partner. The company name was changed to S.C. Johnson and Son. Now both men shared the responsibility for running and financing the company.

The owners of a partnership have advantages. Two or more people invest money. Partners share the responsibilities. They help make decisions. But partners in a company must be careful to cooperate. They must get along with one another. Otherwise they may not agree on what the company should accomplish.

A partnership may be formed by a verbal (spoken) agreement between the partners. However, this is not the wisest way to do it. Many legal complications (difficulties) can develop. It is best to have a **legal contract** (agreement) drawn up by a lawyer. The contract should cover specific rights, duties, and responsibilities of each partner.

A Successful Partnership

Steve Briggs and **Harry Stratton** started the Briggs and Stratton Company as a partnership in 1908. Both men invested their money and efforts in an attempt to produce automobiles. It failed. However, they found success in making small gasoline engines.

In the 1930's, many areas outside of town had no electricity. Briggs and Stratton put a gasoline engine on a washing machine. It gained almost instant success in rural areas.

Today, Briggs and Stratton is the nation's biggest producer of small gasoline engines. They can assemble 5000 engines in a little over an hour. That is about 83 engines each minute. One engine is finished every three/fourths second. Many of these engines are used in lawn mowers and mini-bikes. See Figure 8-7.

Corporations

There are many proprietorships and partnerships. But these tend to be smaller companies. The third way that a company can be owned and organized is as a **corporation.** The bigger companies that do most of the business are corporations. Three-quarters of the

Figure 8-7

Briggs & Stratton Corporation

Failure did not stop partners Steve Briggs and Harry Stratton. Although they were not successful in producing automobiles, they kept trying. Today their company is the nation's leading producer of small gasoline engines.

business in the United States is done by corporations. Where does a corporation come from? How and why does it start?

Often corporations begin as small proprietorships or partnerships. These later grow into corporations. Let's see how this can actually take place.

A Successful Corporation

In 1882, **George Parker** was a teenager still in high school in Salem, Massachusetts. George enjoyed playing games with his friends. They formed a club and met often to play games. They played such old favorites as chess, checkers, and dominoes. But there were not many games for them to play in 1882. George invented a new game. He called it "Banking." By playing and testing it, he refined the game. It became the favorite of their small club.

In 1883, one of the club members suggested that if they liked the game so well, other people probably would, too. George tried to sell the game to several companies, but they were not interested in producing games. George decided to produce the games himself. He took $40 of his $50 life savings and found a printing company that would print 500 sets of the game. The remaining $10 would go for selling expenses.

Three weeks before Christmas, 500 sets of the **Game of Banking** were delivered to the Parker home. Each game was in a separate box. Selling them all by Christmas seemed impossible. George's school principal encouraged the ambition and zeal (enthusiasm) of his young student. He gave George a three-week leave from school. George packed as many games as he could carry. Then he headed for the stores of Boston. With great enthusiasm, he demonstrated his game to every store owner he could find. By Christmas he had only two dozen games left. He got back far more than his original $40 investment. He paid his travel expenses and made a profit of about $100 besides.

George Parker continued to invent and market several more games during his last year in high school. When he graduated, he started the George S. Parker Company. It was a proprietorship. He owned and operated it himself. In 1888, the company had grown. George convinced an older brother, Charles to join the company. The company became a partnership. The name was changed to Parker Brothers.

Little did George Parker and his brother know what was to come. Parker Brothers is a company recognized today as the nation's largest seller of games. Their most famous game is **Monopoly**®. They are also responsible for developing **Ping-Pong**®, the **Nerf Ball**®, and jigsaw puzzles.

In 1898, the oldest Parker brother, Edward, joined the company. He became a third partner. The three brothers worked well together. George was most interested in inventing and selling the games. Charles took care of the **financial** (money) affairs of the company. Edward was in charge of the actual production of the games.

In 1901 the Parker Brothers Company became a corporation. The company had first been a proprietorship. Then it became a partnership. Why did George Parker and his brothers decide to form a corporation? Why not just a partnership?

George and his brothers knew that forming a corporation had some definite advantages for their company. In a corporation, the owners can sell **shares** of ownership in their company. These shares are called **stock**. Then the owners no longer "own" the company. It is not necessary for them to supply all the money needed to run and expand the company.

George Parker became president of the corporation. Everyone who bought a share of stock in Parker Brothers became a part owner. This allowed the Parker Brothers to raise more money for their company. Money **invested** in the company (by buying shares of stock) was used to build new buildings and buy new equipment. They could expand the company. See Figure 8-9.

The Corporation — A Separate "Person"

There is another major advantage of a corporation. The laws of the state in which the company exists treat it as a separate person. The company is said to be a **separate entity**. Suppose something went wrong. Suppose the company's products did not sell. There may not be enough money available in the company to pay the bills. In a corporation, the people who formed it would have laws protecting them. They would not have to pay the bills out of their own pockets. They would not have to sell their homes or furniture. The buildings, machines, materials, and everything else belonging to the company would be sold to pay the company **debts** (what is owed). More money may still be owed after all the **assets** (everything the com-

Figure 8-8
The **Game of Banking** was the first game invented by George S. Parker. He and the company he began have developed and produced many games played today. George was still in high school when he invented "Banking." Perhaps you could invent a new game.
Parker Brothers

Parker Brothers **Figure 8-9**
Investments in the company helped Parker Brothers to grow. The original Parker Brothers building in Salem, Massachusetts, has been changed but it still is used today. The Parker Brothers company celebrates 100 years of fun in 1983.

pany owns) have been sold. If that is the case, the company would declare **bankruptcy.** Those who invested in the company or loaned it money would lose that money but not their other money or property. Operating as a corporation, then, is a protection to those who form and run the company.

Forming a Corporation

A corporation is started by a group of people who are interested in forming a new company. Or they may wish to change a proprietorship or partnership to a corporation. In any case, an application called the **Articles of Incorporation** is made out. It is sent to the state capital for approval. Stock is sold. A **board of directors** is elected. The board members set the company policy and determine the goals of the company. They hire a manager, such as a president, to run it. After the company gets started, the **stockholders** (the ones who own the company) hold a yearly meeting. They hear how the company is doing. They also elect the board of directors for the coming year.

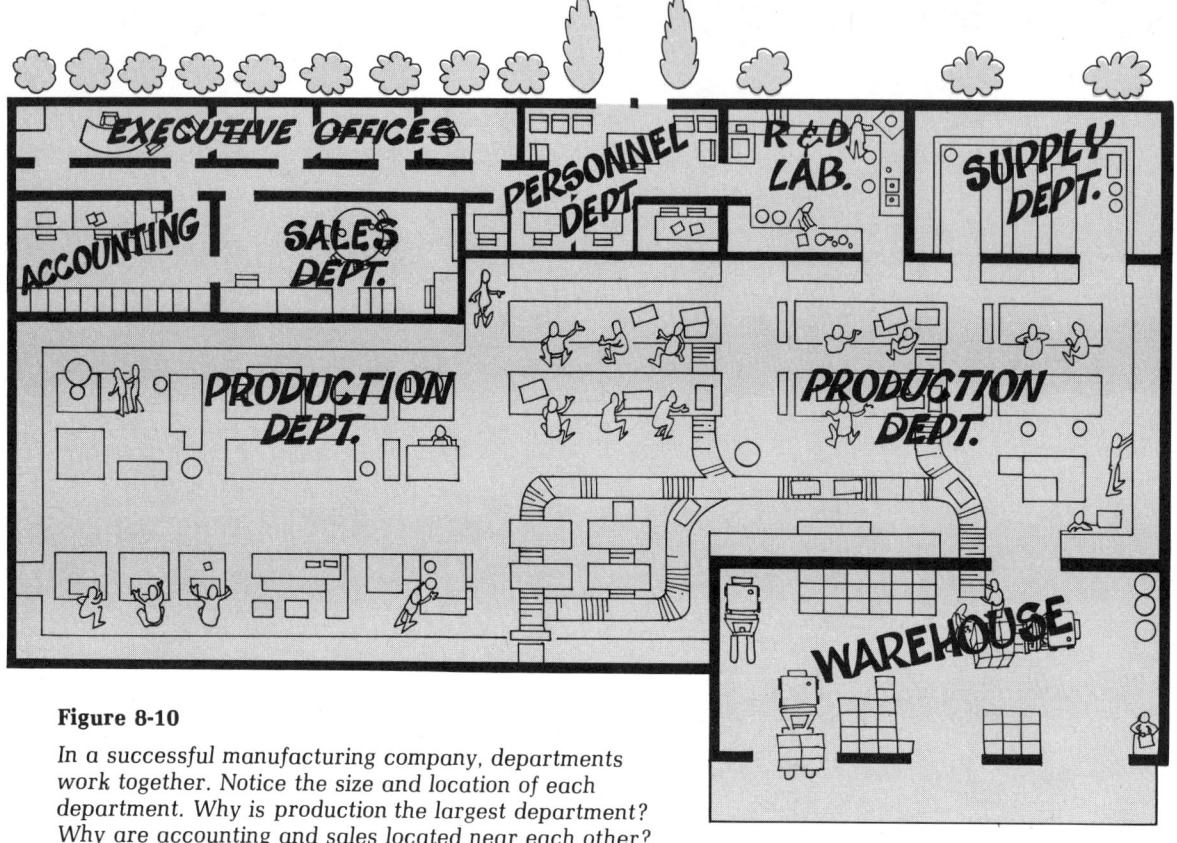

Figure 8-10

In a successful manufacturing company, departments work together. Notice the size and location of each department. Why is production the largest department? Why are accounting and sales located near each other?

How Does a Company Operate?

Imagine flying over a large manufacturing plant that has the roof removed. See Figure 8-10. As you look down inside the huge building, you see people working everywhere. Management people are in offices holding meetings to make decisions about the company. There are secretaries and office workers writing letters and reports and keeping records.

A section of the office area is called the **personnel department.** Here people are applying for jobs in the company, and interviews are being held.

Away from the office area, you can see the **research and development department.** People are experimenting with new materials and products. Designers are making sketches and mockups of new design ideas.

In another, larger area of the plant, much activity is going on. Machines are running. Materials are being processed. Products are being assembled. This is the **production department.** In this department, there is an area where special tools and machines are being made and repaired.

Finished products are stored in a section of the plant or in a nearby building. This storage area is called the **warehouse.** Near the warehouse is the **marketing** or **sales department.** People working here must find out what consumers like and don't like. They need to know where and how to sell the products. They are taking orders for finished products and telling people in the warehouse where to ship them.

This then is a modern manufacturing company. Let's go down inside each of these departments for a moment and see what is taking place.

Management

Management people run the company. To have a successful company, they must be good leaders. Management people must work hard to solve the problems the company faces. These may include how to raise money to buy or rent buildings, where to buy materials and at what price, and how much to pay workers. They pick the products that the company will make.

Management also considers people. They try to find the right people for jobs. They en-

Figure 8-11
Management makes decisions that affect all the departments in the company.

courage workers to do their best and to cooperate with others. Most of all, being a leader in a company takes hard work, determination, and patience. Management people must stick with problems until they are solved.

Administration

Administrative people carry out management decisions. The company president is the **administrator** in charge of the company. The president makes sure that everything is done correctly. However, the president does not make all the decisions. In a corporation, management people such as the board of directors make the major decisions of the company. These decisions may be about building a new factory or making a new product. Then it is up to the president to carry out the decisions of the board and make sure they work. If things are not going well, the president explains the problems to the board. The president does not try to do all this work alone. Depending on the size of the company, the president has one or more vice-presidents. They are responsible for specific parts of the company.

Many companies have a vice-president in charge of each department. There might be vice-presidents in charge of personnel, research and development, production, and marketing or sales. See Figure 8-12. If the company has more than one plant, a vice-president would probably run each facility.

Many support people help in doing much of the administrative work. Secretaries, legal aides, and typists are some of the people who work in the administration offices.

One important job of administration is keeping records. Management people must have information available at all times. It helps them to make the right decisions. How much does it cost to make each item? What products are selling or not selling? How many of a certain item are in the warehouse? How many products should be made to meet the sales demand?

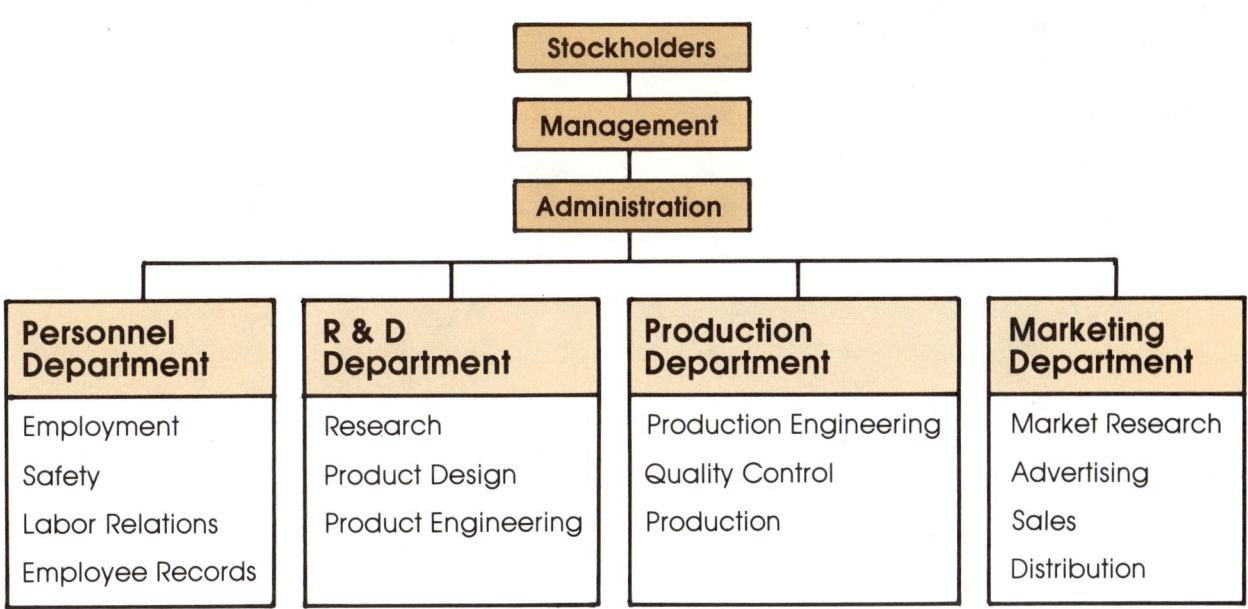

Figure 8-12
This chart shows one way that a company may be organized. Small companies or companies that are divisions of a larger corporation may not have all of the departments shown.

All of this information and much more is needed in order to make correct decisions.

Records must be kept for tax purposes and for making reports to the government. Personal records of each employee are kept. How long did the person work? How much is the worker being paid? How much vacation time has been used?

Keeping records is a big job. Computers can be used for much of it. The company must keep accurate records. Its success and how much profit it makes can be affected by how well the records are kept.

Figure 8-13

People are the most important part of any company. The personnel department works with the people.

Personnel Department

The personnel department works with the people in the company. This department makes sure that the right people are hired to meet the needs of the company. If training is needed, the personnel department sees that it is done. This department may be in charge of safety. It may be responsible for all working conditions. Any disputes (arguments) between workers or about working conditions must be settled. Sometimes workers belong to a labor union. If they do, the personnel department may help **negotiate** (talk over) agreements between workers and management. Questions about working conditions, salaries, and **fringe benefits** (extras) may be settled.

The personnel department is in charge of advancing good workers. Those who do a good job are moved into better positions, usually with more pay. If workers are not needed any longer, they are **laid off**. If they don't do their jobs or can't get along with others, they are fired. Finally, when a worker retires, the personnel department makes sure the worker receives **benefits**. It also gives honors and recognition to employees who have been with the company for many years or have shown outstanding service.

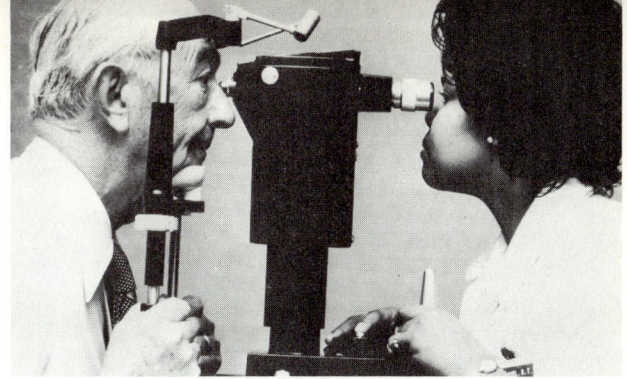

Figure 8-14
Many companies provide health services for their employees.

Research and Development Department

New products and better ways to make products come from research and development. The R & D department designs new products, improves existing products, and finds new and improved ways to make the products.

The R & D department tries to design products that people really want. They design them to be made efficiently. Then the company can make a big profit. Just think what would happen if an R & D department developed an economical and convenient car that could get 100 miles to a gallon of gasoline. Also, if R & D can find improved ways to produce an existing product, the company can save thousands of dollars.

"Boss Ket"

Charles F. Kettering was a pioneer in the field of research and development. He was naturally inclined to "tinker." In spite of poor eyesight, he studied hard and became an engineer. At first he worked in an R & D company that he helped to organize. Then, in 1919, he began doing research for the General Motors Corporation. Soon he was put in charge of the research division. He was head of the General Motors Central Research Laboratory for 27 years.

During his long and productive career he helped to develop many products. Some of these are an electric self-starter for cars, "Ethyl" gasoline, new types of finishes for cars, and a new kind of diesel engine for railroads.

People who worked with him called him "Boss Ket." He was a hard worker and was best at thinking of new ideas. He worked with others to develop the ideas into products.

"Boss Ket" encouraged everyone to think of new and different ways to make life better. He felt that **no one** should be afraid to "try." Charles Kettering recognized that progress requires the creative thinking and hard work of many people.

"What I believe is that, by proper effort, we can make the future almost anything we want to make it."

General Motors Corporation

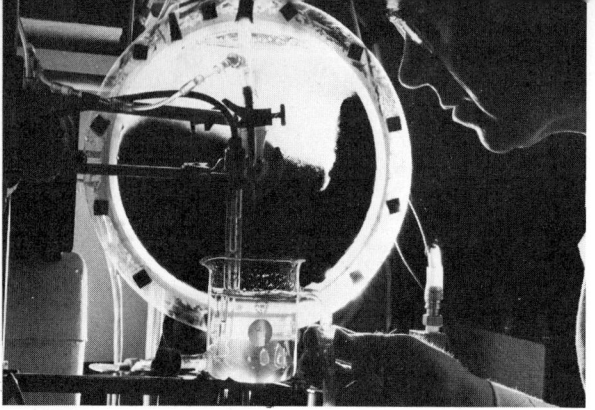

Figure 8-15

The product being developed in this R & D department may someday reduce the amount of water pollution in our country. This researcher is testing a prototype which will purify water which drains from coal mines.

Production Department

The department that is in charge of actually making the products is the production department. Turning materials into components for products, and then assembling the products are very important tasks. The people that run production plan the production system. They make sure each job is done as quickly, as accurately, and as safely as possible.

Poor planning or a mistake in production can cost the company a lot of money. If the right amount of material is not available, production can be slowed or stopped. What would happen, for example, if a company producing yellow refrigerators ran out of yellow paint? The metal parts could be produced but not finished. This would slow or stop many of the production jobs. This is called **downtime**. The workers assembling doors might not have anything to do. If they are idle, the company may still have to pay them. This wastes the company's money. Suppose there were 100 assembly workers and it cost the company $10 per hour for each worker. Then every hour of downtime costs the company $1000. A whole day of downtime would cost $8000. This is why it is so important to plan production before it begins.

Marketing or Sales Department

The marketing or sales department has the responsibility for selling the products. They do consumer surveys to find out what people want. They learn how much people would be willing to pay, and who would be likely to buy the product. This department must have a **marketing plan.** This is the strategy (plan) for how the product will be promoted (advertised) and sold.

Many companies do not do their own advertising. They hire advertising firms to work for them. Catalogs and brochures are printed and sent to stores that may want to order products. This is done many months before the products come off assembly lines. Early orders will give the company an idea of how many of each product to make. Also, products can be shipped as soon as they are finished.

Most larger companies employ sales representatives. These people show samples and pictures to interested customers. Usually, each representative has a specific territory (area) to cover. The territory could be a certain city. Or it could be a much larger area. The western half of the United States could be a territory. Sales representatives try to get orders for many products. They contact customers and show the products to them.

You will learn more about marketing in Chapter 12.

Everyone Must Cooperate

Company people must work together. Workers must cooperate. Departments must cooperate. A company is much like a football team. Everyone on the team has a different job. But all must work together in order to reach the company goal.

Companies of all sizes must operate carefully and efficiently. Years ago people in a small company in Massachusetts found a way to make rubberized clothing. They developed a mixture of rubber and turpentine that could be

Figure 8-16

A marketing plan includes the advertising done to help sell the product.

Figure 8-17
People in all departments work to achieve the company goals.

spread on clothes. Winter was coming. There was a great demand for their product. They rushed to get financing. Then they set up their equipment and began producing and selling their products. Orders poured in. Products were sold as quickly as they could be produced. The owners expanded the plant. It was truly a bonanza.

Then, summer came. Temperatures went up. The new rubberized clothing melted in the heat. Soon the factory was overrun with boxes of melting clothes. Customers sent them back with a demand for the return of the money spent for them. The odor was so strong that the boxes had to be taken out and buried.

What went wrong? This company was very efficient. It had raised the money it needed. The production and sales departments worked well together. One thing was wrong. The company was not careful in selecting its product. The problem was in research and development.

Each department must do its part. And departments in a company must work together and communicate with each other. In a successful company, people work together to reach a common goal.

Looking Ahead

A company must be formed for a good reason. It must be well-organized. The people in the company must do their jobs well and work together.

Do you know
- what is meant by the saying "time is money"?
- if a forklift is used to help people eat meals?

Companies may have the people and equipment needed to manufacture a product. But to produce a good product at the lowest possible cost, plans for production must be made. People must know what they are to do, and machines must be used at the right times. Production planning is a very important part of manufacturing. You will learn more about it in the next chapter.

New Terms

administrator
annual sales
Articles of
  Incorporation
assets
bankruptcy
benefits
board of directors
corporation
debts
downtime
financial affairs
fringe benefits
input-process-
  output system
invested
kept the books
laid off
legal contract
management
marketing
  department
marketing plan
negotiate
partnership
personnel
  department
profit
production
  department
proprietorships
separate entity
shares
stock
stockholders

Study Guide

1. Why are companies formed?
2. What were the important factors that lead to the development of McDonald's as we know it today?
3. What are the advantages of a proprietorship? The disadvantages?
4. What effect did consumer demand have on the development of the Johnson Wax Company?
5. What are the advantages and the disadvantages of a partnership?
6. John Doe wants **to invest** in a certain corporation. What does this mean?
7. How are the bills paid when a corporation goes out of business?
8. Jones and Smith Paper Company decides to become a corporation. What steps must be followed?
9. Explain what these people or departments do within a corporation:
   a. President
   b. Vice-president
   c. Stockholders
   d. Personnel Department
   e. Production Department
   f. Research and Development Department
   g. Marketing or Sales Department
10. In a corporation, what kinds of records are kept? Why is keeping records so important?
11. What may happen if the product or its production is poorly planned?

LET'S GO TO WORK

Starting a Manufacturing Company

Activity: Selecting a Product

Work with your teacher and the other members of your class to select a product for your class to make.

Gathering Product Ideas

Make a list of products that you would like the class to make. Remember, product ideas come from many sources. Look at the Product Previews at the end of this unit. These products may be mass-produced. Review also the Product Previews for Units I and II. Other sources of product ideas include:

- Teacher suggestions,
- Other books, and
- Magazines.

You may also wish to review Chapter 6, **Selecting Products.**

Gathering Information

Answer each of the following questions for each product you have listed. Your answers will help you and the other members of your class to select a product. Your teacher will be able to give you more information.

1. How much time do you have to make the product?
2. How much time is required to make the product?
3. Will the necessary tools and equipment be available?
4. Can you get enough of the right materials when you need them?
5. How would the product be used? By you? At home? As a gift?
6. What other information might help you to make a decision?

Making a Decision

Discuss your product selections and the answers to the questions with the other members of your class. Think about their suggestions and ideas. Which products were selected by the most class members? Which product do you consider best for the class to make?

Class members and teacher may decide together on one product. Or the class may suggest several products and the teacher will make the final decision. After the product has been selected, plans must be made for mass-producing it.

Activity: Selecting a Company Name

Select a company name that reflects something special about the members of your company or class and the product you will make.

Gathering Ideas

Many of the companies that you know were named for the **person** who started the company or who first had the idea for the product. For example, Goodyear Tire and Rubber Company was named for Charles Goodyear, and Ford Motor Company was named for Henry Ford.

Other company names indicate **how** the business was organized. General Motors, for example, includes several names that were once independent companies: Buick, Chevrolet, Oldsmobile, and Cadillac. General Electric includes many manufacturing divisions. They make products as different as light bulbs and refrigerators. The word "general" gives the idea of many different products.

Sometimes companies are named for their **locations.** For example, 3M is the Minnesota Mining and Manufacturing Company.

Your company name might include the name of your school (Metropolis Junior High Manufacturing Company). Or, perhaps, you would like to use the name of your school mascot or a school nickname for the name of your company (Tiger Productions).

Activity: Selecting a Company Trademark

Trademarks are words, groups of words, or symbols used to identify a particular product or business.

Gathering Ideas

Some trademarks are **names.** "Sears" is a name that many people recognize as meaning Sears, Roebuck and Company. Some trademarks are initials. IBM, GE, and AMF are all companies that use initials for trademarks.

Many trademarks include both a **symbol and a name.** The Goodyear trademark includes a winged foot and the name "Goodyear." Still other trademarks are identified by a **symbol** alone. Look at the trademarks shown here. What others can you name that use symbols? Examine some of the trademarks. Why do you think they were chosen?

Designing a Trademark

Design a trademark for your company. First, think about your company name and your product. This is what the trademark will represent. Then think about your design. What are the qualities of a good trademark? Should a trademark attract attention? Are the best trademark designs complicated, or are they simple? How might color be used? Remember, the trademark is a symbol of the company. Design your trademark carefully.

Making the Decision

Consider all of the designs suggested. The class or company can then vote for the design that best represents the company and the product. Other members of the class or company may suggest ways to improve the trademark selected.

Figure 8-18

Companies use trademarks to identify their products. These examples may help you to create your own company trademark.

chapter 9

Planning a Production System

Imagine for a moment that you are the president of a small stereo manufacturing company. You have a suitable building for producing stereos. You have all the necessary equipment and materials. All the workers have been hired. The equipment is at one end of the building, and the materials and components are at the other end. At 8 a.m. the workers come to work. You call a short meeting and tell them that cooperation and effort is needed from everyone. "Let's really manufacture stereos! OK, go to work." What would happen? Do you think they would produce many stereos? Why not?

Let's see what would probably happen. Workers would be carrying materials from one end of the plant to the other looking for places to work. No one would have a certain job or know how to work as efficiently as possible. There would be mass confusion. Why? Because there is no plan, no organization, and no direction. See Figure 9-1.

A production system must be planned and organized. This will help the company to make its products safely and efficiently, and with little waste of materials. Planning will help the company to produce the highest quality product in the least amount of time and at a reasonable cost.

What is Production Planning?

Planning is a special activity that is done before starting a project. Most things in life work out better if they are planned. Production is no exception. Production planning is a process used **before** production starts. It accounts for everything that will be done in the manufacture of a product. The more planning that is done, the smoother the production will run. With planning there is less waste of time, materials, and money.

Types of Production

Before production planning begins, it must be determined whether custom, job-lot, or line production will be used. The type of production chosen will depend upon

- The size of the company,
- How the company is equipped,
- What the product is, and
- How many products will be made.

To custom-produce smaller items, much of the production planning is done by the workers. They decide which tools will be used and when each operation or job will be done. Much more planning must be done to custom-produce large items. See Figure 9-2.

Planning for a job-lot production may be difficult. Since several of the same product will be made, the jobs must be carefully planned. In job-lot production, planning must be done by someone who is familiar with the job shop, tools, and materials. See Figure 9-3.

Line production requires a great deal of planning. **Manufacturing engineers** plan and direct

120 Production

Figure 9-1
What a difference organization makes! A better product is produced at a lower cost.

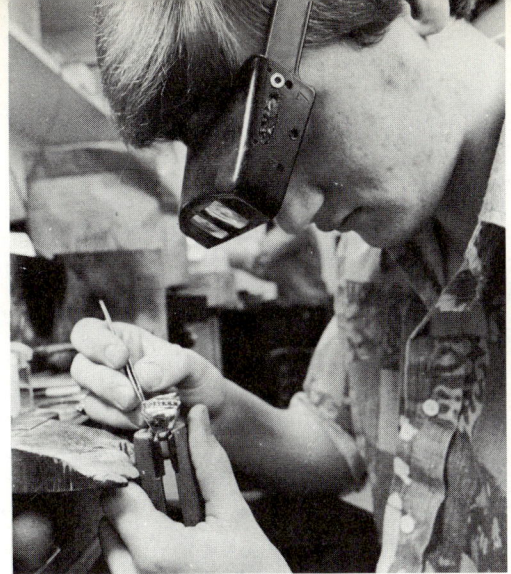

Figure 9-2
This jeweler is custom-producing a ring. He is selecting tools and methods as he goes. More planning is needed when custom-building a ship. Here, a giant crane lowers a tank into a ship.

General Dynamics Corporation

this type of production process. **Manufacturing technologists** may work with the manufacturing engineers. Often they perform many of the same kinds of tasks. Technologists are usually responsible for carrying out the plans and maintaining the system. In a manufacturing plant, some manufacturing engineers and technologists may be called **production engineers.**

In order to plan a line production, engineers must be experts in all the steps that are necessary to make a product. They must know the product, how it is designed, and the materials that will go into it. They must know the number and types of parts and how each type of part can be produced. Engineers plan all the jobs necessary to make the product, where the jobs will be located, and how each operation will be performed. They choose the machines and tools that will be needed. They also decide how materials, components, and finally the product will be moved around the plant during production. See Figure 9-4.

Figure 9-3
Factory equipment must be changed often to produce products in job lots.

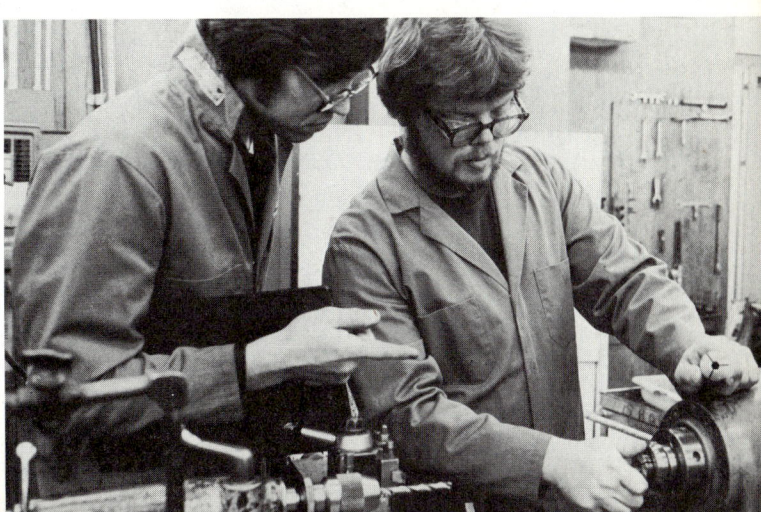

Figure 9-4 Cincinnati Milacron

Engineers use a model of a factory to plan an entire manufacturing system. They know how materials should flow through the plant. They decide where equipment should be set and what tasks the workers should perform.

Production Planning and Product Planning

Production planning is different from product planning. The research and development department plans the product. While this is going on, the people in production planning lay out plans for how the product will be made. Once R & D finishes designing the product, the prototype is built. Production planning may then continue for many months. All the details of how to produce the new item must be worked out.

Why Is Production Planning Important?

We have already seen that a company without a plan will have problems. Workers will be confused and little work will be done. They must know what to do before they can do it.

With good planning, materials are in the right place at the right time. There are enough materials on hand to keep the workers busy. Machines and jobs are set up in the right order. Few mistakes and "reject" parts will be made. Materials are used wisely. Good planning keeps the company's production costs down and raises its profits.

Importance of Costs

Have you heard the saying, "time is money"? This could not be more true than in a manufacturing plant. Consider, for example, a worker in a bicycle manufacturing plant. It costs the company $10 per hour to have this worker on the job. This amount includes the worker's pay, fringe benefits, vacation time, and insurance. These items and others must be included in the cost of the worker. Suppose that the job is running a machine that tightens spokes on a bicycle wheel, and the worker can complete 10 wheels per hour. This means that it costs $1 for the worker to make each wheel.

An engineer, observing this job, has an idea that will make the job go faster. The engineer designs an automatic wrench to tighten the spokes. With this automatic wrench, the worker can do 20 wheels in one hour. Instead of costing the company $1 each time it is done, the cost is now only 50¢. This may not sound like much, but if the company makes 10,000 wheels in one month, the savings is $5000 a month or $60,000 in a year. If the people in production planning develop several new or improved processes like this, they can save the company many thousands of dollars. This is why planning for production is so important. It contributes greatly to the success of a company.

Figure 9-5

If a production engineer can find a more efficient way to complete a task, the company will save money.

The Murray Ohio Manufacturing Co.

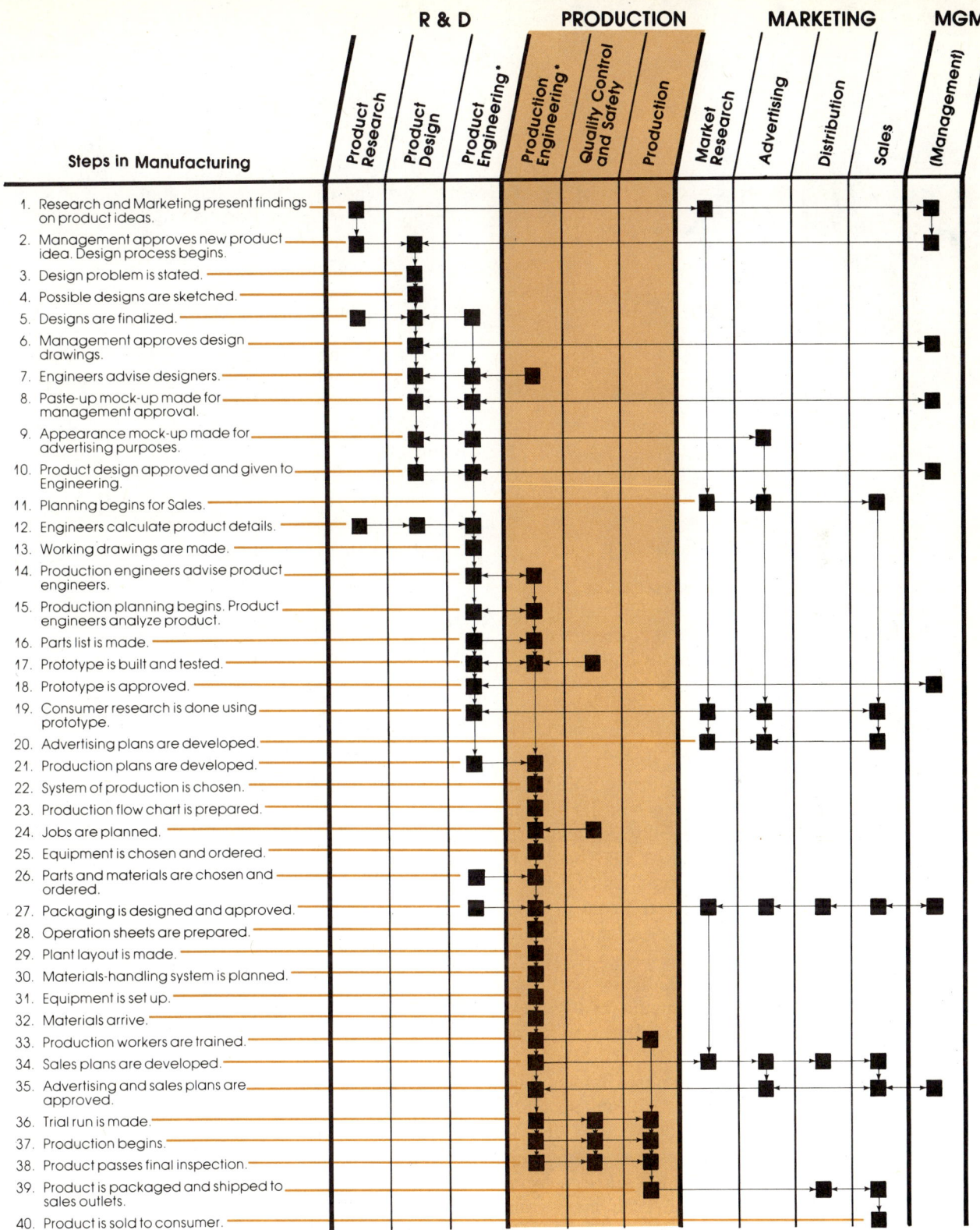

Figure 9-6

Production planning begins before R & D is completed. Production must make certain that the product can be produced. The number of products produced and the amount of time needed to produce them is important to Marketing.

Figure 9-7 The Goodyear Tire and Rubber Co.
The designer and the engineer discuss how the prototype was made. This helps the engineer to plan the mass production of the product.

Production Planning

The first step in production planning is learning how the product is made. Engineers must analyze (look over carefully) the product to see what is required to make it. This is called **product analysis.**

Product Analysis

Engineers want to know exactly what is needed to make the product. Product analysis begins during the earliest stages of the design. Sketches are examined carefully. As mock-ups are put together, the engineers think about how the product can be made. Suggestions are given to the designer that will help to avoid production problems. For example, a designer of electric clocks may want a round design for a new clock-radio. The engineer might know that such a design would require buying costly new equipment and suggest that the design be changed.

There must be a complete **parts list** showing exactly what parts are needed for one product. Some parts, such as nuts, bolts, and screws, will probably be bought from other companies. The parts in which the engineers are most interested are those that must be made and assembled in the plant.

As the prototype of the product is being made, the engineers watch carefully. Although the prototype is usually handmade, they are interested in how each job is done. Final mass production may require a giant machine with automatic controls. However, the **processes** needed to make the product will be the same as those used to make the single prototype. Engineers gather many ideas while watching the prototype being built and may be asked for advice. See Figure 9-7.

The Production Flow Chart

After carefully examining the prototype, the engineers identify the jobs needed to make the product. They also determine the order in which the jobs should be done for the most efficient production. These production factors are shown on a **flow chart.**

Symbols are used on a flow chart. Once you know the symbols, you can read a flow chart and tell what is taking place. Each operation is numbered to show the order in which it must be completed. The points during production when the product should be inspected are also indicated. Suppose the product needs to be stored briefly to allow a finish to dry. Such time periods will show on the flow chart as **delays.** See Figure 9-8.

The flow chart is used to help figure out the amount of time required. How long does it take to do each operation or to move each part around the plant? Where will delays or problems be likely to develop? Remember, planning that is done ahead of time can be a big help toward running a smooth production line.

The flow chart is a very important part of production planning. It gives the engineers and the management people an overall look at what will take place in actual production. The flow chart is used in many planning meetings. The people involved search for the safest and most efficient way to manufacture the product. In making decisions, the main consideration is cost. Which method will cost the least in the long run?

Cost Considerations

The engineers study each operation in order to figure out which machines, tools, and methods are best for the manufacture of the product. The chart in Figure 9-9 shows a study

Planning a Production System 125

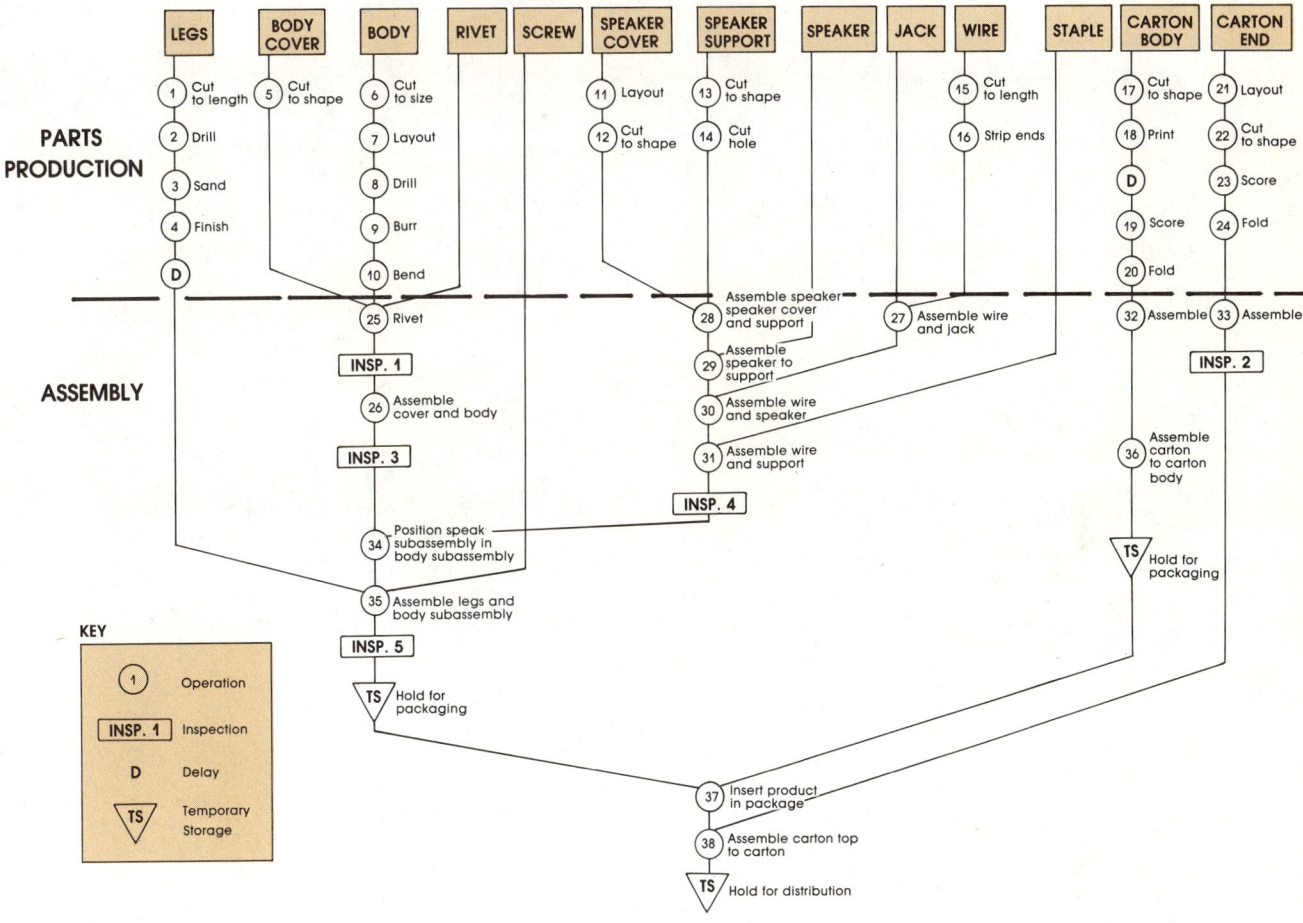

Figure 9-8
Follow the steps of this engineer's flow chart and you'll know exactly how one company makes and packages a stereo speaker.

Method	Setup Cost	Labor Cost	Average Cost, based on number of parts produced				
			1	10	100	1000	10,000
Jig and Drill Press	$ 6.00	$ 5.00	$ 11.00	5.60	5.06	5.00	5.00
Tape-Controlled Drill	$1,000.00	$ 0.50	$1000.50	100.50	10.50	1.50	.60

Figure 9-9
When more products are produced in less time, the cost of producing each product is less.

Operation 30 Assemble wire and speaker

Part speaker subassembly

Production Quantity 500

Stock speaker subassembly, wire subassembly

Tools and Equipment soldering pencil, needle nose pliers, wire cutter

Jigs and Fixtures none

Supplies rosin core solder

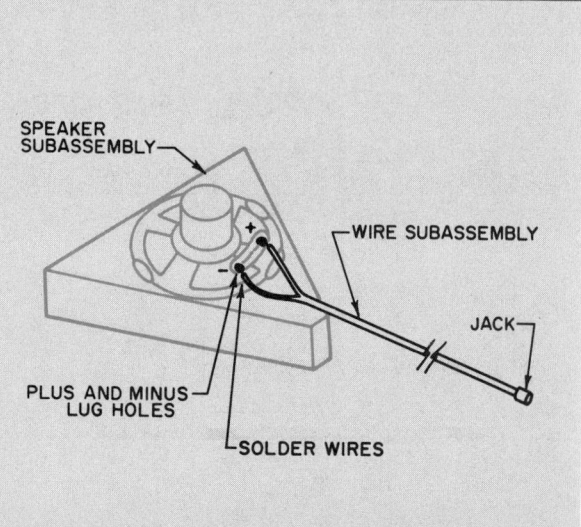

Step Number and Description

1. Insert copper colored lead through lug hole in speaker terminal marked +.
2. Insert silver colored lead through lug hole in speaker terminal marked −.
3. Bend wires leads around terminal lugs.
4. Clip off excess wire sticking through each lug hole and re-bend wires around lugs if necessary.
5. Heat one terminal lug and wire with soldering pencil.
6. Apply solder.
7. Repeat steps 5 and 6 for other terminal and wire lead.
8. Inspect solder joint.
9. Check first subassembly with quality control supervisor.
10. Repeat steps 1 through 8 for production quantity.

Figure 9-10

The worker receives detailed instructions of each operation listed in the production flow chart (See Figure 9-8). Here are the steps involved in Operation 30.

of different methods for drilling holes in the engine of a minibike. The simpler the machine used to drill the holes, the cheaper it is to set up the operation. A simple machine can be set up quickly. This costs less. However, the simpler the machine, the longer it usually takes to do the operation. This means that the labor cost will be higher each time the operation is done.

Notice on the chart how important it is to consider the setup cost and the number of parts that will be made. If only one part is needed, the jig and drill press would cost $11 to use ($6 to set up, $5 for labor to do the job). If the engineer had decided to use the tape-controlled drill, the cost of labor would be much cheaper (50¢) because the machine would go faster. It does more things automatically. Setting up this machine, however, takes much longer and would cost $1000. If only one part is made, the cost of that one part is $1000.50 (the setup cost plus the labor cost). However, if the company needed 10,000 parts, the results would be different.

Using the tape-controlled drill to make 10,000 parts, the least cost to make each part is about 60¢ (50¢ for labor and 10¢ to help pay for the $1000 setup cost). At 60¢ each, the cost for 10,000 parts would be $6000.

Using the jig and drill press, the setup cost for the entire job is only $6.00, **but** the cost of labor is $5.00 for each part. For 10,000 parts,

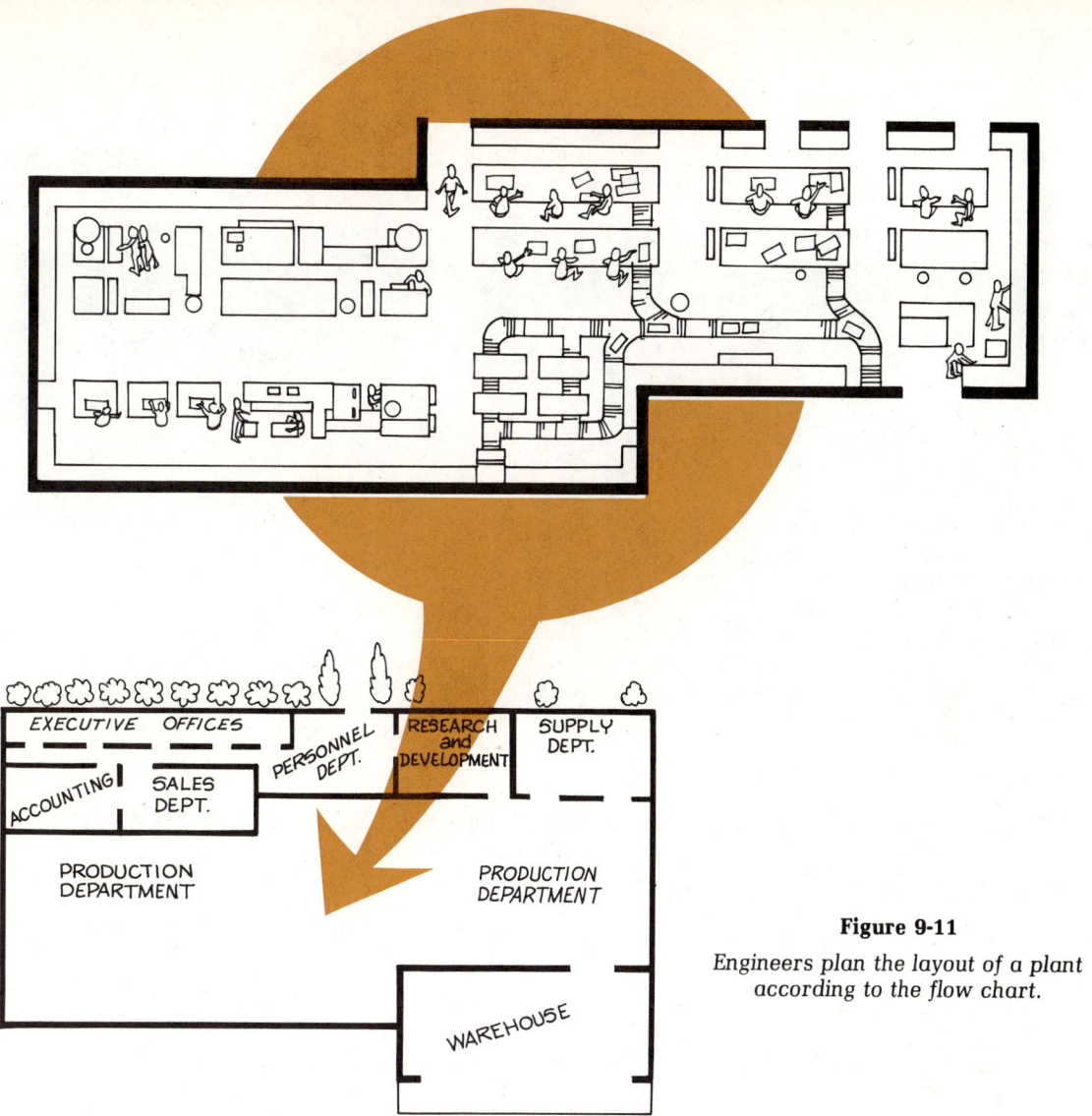

Figure 9-11
Engineers plan the layout of a plant according to the flow chart.

then, the cost would be $50,000 for labor alone. Using the jig and drill press would cost $44,000 more than using the tape-controlled drill. Even though it may cost much more to set up, the tape-controlled drill is much cheaper in the long run. As you can see, it is very important for engineers to study and find the best method for each operation.

Operation Sheets

After carefully studying all possible methods, the engineer makes an **operation sheet** for **each** operation. This sheet will show, step by step, the best way to complete the task. The instructions are **very specific**. As an example, look at Figure 9-10. This is an operation sheet for assembling wire to a radio speaker. The operation sheet refers to Operation 30 on the flow chart in Figure 9-8. By making an operation sheet for each operation, the workers will know the best ways to do their jobs. This will make sure that all workers doing a certain operation will do it in exactly the same way. The operation sheet helps to avoid much confusion and delay once production starts.

The Plant Layout

Using the flow chart as a guide, engineers make a **plant layout.** They decide

■ Where machines will be located,
■ Where the work stations should be placed, and
■ The amount of space needed.

All of this and more must be decided when making the layout.

Figure 9-12
The size of the product determines how it will be moved through the plant. A car requires a mechanical system to move it through assembly.

Figure 9-13
Timing is important. The power equipment for a car is moved along by an overhead chain system. It will meet the car body at the scheduled time.

Often, the plant layout is drawn on paper. However, a model of the plant may be built. Smaller models for all the machines, work stations, and conveyor systems are made. These can be moved around and located in different positions on the larger model until the most efficient plant layout is designed. See again Figure 9-4.

Planning a Materials-Handling System

The simplest way to move a material or a product is to pick it up and carry it. But this is not always possible. The material or product may be too heavy. See Figure 9-12. Also, time that would be spent moving the part by hand from one station to the next might be better spent making products.

Conveyors

Using a conveyor system to make products saves time and effort. **Henry Ford** was one of the first manufacturers to use such a system. When cars were first being made, they were expensive, and most people could not afford them. Production was very slow. To make a car, a frame was placed on the floor. Each part was carried to the frame and assembled to the car. Henry Ford decided that there had to be a better way.

Ford placed the car frame with wheels at one end of his small factory. A track was made for the car to follow. The workers and materials were lined up along the track. Ford tied a rope to the front of the car and extended the rope through the factory and out of the other end. There he tied the rope to a big crank-like setup called a windlass. As each job was completed, the crank was turned. This slowly pulled the car through the plant.

Ford's assembly line was used in 1914. In 1912 (before the assembly line), 170,000 Model T's were made. In 1914 (the first year of the assembly line), 300,000 were produced. Ten years later (1924), nearly two million (2,000,000) were made in one year in 31 assembly plants.

Today, the Ford Motor Company assembly plant in Dearborn, Michigan, has approximately nine miles of conveyor systems. These systems carry many of the millions of parts needed to make over 1000 cars per day.

The **materials-handling system** is designed to meet the production needs of the product. A **conveyor belt** is often used to carry parts around the plant. This is a wide belt that moves. Parts are picked up and laid down as needed by workers. This works much like an escalator in a modern shopping center. Some conveyor systems use overhead chains that look like giant bicycle chains. These chains have hooks on which parts are placed. The chain transports them through the factory. See Figure 9-13. **Timing** is very important. The right part must reach the right place exactly when it is needed.

Other Transporting Methods

Chutes, rollers, and slides are also used to transport materials and products around the factory. Parts are moved by pushing, rolling, or sliding to the next work station. Carts or dollies are often used to roll the products or materials to their proper places. **Forklifts** are used extensively in most manufacturing plants to move large stacks of raw materials, parts, or finished products. See Figure 9-14.

The handling of materials is a very important part of a production system. It must be planned carefully so that production will run smoothly.

When the production planning is completed, the manufacturing company is ready to set up for production. Materials and supplies are purchased. Machines and special tools are set in place. Workers are hired or trained. Many preparations must be made before products are actually turned out of the factory.

Figure 9-14

Two moving systems are being used in this warehouse. A roller system brings boxes of finished goods to the warehouse. A forklift loads them into the truck for shipping.

Looking Ahead

Planning a production system helps to ensure that time, materials, and money will be used wisely.

Do you know

- what "allowance time" means to manufacturing engineers?
- that getting tools and equipment ready to make a new product can sometimes take years?
- how much time would be needed to prepare to produce a new car?
- what happens when a person applies for a job in manufacturing?

In the next chapter you will begin to see how production plans are put into effect as we "get ready to manufacture."

New Terms

delays
flow chart
forklifts
manufacturing engineers
manufacturing technologists
materials-handling system
operation sheet
parts list
plant layout
product analysis
production engineers
work station

Study Guide

1. What production decision must be made before production planning can begin?
2. What is the difference between research and development and production planning?
3. How is product analysis done?
4. How is a flow chart useful in production planning?
5. Make a flow chart for a product you would like to make in your shop.
6. Make an operation sheet for the same product.
7. When manufacturing, is it always the best idea to use the simplest machine? Explain.
8. List some ways that materials and parts can be moved through a factory.

LET'S GO TO WORK

Production Planning

Activity: Making a Shop Layout

A shop layout is a floor plan showing locations of machines and work stations. Making a layout of your shop will help you to develop the most efficient plan for moving materials and parts from one work station to the next.

First think about where to place machines and equipment. Some machines and equipment, such as table saws, planers, and metal lathes, cannot be moved. Others, such as work benches, drill presses, and grinders may be moved easily. On the layout, arrange those that can be moved around those that cannot. Proper arrangement of machines and equipment will help to make the production run quickly and smoothly.

Figure 9-15

Draw your layout of your own shop. Where should equipment be placed to make your assembly line move smoothly?

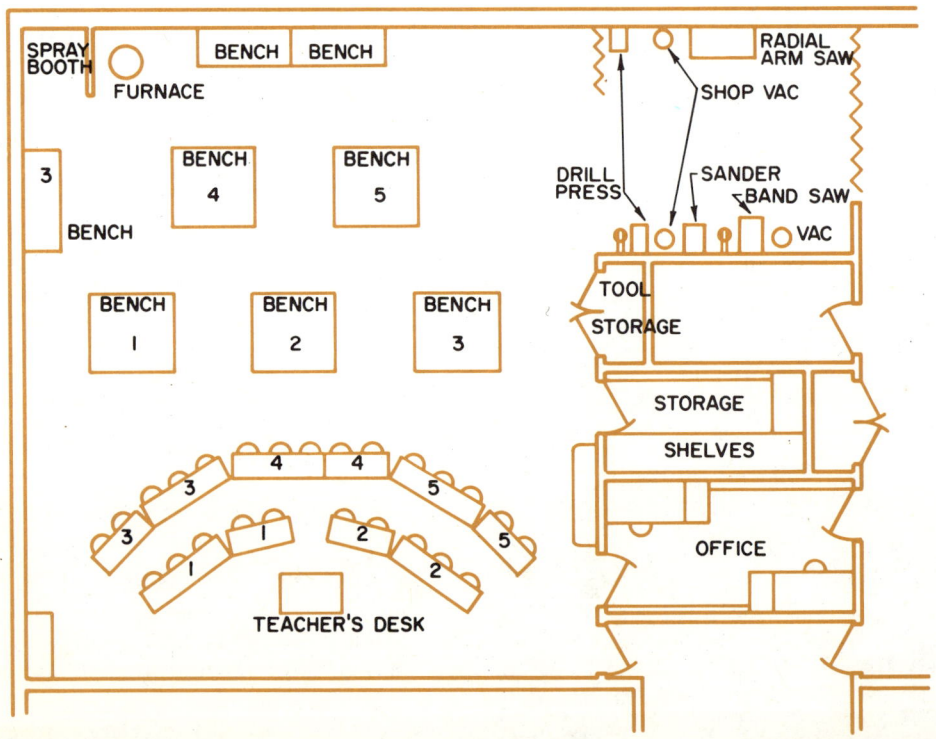

There is a flow chart and a procedure chart provided for each product in this book. These charts can help you to make a shop layout. The procedure chart tells you which tools and processes can be used and the proper sequence (order) for their use.

To make your shop layout drawing:
1. Make a scale drawing of your shop. Use grid paper. Let each square represent a certain area. For example, one square on the paper may equal one square foot of the shop. Measure the shop. Draw as neatly as you can. Include finishing and storage areas.
2. Measure the machines and equipment that cannot be moved. Then draw them in their proper locations. Print the name of each machine near or on your drawing of it.
3. Draw the movable equipment in the locations where they can best be used. Refer to the charts for the order of operations.
4. Number and label the work stations.
5. Draw arrows from station to station showing the path that parts and materials will follow.
6. Show the layout to your teacher. Make any necessary changes. Save this shop layout for later use in activities.

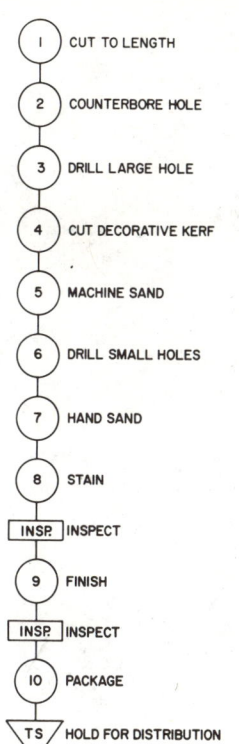

Figure 9-16

Make a flow chart to show all the steps needed to produce a product.

Figure 9-17

Use the shop layout to plan the way in which the product should move through the shop.

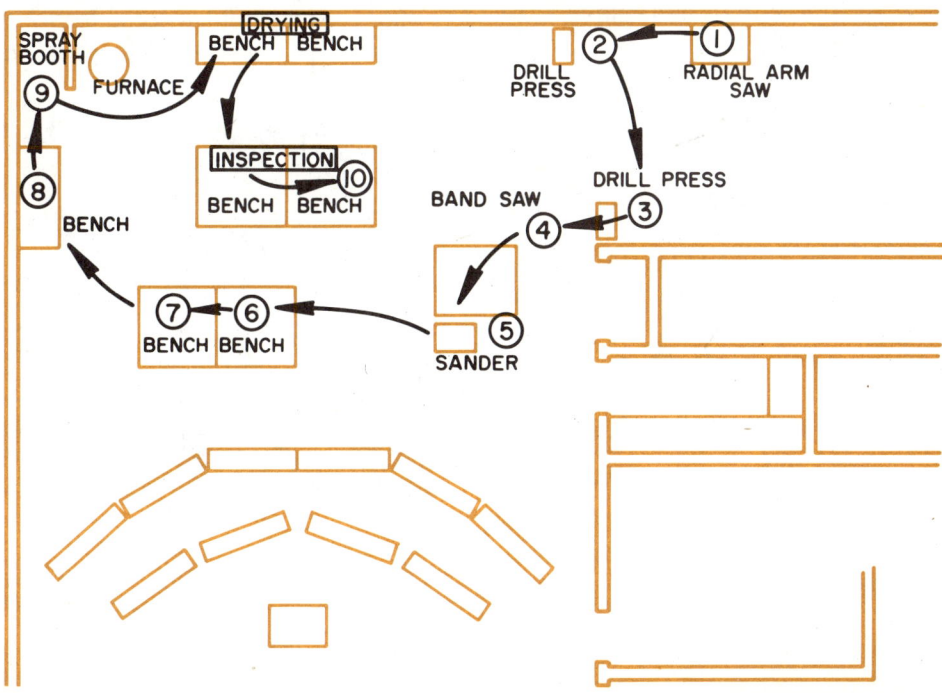

chapter 10

Getting Ready to Manufacture

The manufacturing plant must be made ready according to a plan so that products can be manufactured efficiently. The company has been formed and organized. The product has been designed. A production system has been carefully planned to be as efficient as possible. Now the production plans can be used to organize the manufacturing plant so that production can begin. Many details must be settled. The materials and supplies must be ordered, the time and cost estimated, the machines and equipment set up, and the workers hired and trained.

Ordering Materials and Supplies

The job of the supply department is to make sure that the proper materials and supplies are in the right place at the right time. A whole manufacturing plant could be closed down if the needed items do not arrive on time.

Sources of Supply

When **Soichiro Honda** first began producing Honda motorized bicycles, he got the engines from army surplus communications equipment. Honda sold the motorized bicycles as fast as he made them. Everything went fine for almost two years. Then, suddenly, the whole company stopped. The supply of surplus engines was gone. Because of a supply problem, the small but promising beginnings of the Honda Motor Company looked doomed. What could be done?

Because he was very creative, Honda did not give up. He designed and built his own engines. If he had not been creative, or could not have changed his manufacturing plans, Honda's company might have failed.

Make It or Buy It?

The parts list of the product is carefully studied to determine which parts will be made by the company, and which will be ordered from other companies. The main concern in making this decision is which would cost the least. A company could make its own screws and bolts, but if the machine to make them is expensive, would the company be saving money? Or, for another example, an automobile manufacturer may compare costs for headlights. It may be cheaper to order them from a company that specializes in making headlights.

Preparing to Order

Materials must be ordered for the parts that the company will make in its own plant. These

Figure 10-1
The supply department must keep the right supplies on hand. Then the assembly lines can keep running smoothly.

materials are ordered as standard stock. Standards make ordering and processing materials much easier. People responsible for supplies spend many hours looking through catalogs and telephoning supply houses. They look for the best price and quality for the materials and supplies they need. See Figure 10-2.

When will the supplies arrive? Will they be on hand at the right moment? The time needed between ordering supplies and receiving them is called **lead time.** Supply people must provide enough lead time so that parts will be available when they are needed. Otherwise, a **shortage** may result. Shortages are a big concern for manufacturing companies. Shortages can happen because of bad weather, too much demand for the supply that is available, or work stoppage due to strikes. These factors must be considered when figuring how far ahead to order supplies.

Inventory

Keeping an **inventory** (list) of all supplies is very important. The inventory shows what supplies are in the plant, where they are located and which supplies are getting low. Each time materials or supplies are taken from the warehouse to be used, they are subtracted from the inventory. As new supplies come in, they are added to the inventory. See Figure 10-4. Much of the calculation is now done by computers.

Estimating Production Time

The company needs to know the cost of making a product and the amount of time required to make it. To find these, engineers first study and experiment to see about how long it takes to do each operation. The answers they get are estimates (careful guesses). These times are added together to get the estimated total time needed to manufacture the product. By using estimated time and hourly wages, planners can figure out what the **labor cost** of the product will probably be.

Timing Activities

Let us look at one of **your** normal routines and see how you can estimate the time it takes to do it. Suppose you need to be at school at 8:30 a.m. and must decide what time to get up.

Figure 10-2

Catalogs help persons who order supplies to shop for the best parts at the best prices.

Figure 10-3

Huge rolls of paper arrive at a receiving dock. This standard stock becomes part of the company's inventory until the rolls are needed for production.

Figure 10-4

Computers can help keep track of materials as they arrive or as they are moved to the assembly lines.

Boise-Cascade Corporation

First, make a list of your activities in the morning. Estimate how much time it takes to do each one. Your list may look like the following:

Wash up and brush my teeth	10	minutes
Get dressed	10	minutes
Eat breakfast	15	minutes
Walk to school	15	minutes
Talk with friends and get to class	10	minutes
Total time	60	minutes

You arrived at this total by estimating how long each activity would take. This example shows that you would need to get up no later than 7:30 a.m.

Work Measurement

Determining the amount of time needed to complete a task is called **work measurement.** The most common way to do a work measurement is to watch the operation and time it with a stop watch. This can be done by actually watching someone do the job. If the production has not yet been set up, then a **simulated production** must be used. This should be as much like the actual production as possible. The worker follows the step-by-step instructions shown on the operation sheet while the engineer times the activity with a stopwatch.

What would you do if someone were timing each activity as you were getting ready for school? Probably you would hurry. Would this be an accurate time for getting to school? Experts in industry found that this is what the workers did when timed. They hurried a little and were extra careful. In the actual production, though, workers might be tired, might make a mistake, or just generally work more slowly than when being timed. Extra time was added to make up for this. The extra time is called **allowance time.** Engineers can determine the allowance time by using mathematical figures.

After the time is figured for each operation and allowance time is added, all the operations are added to get the total time needed to produce the product. Now the planner can estimate the cost of the product.

Estimating Costs

If your bicycle were broken, it would probably cost you some money to get it repaired. You might estimate a cost of about $8 for a new bicycle seat, $6 for a new tire, $4 for a tube, and $3 for paint. By adding up all the estimated costs, you would have a total cost estimate for fixing your bike. Of course, this estimate is for parts only.

In industry costs are estimated long before the product is manufactured. The main purpose of estimating costs is to determine the selling price of the product. The company must sell the product for more than it costs to manufacture, or it will not make a profit. The **cost estimator** is the person who figures costs. The estimator finds a total cost estimate by determining three or, in some cases, four major costs:

- Material,
- Labor, and
- Overhead.
- Tooling-up
  (special cases)

The cost of tooling-up is usually included with the overhead. However, when tooling-up is especially costly, it may be considered separately.

Material Costs

Material costs are usually the easiest to figure. By using the parts lists for the product, the cost estimator can figure how much of each

Figure 10-5

This lathe operator and others will be timed to find out how much time is needed to make one cut on this lathe.

Amcar Division, ACF Industries, Inc.

TIME & MOTION STUDY

XYZ Manufacturing Co. Study **106**
Part: **Bracket** Part No. **122606**
Operation: **⑯ Drill Pilot Holes**
Department: **Drilling** Analyst: **Fred Jones**

Step No.	Motion Description	1	2	3	4	5	Average Time (Time in seconds)
1	load part in fixture	16	14	13	10	7	12
2	clamp part in hole	14	14	15	13	14	14
3	drill hole	34	36	33	33	34	34
4	unclamp	10	10	12	8	10	10
5	remove	10	12	12	14	12	12

TOTAL SECONDS **82**
Total in minutes and seconds **1 min. 22 sec.**
Plus 15% allowance for fatigue, breaks and delays **12 sec.**
TOTAL OPERATION TIME **1 min. 34 sec.**

Figure 10-6

Every moment is timed in drilling pilot holes. Notice that allowance time on this form is a percentage of the time needed. Sometimes people are tired and work more slowly. Sometimes they take a break or there is a delay. Engineers must allow for this in their planning.

material is needed and how much it will cost. A little extra material must be added to allow for waste. This includes trimmings, scraps, and other unused materials. Rejected parts are also added in as waste. Even a very careful company will have some parts and products that are not acceptable because of mistakes.

Labor Cost

The cost of the time that workers spend making a product is the **labor cost.** By timing each operation, the total amount of time needed to complete the product is determined. By knowing how much each worker is paid, the labor

Figure 10-7
A company's costs fall into three basic categories:

Labor *Overhead* *Material*

cost for each operation can be figured. These costs are added together in order to find the total labor cost for the product.

Overhead

Many people in a company are not involved in the actual manufacture of the product. Secretaries, product engineers, designers, sales people, janitors, vice-presidents, and others are all necessary in the company. They must be paid with money that comes from the sale of the products. Also, there are costs for building, utilities, office supplies, and advertising. Taxes must be paid. Usually, all costs that are not material or labor costs are put together and called **overhead.**

The Total Cost

The **total cost** of the product is figured by first adding labor and material costs. The cost estimator then uses this figure as a guide to find the overhead and profit that must be included to get a total cost estimate. See Figure 10-8.

When all the figures are in, management holds a meeting in order to make some decisions. These figures give the manager a good idea of what price they must charge for the product. Next they must decide if the selling price is **reasonable.** If they can sell it for more, maybe they can make a better profit. If the price is too high for people to pay, they will need to cut costs.

Figure 10-8

This is how G & S Manufacturing figured the price of their portable fan. Note the reasonable profit they expect to receive on each fan.

PRODUCT COST ESTIMATE

Company, Division: **G & S MFG. — Fan Division**
Project No.: **305 AS**   Date: **6/1**
Product: **Portable Fan**   Production Quantity: **5000**

Part No.	Part Name	Cost/Piece Material	Cost/Piece Labor	Cost/Unit Material	Cost/Unit Labor	Tool Cost
043	rotor	.22	1.22	.22	1.22	375.00
044	blades	.15	.37	.60	1.48	
045	housing	.04	.28	.04	.28	
671	bearings	purchased		.70	—	
046	bracket	.12	.14	.12	.14	
047	cover	.02	.18	.04	.36	
Totals				**1.18**	**3.48**	**375.00**

	Production Cost
Material Cost	1.18
Labor Cost	3.48
Tool Cost per unit	.08
Overhead	.92
Total Manufacturing Cost	5.66
Marketing Expense	1.83
Administrative Expense	.96
Profit	.54
Selling Price	8.99

Lowering Production Costs

When trying to lower costs, the overhead and profit are considered **fixed costs.** They cannot be raised or lowered. The costs that can be lowered are material and labor costs. Maybe the company can buy materials from a different source at a lower price. Perhaps a less costly type of material can be used instead of a more expensive one. If better production techniques are used, materials can be saved because of less waste. The labor cost can be lowered by finding more efficient production methods. These will cut down on the time needed to manufacture the product.

A successful company is concerned about the cost of its product. The lower the costs, the lower the selling price will be. This usually means that the company will sell more products. More sales will mean more profit for the company. For this reason, companies work hard to keep their costs down.

Tooling-Up

Tooling-up means getting the tools and equipment ready for production. This includes preparing all regular machines and tools and making or buying any special tools. For a small company making a simple product, this can take just a few hours. In large companies with very complex products, it may take several years to get all the tools and equipment ready. In automobile manufacturing, for example, companies usually change the design of different models each year. Design changes in the shape of the hood, fenders, top, and windows are very common. **Retooling** (changing the machines) for a new model takes at least a year and a half. If a company designs a completely new car, it takes two to three years to make all the new tools and set up the new production.

Manufacturing companies in the United States have another factor to consider when retooling. The measurement system used is being changed from the U.S. Customary System to the metric system. To make measuring easier and more precise, actual sizes of products and product parts may be changed. For the same reason, standard sizes will probably also be changed. New tools may be needed. Machines must be adapted (changed) or new ones obtained. Many new machines are **dual dimensioned.** This means that they can be adjusted for metric or U.S. Customary dimensions.

AMF

Figure 10-9

The manufacturers of this motorized bike spent 18 months in production planning before mass-producing it.

Figure 10-10

This jig holds a part during welding. Here equipment for a grain elevator is being held for welding.

Most companies are making the changeover gradually. As old machines wear out, they are replaced with metric-dimensioned or dual-dimensioned machines. Often new products are designed with metric measurements, and tooling-up is done accordingly.

Jigs and Fixtures

A manufacturing company usually specializes in a certain type of product, such as toys, electronic equipment, appliances, or furniture. Each company buys equipment that can produce its type of product. Then, as design changes take place or a new product is introduced, the equipment is changed to meet the need of the product.

Jigs and fixtures are used to adapt machines and tools to do specific jobs. A **jig** is used to hold the work and guide the tool to do a certain operation. The jig guides the power tool so that it does the operation the same way each time. See Figure 10-10.

Fixtures hold the material in place while work is being done. Most floor model (stationary) machines or large manufacturing machines use special fixtures to hold materials. These fixtures are usually some type of clamping device that keeps the material from slipping or moving while it is being processed. See Figure 10-11.

Jigs and fixtures must be designed carefully. They should be easy to operate. During production materials must be put in place and removed quickly. Jigs and fixtures must also be durable. They should be able to last through thousands of operations without breaking down under the strenuous (difficult) conditions of production.

Molds and Dies

Many machines are designed to use different molds or dies for making parts. A **mold** is a hollow form. A material is poured, squirted, or forced inside the mold. As the material becomes hard, it is formed into the shape of the mold cavity. A **punch** and **die** is used to press or stamp out parts from metal or plastics. See Figure 10-12. Many sheet metal parts are made

Figure 10-11
Fixtures are usually custom-made to hold a specific part. This fixture permits straight drilling into an angled surface.

Figure 10-13
This die is being cut on a milling machine. The tracer finger on the right follows the pattern. It guides the machine on the left. The new die will be the same as the die used for a pattern.

Figure 10-12
This mold is used to make plastic bottles. The mold is closed over a tube of hot plastic. Air is blown into the mold. The plastic takes the shape of the mold and is hollow on the inside.

This metal fabricating center performs cutting, punching, and tapping processes. With the aid of a computer, the worker can program the machine and let it do most of the work.

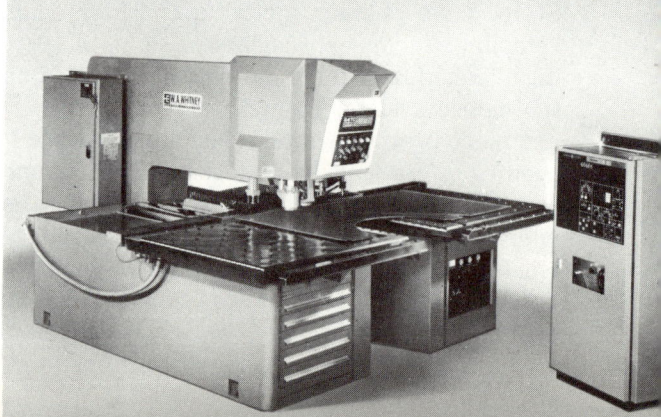

by **die punching**. The part can be cut out and formed into shape in one operation. This is how coins are made.

Some dies or molds can cost less than $100. But complicated dies can cost many thousands of dollars. Tooling-up can be very expensive. For example, Tonka Toys stamps out over 38 million (38,000,000) metal parts each year for 119 different toy designs. For making one part of one toy, such as the cab to a dump truck, the die can be over six feet (2 m) long and cost over $80,000. Each year Tonka Toys introduces about 30 new toy designs. The tooling costs can be over one million five hundred thousand dollars ($1,500,000).

Setting Up

After each tool and machine is built and made ready with all jigs and fixtures, it is tested. Manufacturing engineers and technologists want to make sure that it will hold up under the continuous hard work of production. One breakdown could completely stop production.

Setting up the equipment in the plant can take many weeks or months. Some companies set up the new production in an idle area of the plant. This way the company can continue production of other products. Some companies shut down their plants and lay off workers for several weeks or months. This is done while changes are made and a new production is set up.

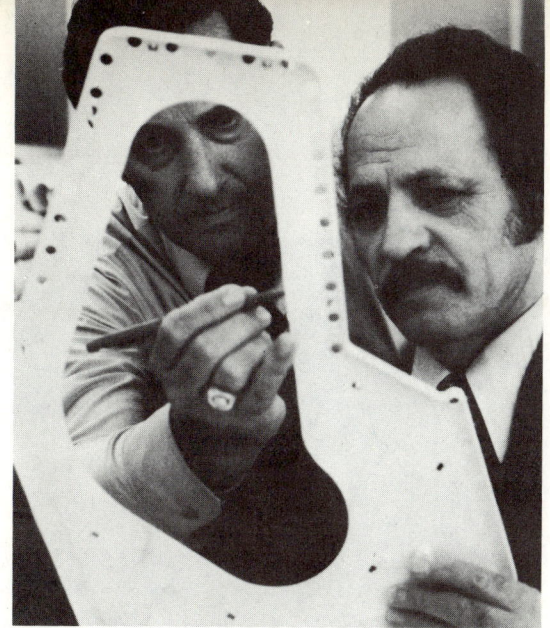

Mattel, Inc.

Figure 10-14

These engineers check the product and tool design before each new toy is produced at this company.

EMPLOYEE REQUISITION

Date: June 23, 1980
Job Title: Drafter   Starting Date: Aug. 1
Requested by: B.L. Jones/Product Engineer   Department: _____
JOB DESCRIPTION: Drawing detail plans from Engineer's sketch.

REQUIREMENTS:
  Academic: High school graduate. Should have vocational drafting courses.

  Skills: General drafting ability. Must be able to make ink drawings. Must have neat lettering.
  Experience: Prefer experience in new product development drafting, especially mechanical & electronics drafting.
JOB SPECIFICS:
  Training: Must attend one week orientation for new employees.
  Salary Range: $250 - $300 per week.
  PERSONAL CHARACTERISTICS: Should be neat, able to follow directions, and get along with other people.

Figure 10-15

The supervisor describes the job opening and the qualifications the person should have. Then the personnel department begins the search.

You're Hired!

In earlier times, many qualified people were not hired for jobs in manufacturing because they were handicapped. **Jayne Baker Spain** owned the Alvey-Ferguson Company in Cincinnati, Ohio. This company engineered and manufactured conveyor systems. In the early 1950's, she began hiring and training handicapped people. The company was helped in several ways. The people in this group came to work on time and were seldom absent from their jobs. Because they liked working, they tried hard and did their jobs well. Most of them stayed with the company for a long time which lowered training costs. They cared about safety and had few, if any, accidents. This caused insurance rates to be lowered. The workers had good attitudes toward their jobs.

Jayne Baker Spain's ways of selecting employees set an example for others. She showed that prejudging people with handicaps is wrong. Employers who had this attitude kept many capable people from developing and using their skills in the manufacturing industry. Now people are judged by what they **can** do, and not by what they **cannot**.

All machines and tools are moved into position and work stations made ready. Conveyor systems are set up and tested. Everything is timed and adjusted. The conveyor system must not go too fast for the operation or slow it down.

Hiring and Training Workers

While tooling-up is being completed, the personnel department is looking for the right people to fill the jobs in production. The personnel department has the task of hiring workers and making sure they are trained to do their jobs correctly.

Job Openings

Each section of a production system has a supervisor in charge of that section. This person knows each job and how it should be performed. When a job opening comes up, the supervisor sends a request to the personnel department for a new worker. The request includes a brief description of the job and the **pay scale** (money the workers will earn). Also listed are **qualifications** (training and experience) needed to do the job.

Using this information, the people in the personnel department begin looking for a person to fill the job. They first look within their own company. Is there someone who wants to change jobs? Is there a qualified person who could be moved up into a better job with more pay? Sometimes no one can be found among the present employees. Then personnel **advertises** for workers. Check your local newspaper and see what types of jobs are available in your community.

Applying for Jobs

Those who see the job notice and are interested get an **application** from the personnel office of the company. After receiving completed applications, the personnel department looks them over carefully. First they must find the people who have the necessary qualifications to do the job. Those who are qualified are usually asked to come in for a **job interview**.

The Job Application

If the personnel department has many qualified applicants, some of them will not be interviewed. Only those people who are best qualified will be selected. This is why care must be taken when filling out a job application. A sloppy or incomplete application tells the company that the applicant doesn't really care and isn't interested in the job.

Even if a company does not have a job opening it will often take applications from people looking for work. Thus, when a job opening comes, qualified people can be immediately called in and interviewed. The opening can be filled as soon as possible.

The Interview

Those who are called in for a job interview are asked many questions. The interviewer wants to get to know the applicant. The interviewer is trying to find out if the applicant would be a good worker, get to work on time, and get along with others.

The applicant should be clean and neat. He or she should learn something about the company and the job before applying. Then the applicant should tell the interviewer how he or she can help the company. Showing an interest and a desire for the job is very helpful.

Before or after the interview, a person in the personnel department may call **references** (people who know the applicant) and ask them questions about the applicant. Personnel people may also check records, such as those from school and previous jobs.

Figure 10-16

The job interview is your chance to convince the company that you are right for the job.

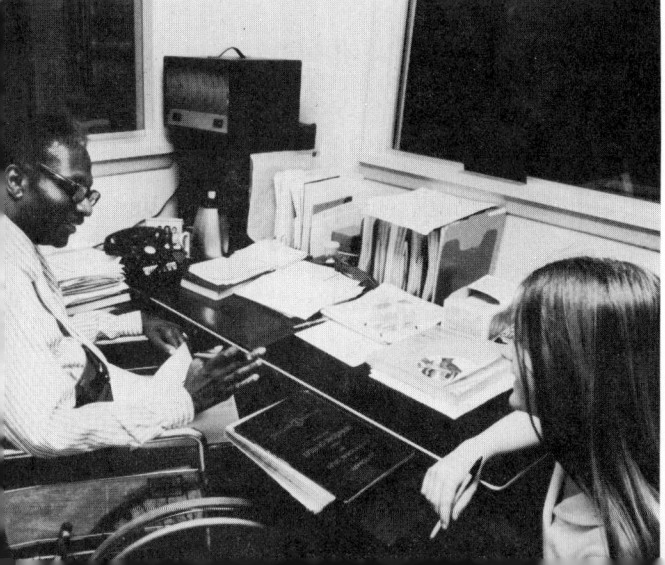

Hiring and Training

After all the information is checked and discussed, the person thought to be the best one for the job is selected. When the person accepts the job, needed papers are filled out. Sometimes the new employee must get a complete physical examination. When all requirements are met, the new worker reports to the supervisor who originally listed the job opening.

In some cases, job training is simple. The new employee is introduced to other employees and shown where to work. In others, weeks or months of specialized training may be necessary. The important thing is that the worker be trained to do the job in the best possible way. See Figure 10-17. The employee is often given a copy of the **operation sheet** in production work. This shows the step-by-step instructions for doing the job.

Unions

Employee unions exist in all the states of the United States of America. These unions include all the trades connected with manufacturing operations. Toolmakers, electricians, and woodworkers are only a few of the skilled people that have unions.

Unions are **collective bargainers** for their members. This means that the individual does not have to negotiate (work out an agreement)

Figure 10-17

Many companies prepare new workers for their jobs by holding special classes.

Figure 10-18 UAW Solidarity

The union steward works with other members of the union and often represents them in meetings with management.

with the company for better wages and working conditions. The union negotiates with the company for all union members.

New employees may need to join the labor union that represents the workers in the manufacturing plant. But in twenty states **compulsory** (required) membership is forbidden by law. The worker has the choice of joining the union or not joining it. If a worker decides to join the union, the **union steward** (representative) will discuss the union benefits and dues with the worker. This is done before the worker joins the union. See Figure 10-18.

Successful Workers

The success of any worker is largely determined by how much effort the worker puts into the job. A worker who tries hard to do a good job is far more likely to get advancements and better pay. Good, dedicated workers help to make the company successful.

Looking Ahead

You can see how much is involved in preparing to manufacture a product. Materials and machines are made ready to use. Time and costs are estimated. Workers are hired and trained. Production can begin.
Do you know
- that all parts need not be **exactly** the same size to be interchangeable?
- why most automobile components are made in the Midwest but assembled elsewhere?

The actual production of a product may require a lot of steps or only a few. But production is always interesting. In fact, it may be fascinating, as you will see in the next chapter.

New Terms

advertise	lead time
allowance time	material costs
application	overhead
collective bargainers	pay scale
	punch and die
cost estimator	qualifications
die punching	references
dual dimensioned	retooling
fixed costs	shortages
fixtures	simulated production
inventory	
jig	total cost
job interview	union steward
labor cost	work measurement

Study Guide

1. Why is it important to be able to change manufacturing plans when necessary?
2. What factors must be considered when deciding how far ahead to order supplies?
3. Select a product that you can make in your shop. Make a list of materials needed. Trade lists with another student to see how material needs are different for different products. Can they also be different for the same product?
4. For the product in Number 3, estimate the total time needed to produce each in your shop.
5. Why must a company be very careful when estimating costs? Which costs are considered?
6. Your company has a high **overhead**. What does this mean? Which costs are involved?
7. How does the cost estimator figure total cost?
8. What are some ways to lower production costs?
9. What determines how long it will take a plant to tool-up?
10. Why are jigs and fixtures important to the manufacturing process?
11. How does a manufacturing plant set up for producing a new product?
12. What are the steps in finding, hiring, and training workers?

LET'S GO TO WORK

Getting Ready for Production

Activity: Taking a Performance Test

Sometimes performance tests are used to help place workers in certain jobs. Performance tests can be used to determine a worker's skill and ability. Each of you will take at least one of the following performance tests. How well you score could determine your job during production.

Test A.

    Needed: Five machine bolts,
                     1/4" dia. × 1", with nuts
                     Stopwatch or timer

1. Place the bolts and nuts on the table in front of you. Ask someone to be the timekeeper for you.
2. At the starting signal, begin threading a nut onto a bolt. Thread the nut all the way to the bolt head.
3. Do the same for the other four bolts and nuts. Have the timekeeper stop the timer when you finish the last one. Your score is the number of seconds it took you to do all five nuts and bolts. Try to get a low score.

READY, SET, GO!

Test B

Needed: One special test board
Nuts of several sizes
Stopwatch or timer

1. Place the test board and nuts on the table in front of you. Ask someone to be the timekeeper for you.
2. At the starting signal, begin taking nuts from the tray and placing them on the posts that they best fit. Place one nut on each post. Have the timekeeper stop the timer when you have finished. Your score will be the number of seconds it took you to properly place all of the nuts onto the posts. Try to get a low score.

READY, SET, GO!

Test C

Needed: One board
Try square
Crosscut handsaw
Pencil

1. Ask someone to be the timekeeper for you. The timekeeper should start the timer when you pick up the square.
2. Draw two parallel lines 1/8" or 3 mm apart all the way around the board.
3. Clamp the board in a vise or to the workbench.
4. Saw through the board keeping the saw blade between the two lines. Have the timekeeper stop the timer when the cut is completed. Your score is the number of seconds it took you to measure, mark, and saw the board. If your saw cut is outside the line at any point add 60 seconds to your score. Try to get a low score.

READY, SET, GO!

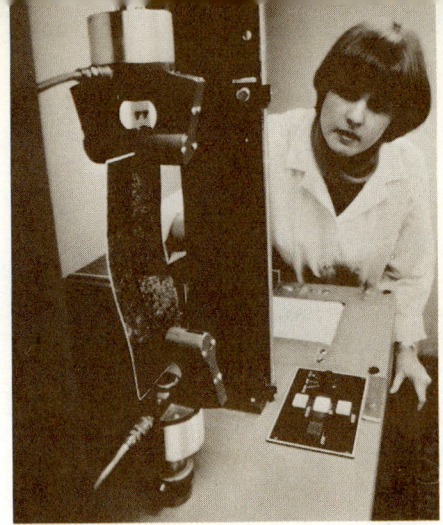

chapter 11

Producing the Products

Let production begin! Every day thousands and thousands of products roll off the assembly lines and out of the production plants of American manufacturing companies. Our manufacturing system has given the U.S. one of the highest standards of living in the world. There are more goods (products) available to more people at lower prices than in any other nation. The key is **productivity.** Productivity simply means how fast and how accurately products are made. A well-planned and well-organized production system contributes to high productivity. To reach this goal, dedicated, hardworking employees are needed.

Where Does Production Begin?

As you have learned, product production starts with raw materials. Iron ore, trees, cotton, and petroleum are a few of the raw materials that nature provides for us. Raw materials are converted into standard stock. Some manufacturers specialize in converting raw material into standard stock. This type of processing is called **primary manufacturing.** Sheets of metal, stacks of lumber, rolls of cloth, drums of liquid plastic, and large containers of ceramic clay are all standard stock. The standard stock is shipped to manufacturing plants. The factories then begin changing it into products.

A product can be made from one single part or component or from many components. Some components are assembled to others to form **subassemblies** (partly finished products). An automobile engine and the speaker for a radio are subassemblies. Then, components and subassemblies are combined to make **assemblies.** The thousands of products we see and use every day are assemblies. For example, a bicycle is an assembly. Components, such as the handle grips, and subassemblies, such as the seat and frame, are combined to make the finished bicycle.

As we have seen in this unit on production, much planning is necessary in order to change raw materials into finished products. Now let's see what must be done to manufacture those products. Most importantly the company must make sure that everything is done correctly and well.

Allis-Chalmers Corporation

Figure 11-1

High productivity helps to keep prices down. Making products quickly and well saves time, materials, and money.

Figure 11-2 United Technologies
This company manufactures propellers for an aircraft company. The planes are assembled in another plant.

Reynolds Metals Company **Figure 11-3**
A worker inspects alumina as it arrives at a plant. In processing, this raw material will become aluminum.

Quality Control

The quality of a product is simply how well it is made. If a product is poorly designed or produced, there is a lot of waste. The company has wasted raw materials. Since the supply of all of the earth's raw materials is limited, this is a very important loss. Many raw materials cannot be replaced. Others, such as trees, take time to grow.

A poor product means that the company has also lost money. It has paid wages for supervising, handling, processing, and inspecting the product. The cost of wasted materials, scrap, or reject parts must be added to the cost of the products to be sold. All of this increases the price to the consumer. Higher prices may cause customers to buy from **competitors** (others who make the same products). The company may go out of business. Having a system to control the quality of the products is an important part of manufacturing.

Purpose of Quality Control

The purpose of **quality control** is to make sure that the product is good enough to be sold. Often, certain people are specifically in charge of quality. But everyone in the plant must take steps to see that a quality product is made.

Inspection is an important step in quality control. Employees should check their work to make sure that it is done correctly. When a problem or a mistake is found, it should be **reported**. It may be reported to a supervisor or an inspector, or to a person who specializes in quality control. Any problems or mistakes will need to be corrected. The faster that problems are resolved (taken care of) and changes are made, the more smoothly the production will run. This will probably also save money for the company.

Inspecting

Inspections are done during three different stages of the production:
- Preparation for production,
- During production, and
- After the product is finished.

The ways in which inspections are done depends upon the type of product and the processes involved.

Material Inspection

When standard stock or raw material are delivered to the manufacturing plant, they must be inspected. See Figure 11-3. Are we getting what we paid for? Is it the right size? the right amount? the right quality? A company cannot make quality products with poor materials.

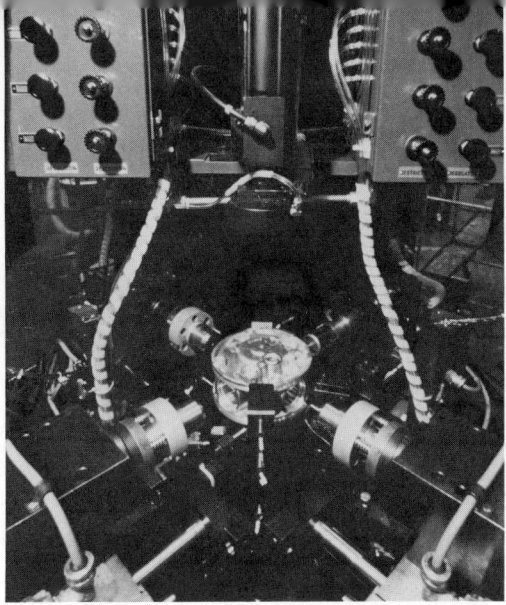

Figure 11-4

Shown here is an automatic gage, a measuring device that can inspect parts at mass-production speeds.

Bendix Corporation

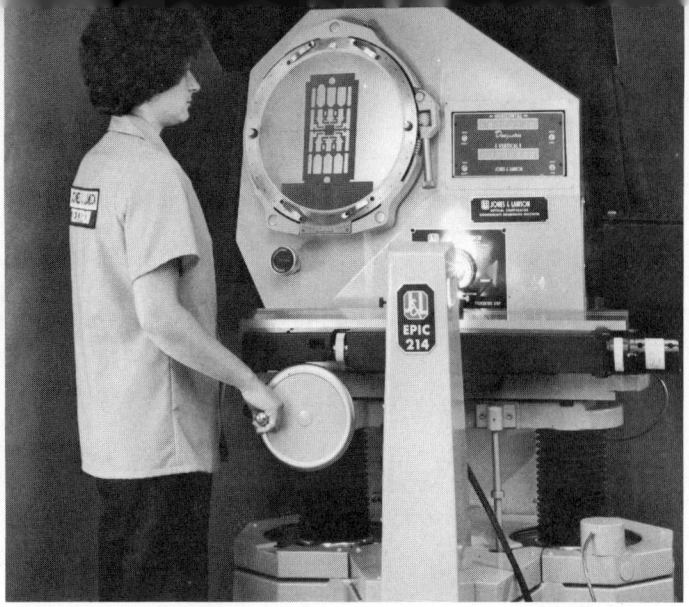

Figure 11-5

An optical comparator is used to inspect small items. It works much like an overhead projector. A reflection of the item is enlarged on the screen. The inspector can then see if the item is made correctly.

The Procter and Gamble Corporation makes many household and laundry products. Materials arriving at the manufacturing plant are carefully inspected. For example, the company buys cardboard boxes for laundry detergent. These are folded flat for shipment to the plant by rail. Many carloads are sent at one time. Before the shipment is accepted, some of the boxes are taken to the testing lab. Here they are tested for strength, color of printing, and even smell. If these boxes meet the quality standards, the shipment is accepted. The boxes are taken to the production line and filled with detergent. Then they are shipped to customers. As you can see, a company that is concerned about quality carefully inspects **all** materials when they are received.

Inspection During Production

The best time to make sure that the parts or products are being made properly is during production. This is usually called **in-process inspection.** There are several reasons for inspecting during production. Some of these are

- To check for worker errors,
- To ensure that critical processes have been done correctly, and
- To check for product defects (flaws) caused by wear on machines and tools.

Even after special training, a worker might make a **mistake.** Imagine a part for a space shuttle made from a special piece of metal that costs $500. Suppose a machine operator had to drill several holes in that metal. If even one hole was drilled in the wrong place, the whole piece would have to be scrapped. How many mistakes like that could the company afford?

A **critical process** such as chrome-plating requires special inspection. If a part should be bright and shiny and it's not, the customer may not want to buy the product.

Errors in products may be caused by **wear** on tools and machines. For example, after drilling 800 holes, a twist drill may be slightly smaller in diameter due to wear. This could affect the parts that must fit into those holes. In-process inspections catch these and other errors.

Inspections are not usually made after every operation. This would require too many inspectors. It would be too costly. Usually inspections are made at **key** times during the production run. When a **new operation** is started, the work being done in that operation is inspected. When a **new machine operator** starts to work, that person's work is inspected. A part or assembly may be inspected after **a series** of similar operations have been done. For example, the block (the main part) of a lawnmower engine would be inspected after all holes have been drilled rather than after each hole. If a part is to be painted, it will need to be inspected both **before and after the painting** operation. Parts that are to be assembled need to be inspected **before assembly.** Companies must make certain that products are made correctly.

Use of Measuring Devices

As product parts are being produced, workers and inspectors use measuring devices to check for quality. Size, strength, weight, roughness, and even color can be measured.

Can parts be made exactly the same size, the same strength, or the same color? Perhaps, but most often not all "identical" parts are **exactly** the same. A small amount of error is allowed. This error is called **tolerance.** Tolerance is the amount that a part can be greater or less than the desired measurement and still be used. For example, a certain part may need to be 50 mm long. The working drawing may show this as 50 mm ± 1 mm (50 millimeters plus or minus one millimeter). This means that the part is still acceptable and will work if it is one millimeter longer (51 mm) or one millimeter shorter (49 mm) than 50 mm. A part that was any bigger or any smaller would be unacceptable. It would have to be changed or rejected.

To produce parts within these tolerances, workers and inspectors use special measuring devices often called **gages.** These measuring devices are usually designed so that the person using them has only to compare the part to the gage. If the part matches the gage, it is accepted. If it doesn't match, it is rejected. A "go and no-go" gage is a good example of this kind of measuring device. If the part being checked fits into the gage it is "go" (acceptable). If the part doesn't fit into the gage or goes into the gage too far, the part is "no-go" (not acceptable). Other kinds of measuring devices are used to measure roughness or smoothness, flatness, and even hardness.

Visual Inspection

The quality of the product can also be checked by **visual inspection** (using your eyes). Inspection for cracks, dents, and scratches is done this way. How well the parts fit together is also checked in this manner. Some products are inspected by machines. Welded parts may be checked by X rays to make sure that the weld is fused into the metal and that there are no "air pockets."

Role of the Inspector

A production system may have several people who are assigned as inspectors. They constantly check the quality of the work being done. Without an inspector, a worker may do a job slightly wrong all day without knowing it. One or two comments from an inspector can help the worker do a better job.

Fred Kenderson, PPG Industries

Figure 11-6

Silverware inspectors look for scratches or changes in color. Flaws must be found and corrected.

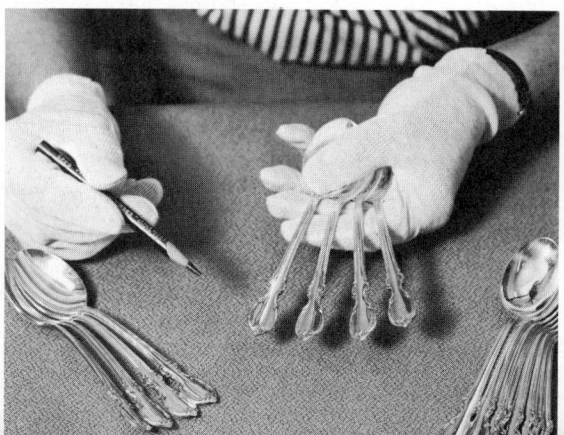

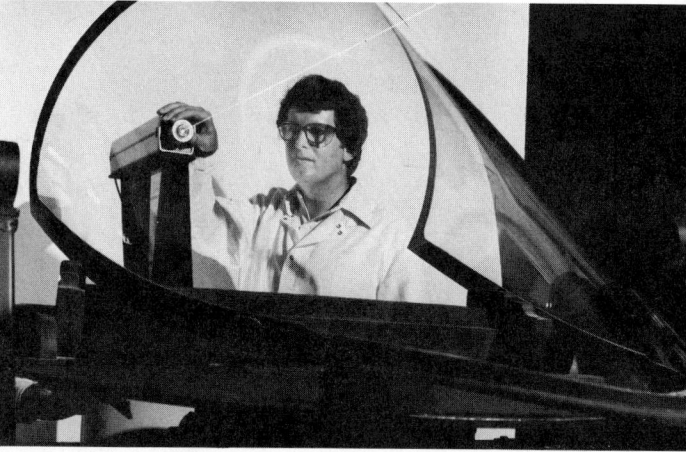

Figure 11-7

This worker is using a laser beam to inspect the clearness of an aircraft windshield. Pilots must be able to see through it easily.

Inspectors may examine the work being done at the work stations or on the line. Sometimes there is a special area where materials or products are taken to be inspected. Inspectors keep records of what they inspect and the results of their inspections.

Final Inspection

Before the final product is packaged and shipped, there is a **final inspection.** This inspection makes sure that the product looks and works as it should.

In some instances, products are tested as they are finished. See Figure 11-7. In an automobile assembly plant, every car is inspected and tested. The motor is started, and every knob, switch, and light is turned on and off to make sure that it works.

It is not always possible to test every product. How could each aspirin tablet be tested when millions are manufactured at one time? Testing each one would be nearly impossible and also very expensive. This is when **sampling** is used. From time to time several products are inspected. If these products pass inspection, then it is assumed that the others will, too. Without proper inspection, thousands of products with defects could be made before an error was found. See Figure 11-8.

Figure 11-8

Every 10 minutes, this inspector selects a jar of powdered drink from the production line as a sample. She checks its weight. She also checks the amount of pressure it takes to open the lid and the tightness of the jar's inner seal.

Correcting Problems

When a problem is discovered, the inspector must find the cause to see that it is corrected. Maybe someone is not doing a job correctly and needs additional instructions. A machine may not be working correctly, or a jig may be wearing out. Someone may be handling a part in such a way that it gets scratched each time it is moved. Whatever the problem, it must be found and quickly corrected.

The job of the quality control department is to find problems and to correct them. However, quality control should also help workers avoid problems.

The Trial Run

Will the system work? Before bringing all the workers in and running the production, a trial run is made. The trial run is very important. It is the first time the new production system will be used. All of the individual parts of the system have been studied and worked out during planning. But the whole system has not been run all together.

The main goal of the trial run is to "debug" the production system. **Debugging** is finding and correcting all the little problems that can cause delays in the actual production. Maybe a worker has to lean over too far in order to place a part on the conveyor system. This movement makes the worker uncomfortable. The worker's production slows down. By doing a trial run, problems such as these can be discovered and corrected before production starts.

Figure 11-9

In a trial run, an engineer may find a better way to assemble a product. Sometimes simply placing tools closer to the worker can save steps. Making changes such as these is called debugging.

Timing must be set up for the whole production system. The speed of the conveyor system must not be too fast or too slow. If it is too fast, workers will not be able to keep up. They will hurry and make mistakes, or parts will stack up and not get processed. If the conveyor is too slow, time will be wasted as workers wait for the parts to come to them.

The trial run may also be used as a time for the **final training** of the workers. Supervisors can give each worker specific instructions on the best way to do the job. The trial run is the last chance to refine the production system before actual production begins.

Now the system is in order. The materials and supplies are in place. The workers are on the job. Production can begin.

The Production Run

After months or even years of careful planning, production starts. Even with the best planning, it will take several days for workers to learn their jobs well. As they do, the system will move more smoothly and quickly.

Producing Components

The first step in producing products is making the components. Some are made in the same manufacturing plant. For example, the frame and push handle for a power lawn mower can be made by the lawn mower manufacturer. However, it is easier to buy some parts from other companies. The nuts, bolts, screws, and even the engines to run the mowers are usually purchased from companies that specialize in making these items.

There are still other components that the manufacturers may not be able to buy or make economically. These parts can be specially made by other companies. For example, the lawn mower manufacturer may have a wheel manufacturer make all the wheels for the lawn mowers.

Parts start out as some form of standard stock. This is changed by separating or forming it into components. Sheets of metal may be cut into shape and then drawn (stamped) into a bicycle fender. Stacks of lumber might be cut into components for skis. Plastic pellets may be

Figure 11-10 Reynolds Metals Company

Companies may make or buy components for their products. This company sells electric motor parts to other companies. The parts will be used in the manufacture of industrial machinery.

Figure 11-11

These are front plates of television tubes. In another part of the plant, they will be assembled with other components into television sets.

RCA Corporation

melted and forced into molds to make phonograph records. Clay may be placed into molds for plates and bowls.

Preparing for Assembly

Components are usually made and stored until enough parts are completed to begin assembling the products. In most large factories, components are made in special areas of the plant and delivered to the assembly area. For example, the Tonka Toys manufacturing company has one area in the plant that is called the press room. Large presses in this area stamp out thousands of metal parts for toys. These parts are stored in large bins until all the parts for a toy are completed. As in most companies, many of the parts are made by Tonka Toys. Some parts are made by other companies, specifically for Tonka Toys. Still other parts are bought from outside suppliers. All components must be ready for the assembly line before assembly of the product begins.

Sometimes components are shipped hundreds or even thousands of miles to the place where they are assembled. Most automobile manufacturers make their components in the Midwest. This is near the supply of iron ore and other raw materials. Fenders, doors, car bodies and engines are made in the Midwestern factories. Some cars are assembled in these factories. But shipping components is cheaper than shipping completed cars. Therefore, automobile assembly plants are located on the west coast, in the South, and in the East. These plants do not make components for cars. They assemble the components shipped to them by railroad from the Midwest.

Assembling Products

Let's take a look at assembly line production. Supply people are delivering parts to the proper locations on the line. Parts are being delivered by a conveyor system. Carts full of components are pushed or pulled into place. **Pallets** (flat platforms for stacking parts or products) or bins of components are carried to the line by forklifts. See Figure 11-12.

At the work stations, workers assemble the parts to the product. The completed product is

Figure 11-12
Components are placed on pallets and in bins. Then they are moved to the assembly lines.

Figure 11-13
Parts needed to complete assembly of cars on the line are ordered through the computer, which stores inventory information.
General Motors Corporation

152 Production

Figure 11-14

On one assembly line, the car body is assembled. Here it is being sanded to prepare it for painting.

Figure 11-16

A hoist lowers the car's body onto its rear axle.

Figure 11-15

On another line, the engine is assembled. Here it is being joined to the transmission (gears).

Figure 11-17

At the end of the line, the cars go through several checks before being shipped to dealers.

Chevrolet Motor Division,
General Motors Corporation

then placed on another conveyor and moved to inspection and packaging.

Controlling Production

Controlling production can be complex. Parts must arrives at the assembly plant at the right time. They must be moved to the right place at the right time during production. A computer may be needed to schedule the movement of parts. Automobile assembly is a good example of using a computer to control production. There are many ways it can be done. Any one of them would be fascinating to watch.

The Orders

Cars can be made according to special orders. When placing the order, consumers consider the many options available. **Options** are equipment choices that can be made. The buyer may also choose the color of the car and the type of trim. Engine size and the number of doors may be selected. Comfort items are often optional. These include such items as the radio, stereo, air conditioning, types of seats, and other special features. Buyers may order a wide variety of features for a car.

Orders are sent to the assembly plant. Information from the orders is entered into the computer. It does not matter if an order for a sedan is followed by an order for a station wagon. It does not matter if many options are ordered or only a few. Different types of cars can be made one after the other.

The Schedule

The computer makes a schedule for each work period. The schedule lists what parts are needed at each work station to make the car as ordered. It also gives the sequence (order) in which the parts will be used.

As you know, parts may be made or assembled in factories miles away. Many are shipped regularly to the assembly plant. Some are ordered as required. The computer keeps track of the supply of parts on hand. By scanning (reviewing electronically) the information in the new orders, it can give the kinds and numbers of parts needed. It can also make necessary changes in regular orders. Computers can even scan past orders to predict what will be needed in the future.

Components, such as seats, are most often made or assembled in the same plant. As in other manufacturing plants, a number of operations are going on at the same time. Most of these use assembly lines.

The Assembly Lines

Many conveyor systems are used. Some carry parts from receiving or storage areas. Others carry parts from the areas in which they are made. All parts are taken to the points on the major assembly lines where they are needed. Major assembly lines are those for the body, the frame, the engine, and the line where trim is attached.

On one major assembly line, the body is assembled. Doors, fenders, hood, and trunk lid are all attached. Then the body is painted. See Figure 11-14.

At the same time, engine assembly operations take place on a different line. Any options related to the engine, such as the air conditioner, are built in. See Figure 11-15.

Meanwhile, at about the speed of a slow walk, the car frame moves slowly along another assembly line. Wheels are assembled and attached.

Now the amazing timing of the assembly line can be seen. The engine and the frame for which it was made reach the same work station at the same time. The engine with the connecting power drive is put into place on the frame. Then, at another work station, the right car body for that combination of frame and engine is there to meet it. The car body is lowered onto the frame and bolted into place. See Figure 11-16.

Finishing touches are applied. The car is complete and ready to be tested. See Figure 11-17.

Corrections During Assembly

This process is complicated, even when things go smoothly. But what happens if something goes wrong? Suppose a mistake is made in painting the body and the body must be repainted. In a situation such as this, the computer sends signals to hold back the components for that certain car. The assembly of

Figure 11-18
Finished products are often stored before being shipped to buyers.

Figure 11-19
These jars are moved on a high speed conveyor belt. Then they are automatically placed into boxes. Packaging helps to protect the breakable glass containers.

other cars in the system continues with no delay. When the necessary correction has been made, the computer again relays instructions. The assembly process for that car begins again from the point at which it was interrupted.

Careful planning and timing are vital in keeping a system such as this running smoothly for long periods of time. Computers are valuable tools in helping to control production.

Working Together

People are necessary in every production system. A successful production is a cooperative effort. Everyone must work together in order to see that each job is done safely and correctly. If each person observes all safety rules and does the best job possible, then the company can produce a quality product.

Inventory Control

Inventory is a list of the amounts of materials owned by the company. All materials must be used wisely. **Inventory control** is a method for making sure that this is done. Inventory control involves keeping track of all the materials that are used by the company. As you know, this may be done by computer.

Categories of Inventory

A manufacturer's inventory is made up of several categories:
- Raw materials,
- Purchased parts,
- Work in process, and
- Finished goods.

Basic materials used in the manufacture of parts are classified as **raw materials.** Standard stock is included in this classification. These materials are vital to manufacturing. If the company runs out of basic materials, parts cannot be made. Without parts, production cannot continue.

Purchased parts are those that are bought from another manufacturer ready for use. Nuts, bolts, motors, bearings, and light bulbs are examples of purchased parts.

When materials are being used in production, they are classified as **work in process.** This means that they are at a stage of production between raw materials and the finished products. Any time work is being done on materials they are classified as work in process.

Finished goods are products that have already been made. A manufacturer must know exactly how many products are finished. Most finished products must be stored before shipping to the buyer. See Figure 11-18.

Inventory Records

Consider, for example, a company that wants to produce 500 cassette tape recorders. It is important to know the inventory in each category. Plastic pellets, sheet metal, wire, and paper are some of the basic materials needed. What would happen if there were not enough of any one of these materials?

Many parts are purchased. Motors, electronic parts, screws, and speakers are ordered from other manufacturers. Lists are kept of the number used and the number on hand.

As plastics, metal, and wire are made into cases, battery holders, and power cords, records are kept to let the company know how many parts are produced each day. If the records show that 300 plastic cases are completed, how many more must be made?

It is also important to know the number of finished products. What if the inventory records show that only 475 tape recorders are finished? What would need to be done?

Packaging and Storing

Important parts of manufacturing are packaging the product and putting it into storage. Even for a simple product, such as toothpaste, packaging plays an important role. The store that sells toothpaste receives it in cases. Each case usually contains 24 tubes. The cases are usually medium-sized, brown, fiberboard (cardboard) boxes. Toothpaste is sold in individual boxes. If you bought toothpaste, you would probably throw away the box, because the toothpaste is in still another container, a tube. As you can see, even for a common product like toothpaste, three containers are needed to get the product from the manufacturer to your toothbrush.

Most manufacturing companies do not make their own packages. They order them from manufacturing companies that specialize in producing containers. Tonka Toys company spends over three million dollars ($3,000,000) for containers for their toys each year.

Why Package a Product?

There are several reasons why a product is packaged:

■ Protection,
■ Storing and shipping,
■ Holding or containing,
■ Identification, and
■ Display and advertising.

Who Are Trade and Mark?

In 1866, the Smith brothers, William and Andrew, inherited the cough drop business that their father had begun. At this time, cough drops were sold from glass jars. The brothers placed pictures of themselves on their jars as a unique **trademark***. Still, it was easy for druggists or sales representatives to fill Smith Brothers jars with cough drops made by other manufacturers. To protect their good business reputation and to make certain their customers were not deceived, the brothers began in 1872 to package their cough drops in paper boxes. **This packaging method was new to the manufacturing industry.** Every box was marked with the Smith Brothers' distinctive trademark. The brothers realized that a trademark is only as valuable as its user's reputation is good.

* A **trademark** is any word, group of words, or symbol which a maker or seller puts on products to distinguish the products from those made or sold by others.

156   Production

Figure 11-20

Look at these packages. At a glance, you can tell a lot about the product: what the product is, how large it is, and how to safely use it.

Figure 11-21

When needed, products can be moved from storage to wholesale or retail stores.

A product must be **protected** during shipping and handling. A company can lose a lot of money when products are damaged. For example, a few years ago, The Firestone Tire and Rubber Company lost money because tire molds rusted while being transported overseas by ship. In 1972, **Grace M. Fagan,** an employee of the company, suggested a packaging system to prevent this. She suggested that the molds be lightly covered with oil and sealed within a heavy plastic cover. This system proved to be effective for protecting the molds from the salty air and the moisture. The company awarded her the exceptional amount of $10,000 for this suggestion. In time, her system will save the company a far greater amount of money.

A container also protects a product from getting bumped, dented, scratched, broken, or dirty. What would your toothpaste tube look like by the time it got to you if it were not shipped in a box?

Packages are used for **storing and shipping** products. Can you imagine trying to stack unpackaged basketballs and footballs in a delivery truck? There would be similar problems when storing the balls. A package makes stacking and moving the product much easier.

Packages may be needed to hold the product or parts of the product in place. If the product has many parts, such as a chess or checkers set, a package is needed to keep the parts together. Liquid products, such as milk or shampoo, must be placed in **containers.** Without containers, how would you carry or store them?

A product can be **identified** on the container. By quickly reading the package, you can find out the size, color, model number, or many other facts about the product. See Figure 11-20.

A package is helpful for **displaying and advertising** the product. Think about the picture on a box containing a plastic model of a hot rod. Doesn't it make the product look interesting!

Packaging is an extremely important part of producing and selling products. Most products are packaged. Imagine the number and the different kinds of packages that are needed! Over 50 billion dollars ($50,000,000,000) are spent for packaging each year.

Storage

After packaging, storage is the final step before shipping products to **wholesale or retail outlets** (stores that sell the products to consumers). If there is an immediate demand for the product, it may not be stored. It may go directly to trucks or to railroad cars and then be shipped to stores. But for most products, a time of storage is needed before shipping the products.

Some products are **seasonal.** Christmas ornaments and patio umbrellas are examples of seasonal products.

Some companies want a supply of extra products in storage. If an order comes for these products, they can be shipped immediately. In this way, when wholesale or retail stores need more products, the company has them on hand. They will not have to wait for them. If stores are asked to wait, they may go to another company.

Looking Ahead

Having manufactured, packaged, and stored the products, the company must be concerned about selling them.
Do you know
- how you may have already been involved in marketing?
- how a sales forecast and a weather forecast are alike?
- which is better: breaking even or making profit?

Like the other activities of a company, marketing begins before the product is available. It, too, must be planned. In the next chapter you will see how companies market their products.

New Terms

assemblies
competitors
critical process
debugging
final inspection
finished goods
gages
in-process inspection
inspection
inventory control
pallets
primary manufacturing

productivity
purchased parts
quality control
retail outlets
sampling
seasonal
subassemblies
tolerance
trademark
visual inspection
wholesale outlet
work in process

Study Guide

1. Give examples of primary manufacturing.
2. What is the difference between a subassembly and an assembly? Give examples.
3. Why do inefficient production methods increase the price of a product?
4. What is the main purpose of a quality control system?
5. How are materials inspected before production?
6. Describe the inspection done during production.
7. What does 45 mm ± 1 mm mean on working drawings?
8. Thousands of bolts are manufactured at one time. Since each one cannot be tested, how is inspection done?
9. What is the purpose of a trial run?
10. Would the procedure for making a "special order" product in a manufacturing plant be different from the procedure used in your shop? Explain.
11. Why is inventory control important? What are the categories of inventory?
12. Suggest a way to package a product that you have already made in your shop. What would be the purpose or purposes of this packaging?

LET'S GO TO WORK

Producing the Product

Activity: Making the Trial Run

The complete production system must be tested before making a large quantity of products. As you know, this is called a **trial run**. During the trial run you will learn your job and help to make any corrections needed in the system. Procedures and equipment must be checked. Machines may need to be adjusted. Sometimes workers must be assigned to different jobs where they can do better work. The trial run will help you to produce good products. Make several trial runs, if necessary. The system must be debugged before production starts.

During the trial run:
1. Go to your assigned work station when your teacher tells you.
2. Pay close attention. Materials for two or three products will move through the stations. Your teacher will instruct you in the proper way to do your job.
3. Remain at your station until the materials and components move through **all** of the stations. Listen to the instructions for all stations. Your job may be changed to one of these other stations before production is completed.
4. Ask any questions that you have about your job or how the system will work.
5. After the trial run, try to think of ways to solve any problems that were discovered.

Activity: Producing the Product

Production can begin! In this activity, a quantity of the product selected for production will be produced. Each job must be done well to produce a good product.

Know what the parts should look like when you get them from the station before yours. Tell the teacher (or the student assigned to be the inspector) if the work does not look right. Then someone can correct whatever may be wrong.

Do your best to be accurate in your work. Each of you is responsible for the product made. Try hard, and cooperate with others. Then the products you produce will be good ones.

Remember:
1. Supplies will need to be delivered. You will need enough for each day's production run.
2. Inspect parts carefully. Report to the teacher or inspector any errors that you find.
3. Keep an accurate record of
   - Inspected parts,
   - Amounts of materials used, and
   - The number of parts or products completed.
4. Store completed parts or products.
5. Report to your teacher for a new assignment when your job is completed.

Begin production when your teacher tells you.

chapter 12

Marketing: Profit or Loss

Have you ever tried to sell anything, such as greeting cards or newpaper subscriptions? Perhaps you have sold candy to raise funds for a school activity. Or maybe you sold some possession that you didn't want anymore to a friend. If you have sold anything, then you have been directly involved in marketing. When you have bought things, you have probably been influenced by marketing.

Marketing

All the activities involved in selling products are called **marketing.** Marketing does not start when the products are finished. It begins long before that. When a designer or an inventor has a new idea for a product, one of the first questions asked is, "Will people buy it?" Marketing is involved from the first stage of planning the product up to the time of its sale and use. Even after the sale there are services to the product that the consumer may expect.

There are four basic steps or activities that must be carried out in order to market a product successfully:
- Market reseach,
- Advertising,
- Sales, and
- Distribution.

Large manufacturing companies often have their own marketing departments that take care of all the areas of marketing. However, most manufacturing companies cannot afford to do all the marketing activities themselves. They may take care of their own sales and distribution but not the market research and advertising. Companies that specialize in these areas are often hired to meet these needs.

Market Research

The goal of **market research** is to find out what people will buy, how much they will buy, and at what price. Market research will find out how the consumers feel about products and product needs. See Figure 12-2. The information gathered revolves around the four "P's": product, price, promotion, and place.

The company will need to know consumers' feelings and opinions about such things as the size, shape, and color of the **product.** How much will people pay for it? Answers to many questions must be gathered before a company can determine the selling **price** for the product. Market research will tell the company the best way to **promote** its product. What is the best method for advertising? Finally, **place** refers to where and how the company should distribute the product to the consumers for best results.

Market research will also show the company the best way to sell its product. From this information the company can make a **sales forecast.** A sales forecast is somewhat like a weather forecast. The weather forecast predicts what the weather will be. A sales forecast predicts what the sale of the product will be. It is important to have an accurate forecast. If the weather report says it will be a sunny day tomorrow, your family may plan a picnic. If it

Research

Sales

The Gillette Company   Advertising

Distribution   The B.F. Goodrich Company

Figure 12-1
There are four major marketing activities.

Figure 12-2
Consumer opinion is important to companies.

Figure 12-3
Once a customer has tested a product, he or she is more likely to buy it.

Marketing: Profit or Loss

Overall Product Preference

	Percent Preferred
Product A	65
Product B	25
No Preference	10

Reasons for Preferring Product A

Taste	**55%**
Better taste	40
Taste family would like	12
Not as sweet tasting	10
Texture	**43%**
Does not crumble as easily	25
Just the right chewiness	19
Appearance	**34%**
Looks good to eat/looks more appetizing	28
It's thicker	10
It's smaller	5
Miscellaneous Comments	
Is better for me/family	9
Probably would cost less	5
Doesn't have nuts in it	4
More convenient to use	2

Figure 12-4

To find out why products are liked or disliked, companies may ask consumers to choose between two similar products.

rains, your plans will be ruined. If a company forecasts sales of 600,000 products and it can only sell 300,000, its plans will be ruined. In this case, however, the loss will be more serious than a missed picnic. Thousands of dollars will be lost, and the company could be forced to go out of business.

If market research indicates that there is a big demand for the company's product, then the company can forecast good sales and plan a large production. This information can be very helpful for raising **capital** (money) to run the company. People will be more likely to invest and to buy stock in a company that has a demand for its products.

Many ideas for new products come from market research. Market researchers work closely with the R & D departments of large companies.

Advertising

Advertising tells people about the product. A company usually takes its product to an **advertising agency** for promotion. People here find the best way to attract attention to the product. They will be very interested in the information found by market research. Who will buy the product? What age group? Where are the consumers located? All of this will help to determine the kind of advertising the company will use. If market research shows that young people buy skateboards, would a company want to advertise by putting posters in retirement homes?

Advertising begins as soon as appearance or hard mock-ups are completed. Because these mock-ups look like the finished product, they can be used for advertising. Many of the pictures you see of products, particularly in catalogs, are actually only mock-ups. By using mock-ups, the company can complete its catalog and start advertising without waiting for products to be produced. Many orders for products can be sent in before production starts. Then the products can be shipped to customers as soon as they are finished.

The advertiser has to decide what to say and how to say it. The consumer's attention must be drawn to the product. In order to do this, advertisers use several **media** (means of expressing ideas). The most common advertising media are:

- Newspapers,
- Magazines,
- Radio,
- Television, and
- Billboards.

Television receives the most money from advertising, followed by newspapers.

Newspapers: Immediacy

In the United States today, over half of any newpaper's space is used for advertising. Since newspapers have many readers, advertisers can reach many people.

Newpaper ads may be published quickly. This is a plus for **retailers** (stores that sell to individuals). For example, an ad for air conditioners can be published immediately after a city's first hot day.

Newspaper ads are portable. People who want to buy these advertised air conditioners can take the ads with them to the store.

Figure 12-5

Denim products appeal to young consumers. As a manufacturer, what would your advertising approach be?

Figure 12-6

An advertising message reaches many people quickly in a newspaper ad.

Magazines: Permanency

Newpapers are generally sold only within their local areas. Most magazines, however, have a national circulation (distribution). This means that one edition of a magazine will probably reach more consumers than one edition of a newpaper. Most newpaper ads are in black and white. Ads in magazines are often in color. Generally, these are very dramatic. Color newspaper ads are usually not as good as those in magazines. Less money is spent for them because newspapers are read within one day. Magazines are read leisurely (over a long period of time). They may be kept around homes or offices for several weeks. See Figure 12-7.

Advertisers can choose from many types of magazines. Some magazines, such as **Time,** are read by many kinds of people. Others, like **Seventeen,** are written for one group, in this case teens.

Radio: Drama

Radio does not allow the consumer to see the product. Ads on the radio must rely on music, special sound effects, and the human voice to sell products. See Figure 12-8.

Consumers can do other things while listening to the radio. In fact, radio ads cost the most during rush hours when people are driving to

Figure 12-7

A consumer who wants the product is a consumer who will buy it. Magazine ads try to influence what you want through colorful and dramatic ads.

Figure 12-8
A radio ad causes you to use your imagination. You may buy the product the next time you see it.

The Gillette Company

Figure 12-9
The message is the same: buy this product. But there are many ways a television ad can sell the product. Words, people, and scenes are planned carefully.

work. The large number of ads on the radio can be a disadvantage to advertisers. A person who hears many ads while driving to work, may find it hard to remember any one of them.

By carefully placing an ad at the right time of day, an advertiser may reach the people most likely to buy the product. Radio ads are picked for the type of program. For example, an anti-acne cream would be advertised on a program for teens.

Television: Action

When watching a television commercial (advertisment), consumers see the product while hearing about it. The product may be demonstrated. As with many magazine ads, the product may be shown in natural color. TV ads can also use motion in demonstrating the product.

Television backgrounds can easily be changed to show different times of the day or the year, and different parts of the world. This may help the consumer to identify a package or a trademark.

Television ads offer more appeal than do radio ads. It is easy for consumers to identify with a TV commercial. People can place themselves in the action.

Figure 12-10
Billboards get only seconds of your attention. Messages must be brief and effective.

Billboards: Impact

Most billboards are used by those advertisers whose products are sold in stores across the country. Messages on billboards must be short. People passing by will see the sign for only a few seconds. See Figure 12-10.

Billboards are large and can be seen easily. The size emphasizes (adds importance to) the product or trademark shown. Billboards are often set up along highways near retail outlets to attract consumers who are passing by. Over a period of time, billboards may be seen by thousands of people.

Other Means of Advertising

Advertising is done in many different ways. Sometimes people are hired to do spectacular stunts, such as skydiving, to draw attention to a product. Companies may sponsor special events. For example, the Colgate-Palmolive Company has sponsored The Colgate-Dinah Shore Winners Circle golf tournament for many years.

An ad can be placed on other kinds of products, such as ball-point pens, matchbooks, t-shirts, or glass tumblers. It may even be written in lights high in the sky on a blimp.

A Special Way to Advertise

Before its defeat in World War I, Germany had many blimps (airships). After the war, Germany was ordered to stop making them. The engineers who built blimps were out of work. Goodyear Aircraft Corporation hired many of these engineers to make helium-filled blimps, marked with the Goodyear name. During World War II, the blimps went "back to war" on antisubmarine patrols.

Today, Goodyear blimps are used for advertising. See Figure 12-11. At night the blimps act out a "Super Skytacular." This effect is achieved with light bulbs. Each Skytacular billboard has 3780 blue, green, red, and yellow bulbs. These bulbs light up to spell out a message in a continuing ribbon of words.

Goodyear's blimps have been so successful that a competitor states, "We're the ones without the blimp."

Many unusual and interesting methods may be used to attract attention to a company's product. But remember, the main goal of advertising is to cause more people to buy the product. If advertising is not well-planned and more people do not buy the product, then money is wasted. Billions of dollars each year are spent on the thousands of ads we see around us each day. Advertising is an important part of marketing a product.

Sales

The company must sell its products. Selling is simply convincing people to buy the product. Usually one person in the company is in charge of sales. Most often this person is the vice-president of sales or a **sales manager.** Unless the company is very small, it will probably have a sales department.

Figure 12-11

The Goodyear Blimp is a good example of special advertising.

Figure 12-12

Sales representatives and buyers for retail stores help to get the products to the consumers.

There are many jobs to be done in the sales department. The most important job is selling the product. This is done by **sales representatives.** They present the product to future customers. Usually they are given a territory. It is their responsibility to contact as many people as possible and to sell all the products they can within that area.

Many people in sales work on a **commission.** This means they get a certain **percentage** of what they sell. The commission is an **incentive** (encouragement) to sell more products. Good sales representatives are very important to the company. They are the ones who show the product to the customers and convince them to buy it. They may sell to individuals or to retailers. See Figure 12-12.

Distribution

The products must be distributed to the consumer. The product reaches the consumer through **marketing channels.** The first step in the marketing channel is the manufacturer. In some cases, the products are shipped directly from the manufacturer to the consumer. This is usually done with custom-produced products, such as handcrafted items. Sometimes it is possible (depending on where you live) to buy items directly from **factory outlets.** However, for most products, the consumer does not go directly to the manufacturing company to buy the product.

The most common marketing channel is from the manufacturer to the wholesaler to the retailer and, finally, to the consumer. **Wholesalers** buy large quantities of products at cheaper prices. Several railroad cars full of finished products may be sent to one wholesaler. The wholesaler, in turn, raises the price and then sells the products in smaller **lots** (amounts) to retailers. The retailers again raise the price and sell the products to the consumers. This practice of raising prices from one sales outlet to another is called **marking up** the product.

Whichever method of distribution is used, the products must be shipped. They may be loaded on trains, boats, airplanes, or trucks to be transported to the customers. The customer may be a wholesale outlet, a retail store, or an individual consumer.

E.I. du Pont de Nemours & Company

Figure 12-13

This carpeting went from manufacturer to wholesaler to retail store before reaching this potential buyer.

Figuring Profit or Loss

Will the company make money? This is the big question that concerns every company. If a company makes money, then it can continue to grow and expand. If the company does not make money, it will develop problems. It might end up going out of business.

In order for companies to know if they are making a **profit** or losing money, they must keep accurate records. Each year most companies put these records together in an annual report. The **annual report** is usually written in an attractive booklet (much like a magazine). A copy is sent to each of the stockholders. This is the report to the stockholders of what took place in their company over the past year.

Keeping Accurate Records

Determining the cost of the products made during the year is different from figuring cost estimates (Chapter 10). A cost estimate is just that, an estimate or close guess of how much the product will cost. It is figured before the product is made. Determining the real cost of the product is no longer an estimate. It is the actual cost. This is done by keeping records of exactly how much was spent and for what.

When all the records are checked and totaled, the management of the company can see exactly how much it has spent to make a certain product, to advertise it, and to ship it. The records also show the overhead cost of running the company for a whole year.

Figure 12-14 Association of American Railroads

Truck trailers are loaded to ride "piggyback" on rail cars. Then they are hitched to a truck cab to finish delivering the goods.

Bill Osmun, Air Transport Association **Figure 12-16**

Air transportation of goods is fast. These automobile parts, loaded on racks inside a jet freighter, can be shipped tonight and used in a production line tomorrow. How do they get from the airport to the consumer?

Figure 12-15 Boise Cascade Corporation

Seventy-five percent of all the nation's freight is carried at least part of the way by trucks. However, all of the final distribution of finished products is handled by trucks. Ships, trains, and airplanes cannot deliver to retail outlets.

Lykes Bros. Steamship Company **Figure 12-17**

A ship the size of three football fields can carry a lot of goods. Once on land, these containers are fastened to truck trailers for fast delivery.

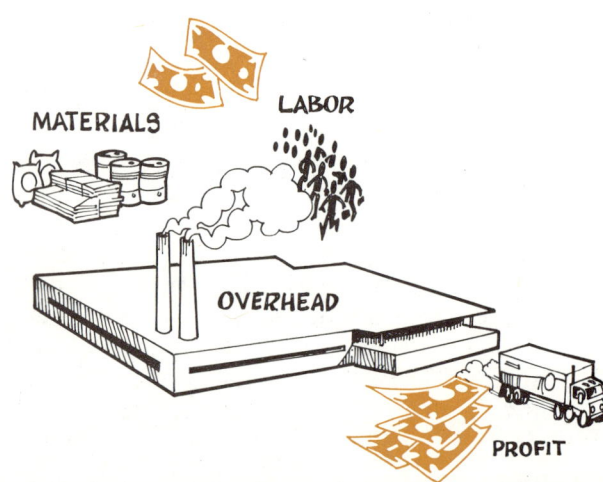

Figure 12-19

The company must spend money on labor, materials, and overhead, and still make a profit when the products are sold.

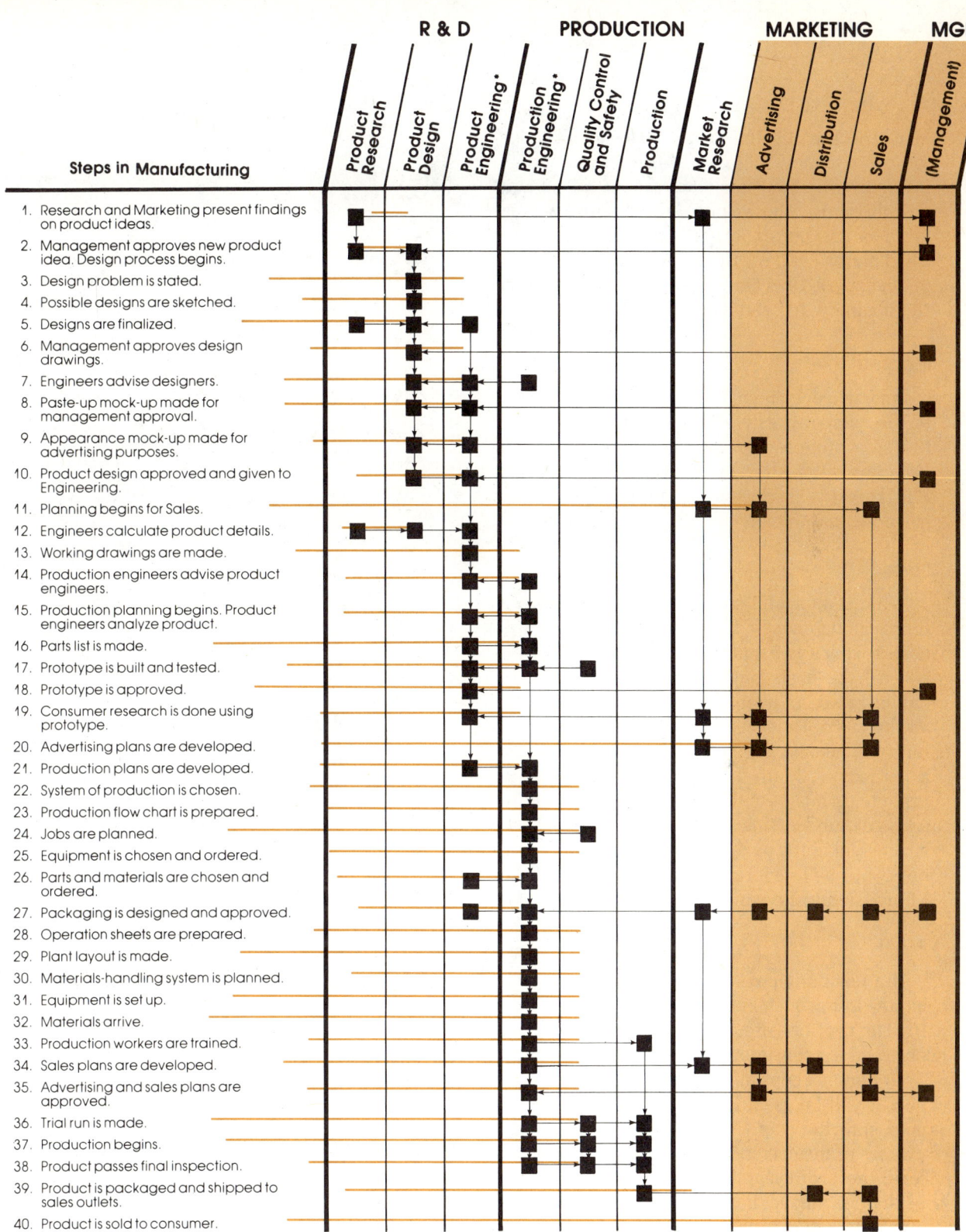

Figure 12-18

Marketing is involved from the time that the product is only an idea to the time of the final distribution.

Costs

Accurate records will help the company keep track of all the **costs**. Costs are monies (plural of money) that go out from the company. Records will show how much money was spent on materials. The amount of money spent on labor is another expense, and overhead is the third major cost. All the basic costs are shown in Figure 12-19. This money must go out from the company to keep it in business.

Income

Income (revenue) is the money that comes into the company. The source of income is from the sale of products. The more products the company sells, the more **revenue** it receives. The key to a successful company is to have more income than costs.

Loss

What if a careful check of all records shows there is no money left after paying all the costs? There is no profit. If the company can pay all the costs but has nothing left, this is called **breaking even**. The income is the same as the costs. However, if the costs are greater than the income, the company suffers a **loss**. It loses money. Many large companies can continue to do business through a year or two with a loss of money. This is because they have been successful in the past. But most companies cannot continue to operate for very long at a loss. Many go out of business quickly.

Profit

If a company pays all the costs and there is money left over then the company makes a profit. The profit belongs to the owners of the company. In a proprietorship, there is one owner. That person keeps all the profit and determines how it will be used. In a partnership, the partners split the profit. The stockholders are the owners of the company in a corporation. Therefore, the profit legally belongs to them. It can be divided among the stockholders based on the number of shares each stockholder owns.

Wise company owners will not keep all the profits for themselves. Most owners take some of their profits and use them to improve or expand the company. New equipment is purchased, buildings are made larger, or sometimes a new product is developed.

Figure 12-20

By examining this statement, stockholders know exactly how much income this company made in 1980. Some of the profit this year went to stockholders as a dividend (share of the money). Some profit was spent on a building addition.

XYZ Manufacturing Company, Inc.

INCOME STATEMENT — YEAR 1980

Net Sales		$11,000,000
Cost of sales and operating expenses		
Cost of goods made	$8,200,000	
Marketing and administrative expenses	1,400,000	
		9,600,000
Gross Profit		$ 1,400,000
Less: federal income tax	$ 480,000	
Less: state and local tax	358,000	
		838,000
Net Profit		$ 562,000

ACCUMULATED RETAINED EARNINGS STATEMENT (EARNED SURPLUS) — YEAR 1980

Balance January 1, 1980		$ 1,315,000
Net profit for the year		562,000
Total		1,877,000
Less: dividends paid to stockholders	$ 120,000	
Less: building addition	250,000	
		370,000
Balance December 31, 1980		$ 1,507,000

Figure 12-21
Newsday, Inc., The Times Mirror Company

A successful business can use some of the profits to expand the company. This creates more jobs and workers will be able to buy more goods. Healthy companies can help keep the economy healthy.

In a corporation, the total profit made is called the **gross profit.** From this, the company pays its taxes to the federal government as well as state and local taxes. The money that the company keeps is called **net profit.** Some of this profit may be paid to the stockholders as a **dividend.** Part of the net profit may be put back into the company for growth. See Figure 12-20.

The Importance of Profit in the Manufacturing Industry

A well-run company should make a profit and keep on earning a good profit over a long period of time. This means the company should improve or at least maintain its sales each year. A company that consistently makes a profit is known as a **healthy company.** People want to invest and buy stock in the company. Banks are more willing to lend money for new machines or for tools. A healthy company is willing to try new ideas. It has money to cover losses if the ideas don't work.

Profit that is put into new machines or into improved processes will shorten production time and lower the company's costs. Then the prices of products can be lowered. As companies expand, new employees are hired. This helps supply jobs for local workers.

Healthy companies help the economy of the area in which they are located. The **economy** is the system of producing, selling, and buying products and services. Healthy companies also

Figure 12-22
To meet competition, companies work hard to improve their products. Michael Cochran helped to develop the microcomputer for his company. This invention improved the company's product, calculators. The silicon chip is inside the plastic case that Cochran is holding.

help the economy of our country. They buy products and services and sell products and services. They pay workers. They pay taxes. They keep the system going.

The American Economic System

The economic system in America is called **capitalism.** This means that the factories and all that they contain are owned by individuals or by companies. It means, also, that the raw materials and finished products can be owned by individuals. With private ownership, businesses are free to produce whatever they wish. They compete with other businesses making the same product. This competition generally brings improved products to the marketplace. Consumers benefit because products are better and cost less.

Supply and Demand

Pocket calculators cost over $100 when they were first introduced. Now you can buy one for just a few dollars. Why is this so? When pocket calculators were first developed, there were only a few companies making them. The amount of products made is called the **supply**. More and more people wanted the calculators. There was a great **demand**. As other companies recognized the increased demand, they began making calculators. The price went down as the supply increased. Improvements were made on the calculators. Companies tried hard to make a better product than their competitors. See Figure 12-22.

Capitalism encourages profit-making, and profits encourage people to invest in successful businesses. Companies grow. There is more work for people. Workers who earn good wages and salaries are able to buy more, to travel more, and to afford better health care and education. All of these things improve the way a person can live.

Free Enterprise

In some foreign countries, the factories and businesses are owned by the government. There is little or no competition among the companies, and sometimes no improvements are made in the products. In some of these countries, there are few products available to the consumer. For example, telephones and automobiles are very expensive. Few people can afford them. There are not as many jobs, then, in these industries as there are in capitalist industries.

Here in the United States, and in some other countries, the industrial system is called **free enterprise.** This means that we are free to start up almost any kind of business we wish and to make nearly any kind of product we wish.

In "free enterprise" countries, individuals and businesses are encouraged to create new products. The government grants **patents** to inventors. Patents protect their ideas. A product that receives a patent cannot be copied. Improvements on products can be patented as well. Products are sometimes protected by many patents. Each patent shows what an individual or company research team has invented or how the product was improved.

An individual or a company can allow others to use the patent. This is called **licensing.** When a license is granted, the user pays the inventor for the use of the patent.

Thousands of patents are granted by the United States government every year. A person

Allis-Chalmers Corporation

United Technologies

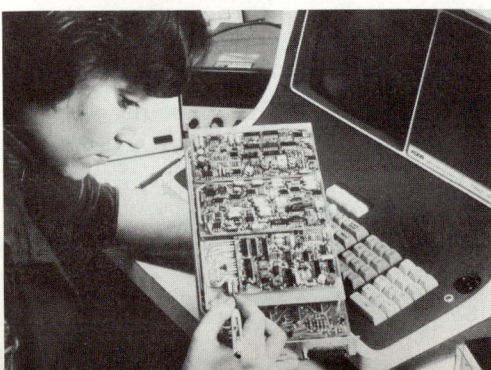

Figure 12-23

What is free enterprise? It is the freedom to start a business, the freedom to create new products, the freedom to work.

General Electric Company

of any age can receive a patent. If you can think of a new product or a new way to make a product or part of a product, you can receive a patent. Remember reading about Margaret Knight in Chapter 4? She could have received a patent at age twelve. Perhaps you will get an idea for a product during this course. You may be able to patent your idea.

Laws and Free Enterprise

Free enterprise doesn't mean that an individual can make a dangerous product. For example, years ago paint contained lead as an ingredient. Toys finished with this kind of paint made children sick. When this was brought to the attention of the toy manufacturers, they asked the paint companies to develop non-toxic paint (paint that is harmless).

There are government and private consumer protection agencies. The people working for the agencies keep records of accidents caused by some parts in machines, cars, and the like. When a part does not do the job it is supposed to do, the manufacturer will "recall" it. That is, the customers will be contacted by the manufacturer. They will be told to take the item to a dealer for free repair.

The government makes rules for many products, but these limitations are made to protect the consumer and the worker. They are not made simply to keep someone out of business.

There are government rules regarding pollution. There are laws that prevent certain kinds of industry from locating in certain places, where the homes of workers and others would be endangered. There are other rules or laws that govern the weight and size of vehicles that can be driven on local roads. There are frequent disagreements about rules. But remember, **we** are the people who make the rules, and **we** can change them. That is what is really **free** about free enterprise.

Looking Back

This, then, is manufacturing. Yesterday, today, and tomorrow, ideas become realities, and people shape industries and lives while shaping products. And you — you have become a manufacturer yourself. You are a part of it all.

New Terms

advertising agency
annual report
breaking even
capital
capitalism
commission
costs
dividend
economy
factory outlets
free enterprise
gross profit
healthy company
incentive
income
loss
lots
market research
marketing channels
marking up
media
net profit
patents
percentage
profit
promotion
retailers
revenue
sales forecast
sales manager
sales representative
supply and demand
wholesaler

Study Guide

1. What is marketing?
2. What are the four "P's" of market research?
3. Why does a company have to be very careful when making a sales forecast?
4. Suppose you are given money to advertise a product made in your shop. What will you do to advertise the product?
5. If you were selling a product, would you want to work on commission? Why or why not?
6. Describe the most common marketing channel.
7. How does a company determine the cost of making a certain product?
8. Are a company's gross profit and its net profit the same thing? Explain.
9. Businesses in country X are owned by the government. Is this free enterprise? Why or why not?
10. Suppose you invented a new product or a new way to produce a product. Why would you apply to the government for a patent?
11. How does the system of free enterprise affect the work that you do in your school shop?

LET'S GO TO WORK

Marketing

Activity: Packaging the Product

Packaging is important to marketing. An attractive package is one of the best ways of advertising. Packaging is also very important in preparing a product for distribution. Information about the product is printed on the container.

Packages are designed for the type of product inside. Some products are enclosed in a plastic bag. Some have only a label attached to provide special information about the product. Other products need boxes or special wrappings.

Planning for Packaging

With other members of your company or class, decide how your product should be packaged. Think about the types of packaging which might be best for your product. What materials are available? How much do they cost? Your teacher will help you to find the information that you need for designing and making the package.

Decisions to be made:
1. The kind of packaging.
   - A clear plastic bag with a label?
   - A cardboard box?
   - Labeling only (no container)?
   - Plastic blister pack?
   - Other?
2. The design for the package.
   - What size must it be?
   - What shape?
   - How will you label or decorate it?
3. How to produce the package if it has to be made.
   - How many workers will be needed?
   - What equipment and supplies are necessary?
4. How to organize workers to place the product into the package.
   - Should each student completely package a product?
   - Should small groups of students package the product on an assembly line?

Packaging the Product

1. Make a detailed drawing of the package. Show the size (dimensions) of the package, the information that will be on the package, and any special features.
2. Make a sample package. Sometimes design changes are necessary.
3. Examine and discuss the packages designed by the members of your company or class. Select the package design that is best for your product.
4. Decide which method should be used to produce the package.
5. Make (or help to make) as many packages as are needed for the products.
6. Package the products.

Figure 12-24

Working together on an assembly line can speed up packaging.

Unit III PRODUCT REVIEW

Note: Before making any of the products presented in Unit IV, read **Workers and Safety** and **Understanding Product Plans**, pages 52 through 64.

Product		Material	Level	Time	Page
WHISTLE Use this while you work.		Wood	★	★	178
JUMP-A-PEG Test your IQ.		Wood	★	★	182
SALT AND PEPPER SHAKERS These are always in season.		Wood	★	★★	184
SUPER Q Three games in one.		Wood	★	★	186
NOTE HOLDER A nice gift.		Wood	★	★★	188

(continued next page)

Product		Material	Level	Time	Page
MARBLE DROP Don't lose your marbles.		Wood	★	★	190
HANDY DANDY TENNIS One-handed table tennis.		Wood	★	★★	196
PAPER TOWEL STAND Stands on its own.		Wood	★★	★★★	198
MARBLE SHOOT A long shot.		Wood	★★	★★★	206
PLANT STAND A grand stand for indoor gardeners.		Wood	★★	★★★	210

PLANTER You'll really dig this one!	Metal	★★	★★★	244
TARGET GAME Right on!	Wood Metal	★★	★★★★	216
GIZMO A three-way winner.	Plastic Wood Metal	★★	★★★	260
COASTER SET See if you can coast through these.	Wood	★★★	★★★★	212

(continued next page)

Product	Material	Level	Time	Page
ROCKET The sky's the limit!	Paper	★★★	★★★	282
GUMBALL MACHINE Pull the handle and get a treat.	Wood Plastic	★★★	★★★★	228
TOOLBOX Getting it all together.	Metal	★★★	★★★★	248
DESK LAMP Light up your life.	Metal Wood	★★★	★★★★	252

UNIT IV

PRODUCT PLANS

Whistle

DIMENSIONS

Dimension Symbol	Metric mm	Cust. in.
A	75	3
B	32	1-1/4
C	16	5/8
D	8	5/16
E	10	3/8
F	20	3/4
G	7	9/32
H	2	1/16
I	6	1/4
J	60	2-3/8
K	50	2

PARTS AND MATERIALS

Qty.	Part	Size	Material
1	Body	C×B×A	Wood
2	Plugs	3/8" dia. ×F Wood Dowel	

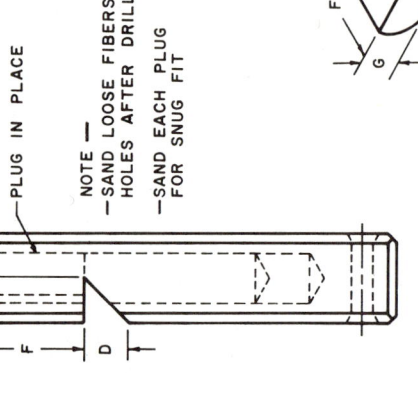

PLUG DETAIL
2 REQ'D.

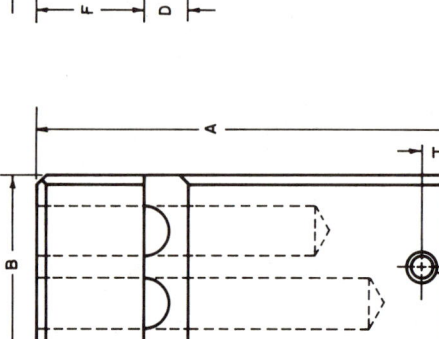

BODY DETAIL

Whistle

PROCEDURE

Operation Number	Operation	Tools & Equipment	Topics	Notes
BODY				
B-1	Cut to size	Handsaw or power saw	5	
B-2	Drill sound holes	Drill press, 3/8" twist drill	6, 25	A drilling fixture may be used.
B-3	Sand holes	Abrasive paper	15	Sand inside holes to remove loose fibers.
B-4	Drill chain hole	Hand drill or power drill, 3/16" twist drill	6, 25	A drilling fixture may be used.
B-5	Countersink hole	Hand drill or power drill, countersink	6	
B-6	Chamfer edges	Belt sander or disk sander or block plane or file	7 or 15	
B-7	Cut notch	Band saw or backsaw	5	
B-8	Sand	Abrasive paper	15	
PLUGS				
P-1	Cut to length	Miter saw or backsaw or coping saw	5	
P-2	File flat side	File	15	
P-3	Sand	Abrasive paper	15	Sand plugs until they fit snugly in holes.
ASSEMBLY				
W-1	Assemble plugs into body			Press plugs into holes.
W-2	Finish	Spray or brush	23	Do not get finish into sound holes.

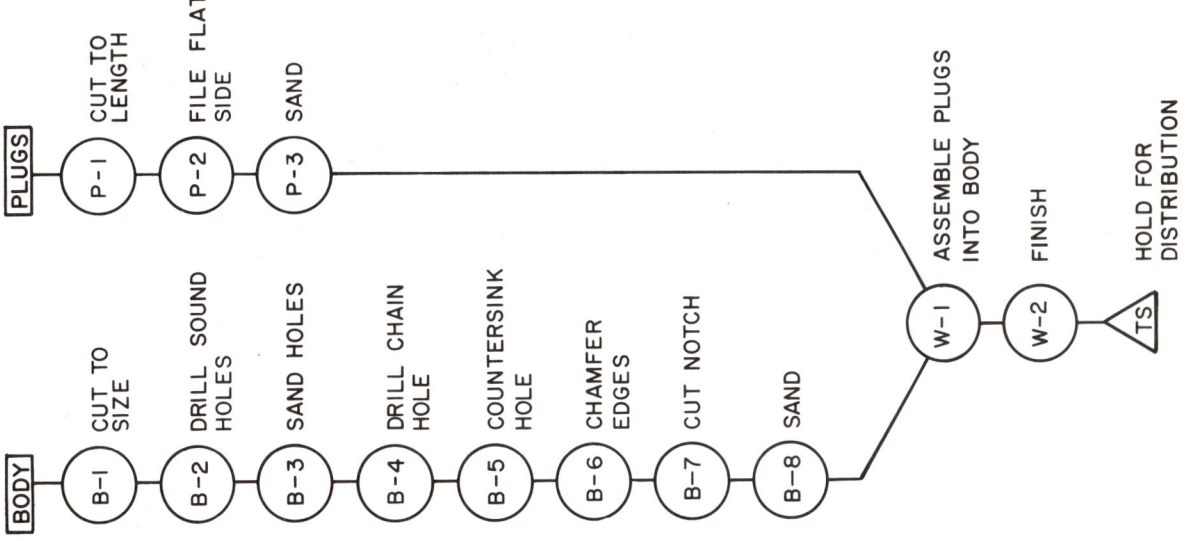

Xlyobox

PARTS AND MATERIALS

Qty.	Part	Size	Material
1	Top and Bottom	F x E x H	Wood
2	Sides	F x K x H	Wood
2	Ends	F x K x I	Wood
1	Stick	3/8" dia. x L Wood Dowel	
1	Ball	1" dia.	Rubber
			Glue
			Finish

DIMENSIONS

Dimension Symbol	Metric mm	Cust. in.
A	265	10-1/2
B	215	8-1/2
C	165	6-1/2
D	58	2-1/4
E	140	5-1/4
F	20	3/4
G	33	1-1/4
H	460	18-1/2
I	108	4-1/2
J	9	3/8
K	100	4
L	300	12

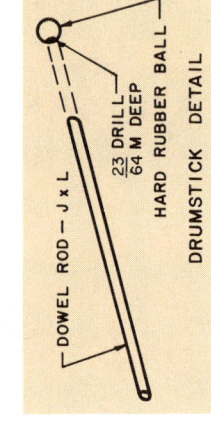

DOWEL ROD — J x L
23/64 DRILL M DEEP
HARD RUBBER BALL
DRUMSTICK DETAIL

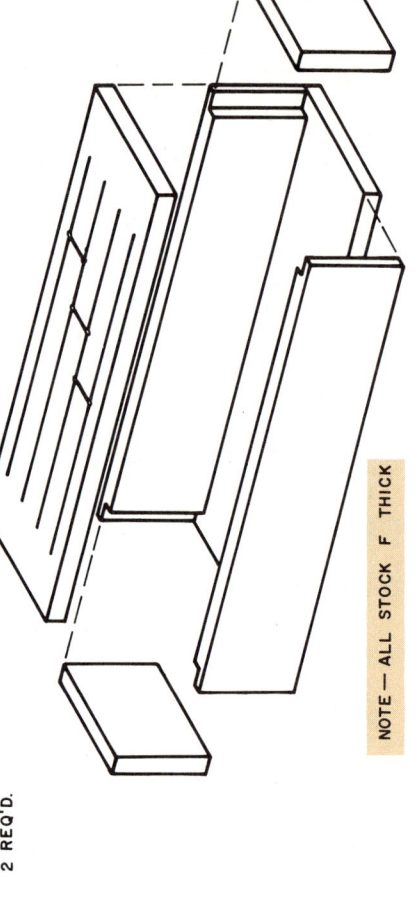

NOTE — ALL STOCK F THICK
ASSEMBLY DETAIL

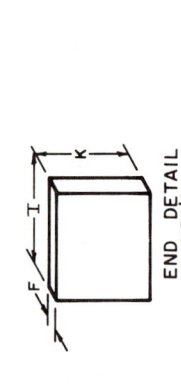

END DETAIL
2 REQ'D.

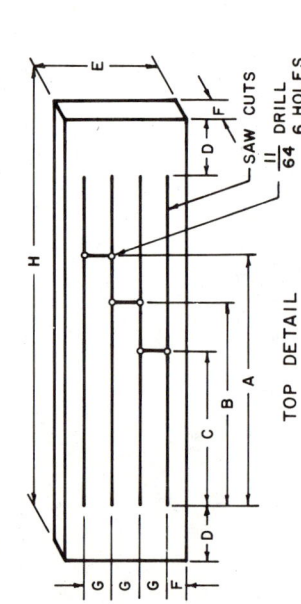

SAW CUTS
11/64 DRILL 6 HOLES
TOP DETAIL

BOTTOM DETAIL

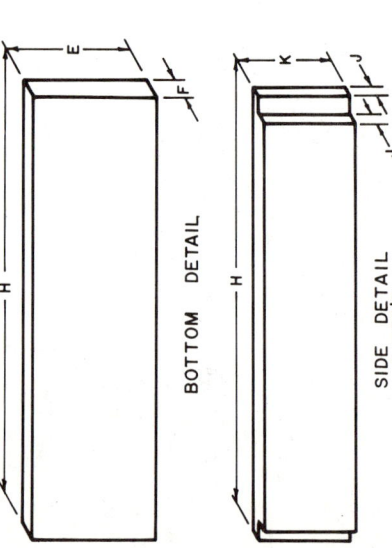

SIDE DETAIL
2 REQ'D.

Xylobox

PROCEDURE

Operation Number	Operation	Tools & Equipment	Topics	Notes
TOP				
T-1	Cut to size	Handsaw or power saw	5	
T-2	Drill holes	Hand drill or portable electric drill or drill press, 11/64" twist drill	6	
T-3	Cut slots	Coping saw or table saw or saber saw or scroll saw	5	
BOTTOM				
B-1	Cut to size	Handsaw or power saw	5	
SIDES				
S-1	Cut to size	Handsaw or power saw	5	
S-2	Rabbet ends	Backsaw or table saw with a dado blade	5	Parts could be butt-joined instead of rabbeted.
ENDS				
E-1	Cut to size	Handsaw or power saw	5	
ASSEMBLY				
X-1	Assemble ends and sides	Bar clamps or hand-screw clamps	11, 24	Glue and clamp.
X-2	Assemble top and bottom to box	Bar clamps or hand-screw clamps	11, 24	Glue and clamp.
X-3	Sand	Finishing sander, abrasive paper	15	
X-4	Finish	Spray or brush	23	
DRUMSTICK				
DS-1	Cut dowel to length	Miter saw or coping saw or backsaw	5	
DS-2	Drill ball	Portable electric drill or drill press, 23/64" twist drill	6	
DS-3	Assemble ball to dowel		11	Use glue.

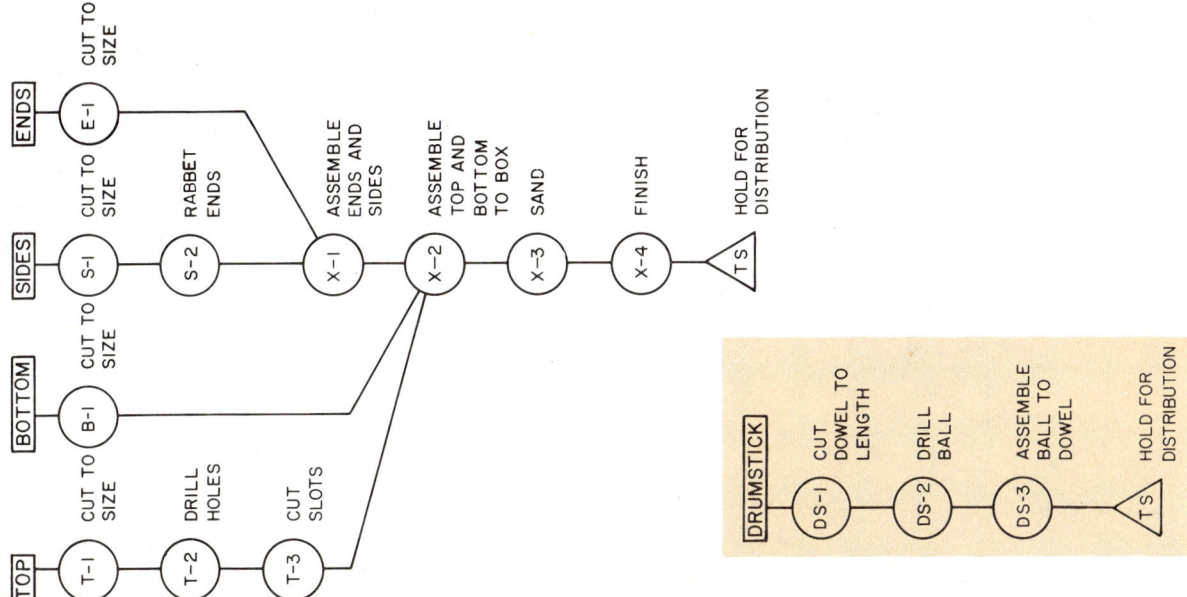

Jump-a-Peg

PARTS AND MATERIALS

Qty.	Part	Size	Material
1	Board	E×A×A	Wood
14	Pegs	1/4" dia. ×G	Wood
	Finish		

DIMENSIONS

Dimension Symbol	Metric mm	Cust. in.
A	127	5
B	13	1/2
C	18	11/16
D	10	3/8
E	20	3/4
F	3	1/8
G	25	1

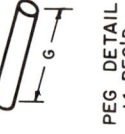

PEG DETAIL
14 REQ'D.

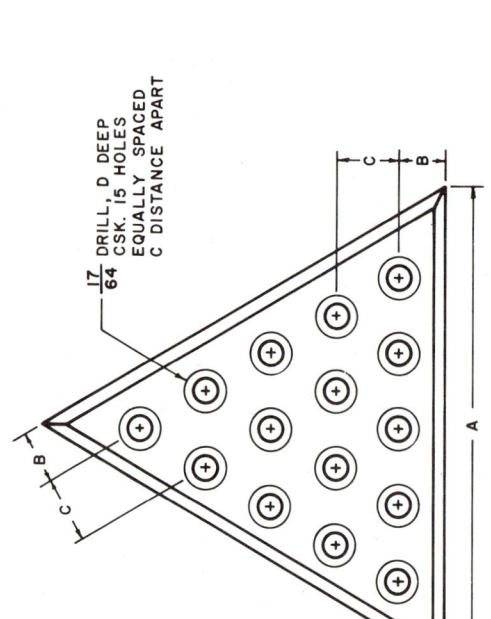

17/64 DRILL, D DEEP
CSK. 15 HOLES
EQUALLY SPACED
C DISTANCE APART

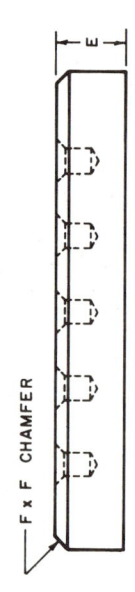

F × F CHAMFER

BOARD DETAIL

Jump-a-Peg

PROCEDURE

Operation Number	Operation	Tools & Equipment	Topics	Notes
BOARD				
B-1	Cut to size and shape	Handsaw or power saw	5	
B-2	Drill holes	Portable electric drill or drill press, 17/64" twist drill	6	A drilling fixture may be used.
B-3	Countersink holes	Portable electric drill or drill press, countersink	6	
B-4	Chamfer edges	File or hand plane or disk sander or belt sander or Uniplane	7 or 15	
B-5	Sand	Disk sander or belt sander or finishing sander, abrasive paper	15	
B-6	Finish	Spray or brush	23	
PEGS				
P-1	Cut to length	Miter saw or coping saw or backsaw	5	
P-2	Sand	Abrasive paper	15	
P-3	Finish	Spray or brush	23	Stain pegs a different color from base.
ASSEMBLY				
JAP-1	Insert pegs into base			

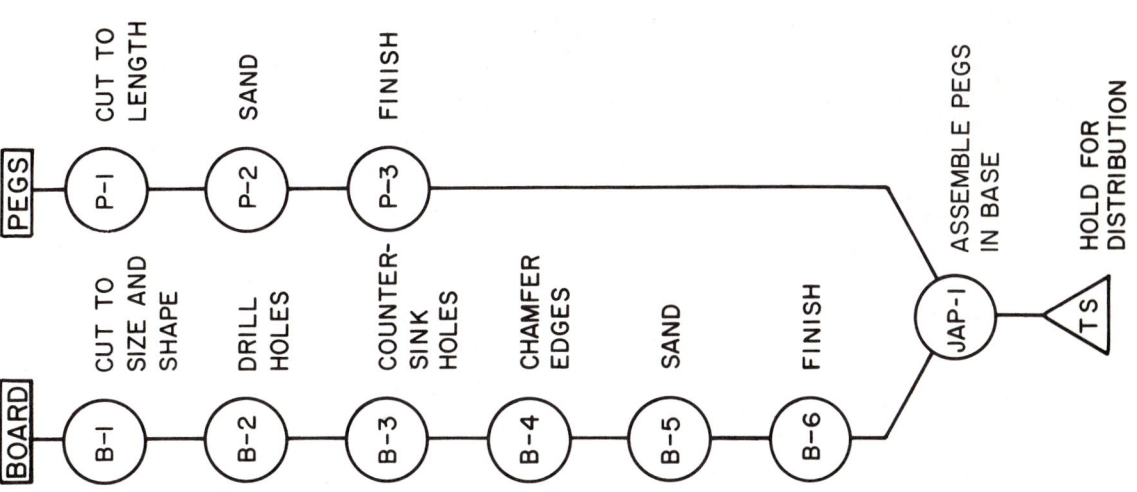

Salt and Pepper Shakers

PARTS AND MATERIALS

Qty.	Part	Size	Material
2	Shakers	A×A×C	Wood
2	Stoppers	#4	Cork
2	Screws PH	#6×3/8"	
	Finish		

DIMENSIONS

Dimension Symbol	Metric mm	Cust. in.
A	32	1-1/4
B	12	1/2
C	75	3
D	3	1/8
E	8	5/16
F	70	2-3/4

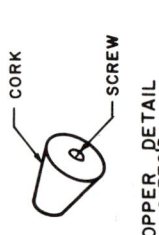

STOPPER DETAIL
2 REQ'D.

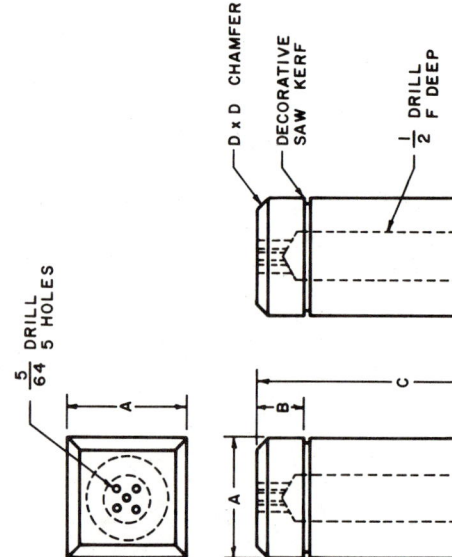

SHAKER DETAIL
2 REQ'D.

Salt and Pepper Shakers

PROCEDURE

Operation Number	Operation	Tools & Equipment	Topics	Notes
SHAKERS				
S-1	Cut to size	Handsaw or power saw	5	
S-2	Counterbore	Drill press, 7/8" Forstner bit or 7/8" spade bit	6, 25	A drilling fixture may be used.
S-3	Drill hole in bottom for filling	Drill press, 1/2" twist drill	6, 25	A drilling fixture may be used.
S-4	Cut kerf	Handsaw or band saw	5, 25	A sawing fixture may be used.
S-5	Chamfer top edges	Sanding block or belt sander or disk sander	15, 25	A sanding fixture may be used.
S-6	Drill holes for pouring	Hand drill or portable electric drill or drill press, 5/64" twist drill	6, 25	A drilling jig may be used.
S-7	Sand	Finishing sander, abrasive paper	15	
S-8	Label			A rubber stamp or a stencil could be used.
S-9	Finish	Spray or brush	23	
STOPPERS				
ST-1	Drill pilot hole	Hand drill, 1/16" twist drill	6	Drill hole in large end of cork.
ST-2	Install screw	Screwdriver	3, 9	Do not over-tighten the screw. Leave a "fingernail" space.
ASSEMBLY				
SPS-1	Assemble stopper in shaker			

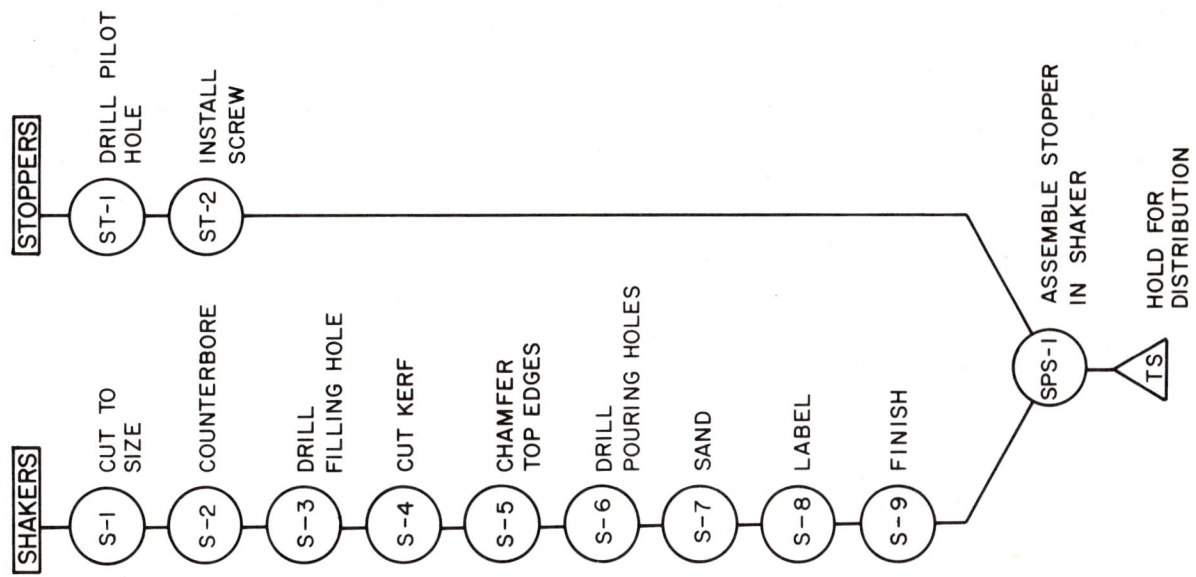

Super Q

SUPER Q GAMES

Super Q
1. The game is started with pegs in all of the holes except the center hole.
2. The moves are made by jumping over a peg and into an open hole.
3. Remove the peg that you jump over.
   Note: The jumps can be made in any direction, even diagonally.
4. The game is continued until you cannot make any more jumps.
5. You then count the number of pegs you have left. This number will be your score.

4 or more left	Try again.
3 left	You can do better.
2 left	Good.
1 left	Excellent!
1 left in the center hole	You are a genius!

Tic-Tac-Toe
Use the square of nine holes in the middle of the block to play tic-tac-toe.

Super Pick Up
1. Place only **18** pegs in the holes.
2. Two people play the game. The object is to take turns removing pegs from the board. The person who picks up the last peg is the loser.
3. Each player must pick up either one, two, or three pegs at a time.
4. Take turns being the first to pick up the peg or pegs.

Note: There is a clever system to winning this game. See if you can figure it out.

PARTS AND MATERIALS

Qty.	Part	Size	Material
1	Board	B×A×A	Wood
20	Pegs	1/4" dia. ×D	Wood Dowel
			Finish

DIMENSIONS

Dimension Symbol	Metric mm	Cust. in.
A	127	5
B	20	3/4
C	3	1/8
D	25	1
E	12	1/2

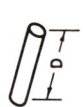

PEG DETAIL
20 REQ'D.

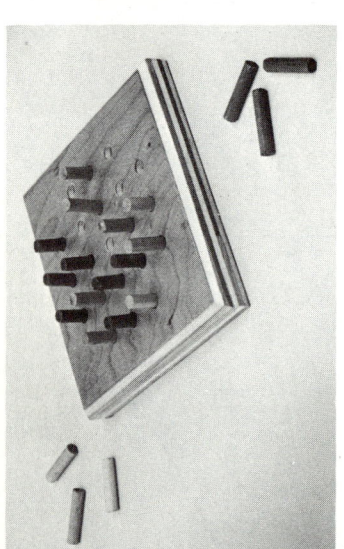

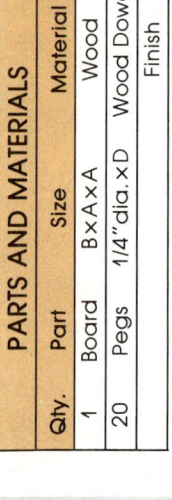

$\frac{9}{32}$ DRILL — E DEEP
21 HOLES

C × C CHAMFER

BOARD DETAIL

Super Q

PROCEDURE

Operation Number	Operation	Tools & Equipment	Topics	Notes
BOARD				
B-1	Cut to size	Handsaw or power saw	5	Use stop block.
B-2	Drill holes	Hand drill or portable electric drill or drill press, 9/32" twist drill	6, 25	A drilling fixture may be used, or mark using a template.
B-3	Chamfer edges	File or hand plane or disk sander or Uni-plane	7 or 15	
B-4	Sand	Belt sander or disk sander or finishing sander, abrasive paper	15	
B-5	Finish	Spray or brush	23	
PEGS				
P-1	Cut to length	Miter saw or coping saw or backsaw	5	Cut 20 pegs for each game.
P-2	Sand	Abrasive paper	15	
P-3	Finish	Spray or brush	23	Stain 10 pegs light. Stain 10 pegs dark. Pegs could be painted the school colors.
ASSEMBLY				
SQ-1	Assemble pegs into base			

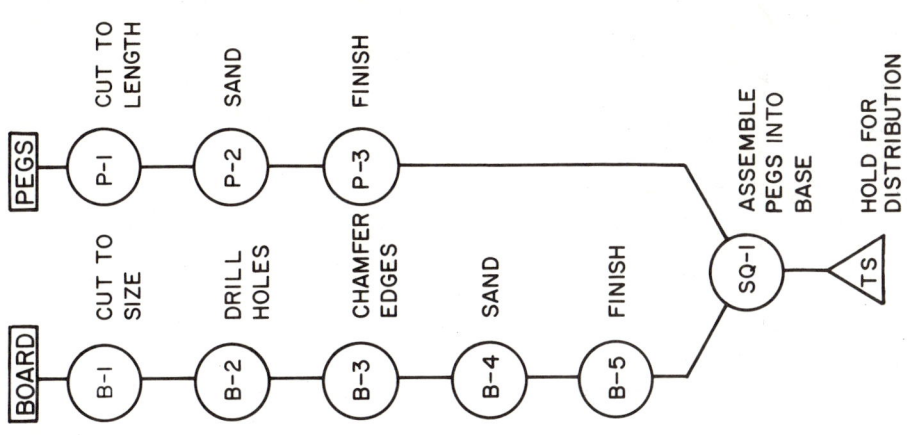

Note Holder

PARTS AND MATERIALS

Qty.	Part	Size	Material
1	Base	C×A×A	Wood
1	Pedestal	C×B×A	Wood
1	Top	1/8"×G×F	Wood
1	Clothespin		
2	Screws FH	#6×1-1/4"	Finish

DIMENSIONS

Dimension Symbol	Metric mm	Cust. in.
A	90	3-1/2
B	75	3
C	20	3/4
D	60	2-1/4
E	35	1-3/8
F	84	3-1/4
G	25	1
H	12	1/2
I	50	2
J	44	1-3/4
K	3	1/8

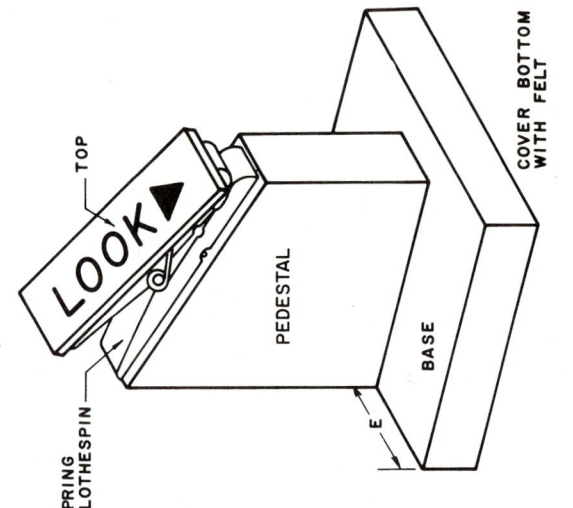

ASSEMBLY DETAIL

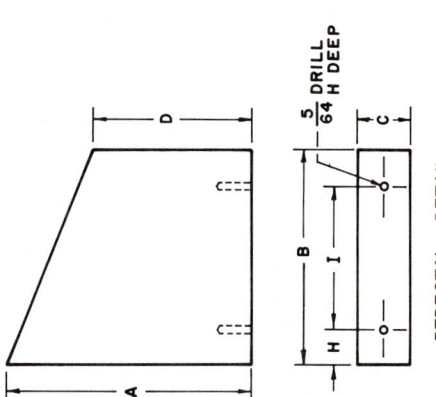

TOP DETAIL

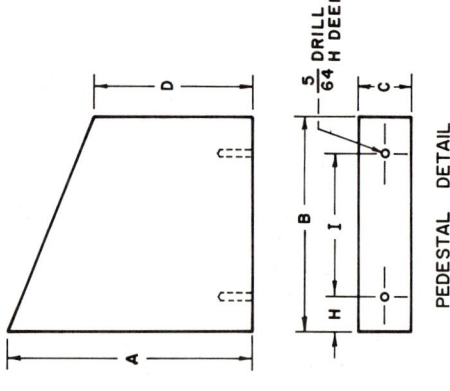

PEDESTAL DETAIL

BASE DETAIL

Note Holder

PROCEDURE

Operation Number	Operation	Tools & Equipment	Topics	Notes
PEDESTAL				
P-1	Cut to size and shape	Handsaw or power saw	5	
P-2	Drill pilot holes	Hand drill or portable electric drill, drill press, 5/64" twist drill	6	Could use base holes as a guide and drill at assembly (match drill).
P-3	Sand	Belt or disk sander or finishing sander, abrasive paper	15	
BASE				
B-1	Cut to size	Handsaw or power saw	5	
B-2	Drill holes	Hand drill or portable electric drill, drill press, 9/64" twist drill	6, 25	A drilling fixture may be used.
B-3	Countersink holes	Hand drill or portable electric drill, drill press, countersink	6	
B-4	Sand	Belt sander or disk sander or finishing sander, abrasive paper	15	
TOP				
T-1	Cut to size	Handsaw or power saw	5	
T-2	Sand edges	Abrasive paper	15	
T-3	Print	Screen stencil equipment or stencil and spray equipment	23	Could print paper label and glue it to top, or mark with a felt tip pen.
ASSEMBLY				
NH-1	Assemble pedestal to base	Screwdriver	3, 9, 11	Use glue.
NH-2	Assemble clothespin to pedestal		11	Use hot melt glue or another adhesive.
NH-3	Assemble top to clothespin		11	Use hot melt glue or another adhesive.
NH-4	Finish	Spray or brush	23	

Marble Drop

DIMENSIONS

Dimension Symbol	Metric mm	Cust. in.
A	140	5-1/2
B	64	2-1/2
C	20	3/4
D	25	1
E	12	1/2
F	35	1-3/8
G	22	7/8
H	46	1-7/8
I	14	9/16
J	32	1-1/4
K	15	5/8
L	10	3/8
M	178	7

PARTS AND MATERIALS

Qty.	Part	Size	Material
1	Base	C×B×A	Wood
2	Ends	C×D×B	Wood
2	Rods	1/8" dia. ×M	Metal Rod
4	Finishing Nails	4d	
1	Marble		
			Glue
			Finish

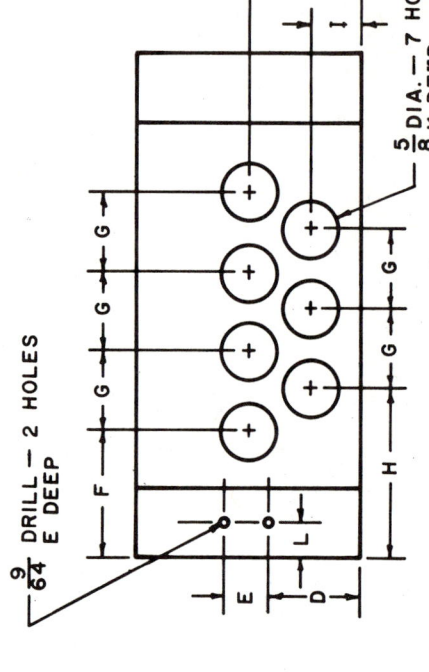

LAYOUT DETAIL

$\frac{5}{8}$ DIA. — 7 HOLES, K DEEP

$\frac{9}{64}$ DRILL — 2 HOLES, E DEEP

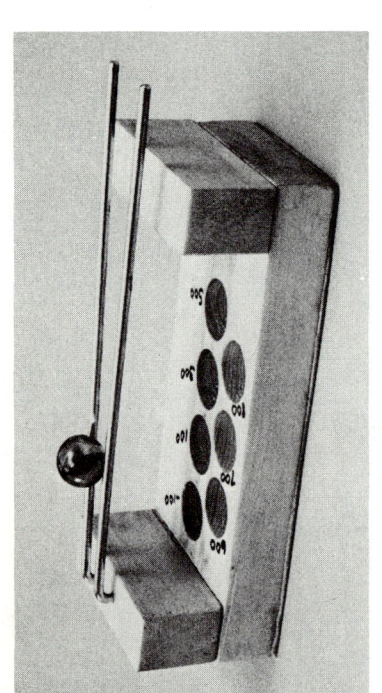

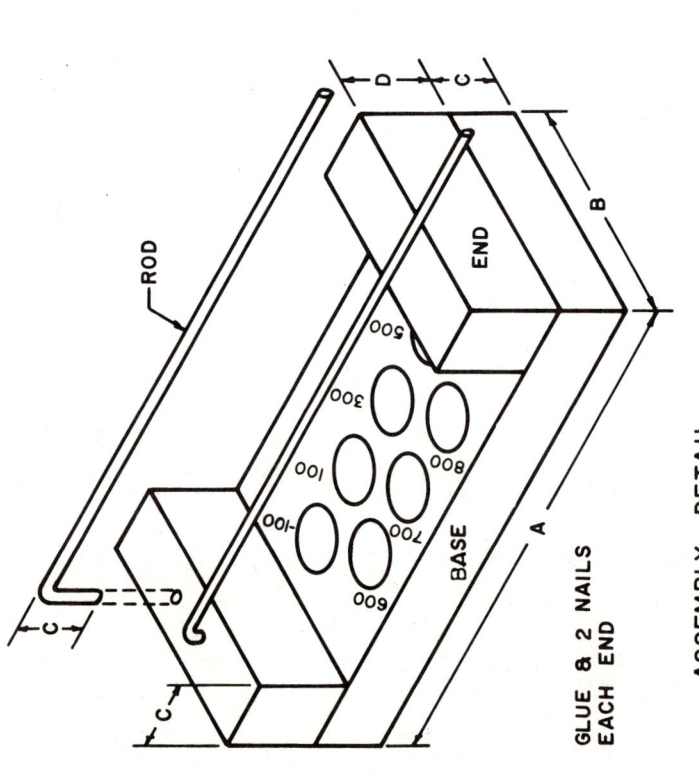

ASSEMBLY DETAIL

GLUE & 2 NAILS EACH END

Marble Drop

PROCEDURE

Operation Number	Operation	Tools & Equipment	Topics	Notes
ENDS				
E-1	Cut to size	Handsaw or power saw	5	Use stop block for cutting to length.
E-2	Drill rod holes	Hand drill or portable electric drill or drill press, 9/64" twist drill	6, 25	A drilling fixture may be used. Drill only one end piece.
E-3	Sand	Belt sander or finishing sander or abrasive paper	15	
BASE				
B-1	Cut to size	Handsaw or power saw	5	
B-2	Bore marble holes	Portable electric drill or drill press or bit brace, 5/8" Forstner bit or 5/8" spade bit	6, 25	A drilling fixture may be used. Do not bore completely through base.
B-3	Sand	Belt sander or finishing sander or abrasive paper	15	
B-4	Print	Felt tip pen, stencil	23	Use scoring numbers as on plan or make up your own.
RODS				
R-1	Cut to length	Wire cutters	3	
R-2	Bend	Vise, pliers	3, 19	A bending jig may be used.
ASSEMBLY				
MD-1	Assemble ends to base	Hammer	3, 11	Glue and nail.
MD-2	Finish	Spray or brush		
MD-3	Insert rods		23	
MD-4	Add marble			

Pet Rooster

DIMENSIONS

Dimension Symbol	Metric mm	Cust. in.
A	12	1/2
B	140	5-1/2
C	20	3/4
D	685	27
E	240	9-1/2
F	6	1/4
G	90	3-1/2
H	125	5
I	75	3
J	150	6
K	65	2-1/2
L	405	16

PARTS AND MATERIALS

Qty.	Part	Size	Material
Pet Rooster			
1	Head	C x G x H	Wood
1	Body	C x I x J	Wood
2	Feet	C x K x K	Wood
2	Handles	A x C x E	Wood
1	Legs	1/4" dia. x L	Rope
1	Neck	1/4" dia. x H	Rope
	String	8'	Monofilament Nylon Line
2	Wire Brads 18 ga. x 1/2"		
	Feathers		
			Paint
Perch			
2	Ends	C x B x B	Wood
1	Post	1/2" x 27"	Wood Dowel
			Finish

PROCEDURE

Operation Number	Operation	Tools & Equipment	Topics	Notes
FEET				
F-1	Cut to size and shape	Coping saw or band saw or scroll saw	5	
F-2	Drill hole	Hand drill or power drill, 1/4" twist drill	6	
F-3	Smooth edges and sides	File or abrasive paper	15	
F-4	Paint	Brush	5	See suggested colors and guidelines on pattern detail.
HEAD				
H-1	Cut to size and shape	Coping saw or scroll saw or band saw	5	
H-2	Drill hole	Hand drill or power drill, 1/4" twist drill	6	
H-3	Smooth edges and sides	File or abrasive paper	15	
H-4	Paint	Brush	23	See suggested colors and guidelines on pattern detail.

PATTERN DETAIL

Pet Rooster

BODY				
B-1	Cut to size and shape	Coping saw or scroll saw or band saw	5	
B-2	Drill holes	Hand drill or power drill, 1/4" twist drill and 3/32" twist drill	6	Attach feathers in 3/32" dia. hole.
B-3	Smooth edges and sides.	File or abrasive paper	15	
B-4	Paint	Brush	23	See suggested colors and guidelines on pattern detail.
B-5	Attach feathers		11	Use a small amount of glue.
NECK				
N-1	Cut to length	Knife	4	Neck should be 5" (127 mm) long.
LEGS				
L-1	Cut to length	Knife	4	One piece of rope 16" (405 mm) long will make both legs. Do **not** cut into two pieces.
STRINGS				
S-1	Cut to length	Knife or scissors	4	Use one 15" (380 mm) string for head, one 19" (480 mm) string for body and two 28" (710 mm) string for feet.
HANDLE				
HA-1	Cut to size	Handsaw or power saw	5	
HA-2	Cut dadoes	Backsaw and chisel or table saw with a dado blade	5, 7	Dado should be in center of each crosspiece.
HA-3	Drill holes	Hand drill or power drill, 1/16" twist drill	6	
HA-4	Sand	Abrasive paper	15	
HA-5	Assemble	Hammer	3, 10, 11	Use glue and wire brads.
HA-6	Finish	Spray or brush	23	
HA-7	Attach strings		11	Tie a knot on one end of each string. Pull the string through the handle. Glue in place.

(continued next page)

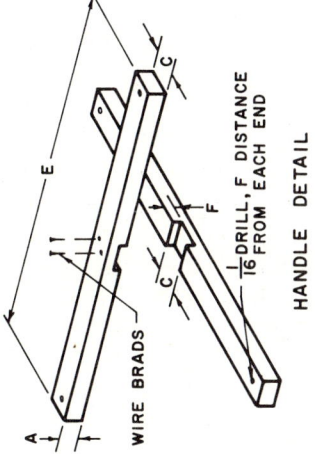

HANDLE DETAIL

Pet Rooster (cont.)

			ASSEMBLY	
PR-1	Assemble neck to head and body		11	Use a small amount of glue to secure the rope ends in hole.
PR-2	Assemble legs to body		11	Leg rope should be pulled through body hole to center position. Use glue to make it stay.
PR-3	Assemble feet to legs		11	Use glue to make them stay.
PR-4	Assemble strings to body, head, and feet	Staple gun or hammer	3 or 10	Staple string ends at locations shown on pattern detail. Could use tacks.

Perch

			ENDS	
E-1	Cut to size	Handsaw or power saw	5	
E-2	Drill hole	Portable electric drill or drill press or bit brace, 1/2" drill or 1/2" bit	6	
E-3	Round corners	Belt or disk sander or file, abrasive paper	15	
E-4	Sand	Finishing sander, abrasive paper	15	
			POST	
PO-1	Cut to length	Miter saw or backsaw or coping saw	5	
PO-2	Sand	Abrasive paper	15	
			ASSEMBLY	
P-1	Assemble ends to post		11	Use glue. Top and base should be positioned in the same way at either end of the post.
P-2	Finish	Spray or brush	23	

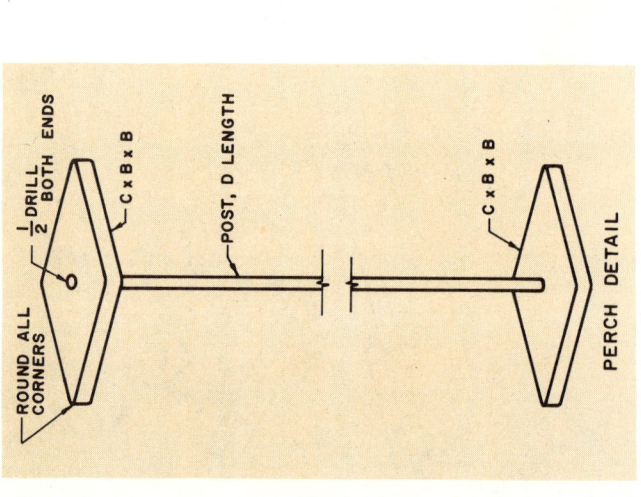

PERCH DETAIL

Pet Rooster (cont.)

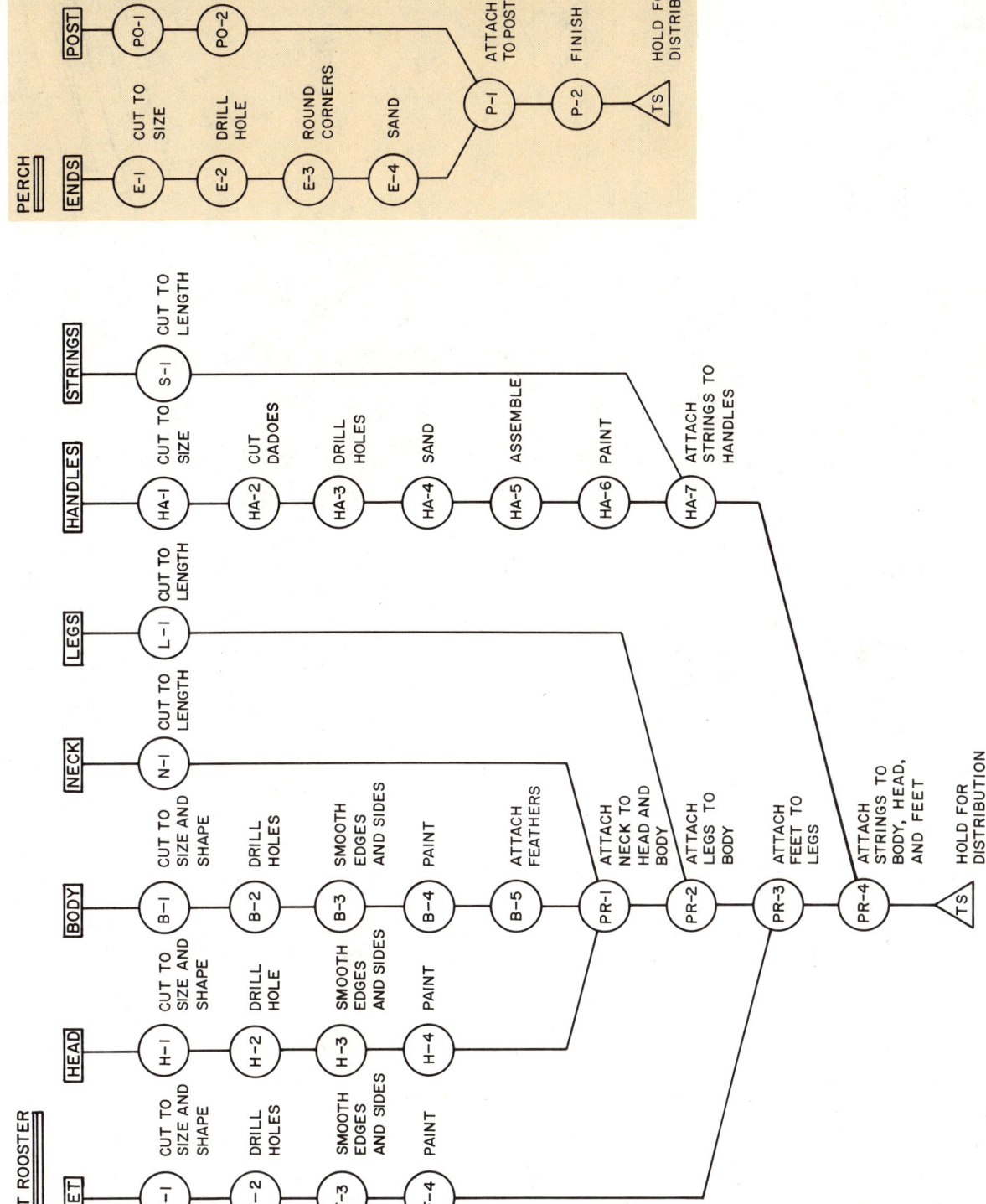

Handy Dandy Tennis

PARTS AND MATERIALS

Qty.	Part	Size	Material
1	Table	1/4" × B × A	Plywood
1	Handle	G × G × H	Wood
2	Posts	1/4" dia. × C	Wood Dowels
1	Net	D × E	Nylon Mesh or Cotton Gauze
1	Table Tennis Ball		
2	Brads	18ga. × 1/2"	
			Glue
			Finish

DIMENSIONS

Dimension Symbol	Metric mm	Cust. in.
A	300	12
B	150	6
C	50	2
D	25	1
E	175	7
F	125	5
G	20	3/4
H	280	11
I	10	3/8

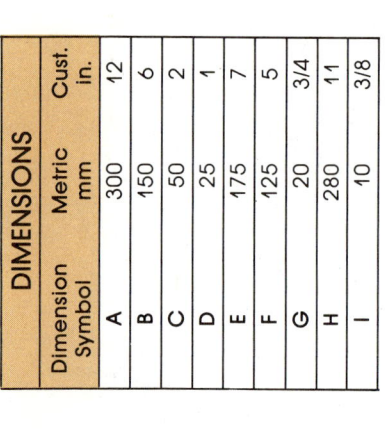

DECORATIVE LINES

DRILL — 2 HOLES — 1/4

BRADS

HANDLE DETAIL

OBJECTIVE — KEEP THE BALL BOUNCING ACROSS THE NET

POST & NET DETAIL

TABLE DETAIL

Handy Dandy Tennis

PROCEDURE

Operation Number	Operation	Tools & Equipment	Topics	Notes
TABLE				
T-1	Cut to size	Handsaw or power saw	5	
T-2	Sand	Finishing sander, abrasive paper	15	
T-3	Finish	Spray or brush	23	Paint top only. Green paint or school colors may be used.
T-4	Add decorative lines	Spray, stencil or screen printing equipment	23	White or school colors may be used
HANDLE				
H-1	Cut to size	Handsaw or power saw	5	
H-2	Round edges	File or abrasive paper	15	Round the handle grip only.
H-3	Sand	Finishing sander, abrasive paper	15	
H-4	Finish	Spray or brush	23	
POSTS				
P-1	Cut to length	Miter saw or coping saw or backsaw	5	
P-2	Sand	Abrasive paper	15	
NET				
N-1	Cut to size	Scissors	4	
ASSEMBLY				
HDT-1	Assemble handle and table	Hammer	3, 10, 11	Use glue and brads.
HDT-2	Drill post holes	Hand drill or portable electric drill or drill press, 1/4" twist drill	6	Drill holes through table and handle.
HDT-3	Assemble posts to handle and table	Mallet	3, 11	Use a small amount of glue for each.
HDT-4	Assemble net to posts	Stapler	10	Fold net to double thickness. Glue and staple net to posts. Trim excess net.
HDT-5	Add ball			

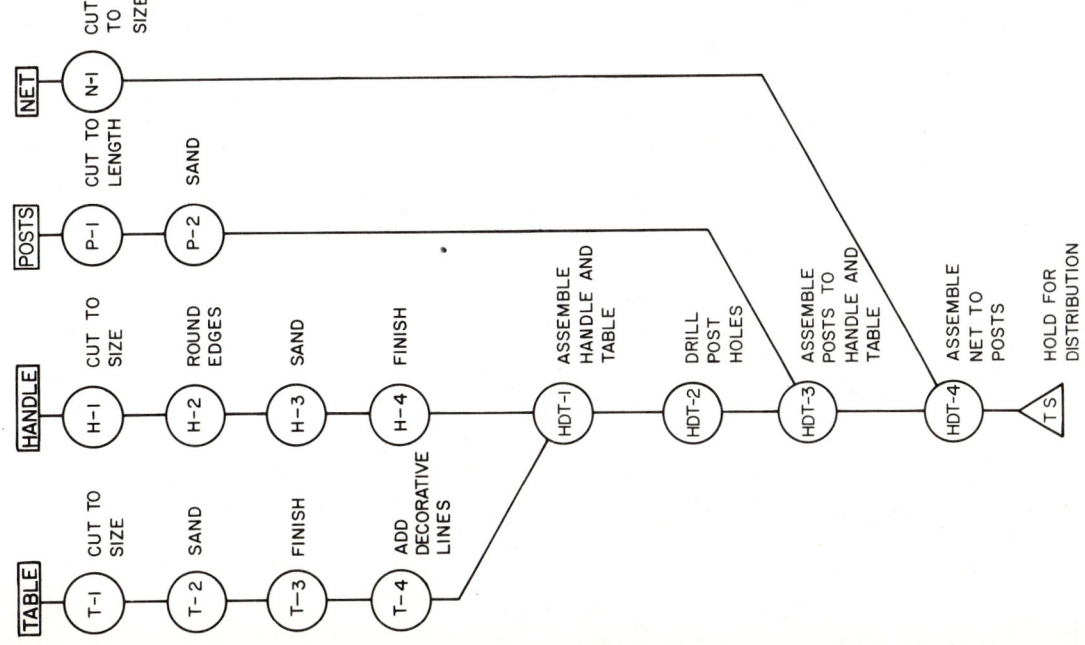

Paper Towel Stand

PARTS AND MATERIALS			
Qty.	Part	Size	Material
1	Base	C×B×B	Wood
1	Top	C×B dia.	Wood
1	Support	C×A×I	Wood
1	Rod	1/2" dia.×H	Wood dowel
1	Cap	1" dia.×E	Wood dowel
4	Screws FH	#8×1-1/2"	
			Glue
			Finish

DIMENSIONS		
Dimension Symbol	Metric mm	Cust. in.
A	64	2-1/2
B	140	5-1/2
C	20	3/4
D	12	1/2
E	25	1
F	230	9
G	10	3/8
H	326	13
I	380	15
J	70	2-3/4
K	305	12-1/8

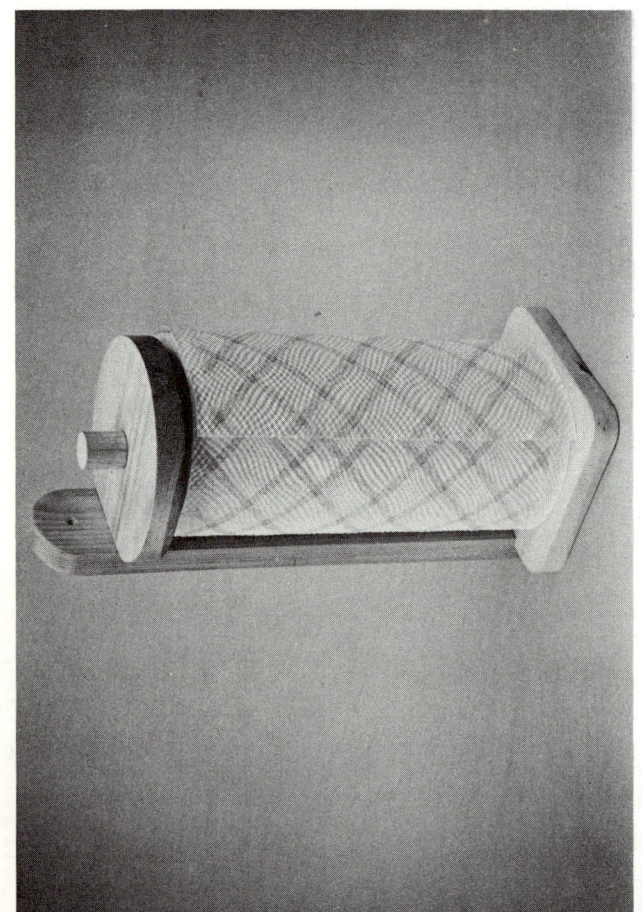

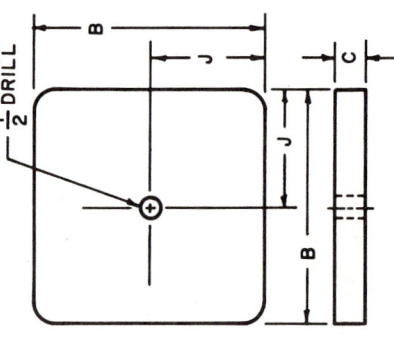

BASE DETAIL

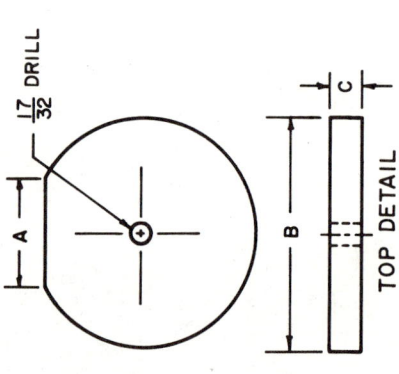

TOP DETAIL

Paper Towel Stand

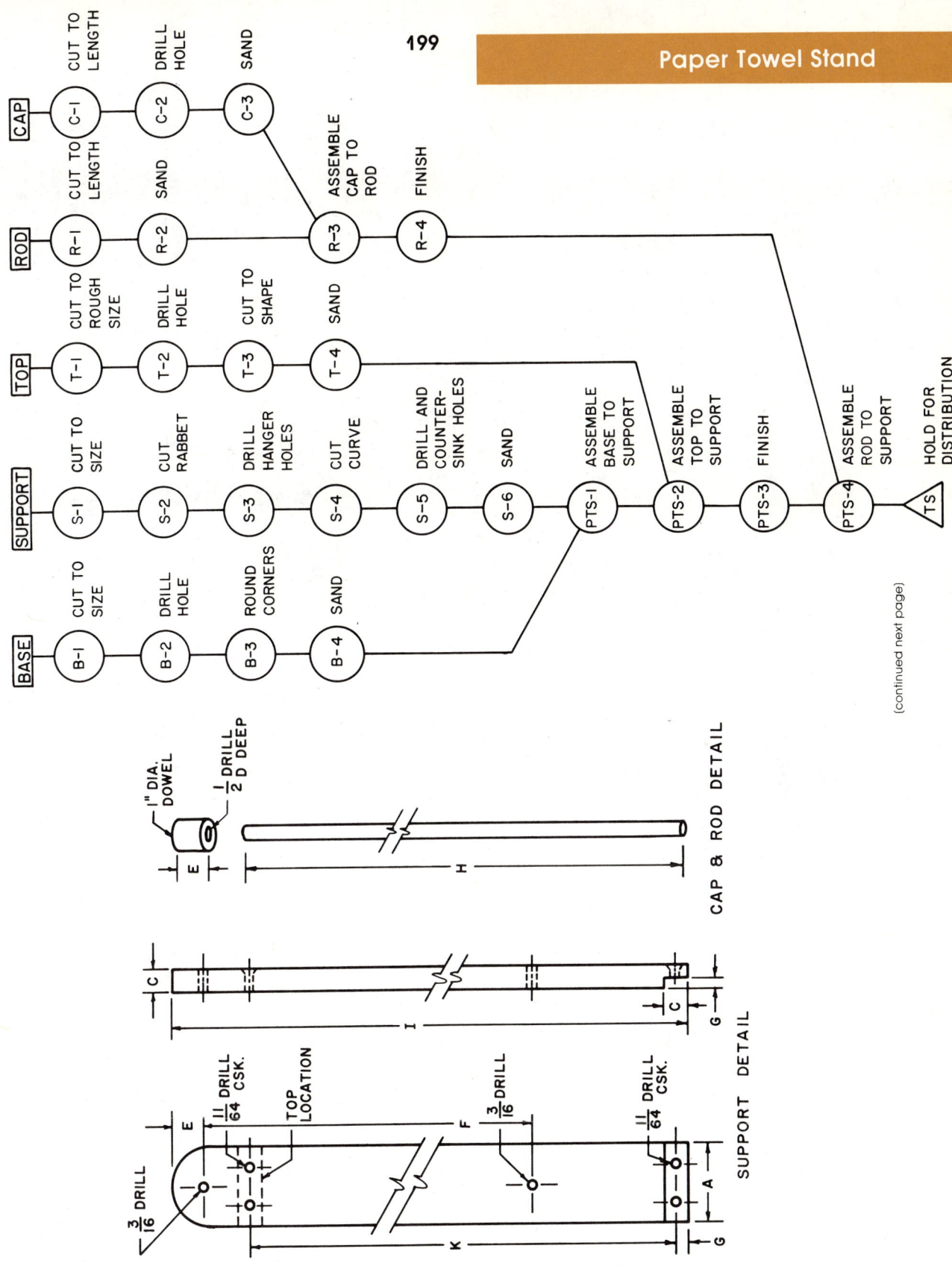

(continued next page)

Paper Towel Stand (cont.)

PROCEDURE

Operation Number	Operation	Tools & Equipment	Topics	Notes
BASE				
B-1	Cut to size	Handsaw or power saw	5	
B-2	Drill hole	Drill press, 1/2" twist drill or 1/2" spade bit	6	
B-3	Round corners	Band saw or belt sander or disk sander	5 or 15, 25	A modified circle cutting jig may be used.
B-4	Sand	Belt sander or finishing sander, abrasive paper	15	
SUPPORT				
S-1	Cut to size	Handsaw or power saw	5	
S-2	Cut rabbet	Backsaw or table saw with a dado blade	5	
S-3	Drill hanger holes	Hand drill or portable electric drill or drill press, 3/16" twist drill	6	
S-4	Cut curve	Band saw or scroll saw or coping saw	5, 25	A circle cutting jig may be used.
S-5	Drill and countersink holes	Hand drill or portable electric drill or drill press, 11/64" twist drill, countersink	6	
S-6	Sand	Belt sander or finishing sander, abrasive paper	15	
TOP				
T-1	Cut to rough size (square)	Handsaw or power saw	5	Cut slightly oversize.
T-2	Drill hole	Drill press, 17/32" twist drill	6	
T-3	Cut to shape	Band saw or scroll saw or coping saw	5, 25	A circle cutting jig may be used.
T-4	Sand	Belt sander or finishing sander, abrasive paper	15, 25	A circle jig for sanding edges may be used.

Paper Towel Stand (cont.)

CAP			
C-1	Cut to length	Coping saw or miter saw	5
C-2	Drill hole	Drill press, 1/2" twist drill or 1/2" spade bit	6, 25	Use fixture to hold while drilling.
C-3	Sand	Abrasive paper	15
ROD			
R-1	Cut to length	Miter saw or backsaw or coping saw	5
R-2	Sand	Abrasive paper	15
R-3	Assemble cap to rod	Mallet	3, 11	Use a small amount of glue.
R-4	Finish	Spray or brush	23
ASSEMBLY			
PTS-1	Assemble base to support	Scratch awl, screwdriver	2, 3, 9	Use scratch awl to make starter holes for screw.
PTS-2	Assemble top to support	Scratch awl, screwdriver	2, 3, 9
PTS-3	Finish	Spray or brush	23
PTS-4	Assemble rod to support		

Candle Holder

PARTS AND MATERIALS

Qty.	Part	Size	Material
1	Candle Holder	B×B×D	Wood
	Finish		
1	Finishing Nail	6d	

DIMENSIONS

Dimension Symbol	Metric mm	Cust. in.
A	75	3
B	90	3-1/2
C	25	1
D	112	4-1/2

NOTE —
MAKE A FULL-SIZE PROFILE OF YOUR OWN DESIGN. HAVE IT APPROVED PRIOR TO STARTING WORK.

SAMPLE PROFILE

HOLDER DETAIL

Candle Holder

PROCEDURE

Operation Number	Operation	Tools & Equipment	Topics	Notes
CH-1	Cut stock to rough length	Handsaw or power saw	5	
CH-2	Turn to size and shape	Wood lathe, lathe tools	7	
CH-3	Sand	Abrasive paper	15	Sand while still on the lathe.
CH-4	Cut to length	Backsaw or miter saw	5	
CH-5	Drive nail in top	Hammer	3, 10	
CH-6	Sharpen head of nail	File or grinder	15 or 16	
CH-7	Finish	Spray or brush	23	

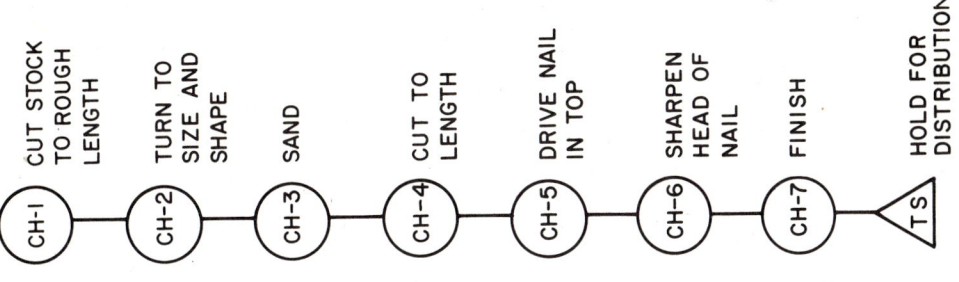

Cutting Board

PARTS AND MATERIALS

Qty.	Part	Size	Material
1	Light Board	C×A×A	Wood
1	Dark Board	C×A×A	Wood
			Glue
			Finish

Note: These materials are enough for two cutting boards.

DIMENSIONS

Dimension Symbol	Metric mm	Cust. in.
A	300	12
B	265	10-1/2
C	20	3/4

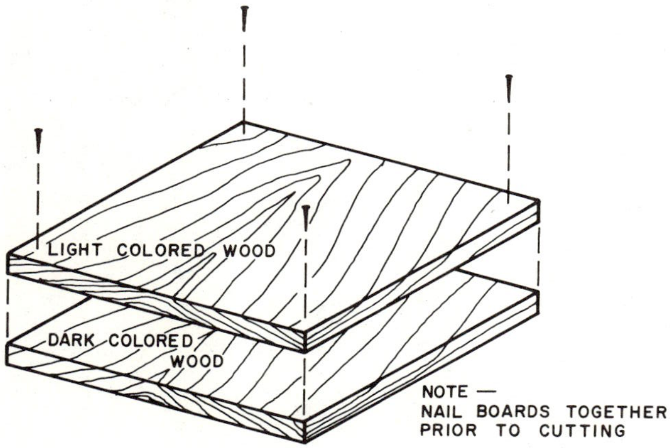

CUTTING DETAIL

TRIM SHADED AREA TO REMOVE NAIL HOLES

BOARD DETAIL

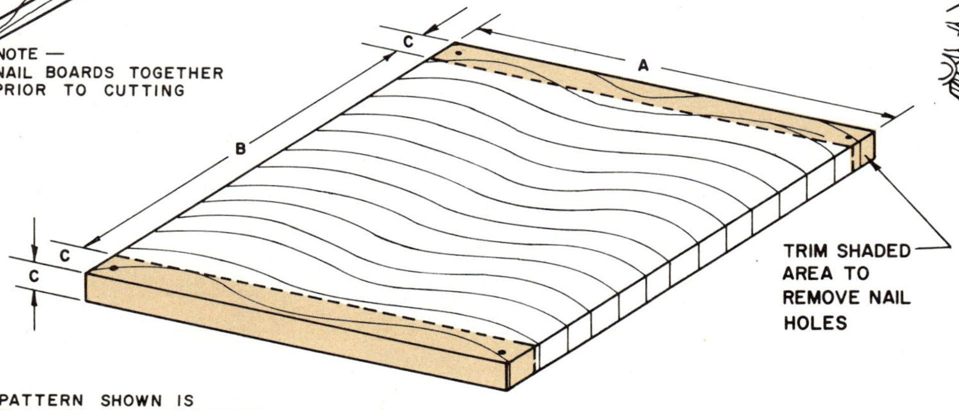

NOTE — CUTTING PATTERN SHOWN IS AN EXAMPLE. YOU MAY SUBSTITUTE YOUR OWN DESIGN.

NOTE — NAIL BOARDS TOGETHER PRIOR TO CUTTING

LIGHT COLORED WOOD

DARK COLORED WOOD

PRE-CUTTING DETAIL

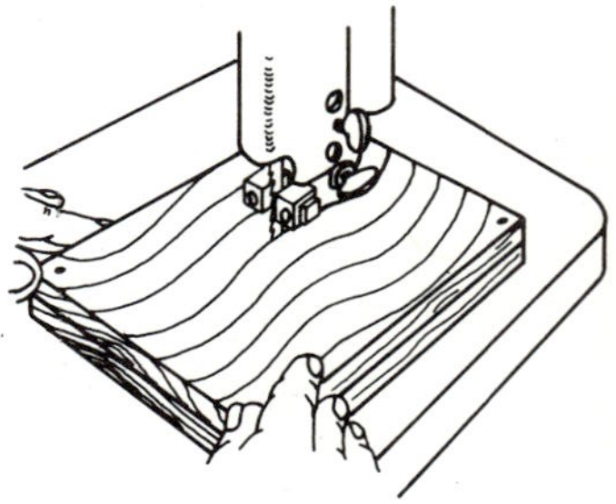

Cutting Board

PROCEDURE

Operation Number	Operation	Tools & Equipment	Topics	Notes
LIGHT BOARD				
LB-1	Cut to rough size	Handsaw or power saw	5	
DARK BOARD				
DB-1	Cut to rough size	Handsaw or power saw	5	
ASSEMBLY				
CB-1	Nail boards together	Hammer	3, 10	Position nails in waste area. Make sure boards are lined up properly.
CB-2	Cut into strips or pieces	Scroll saw or band saw	5	Cut through both boards at the same time. Avoid nails. Keep pieces in order.
CB-3	Assemble strips or pieces	Clamps	11, 24	Alternate color of wood. Glue and clamp.
CB-4	Cut to finish size	Handsaw or power saw	5	Trim two opposite edges to remove pieces with nail holes.
CB-5	Round all edges	File or router or abrasive paper	7, 15	
CB-6	Sand	Finishing sander, abrasive paper	15	
CB-7	Finish		23	Use mineral oil. Rub on and let it soak in.

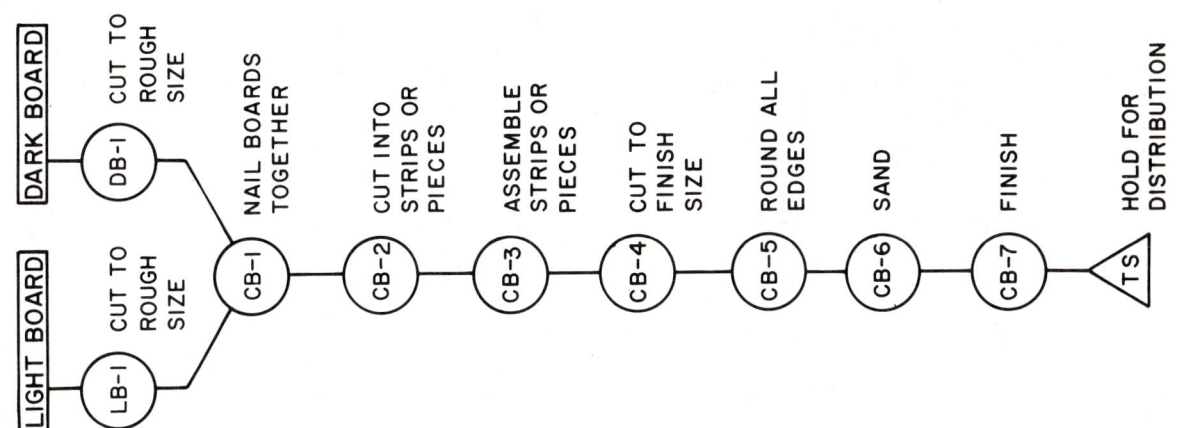

Marble Shoot

DIMENSIONS

Dimension Symbol	Metric mm	Cust. in.
A	292	11-1/2
B	40	1-1/2
C	12	1/2
D	57	2-1/4
E	35	1-3/8
F	20	3/4
G	50	2
H	21	13/16
I	318	12-1/2
J	25	1
K	6	1/4
L	100	4
M	3	1/8

PARTS AND MATERIALS

Qty.	Part	Size	Material
1	Base	C×B×A	Wood
2	Ends	C×B×G	Wood
2	Sides	1/8"×J×I	Plastic
1	Ramp	M×C×J	Wood
2	Guides	1/8" dia. ×G	Wood Dowel
1	Rebound Pad	B×B	Felt
1	Bottom Pad	B×I	Felt
1	Shooter	1/2" dia. ×G	Wood Dowel
12	Nails	18ga. ×3/4"	
1	Rubber Band		
1	Marble		
			Glue
			Finish

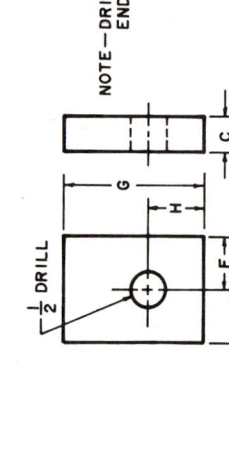

NOTE—DRILL ONE END ONLY

1/2 DRILL

END DETAIL
2 REQ'D.

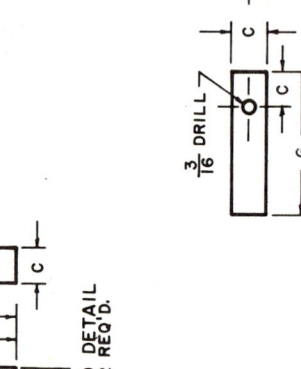

3/16 DRILL

SHOOTER DETAIL

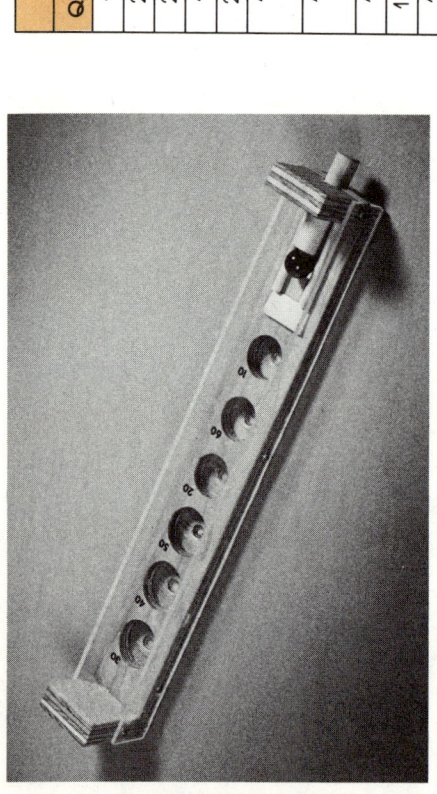

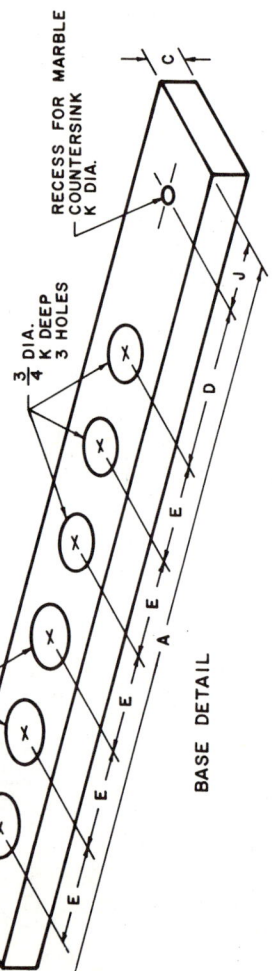

1/16 DRILL 4 HOLES

SIDE DETAIL
2 REQ'D.

RECESS FOR MARBLE COUNTERSINK K DIA.

3/4 DIA. K DEEP 3 HOLES

7/8 DIA. K DEEP 3 HOLES

BASE DETAIL

Marble Shoot

BOTTOM PAD: BP-1 CUT TO SIZE

REBOUND PAD: RP-1 CUT TO SIZE

SHOOTER: SH-1 CUT TO LENGTH → SH-2 DRILL HOLE → SH-3 SAND

SIDES: S-1 CUT TO SIZE → S-2 DRILL HOLES → S-3 SMOOTH EDGES → S-4 POLISH EDGES

GUIDES: G-1 CUT TO LENGTH → G-2 SAND

RAMP: R-1 CUT TO SIZE → R-2 SAND

ENDS: E-1 CUT TO SIZE → E-2 DRILL HOLE → E-3 SAND

BASE: B-1 CUT TO SIZE → B-2 BORE LARGE DIAMETER HOLES → B-3 BORE SMALL DIAMETER HOLES → B-4 COUNTERSINK RECESS FOR MARBLE → B-5 SAND → B-6 PRINT NUMBERS

Assembly: MS-1 ASSEMBLE ENDS TO BASE → MS-2 ASSEMBLE RAMP AND GUIDES TO BASE → MS-3 FINISH → MS-4 ASSEMBLE SIDES TO BASE → MS-5 INSTALL SHOOTER → MS-6 ASSEMBLE PADS TO BASE AND END → MS-7 ADD MARBLE → TS HOLD FOR DISTRIBUTION

(continued next page)

RAMP DETAIL

GUIDE DETAIL 2 REQ'D.

ASSEMBLY DETAIL — RECESS FOR MARBLE, RUBBER BAND, END, SHOOTER, RAMP, GUIDE

Marble Shoot (cont.)

PROCEDURE

Operation Number	Operation	Tools & Equipment	Topics	Notes
BASE				
B-1	Cut to size	Handsaw or power saw	5	
B-2	Bore large diameter holes	Drill press or portable electric drill or bit brace, 7/8" auger bit or 7/8" spade bit or 7/8" Forstner bit	6, 25	A drilling fixture may be used. Do not bore completely through base.
B-3	Bore small diameter holes	Drill press or portable electric drill or bit brace, 3/4" auger bit or 3/4" spade bit or 3/4" Forstner bit	6, 25	A drilling fixture may be used. Do not bore completely through base.
B-4	Countersink recess for marble	Portable electric drill or drill press, countersink	6	
B-5	Sand	Finishing sander, abrasive paper	15	
B-6	Print numbers	Silk screen stencil equipment or felt tip pen	23	
ENDS				
E-1	Cut to size	Handsaw or power saw	5	
E-2	Drill hole	Drill press or portable electric drill or bit brace, 1/2" twist drill or 1/2" auger bit	6, 25	Drill hole in one end only. A drilling fixture may be used.
E-3	Sand	Belt sander or finishing sander, abrasive paper	15	
RAMP				
R-1	Cut to size	Band saw or backsaw or coping saw	5	
R-2	Sand	Abrasive paper	15	
GUIDES				
G-1	Cut to length	Coping saw or miter saw	5	
G-2	Sand	Belt sander, abrasive paper	15	Make a flat side.

Marble Shoot (cont.)

SIDES					
S-1	Cut to size	Handsaw or power saw	5		
S-2	Drill holes	Hand drill or portable electric drill or drill press, 1/16" twist drill	6		Drill at very slow speed.
S-3	Smooth edges	File or abrasive paper	15		
S-4	Polish edges	Buffer	15		
SHOOTER					
SH-1	Cut to length	Coping saw or miter saw or backsaw	5		
SH-2	Drill hole	Hand drill or portable electric drill or drill press, 3/16" twist drill	6, 25		A drilling fixture may be used.
SH-3	Sand	Abrasive paper	15		
REBOUND PAD					
RP-1	Cut to size	Scissors or paper cutter	4		
BOTTOM PAD					
BP-1	Cut to size	Scissors or paper cutter	4		
ASSEMBLY					
MS-1	Assemble ends to base	Hammer	3, 10, 11		Attach ends even with sides and bottom of base. Use glue and nails. Use 2 nails in each end.
MS-2	Assemble ramp and guides to base		11		Use glue. Lower end of ramp should be towards shooter.
MS-3	Finish	Spray or brush	23		
MS-4	Assemble sides to base	Hammer	3, 10		Use nails. Leave 2 nails sticking out slightly for rubber band.
MS-5	Install shooter				Hook rubber band over nails.
MS-6	Assemble pads to base and end		11		Glue pads to bottom of base and inside surface of end.
MS-7	Add marble				

Plant Stand

DIMENSIONS

Dimension Symbol	Metric mm	Cust. in.
A	225	9
B	20	3/4
C	10	3/8
D	38	1-1/2
E	32	1-1/4
F	150	6
G	840	33

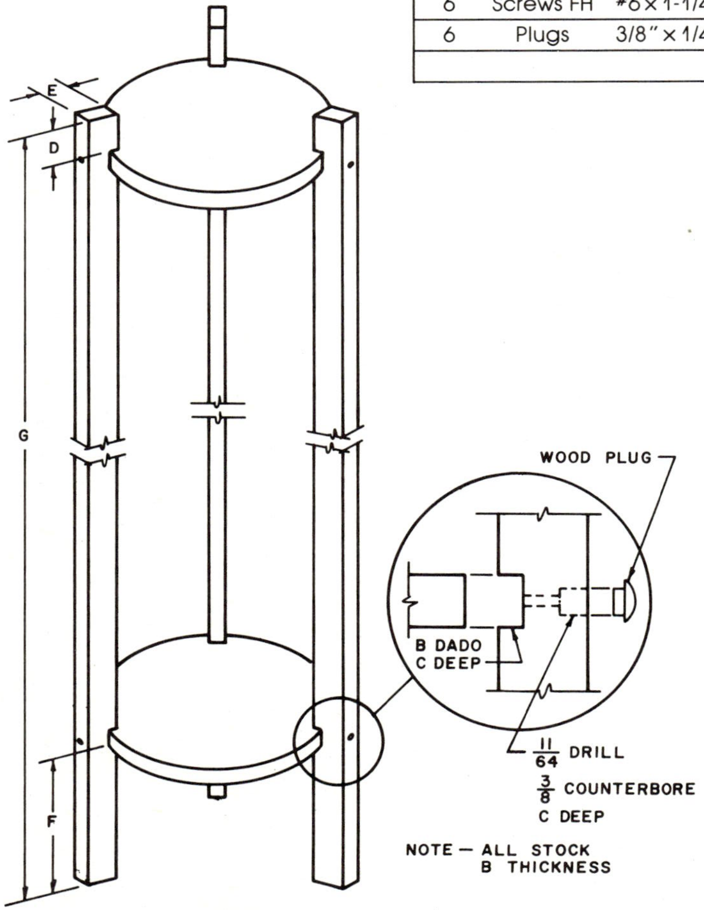

PARTS AND MATERIALS

Qty.	Part	Size	Material
2	Disks	B × A dia.	Wood
3	Legs	B × E × G	Wood
6	Screws FH	#6 × 1-1/4"	
6	Plugs	3/8" × 1/4"	Glue
			Finish

DISK DETAIL
2 REQ'D.

$\frac{3}{32}$ DRILL — 3 HOLES EQUALLY SPACED

WOOD PLUG

$\frac{11}{64}$ DRILL
$\frac{3}{8}$ COUNTERBORE
C DEEP

B DADO
C DEEP

NOTE — ALL STOCK B THICKNESS

ASSEMBLY DETAIL

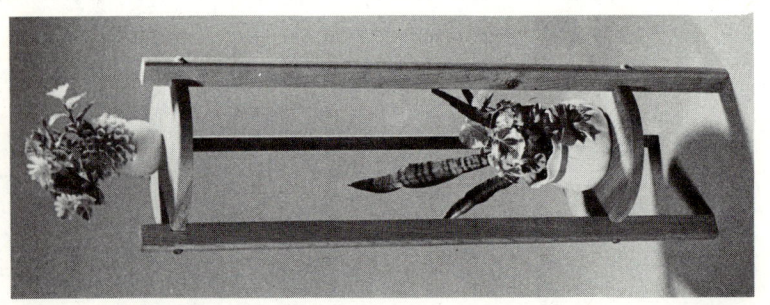

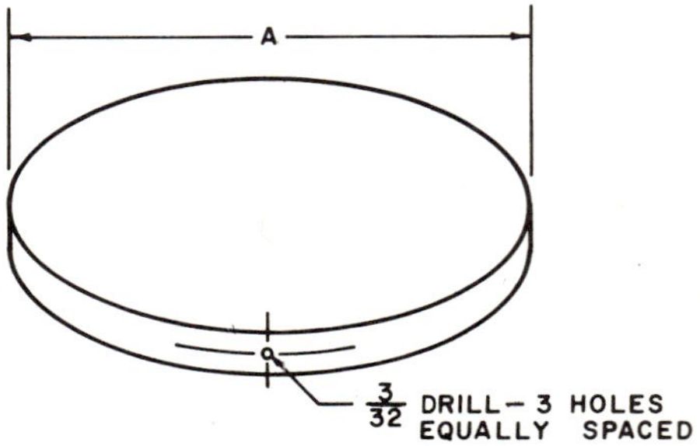

Plant Stand

PROCEDURE

Operation Number	Operation	Tools & Equipment	Topics	Notes
LEGS				
L-1	Cut to size	Handsaw or power saw	5	
L-2	Cut dadoes	Backsaw or table saw or radial arm saw with dado head	5	
L-3	Round edges	Router, 1/4" bead cutter bit	7	Clamp work securely.
L-4	Drill holes	Drill press, 11/64" twist drill	6	
L-5	Counterbore	Portable electric drill or drill press, 3/8" twist drill	6	
L-6	Sand	Finishing sander, abrasive paper	15	
DISKS				
D-1	Cut square disk blanks	Handsaw or power saw	5	Cut slightly over-size.
D-2	Cut circular shape	Band saw or coping saw or scroll saw	5, 25	A circle cutting jig may be used.
D-3	Sand circular shape	Belt sander or finishing sander, abrasive paper	15, 25	A circle sanding jig may be used.
D-4	Round edge	Router, 1/4" bead cutting bit	7	Clamp work securely.
D-5	Sand	Finishing sander, abrasive paper	15	
D-6	Drill pilot holes	Hand drill or power drill, 3/32" twist drill	6	
ASSEMBLY				
PS-1	Assemble legs to disks	Screwdriver	3, 9	Plugs may be made with a plug cutter or purchased.
PS-2	Assemble plugs to legs	Mallet	3, 11	Use a small amount of glue.
PS-3	Sand plugs	Abrasive paper	15	
PS-4	Finish	Spray or brush	23	

DISKS
- D-1 CUT SQUARE DISK BLANKS
- D-2 CUT CIRCULAR SHAPE
- D-3 SAND CIRCULAR SHAPE
- D-4 ROUND EDGE
- D-5 SAND
- D-6 DRILL PILOT HOLES

LEGS
- L-1 CUT TO SIZE
- L-2 ROUND EDGES
- L-3 CUT DADOES
- L-4 DRILL HOLES
- L-5 COUNTERBORE
- L-6 SAND

- PS-1 ASSEMBLE LEGS TO DISKS
- PS-2 ASSEMBLE PLUGS TO LEGS
- PS-3 SAND PLUGS
- PS-4 FINISH
- TS HOLD FOR DISTRIBUTION

Coaster Set

PARTS AND MATERIALS

Qty.	Part	Size	Material
1	Base	B×A×A	Wood
6	Coasters	C×A×A	Wood
2	Posts	1/4" dia. ×H	Wood Dowel
6	Disks	1/16" thick ×G dia.	Cork
			Glue
			Finish

DIMENSIONS

Dimension Symbol	Metric mm	Cust. in.
A	100	4
B	20	3/4
C	12	1/2
D	3	1/8
E	6	1/4
F	50	2
G	70	2-3/4
H	150	6

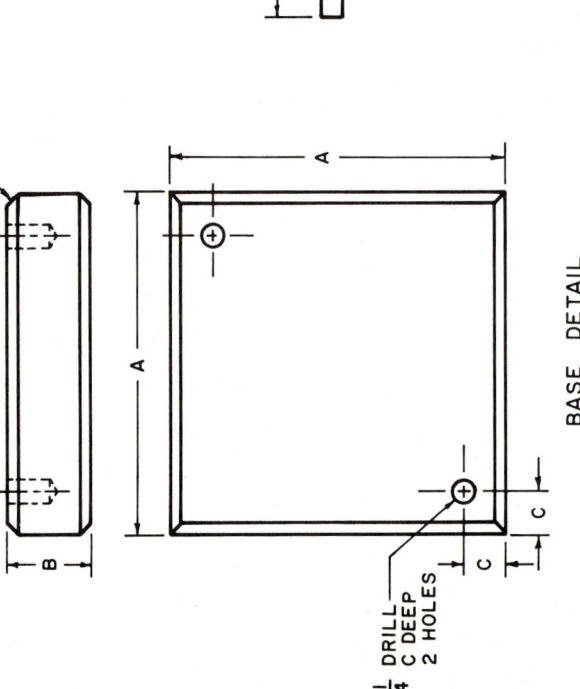

POST DETAIL
2 REQ'D.

BASE DETAIL

Coaster Set

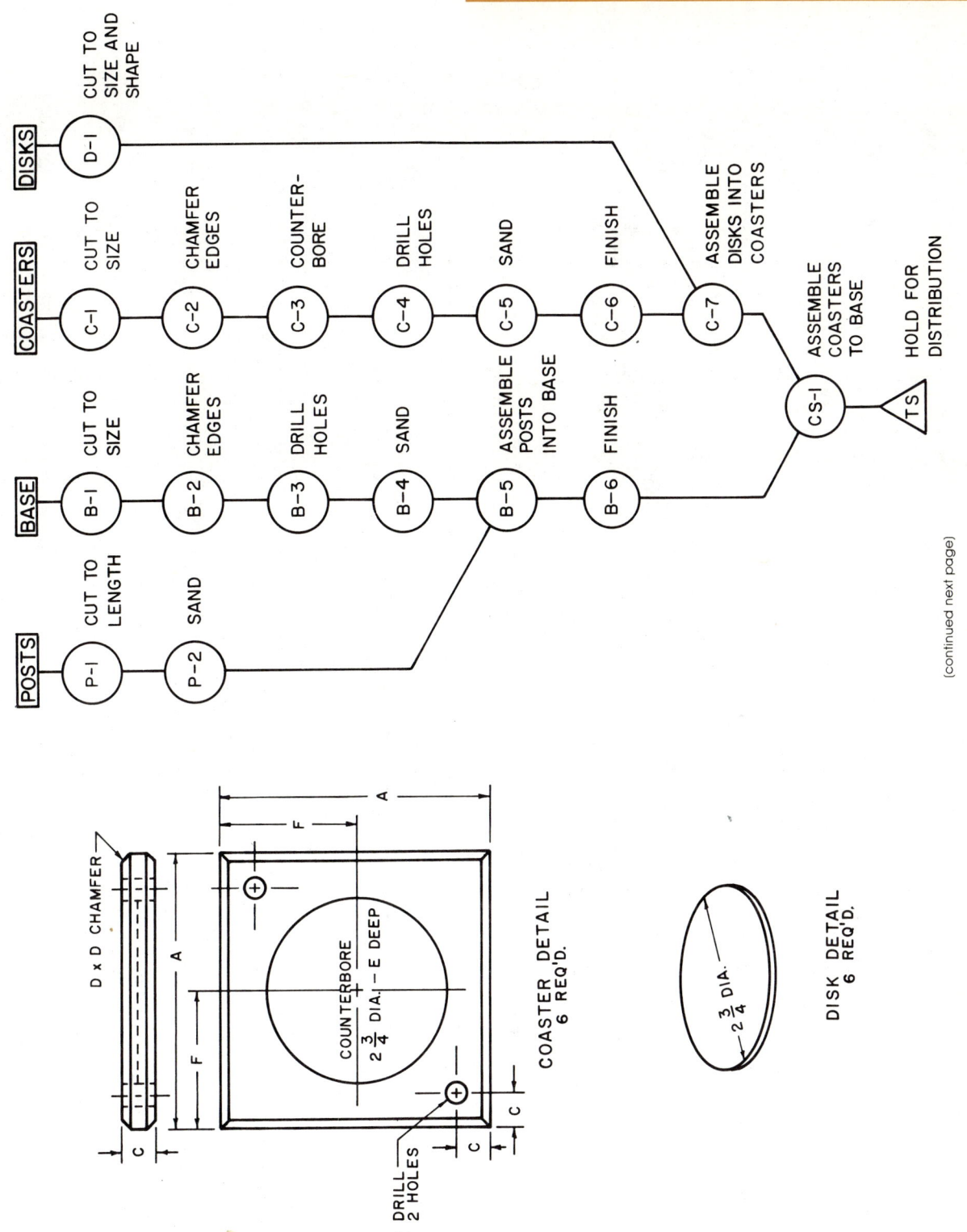

(continued next page)

Coaster Set (cont.)

PROCEDURE

Operation Number	Operation	Tools & Equipment	Topics	Notes
POSTS				
P-1	Cut to length	Miter saw or backsaw or coping saw	5	
P-2	Sand	Abrasive paper	15	
BASE				
B-1	Cut to size	Handsaw or power saw	5	A stop block may be used.
B-2	Chamfer edges	Belt sander or disk sander or file or Uniplane	7 or 15	
B-3	Drill holes	Drill press, 1/4" twist drill	6, 25	A drilling fixture may be used.
B-4	Sand	Finishing sander or belt sander, abrasive paper	15	
B-5	Assemble posts into base	Hammer or mallet	3, 11	Glue pegs in place.
B-6	Finish	Spray or brush	23	
DISKS				
D-1	Cut to size and shape	Knife or scissors	2, 4	Use template for layout.

Coaster Set (cont.)

COASTERS				
C-1	Cut to size	Handsaw or power saw	5	Cut 6 coasters for each set. A stop block may be used.
C-2	Chamfer edges	Belt sander or disk sander or file or Uniplane	7 or 15	
C-3	Counterbore	Drill press, 2-3/4" multi-spur or Forstner bit	6, 25	A drilling fixture may be used. A bit brace and expansive bit may be used. Clamp securely.
C-4	Drill holes	Drill press, 5/16" twist drill	6, 25	A drilling fixture may be used.
C-5	Sand	Finishing sander or belt sander, abrasive paper	15	
C-6	Finish	Spray or brush	23	
C-7	Assemble disks in coasters		11	Carefully glue disks in place.
ASSEMBLY				
CS-1	Assemble coasters to base			Stack 6 coasters on base.

Target Game

DIMENSIONS

Dimension Symbol	Metric mm	Cust. in.
A	203	8
B	133	5-1/4
C	22	7/8
D	29	1-1/8
E	32	1-1/4
F	20	3/4
G	12	1/2
H	75	3
I	42	1-3/4
J	10	3/8
K	25	1
L	216	8-1/2
M	37	1-1/2
N	50	2
O	5	3/16
P	178	7
Q	3	1/8
R	6	1/4

PARTS AND MATERIALS

Qty.	Part	Size	Material
1	Base	G×B×A	Wood
1	End	G×M×B	Wood
1	Ramp	O×G×K	Wood
1	Guide	K×P	Sheet Metal (28 gage)
1	Starting Gate	K×N	Sheet Metal (28 gage)
1	Catapult	3/32"×K×L	Acrylic Plastic
6	Screws PH	4"×1/2"	
3	Wire Brads	16ga.×1"	
1	String		
1	Pad	F×B	Felt
1	Marble		
			Glue
			Finish

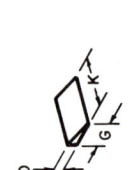

RAMP DETAIL

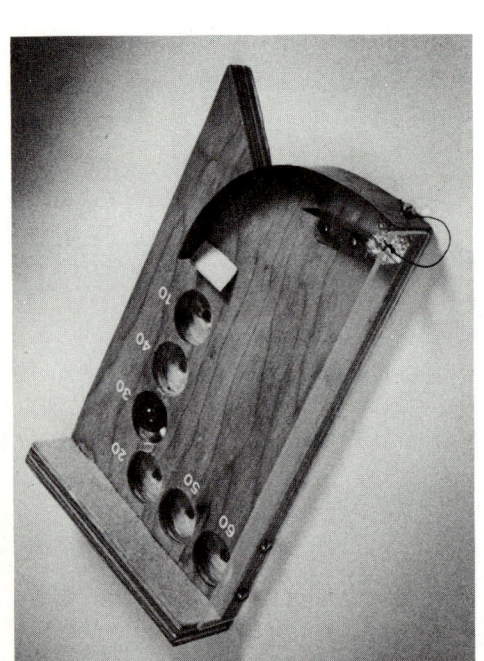

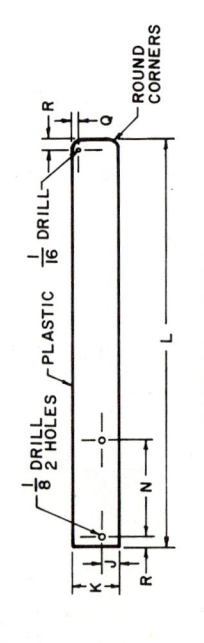

BASE DETAIL

CATAPULT DETAIL

Target Game

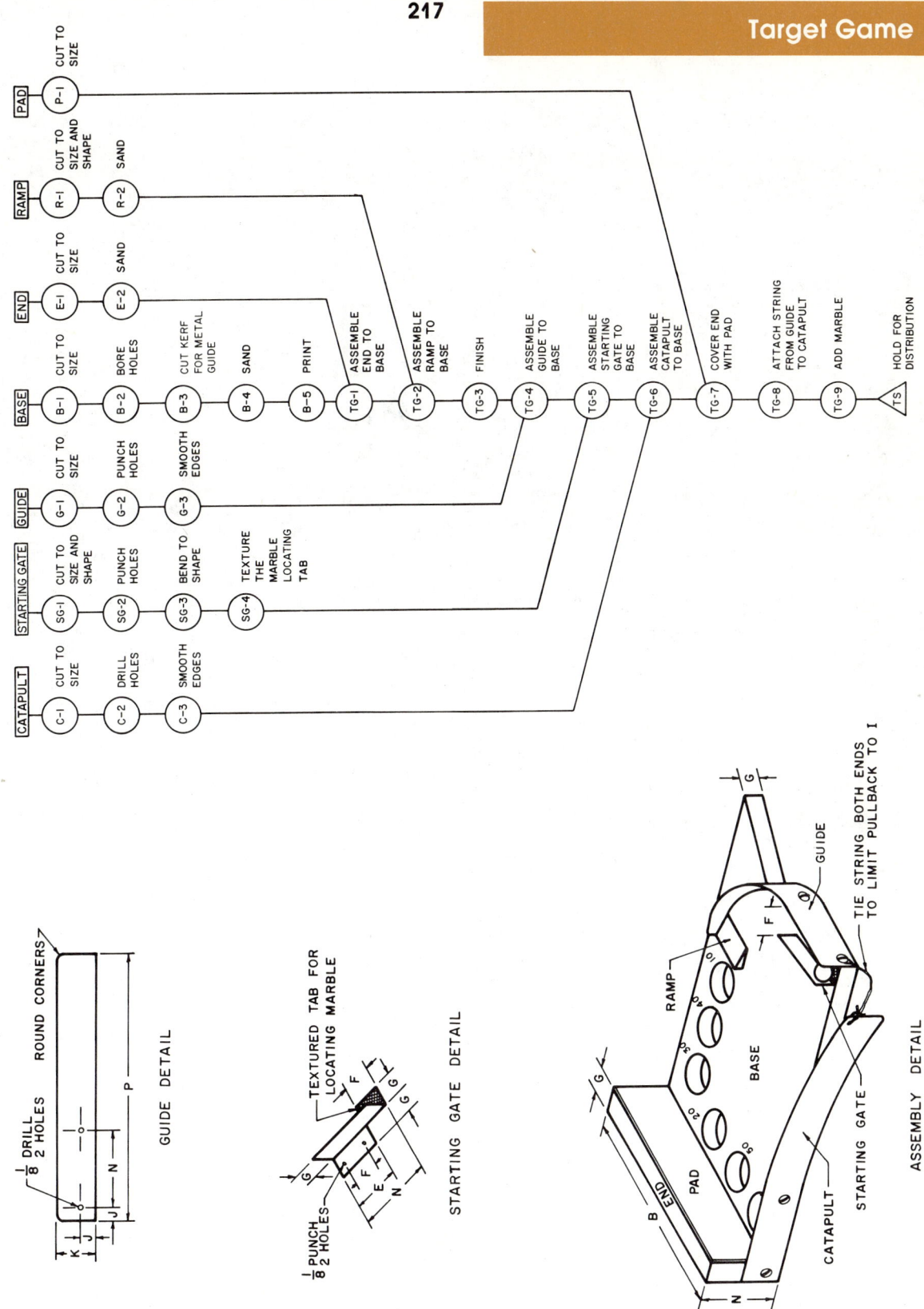

Target Game (cont.)

PROCEDURE

Operation Number	Operation	Tools & Equipment	Topics	Notes
CATAPULT				
C-1	Cut to size	Handsaw or power saw	5	
C-2	Drill holes	Hand drill or portable electric drill or drill press, 1/8" twist drill	6	Use slow speed.
C-3	Smooth edges	File or belt sander, abrasive paper	15	
STARTING GATE				
SG-1	Cut to size and shape	Tin snips or squaring shears	4	
SG-2	Punch holes	1/8" hand punch	4 or 6	Could be drilled.
SG-3	Bend to shape	Hand seamer	19	
SG-4	Texture the marble locating tab	Vise		Squeeze tab in vise to make dents to locate the marble.
GUIDE				
G-1	Cut to size	Tin snips or squaring shears	4	
G-2	Punch holes	1/8" hand punch	4 or 6	Could be drilled.
G-3	Smooth edges	File	15	
BASE				
B-1	Cut to size	Handsaw or power saw	5	
B-2	Bore holes	Drill press or portable electric drill or bit brace, 1" spade bit or 1" Forstner bit	6	A drilling (boring) fixture may be used.
B-3	Cut kerf for metal guide	Band saw or scroll saw or coping saw	5	Cut curve carefully.
B-4	Sand	Belt sander or disk sander or finishing sander, abrasive paper	15	
B-5	Print	Screen stencil equipment or felt tip pen		Print scoring numbers around holes. A design or your school name could be printed.

Target Game (cont.)

PROCEDURE (cont.)

Operation Number	Operation	Tools & Equipment	Topics	Notes
RAMP				
R-1	Cut to size and shape	Miter saw or band saw	5	
R-2	Sand	Abrasive paper	15	
END				
E-1	Cut to size	Handsaw or power saw	5	
E-2	Sand	Belt sander or disk sander or finishing sander, abrasive paper	15	
PAD				
P-1	Cut to size	Scissors		
ASSEMBLY				
TG-1	Assemble end to base	Hammer	3, 10, 11	Use glue and nails.
TG-2	Assemble ramp to base		11	Use glue.
TG-3	Finish	Spray or brush	23	
TG-4	Assemble guide to base	Screwdriver, scratch awl	2, 3, 9	Use scratch awl to make starting hole for screw.
TG-5	Assemble starting gate to base	Screwdriver, scratch awl	2, 3, 9	
TG-6	Assemble catapult to base	Screwdriver, scratch awl	2, 3, 9	
TG-7	Cover end with pad	Scissors	11	Use glue.
TG-8	Attach string from guide to catapult			Do this to avoid breaking the plastic catapult.
TG-9	Add marble			

Desk Caddy

DIMENSIONS

Dimension Symbol	Metric mm	Cust. in.
A	114	4-1/2
B	255	10
C	180	7-1/4
D	57	2-1/4
E	87	3-1/2
F	37	1-1/2
G	42	1-5/8
H	128	5
I	6	1/4
J	3	1/8
K	186	7-1/2
L	35	1-3/8
M	70	2-3/4
N	96	3-7/8
O	50	2
P	22	7/8
Q	82	3-1/4
R	128	5-1/4
S	108	4-1/4
T	43	1-3/4

PARTS AND MATERIALS

Qty.	Part	Size	Material
2	Sides	1×A×B	Wood
2	Lower Shelves	1×A×C	Wood
2	Upper Shelves	1×A×C	Wood
3	Large Dividers	1×A×R	Wood
1	Small Divider	1×A×T	Wood
2	Drawer Fronts	1×O×N	Wood
2	Drawer Backs	1×L×M	Wood
4	Drawer Sides	1×L×A	Wood
2	Drawer Bottoms	1×M×S	Wood
1	Knob		Wood
24	Wire Brads	18ga. × 1/2"	
			Glue
			Finish

Desk Caddy

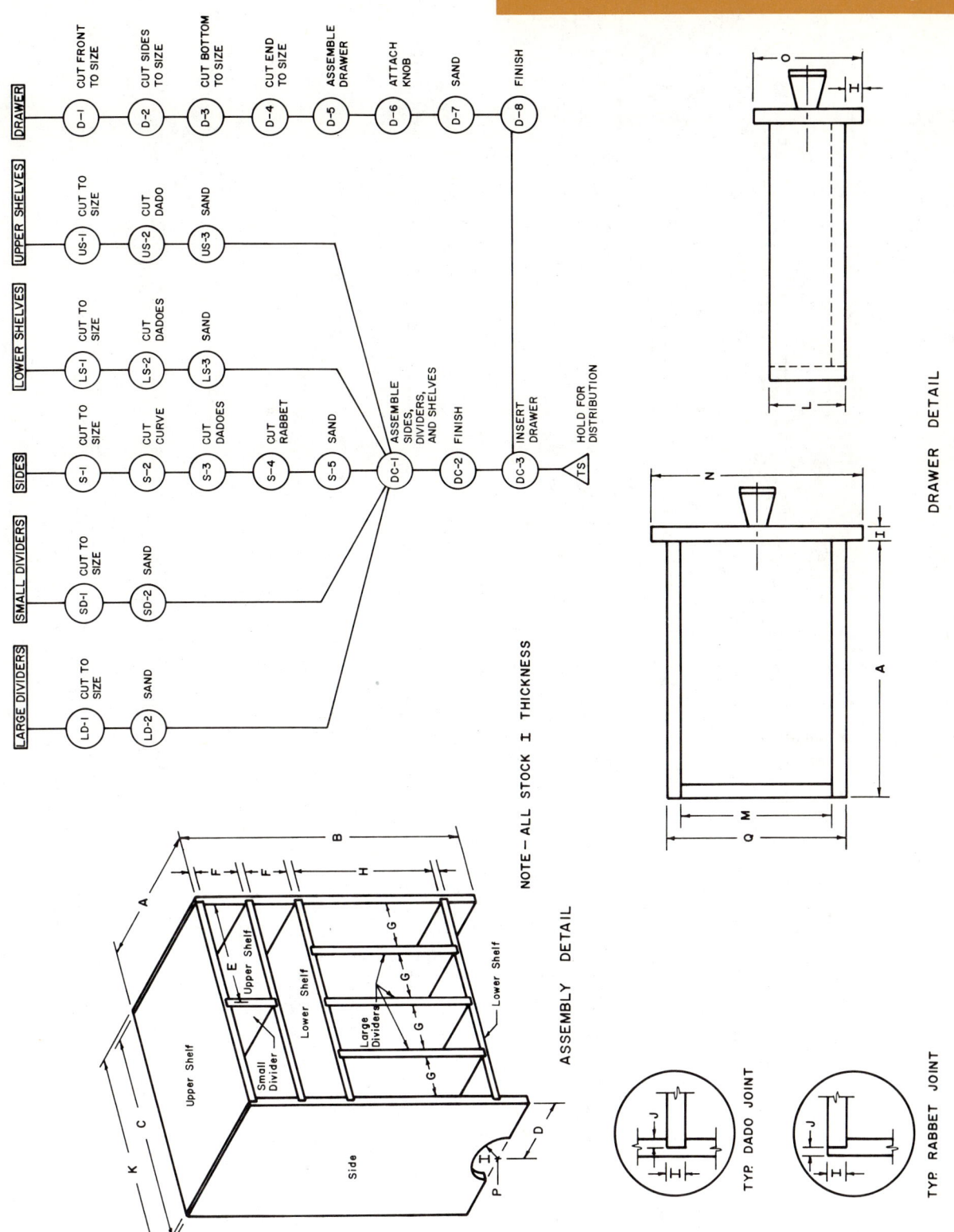

Desk Caddy (cont.)

Operation Number	PROCEDURE			
	Operation	Tools & Equipment	Topics	Notes
	LARGE DIVIDERS			
LD-1	Cut to size	Handsaw or power saw	5	
LD-2	Sand	Finishing sander, abrasive paper	15	
	SMALL DIVIDERS			
SD-1	Cut to size	Handsaw or power saw	5	
SD-2	Sand	Finishing sander, abrasive paper	15	
	SIDES			
S-1	Cut to size	Handsaw or power saw	5	
S-2	Cut curve	Band saw or coping saw or scroll saw	5	
S-3	Cut dadoes	Table saw, dado blade	5	
S-4	Cut rabbet	Table saw, dado blade	5	
S-5	Sand	Finishing sander, abrasive paper	15	
	LOWER SHELVES			
LS-1	Cut to size	Handsaw or power saw	5	
LS-2	Cut dadoes	Table saw, dado blade	5	
LS-3	Sand	Finishing sander, abrasive paper	15	
	UPPER SHELVES			
US-1	Cut to size	Handsaw or power saw	5	
US-2	Cut dado	Table saw, dado blade	5	
US-3	Sand	Finishing sander, abrasive paper	15	

Desk Caddy (cont.)

DRAWER				
D-1	Cut front to size	Handsaw or power saw	5	
D-2	Cut sides to size	Handsaw or power saw	5	
D-3	Cut bottom to size	Handsaw or power saw	5	
D-4	Cut end to size	Handsaw or power saw	5	
D-5	Assemble drawer	Hammer	3, 10, 11	Use wire brads and glue.
D-6	Attach knob	Screwdriver	3, 9	
D-7	Sand	Finishing sander abrasive paper	15	
D-8	Finish	Spray or brush	23	
ASSEMBLY				
DC-1	Assemble sides, dividers, and shelves	Mallet, clamps	3, 11, 24	Use a small amount of glue.
DC-2	Finish	Spray or brush	23	
DC-3	Insert drawer			

Mirror Shelf

DIMENSIONS		
Dimension Symbol	Metric mm	Cust. in.
A	330	13
B	150	6
C	100	4
D	34	1-1/4
E	50	2
F	70	2-7/8
G	36	1-1/2
H	65	2-9/16
I	35	1-3/8
J	116	4-1/2
K	92	3-1/2
L	40	1-9/16
M	5	3/16
N	20	3/4
O	10	3/8
P	110	4-1/4
Q	75	3
R	14	1/2
S	25	1
T	140	5-1/2
U	58	2-1/4

PARTS AND MATERIALS			
Qty.	Part	Size	Material
1	Back	N×B×A	Wood
1	Shelf	N×J×O	Wood
1	Brace	M×E×Q	Wood
1	Mirror	C×B	Mirror
3	Screws FH	#6×1-1/4"	
			Glue
			Finish

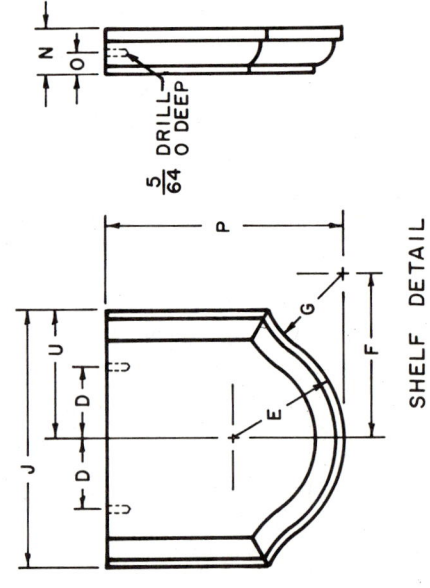

SHELF DETAIL

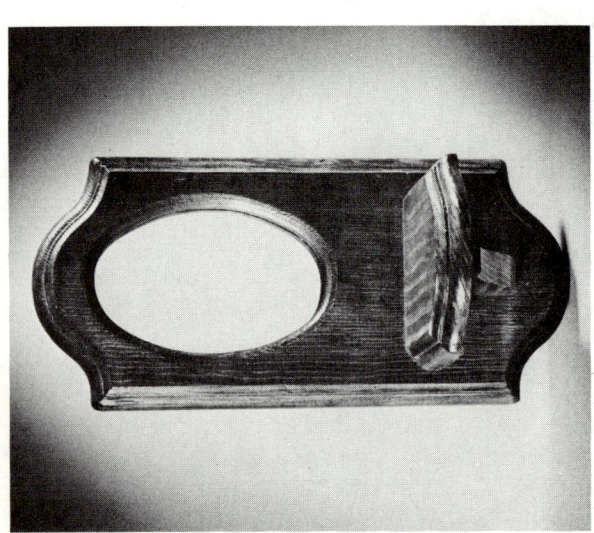

Mirror Shelf

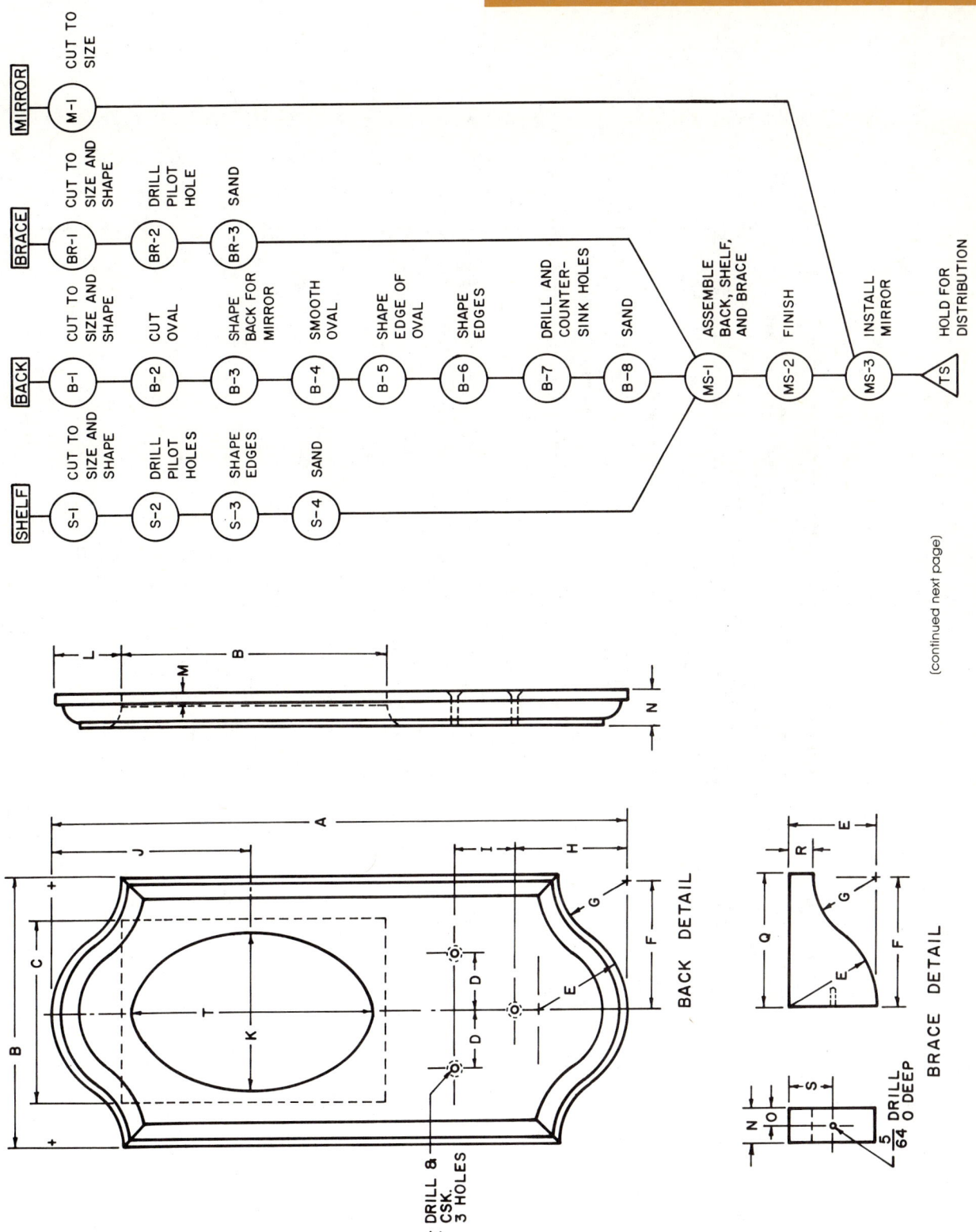

(continued next page)

Mirror Shelf (cont.)

PROCEDURE

Operation Number	Operation	Tools & Equipment	Topics	Notes
SHELF				
S-1	Cut to size and shape	Handsaw or power saw, scroll saw or band saw	5	
S-2	Drill pilot holes	Hand drill or portable electric drill, 5/64" twist drill	6	Could use back holes as a guide and drill at assembly (match drill).
S-3	Shape edges	Router, router bit	7	Use router bit of your choice.
S-4	Sand	Finishing sander, abrasive paper	15	
BACK				
B-1	Cut to size and shape	Handsaw or power saw, scroll saw or band saw	5	
B-2	Cut oval	Saber saw or scroll saw	5	
B-3	Shape back for mirror	Mallet and chisel or router and straight router bit	7	
B-4	Smooth oval	File or sanding drum or abrasive paper	15	
B-5	Shape edge of oval	Router, router bit	7	Use a rounding router bit.
B-6	Shape edges	Router, router bit	7	Use the same router bit as used for shelf.
B-7	Drill and countersink holes	Hand drill or portable electric drill or drill press, 9/64" twist drill, countersink	6	
B-8	Sand	Finishing sander, abrasive paper	15	

Mirror Shelf (cont.)

BRACE				
BR-1	Cut to size and shape	Handsaw or power saw or scroll saw or band saw	5	
BR-2	Drill pilot hole	Hand drill or portable electric drill, 5/64" twist drill	6	Could match drill at assembly.
BR-3	Sand	Sanding drum or abrasive paper	15	
MIRROR				
M-1	Cut to size	Glass cutter, square or straight edge	2, 17	Practice on a scrap piece first.
ASSEMBLY				
MS-1	Assemble back, shelf, and brace	Screwdriver	3, 10, 11	Use glue and screws.
MS-2	Finish	Spray or brush	23	
MS-3	Install mirror	Glue gun	10 or 11	Use hot melt glue. May be installed using brads.

Gumball Machine

DIMENSIONS

Dimension Symbol	Metric mm	Cust. in.
A	90	3-1/2
B	178	7
C	12	1/2
D	45	1-3/4
E	32	1-1/4
F	3	1/8
G	38	1-1/2
H	20	3/4
I	6	1/4
J	5	3/16
K	120	4-3/4
L	100	3-7/8
M	10	3/8
N	14	9/16
O	114	4-1/2
P	140	5-1/2
Q	25	1
R	68	2-5/8

PARTS AND MATERIALS

Qty.	Part	Size	Material
2	Sides	C×A×B	Wood
1	Back	C×A×P	Wood
1	Top	H×A×O	Wood
1	Bottom	C×A×A	Wood
1	Drawer	H×A×A	Wood
1	Spacer	I×A×R	Wood
1	Front	1/8"×K×L	Acrylic Plastic
1	Deflector	A×D	Sheet Metal (26 gage)
1	Peg	1/4" dia. ×C Wood Dowel	
1	Knob		
2	Screws PH	#4 × 3/8"	
12	Wire Brads	18ga. × 1"	
			Glue
			Finish

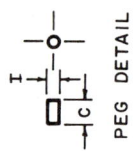
PEG DETAIL

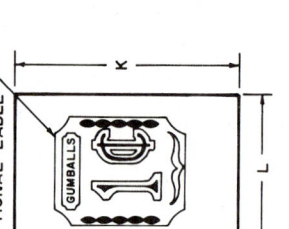
FRONT DETAIL

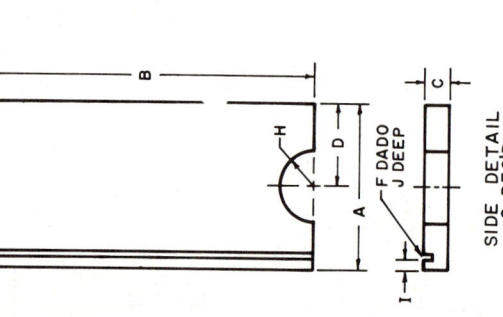
BACK DETAIL

SIDE DETAIL
2 REQ'D.

Gumball Machine

LID: L-1 CUT TOP AND SPACER TO SIZE → L-2 SAND → L-3 ASSEMBLE SPACER TO TOP → L-4 FINISH

FRONT: F-1 CUT TO SIZE → F-2 SMOOTH EDGES

DEFLECTOR: D-1 CUT TO SIZE → D-2 PUNCH HOLES → D-3 BEND TO SHAPE

DRAWER: DR-1 CUT TO SIZE → DR-2 CUT SLOT → DR-3 BORE HOLE → DR-4 SAND → DR-5 FINISH → DR-6 INSTALL KNOB

PEG: P-1 CUT TO LENGTH → P-2 SAND

BOTTOM: BO-1 CUT TO SIZE → BO-2 DRILL HOLE → BO-3 CUT CURVE → BO-4 SAND → BO-5 FINISH

BACK: B-1 CUT TO SIZE → B-2 CUT RABBET → B-3 SAND → B-4 FINISH

SIDES: S-1 CUT TO SIZE → S-2 CUT GROOVE → S-3 CUT CURVE → S-4 SAND → S-5 FINISH

GM-1 ASSEMBLE SIDES, BOTTOM, AND BACK → GM-2 INSERT PEG IN BOTTOM → GM-3 INSTALL DRAWER → GM-4 INSTALL DEFLECTOR → GM-5 INSTALL FRONT → GM-6 ASSEMBLE LID TO GUMBALL MACHINE → TS HOLD FOR DISTRIBUTION

(continued next page)

LID DETAIL

DEFLECTOR LAYOUT — $\frac{1}{8}$ PUNCH 2 HOLES, BEND LINE

ASSEMBLY SECTION — LID, DEFLECTOR 120°, DRAWER, BOTTOM, PEG

BOTTOM DETAIL — $\frac{1}{4}$ DRILL 1 DEEP

DRAWER DETAIL — $\frac{1}{2}$ SLOT 1 DEEP, $\frac{3}{4}$ DIA.

Gumball Machine (cont.)

PROCEDURE

Operation Number	Operation	Tools & Equipment	Topics	Notes
SIDES				
S-1	Cut to size	Handsaw or power saw	5	
S-2	Cut groove	Table saw, dado blade	5	
S-3	Cut curve	Scroll or band saw or coping saw	5	
S-4	Sand	Sanding block	15	
S-5	Finish	Spray or brush	23	
BACK				
B-1	Cut to size	Handsaw or power saw	5	
B-2	Cut rabbet	Table saw, dado blade	5	
B-3	Sand	Finishing sander, abrasive paper	15	
B-4	Finish	Spray or brush	23	
BOTTOM				
BO-1	Cut to size	Table saw or radial arm saw or miter saw	5	
BO-2	Drill hole	Drill press, 1/4" twist drill	6, 25	A drilling fixture may be used.
BO-3	Cut curve	Scroll saw or band saw or coping saw	5	
BO-4	Sand	Finishing sander, abrasive paper	15	
BO-5	Finish	Spray or brush	23	
PEG				
P-1	Cut to length	Handsaw or miter saw	5	
P-2	Sand	Abrasive paper	15	Sand ends only.
DRAWER				
DR-1	Cut to size	Handsaw or power saw	5	
DR-2	Cut slot	Router, 1/2" straight router bit	7, 25	Mount router in router table or make a fixture.
DR-3	Bore hole	Drill press, 3/4" spade bit	6, 25	A drilling fixture may be used.
DR-4	Sand	Finishing sander, abrasive paper	15	
DR-5	Finish	Spray or brush	23	
DR-6	Install knob			

Gumball Machine (cont.)

DEFLECTOR			
D-1	Cut to size	Tin snips or squaring shears or aviation snips	4
D-2	Punch holes	1/8" hand punch	4
D-3	Bend to shape	Hand seamer or bar folder	19
FRONT			
F-1	Cut to size	Handsaw or power saw	5	Use a fine tooth blade.
F-2	Smooth edges	File and abrasive paper	15	A design could be printed on the plastic.
LID			
L-1	Cut top and spacer to size	Handsaw or power saw	5
L-2	Sand	Finishing sander, abrasive paper	15
L-3	Assemble spacer to top	Hammer	3, 10, 11	Use glue and wire brads.
L-4	Finish	Spray or brush	23
ASSEMBLY			
GM-1	Assemble sides, bottom, and back	Hammer	3, 10, 11, 25	Use glue and wire brads. An assembly fixture may be used.
GM-2	Insert peg in bottom	Hammer or mallet	3	Peg should fit tightly.
GM-3	Install drawer		
GM-4	Install deflector	Screwdriver	3, 9
GM-5	Install front		
GM-6	Assemble lid to gumball machine		

Name Badge

PARTS AND MATERIALS

Qty.	Part	Size	Material
1	Badge	A × A	Sheet Metal (28 gage)
			Paint
			Label

DIMENSIONS

Dimension Symbol	Metric mm	Cust. in.
A	75	3
B	20	3/4
C	12	1/2
D	6	1/4

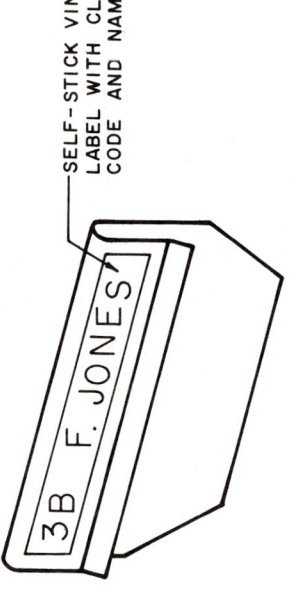

SELF-STICK VINYL LABEL WITH CLASS CODE AND NAME

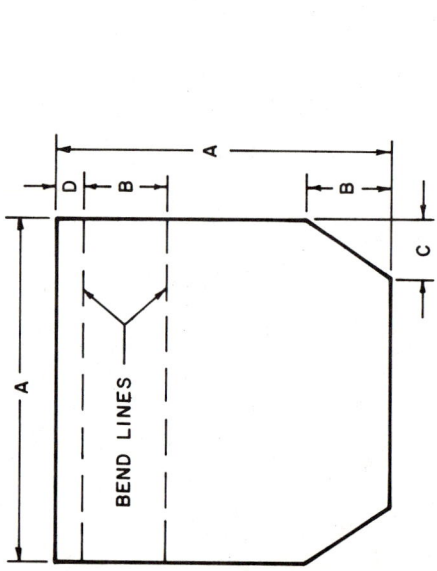

BADGE LAYOUT

Name Badge

PROCEDURE

NAME BADGE

Operation Number	Operation	Tools & Equipment	Topics	Notes
NB-1	Cut to size and shape	Tin snips or squaring shears or aviation snips	4	
NB-2	Smooth edges	File	15	
NB-3	Bend to shape	Hand seamer or box and pan brake	19	
NB-4	Paint	Spray	23	
NB-5	Label	Self-stick label tape		Labels may be written with felt tip pen and covered with transparent tape.

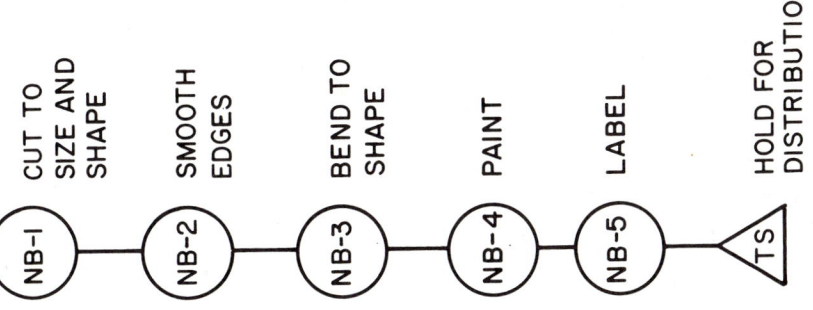

NB-1 CUT TO SIZE AND SHAPE
NB-2 SMOOTH EDGES
NB-3 BEND TO SHAPE
NB-4 PAINT
NB-5 LABEL
TS HOLD FOR DISTRIBUTION

Bike Beverage Holder

DIMENSIONS

Dimension Symbol	Metric mm	Cust. in.
A	230	9
B	45	1-3/4
C	38	1-1/2
D	64	2-1/2
E	16	5/8
F	6	1/4
G	241	9-1/2
H	12	1/2
I	50	2
J	32	1-1/4

PARTS AND MATERIALS

Qty.	Part	Size	Material
1	Frame	C×A	Sheet Metal (24 gage)
1	Band	I×G	Sheet Metal (24 gage)
1	Machine Screw RH	#8-32	
1	Nut	#8-32	
	Paint		

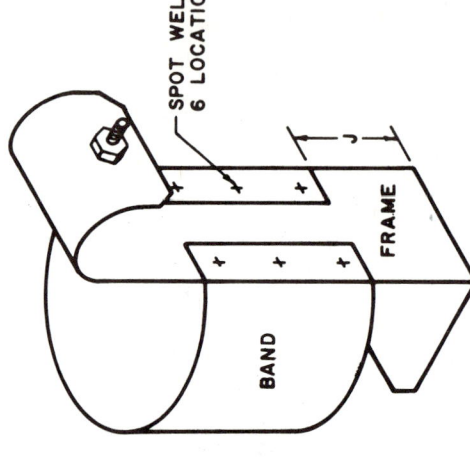

ASSEMBLY DETAIL

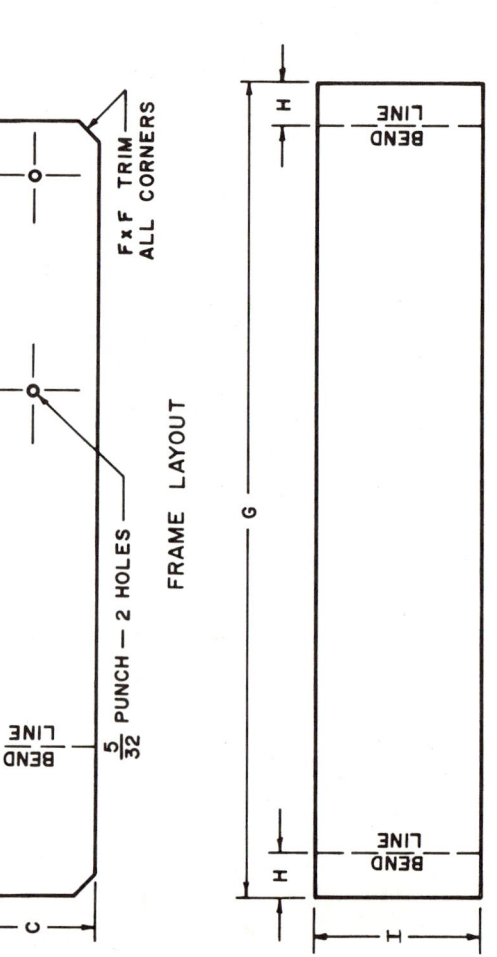

FRAME LAYOUT

BAND LAYOUT

Bike Beverage Holder

PROCEDURE

Operation Number	Operation	Tools & Equipment	Topics	Notes
BAND				
B-1	Cut to size	Squaring shears, tin snips or aviation snips	4	
B-2	Form curve	Slip roll former or mallet and stake	3, 19	Form metal around stake.
B-3	Bend angles	Box and pan brake or hand seamer	19	
FRAME				
F-1	Cut to size	Squaring shears or tin snips or aviation snips	4	
F-2	Punch holes	5/32" hand punch	4 or 6	Hand drill or power drill could be used.
F-3	Form curve	Box and pan brake	19	
F-4	Bend angle	Slip roll former or mallet and stake	19	
ASSEMBLY				
BBH-1	Assemble band to frame	Spot welder or soldering copper	14 or 12	Pop® rivets could be used.
BBH-2	Paint	Spray	23	
BBH-3	Install fastener	Screwdriver	3, 9	

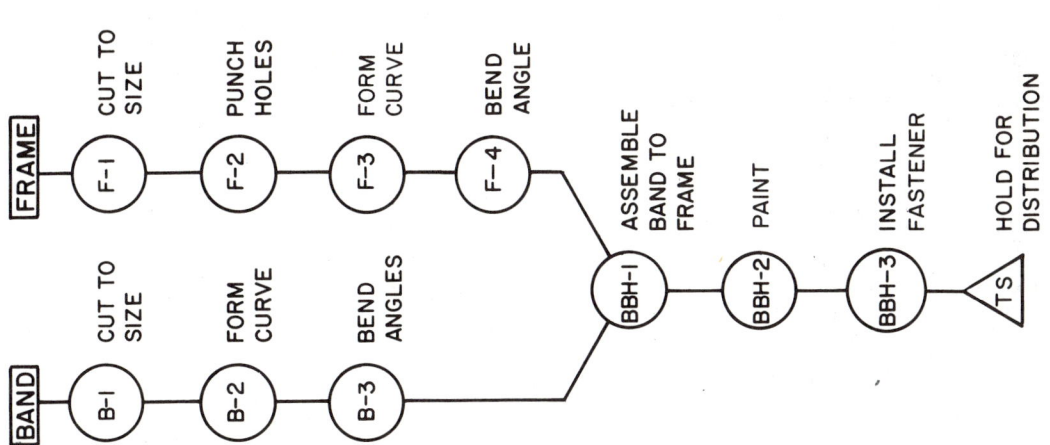

Belt Buckle

DIMENSIONS

Dimension Symbol	Metric mm	Cust. in.
A	75	3
B	50	2
C	25	1
D	6	1/4
E	62	2-1/2
F	38	1-1/2
G	12	1/2
H	19	3/4
I	16	5/8
J	10	3/8
K	95	3-3/4
L	70	2-3/4

PARTS AND MATERIALS

Qty.	Part	Size	Material
1	Insert	E × F	Copper Metal Sheet (16 gage)
1	Hook	1/8" dia. × K	Metal Rod
1	Loop	1/8" dia. × A	Metal Rod
			Brass Rod
			Glass Powder
2	Sides	3/4" × 3/4" × K	Steel
2	End	3/4" × 3/4" × L	Steel

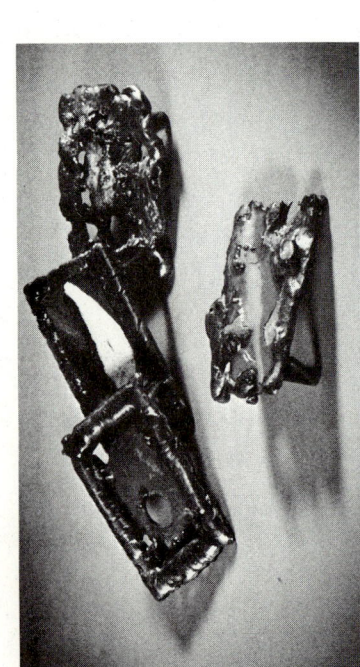

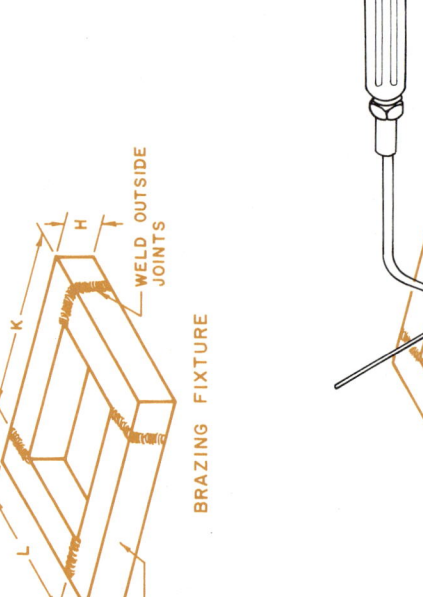

BUCKLE DETAIL

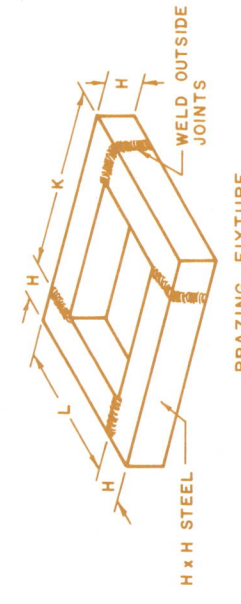

BRAZING FIXTURE

WELD OUTSIDE JOINTS

H × H STEEL

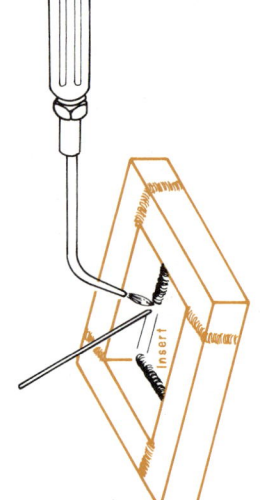

FORMING BRASS FRAME

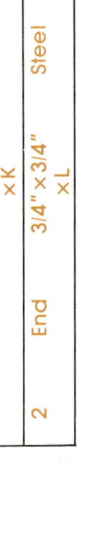

INSERT DETAIL

HOOK DETAIL

LOOP DETAIL

ASSEMBLY DETAIL

Belt Buckle

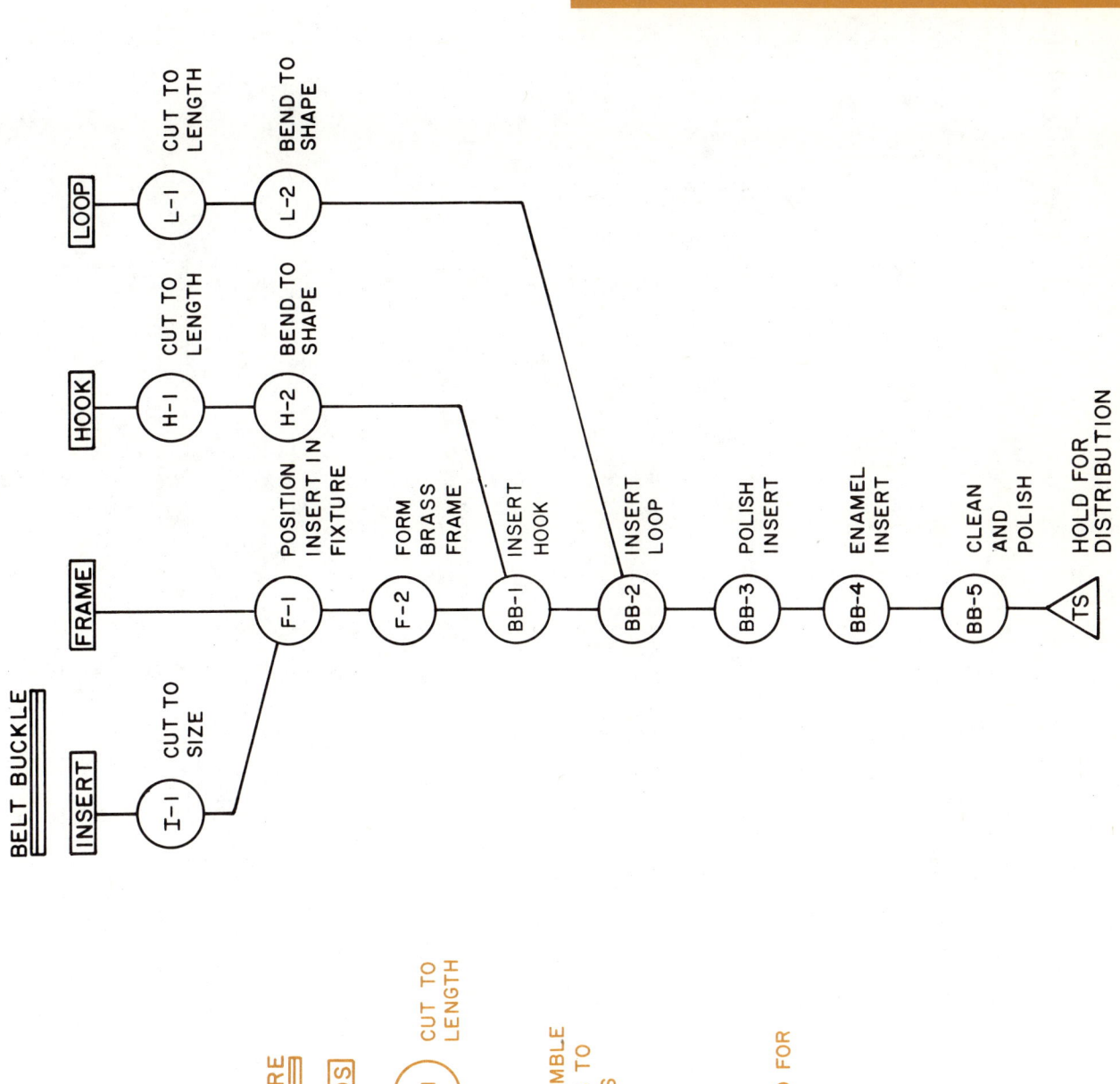

Belt Buckle (cont.)

PROCEDURE

Belt Buckle Fixture

Operation Number	Operation	Tools & Equipment	Topics	Notes
SIDES				
S-1	Cut to length	Hacksaw or metal-cutting band saw	5	Cut two pieces.
ENDS				
E-1	Cut to length	Hacksaw or metal-cutting band saw	5	Cut two pieces.
ASSEMBLY				
FX-1	Assemble by welding ends to sides	Oxyacetylene torch	13	Be sure that fixture is square.

Belt Buckle

Operation Number	Operation	Tools & Equipment	Topics	Notes
INSERT				
I-1	Cut to size	Tin snips or squaring shears	4	
FRAME				
F-1	Position insert in fixture			Center insert in fixture.
F-2	Form brass frame	Oxyacetylene torch, brazing fixture	13	See drawing.

Belt Buckle (cont.)

HOOK				
H-1	Cut to length	Wire cutters or bolt cutters or hacksaw	3 or 4 or 5	
H-2	Bend to shape	Vise, hammer or pliers or bending jig	3, 19	
LOOP				
L-1	Cut to length	Wire cutters or bolt cutters or hacksaw	3 or 4 or 5	
L-2	Bend to shape	Vise, hammer or pliers or bending jig	3 or 19	
ASSEMBLY				
BB-1	Insert hook	Oxyacetylene torch, pliers	3, 13	Melt small puddle in frame and insert hook.
BB-2	Insert loop	Oxyacetylene torch, pliers	3, 13	Use the same procedure as for hook. Insert only one end at a time.
BB-3	Polish insert		15	Use fine steel wool.
BB-4	Enamel insert	Kiln or oxyacetylene torch	23 or 13	
BB-5	Clean and polish	Wire brush, buffer	15	Clean and polish frame and insert. Avoid scratching enameled parts.

Charcoal Tongs (Easy)

PARTS AND MATERIALS

Qty.	Part	Size	Material
1	Handle 1	1/8" × 1/2" × G	Wrought Iron
1	Handle 2	1/8" × 1/2" × H	Wrought Iron
1	Rivet RH	1/8" × 1/2"	
			Paint

DIMENSIONS

Dimension Symbol	Metric mm	Cust. in.
A	250	9-3/4
B	230	9
C	95	3-3/4
D	70	2-3/4
E	28	1-1/8
F	20	3/4
G	432	17
H	482	19
I	25	1

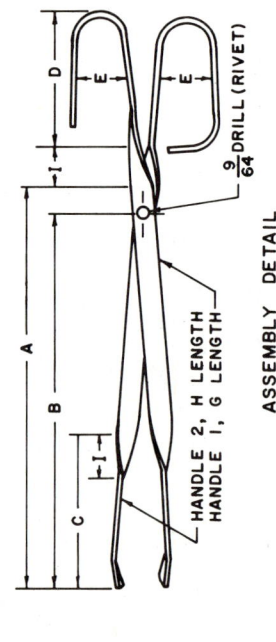

TWIST DETAIL — HANDLE END UP. TWISTING SPACE SHOULD BE AT LEAST TWICE THE WIDTH OF METAL.

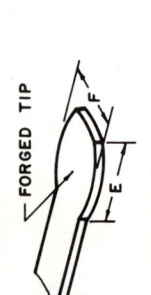

ASSEMBLY DETAIL — 9/64 DRILL (RIVET), HANDLE 2, H LENGTH; HANDLE 1, G LENGTH

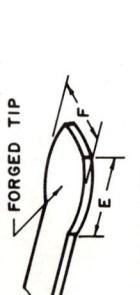

TIP DETAIL — FORGED TIP

Charcoal Tongs (Easy)

PROCEDURE

Operation Number	Operation	Tools & Equipment	Topics	Notes
CT-1	Cut handles to length	Hacksaw or throatless shears	5	Cut handle 1 to G length. Cut handle 2 to H length.
CT-2	Forge tips to rough shape	Oxyacetylene torch or furnace, hammer, anvil	13 or 3, 19, 22	
CT-3	File or grind to finish shape	File or grinder	15 or 16	
CT-4	Bend handles to shape	Bending jig	19	Notice difference in handle shapes.
CT-5	Clamp and then drill hole	Clamp or drill press vise, drill press, 9/64" twist drill	6, 24	Align handles and drill both at the same time.
CT-6	Assemble by riveting handles together	Hammer, rivet set	3, 10	
CT-7	Twist tongs to line up properly	Vise, wrench	3, 19	See drawing.
CT-8	Paint	Spray	23	

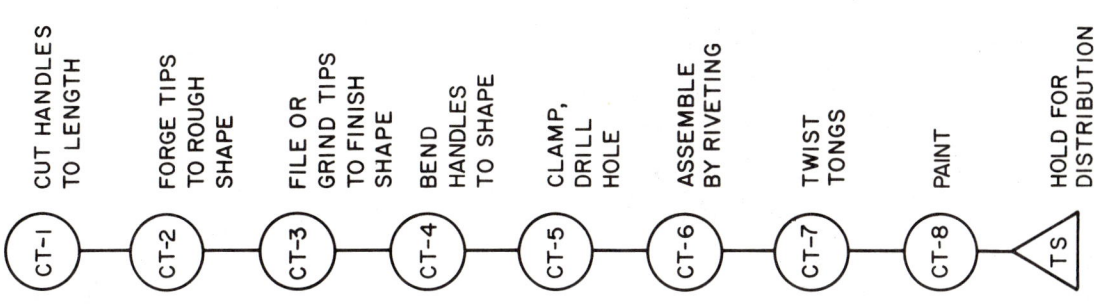

Charcoal Tongs (Difficult)

PARTS AND MATERIALS

Qty.	Part	Size	Material
2	Tongs	1/8" × 3/4" × F	Wrought Iron
2	Handles	1/4" dia. × E	Wrought Iron
1	Rivet	1/8" dia. × 3/8"	
			Paint

DIMENSIONS

Dimension Symbol	Metric mm	Cust. in.
A	266	10-1/2
B	250	9-3/4
C	90	3-1/2
D	28	1-1/8
E	230	9
F	300	12
G	38	1-1/2

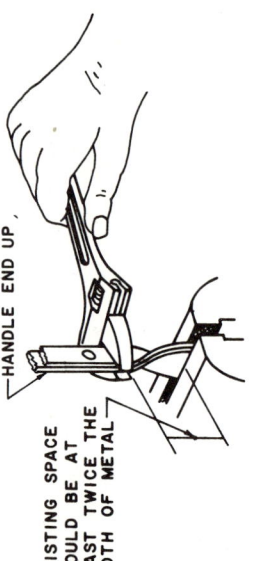

TWIST DETAIL

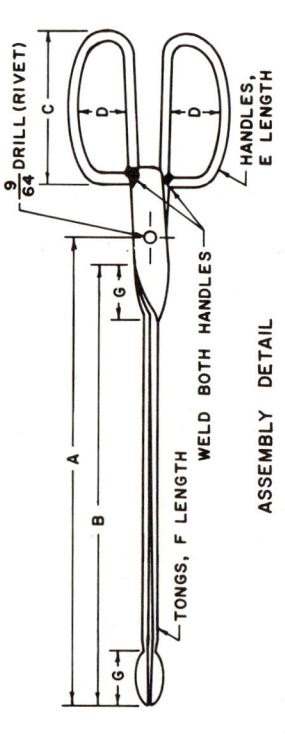

ASSEMBLY DETAIL

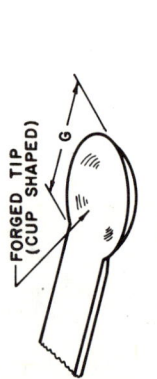

TIP DETAIL

Charcoal Tongs (Difficult)

PROCEDURE

Operation Number	Operation	Tools & Equipment	Topics	Notes
TONGS				
T-1	Cut to lengths	Hacksaw or throatless shear	5	
T-2	Forge tips to rough shape	Oxyacetylene torch or furnace, hammer, anvil	13 or 3, 19, 22	Use ball peen end of hammer.
T-3	File or grind to finish shape	File or grinder	15 or 16	
T-4	Clamp and then drill hole	Clamp or drill press vise, drill press, 9/64" twist drill	6, 24	Align tongs and drill both at the same time.
T-5	Assemble by riveting tongs together	Hammer, rivet set	3, 10	
T-6	Twist tongs to line up properly	Vise, wrench	3	See drawing.
HANDLES				
H-1	Cut to length	Hacksaw	5	
H-2	Bend to shape	Vise, bending jig	19	
ASSEMBLY				
CT-1	Weld handles to tongs	Oxyacetylene torch	13	Could be brazed.
CT-2	Paint	Spray	23	

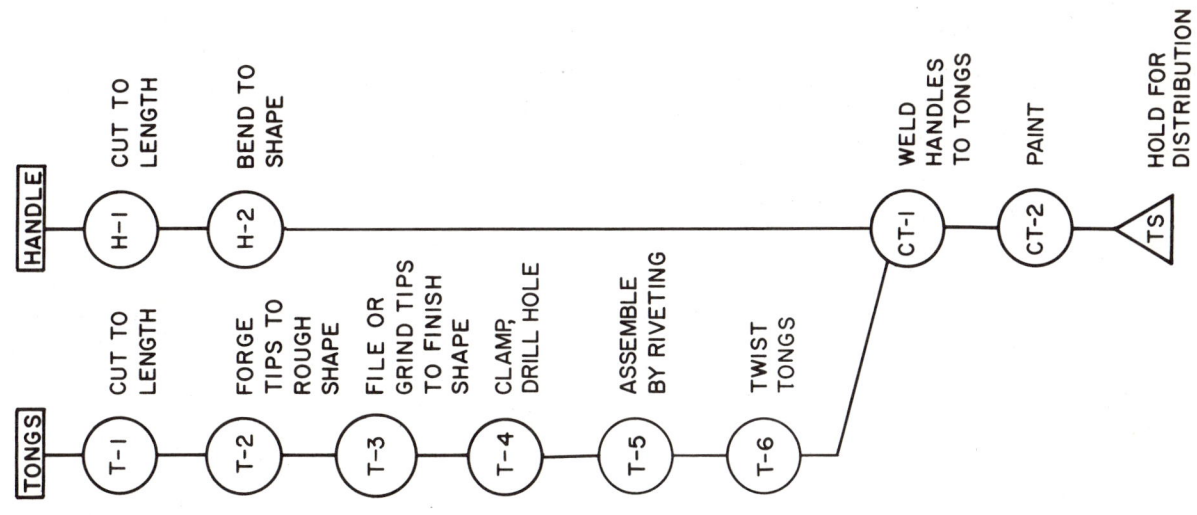

Planter

PARTS AND MATERIALS

Qty.	Part	Size	Material
3	Legs	1/8" × 1/2" × D	Wrought Iron
1	Ring	1/8" × 1/2" × A	Wrought Iron
3	Rivets RH	1/8" dia. × 3/8"	
			Paint

DIMENSIONS

Dimension Symbol	Metric mm	Cust. in.
A	243	9-3/4
B	25	1
C	81	3-1/4
D	406	16
E	12	1/2

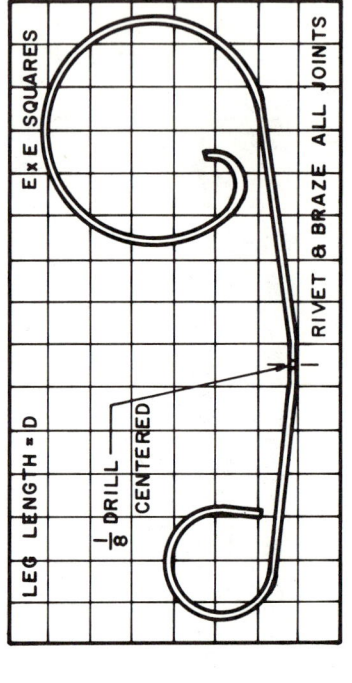

LEG LENGTH = D   E × E SQUARES
1/8 DRILL CENTERED
RIVET & BRAZE ALL JOINTS
LEG DETAIL
3 REQ'D.

1/8 DRILL 3 HOLES
RING LAYOUT

Planter

PROCEDURE

Operation Number	Operation	Tools & Equipment	Topics	Notes
LEGS				
L-1	Cut to length	Throatless shear or hacksaw	4 or 5	
L-2	File ends smooth	File or grinder	15 or 16	
L-3	Bend scroll shapes	Bending jig or Universal® bender or Di-acro® bender	19	
L-4	Drill holes	Drill press and drill press vise or portable electric drill, 1/8" twist drill	2, 6	Center punch before drilling.
L-5	Bend angle	Vise	19	A bending jig could be made. Bend may be made by hand.
RING				
R-1	Cut to length	Throatless shear or hacksaw	4 or 5	
R-2	Drill holes	Drill press and drill press vise or portable electric drill, 1/8" twist drill	2, 6	Center punch before drilling.
R-3	Bend to shape	Bending jig or Universal® bender or Di-acro® bender	19	
R-4	Braze ends together	Oxyacetylene torch	13	May be welded.
ASSEMBLY				
P-1	Rivet legs to ring	Hammer, rivet set, anvil	10, 13	May be welded.
P-2	Braze legs to ring	Oxyacetylene torch	13	Make sure legs are lined up properly.
P-3	Paint	Spray or brush	23	

RING: R-1 CUT TO LENGTH — R-2 DRILL HOLES — R-3 BEND TO SHAPE — R-4 BRAZE ENDS TOGETHER

LEGS: L-1 CUT TO LENGTH — L-2 FILE ENDS SMOOTH — L-3 BEND SCROLL SHAPES — L-4 DRILL HOLE — L-5 BEND ANGLE

P-1 RIVET LEGS TO RING — P-2 BRAZE LEGS TO RING — P-3 PAINT — TS HOLD FOR DISTRIBUTION

Screwdriver

PARTS AND MATERIALS

Qty.	Part	Size	Material
1	Handle	3/4" dia. × H	Metal Rod
1	Blade	1/4" dia. × G	Metal Rod

DIMENSIONS

Dimension Symbol	Metric mm	Cust. in.
A	20	3/4
B	95	3-3/4
C	3	1/8
D	12	1/2
E	6	1/4
F	25	1
G	146	5-3/4
H	112	4-1/2

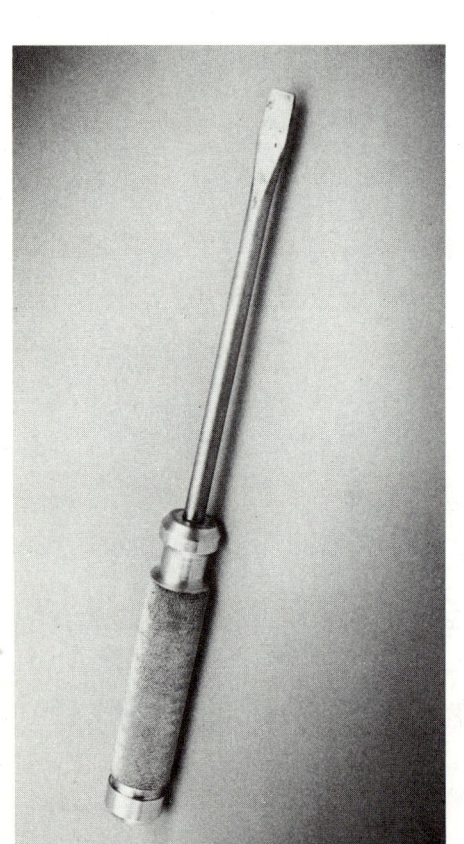

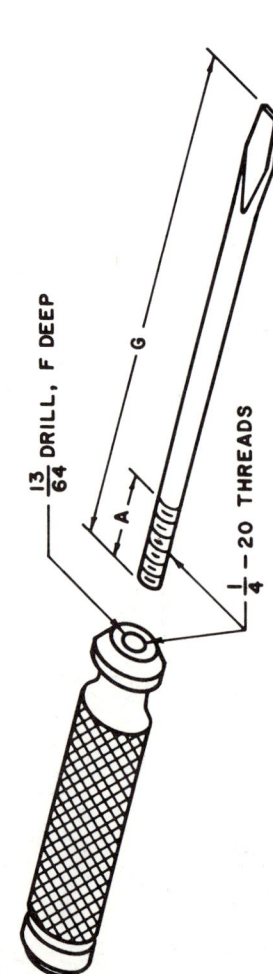

13/64 DRILL, F DEEP
1/4 - 20 THREADS

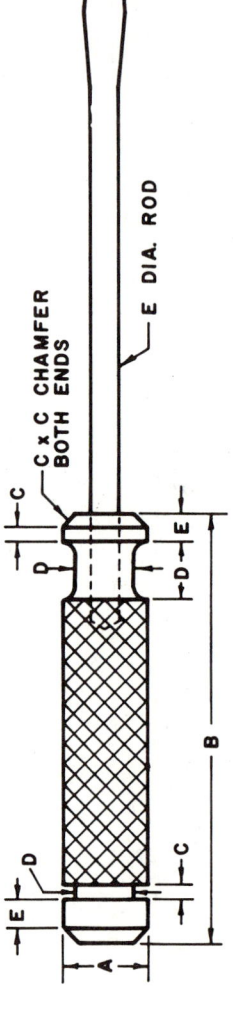

C × C CHAMFER BOTH ENDS
E DIA. ROD

Screwdriver

PROCEDURE

Operation Number	Operation	Tools & Equipment	Topics	Notes
HANDLE				
H-1	Cut to rough length	Hacksaw or metal-cutting band saw	5	
H-2	Face and center-drill each end	Metalworking lathe, facing tool, center drill	8	Use a 3-jaw chuck.
H-3	Turn to shape	Metalworking lathe, cutting tools	8	
H-4	Knurl	Metalworking lathe, knurling tool	8	Use back gears.
H-5	Polish	File or abrasive cloth	15	
H-6	Drill hole	Metalworking lathe, 13/64" twist drill	8	Mount the handle in a 3-jaw chuck and the twist drill in the tailstock.
H-7	Tap threads	Tap wrench, 1/4-20 tap	9	Leave handle mounted in chuck to tap threads. Lathe should not be turned on.
H-8	Cut to finish length and face end	Metalworking lathe, cutoff tool, facing tool	8	Mount in a 3-jaw chuck.
BLADE				
B-1	Cut to length	Hacksaw	5	
B-2	Forge to rough shape	Forging furnace or oxyacetylene torch, hammer, pliers	3, 13, 22	
B-3	File or grind to shape	File or grinder	15 or 16	
B-4	Heat treat tip	Heat treating furnace or oxyacetylene torch	13 or 22	Use oil or water for quenching.
B-5	Cut threads	Die stock, 1/4-20 die	9	
B-6	Clean and polish	Abrasive cloth or steel wool	15	
ASSEMBLY				
SD-1	Assemble blade to handle		9	To hold blade in handle, deform last two threads in blade.

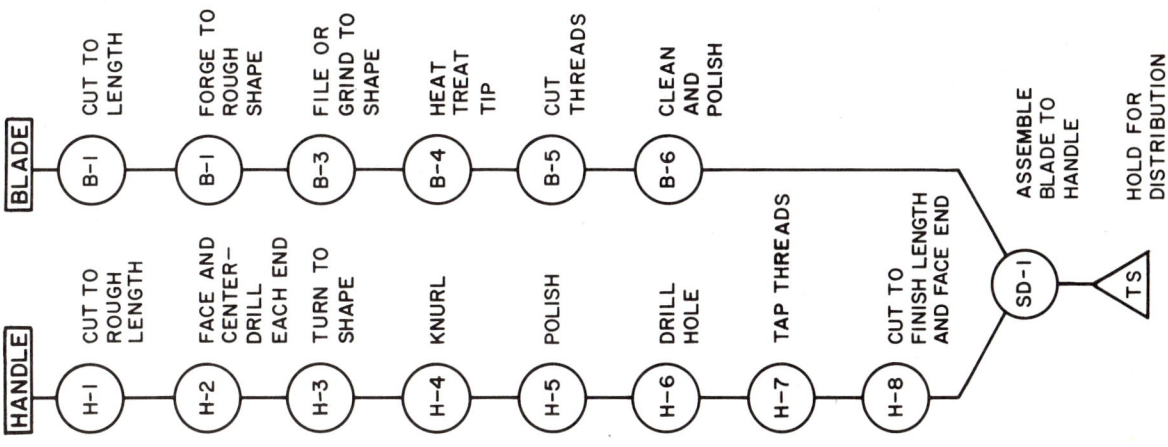

Toolbox

DIMENSIONS

Dimension Symbol	Metric mm	Cust. in.
A	355	14
B	238	9-3/8
C	80	3-1/8
D	22	7/8
E	10	3/8
F	119	4-11/16
G	105	4-1/4
H	33	1-1/4
I	286	11-1/4
J	104	4-1/8
K	67	2-5/8
L	25	1
M	38	1-1/2
N	6	1/4
O	12	1/2
P	150	5-7/8
Q	89	3-1/2
R	36	1-7/16
S	35	1-3/8
T	3	1/8
U	120	4-3/4
V	20	3/4
W	137	5-1/2
X	193	7-3/4

PARTS AND MATERIALS

Qty.	Part	Size	Material
1	Lid	A×B	Sheet Metal (26 gage)
1	Bottom	A×I	Sheet Metal (26 gage)
2	Ends	P×Q	Sheet Metal (26 gage)
1	Handle	1/8" × 1/2" × X	Metal
14	Rivets	1/8" dia. × 1/8"	
2	Hinges	1" × 1"	
1	Hasp Set	1" × 2"	
			Paint

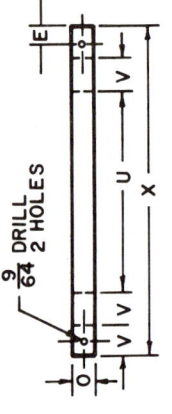

HANDLE LAYOUT

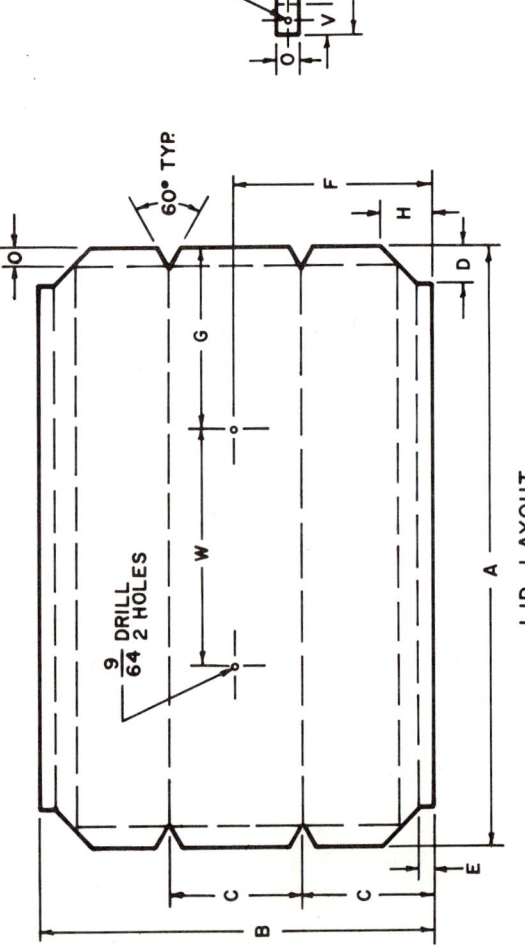

LID LAYOUT

Toolbox

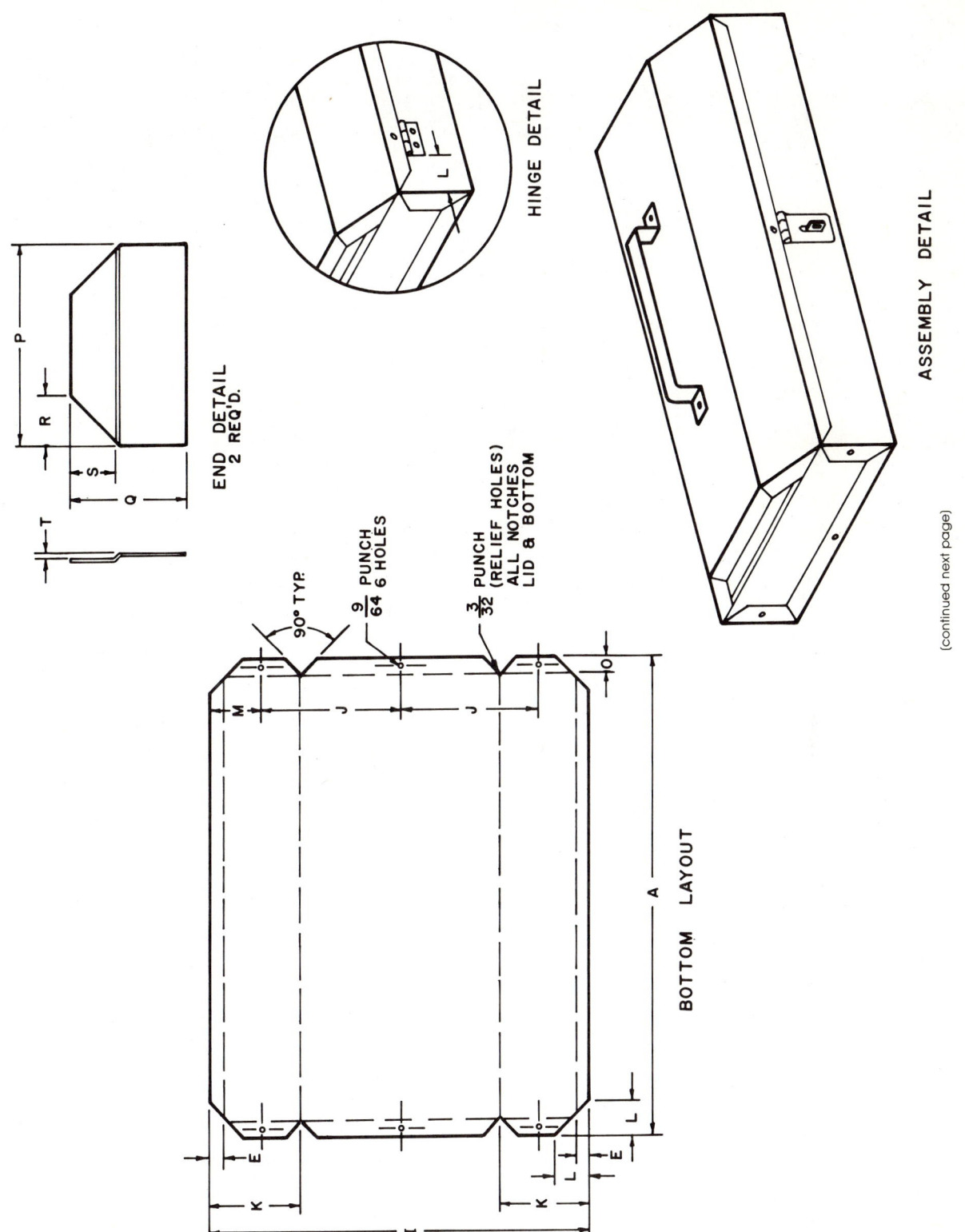

(continued next page)

Toolbox (cont.)

PROCEDURE

Operation Number	Operation	Tools & Equipment	Topics	Notes
HANDLE				
H-1	Cut to length	Hacksaw	5	
H-2	File ends	File or grinder	15 or 16	
H-3	Drill holes	Drill press and drill press vise or portable electric drill, 9/64" twist drill	2, 6	Center punch before drilling.
H-4	Bend to shape	Vise, hammer	3, 19	
H-5	Paint	Spray or brush	23	
LID				
L-1	Cut to size	Tin snips or squaring shears	4	
L-2	Punch and drill holes	3/32" hand punch (relief holes), 9/64" twist drill (handle holes)	4, 6	Drill handle holes with a hand drill or portable electric drill or drill press.
L-3	Cut notches	Tin snips or hand notcher	4	
L-4	Bend and flatten hems	Bar folder or box and pan brake, hammer or mallet and vise	3, 19	
L-5	Bend ends 90°	Bar folder or box and pan brake	19	
L-6	Bend to shape	Box and pan brake	19	Angle is 42°.
L-7	Attach hinges and hasp to top	Riveting tool	10 or 9	Screws could be used.
L-8	Assemble handle to top	Riveting tool	10 or 9	Screws could be used.
ENDS				
E-1	Cut to size and shape	Tin snips or squaring shears	4	
E-2	Bend offset	Bar folder or box and pan brake	19	

ENDS: E-1 CUT TO SIZE AND SHAPE → E-2 BEND OFFSET

BOTTOM: B-1 CUT TO SIZE → B-2 PUNCH HOLES → B-3 CUT NOTCHES → B-4 BEND AND FLATTEN HEMS → B-5 BEND ENDS 90° → B-6 BEND TO SHAPE → B-7 ASSEMBLE ENDS TO BOTTOM

LID: L-1 CUT TO SIZE → L-2 PUNCH AND DRILL HOLES → L-3 CUT NOTCHES → L-4 BEND AND FLATTEN HEMS → L-5 BEND ENDS 90° → L-6 BEND TO SHAPE → L-7 ATTACH HINGES AND HASP TO TOP → L-8 ASSEMBLE HANDLE TO TOP

HANDLE: H-1 CUT TO LENGTH → H-2 FILE ENDS → H-3 DRILL HOLES → H-4 BEND TO SHAPE → H-5 PAINT

TB-1 ATTACH LID TO BOTTOM → TB-2 INSTALL HASP LOOP → TB-3 PAINT → TS HOLD FOR DISTRIBUTION

		BOTTOM		
B-1	Cut to size	Tin snips or squaring shears	4	
B-2	Punch holes	9/64" hand punch and 3/32" hand punch	4 or 6	May be drilled.
B-3	Cut notches	Tin snips or hand notcher	4	
B-4	Bend and flatten hems	Bar folder or box and pan brake, hammer or mallet	3, 19	
B-5	Bend ends 90°	Bar folder or box and pan brake	19	
B-6	Bend to shape	Box and pan brake	19	
B-7	Assemble ends to bottom	Riveting tool	10 or 12 or 14	Could be spot welded or soldered.
		ASSEMBLY		
TB-1	Attach lid to bottom	Riveting tool	10 or 9	Screws could be used.
TB-2	Install hasp loop	Riveting tool	10 or 9	Screws could be used.
TB-3	Paint	Spray	23	If galvanized metal was used, product need not be painted.

Desk Lamp

DIMENSIONS

Dimension Symbol	Metric mm	Cust. in.
A	254	10
B	267	10-1/2
C	5	3/16
D	94	3-11/16
E	143	5-5/8
F	86	3-3/8
G	38	1-1/2
H	62	2-7/16
I	12	1/2
J	75	3
K	6	1/4
L	124	4-7/8
M	25	1
N	215	8-1/2
O	203	8
P	32	1-1/4
Q	44	1-3/4
R	20	3/4
S	10	3/8
T	264	10-3/8
U	112	4-1/2
V	56	2-1/4
W	88	3-1/2
X	56	2-1/4

PARTS AND MATERIALS

Qty.	Part	Size	Material
1	Base	2"×4"×A	Wood
2	Ends	E×F	Sheet Metal (26 gage)
1	Shade	A×B	Sheet Metal (26 gage)
1	Bracket	M×U	Sheet Metal (26 gage)
2	Stems	1/2" o.d. ×T	Metal Tube
1	Socket Standard	6A, 125V	
1	Switch, Single Pole		
1	Pipe Nipple with Lock Nuts	1/8" dia. ×1/2"	
1	Electrical Lamp Cord #18-2, Brown	7'	
1	Plug	73B	
3	Wire Nuts		
1	Felt	A×W	
1	Self-stick Vinyl	18"×14"	Plastic
4	Screws PH	#6×3/8"	
1	Washer (Flat)	3/16"	
2	Machine Screws RH	#6-32×3/4"	
2	Hex Nuts	#6-32	
			White Paint

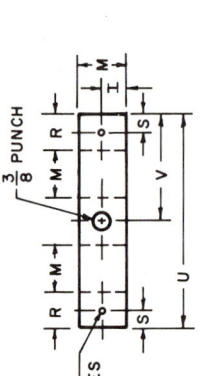

BRACKET LAYOUT

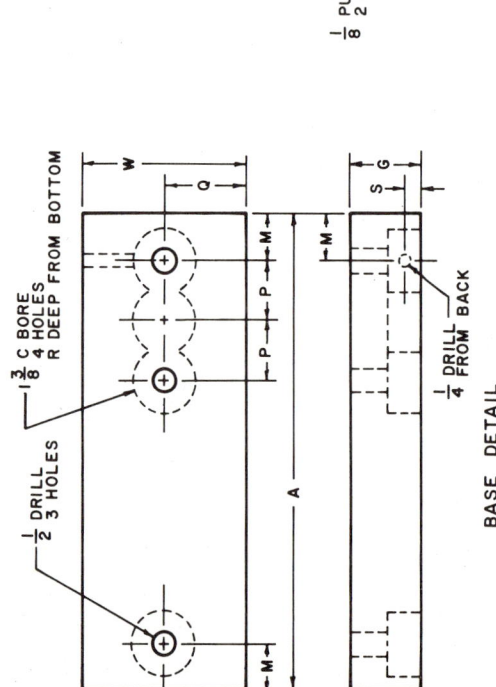

BASE DETAIL

Desk Lamp

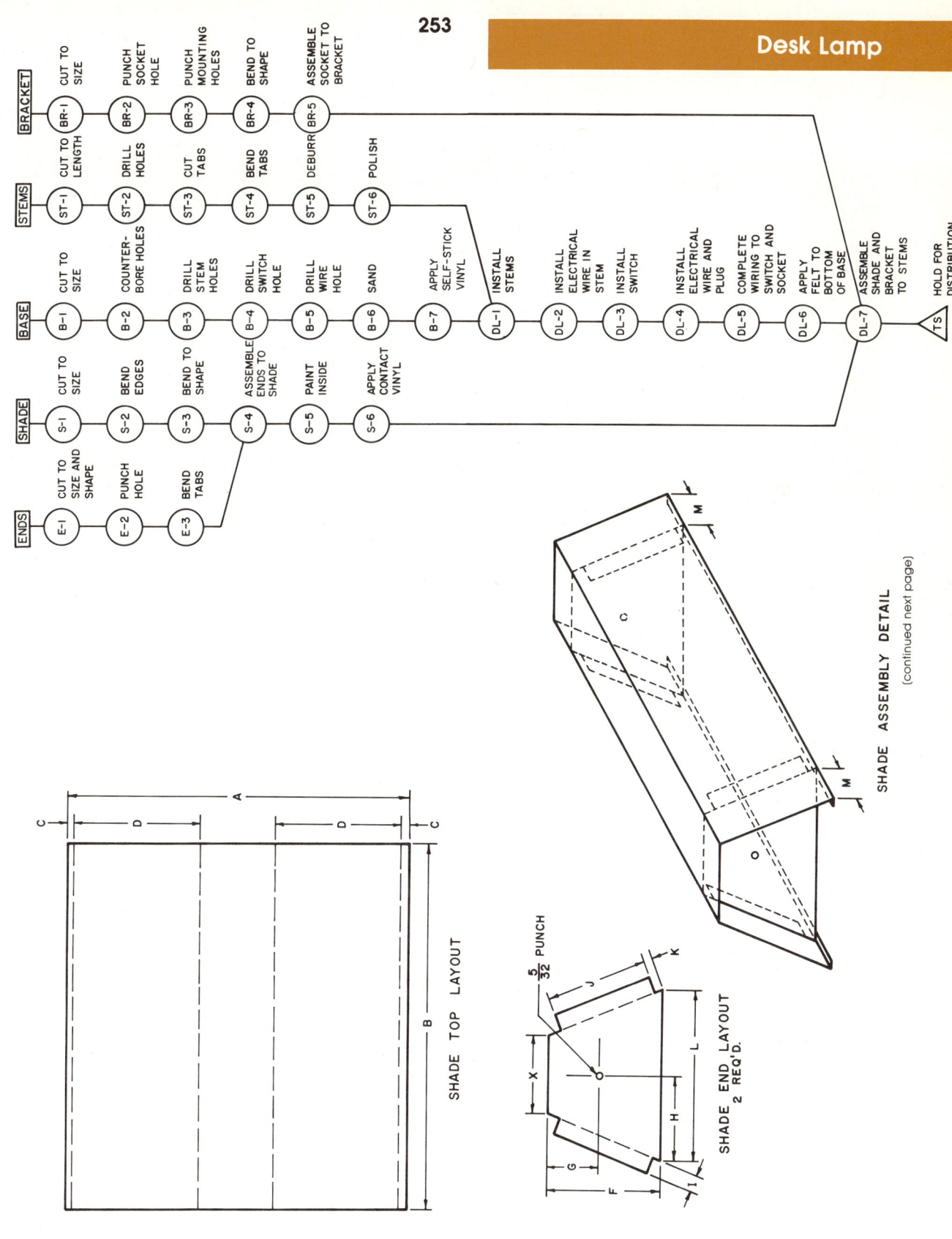

Desk Lamp (cont.)

PROCEDURE

Operation Number	Operation	Tools & Equipment	Topics	Notes
ENDS				
E-1	Cut to size and shape	Squaring shears, tin snips	4	
E-2	Punch hole	5/32" hand punch	4 or 6	Could be drilled.
E-3	Bend tabs	Bar folder or box and pan brake	19	Bend 90°.
SHADE				
S-1	Cut to size	Squaring shears, tin snips	4	
S-2	Bend edges	Bar folder or box and pan brake	19	Bend 90°.
S-3	Bend to shape	Box and pan brake	19	
S-4	Assemble ends to shade	Spot welder	14 or 10	Could be riveted.
S-5	Paint inside	Spray or brush	23	Paint entire under-side (including ends) with white paint.
S-6	Apply self-stick vinyl	Scissors, knife	4	Apply self-stick vinyl carefully to outside of shade.
BASE				
B-1	Cut to size	Handsaw or power saw	5	
B-2	Counterbore hole	Drill press or portable electric drill, 1-3/8" spade bit	6, 25	A drilling fixture may be used.
B-3	Drill stem holes	Portable electric drill or drill press, 33/64" twist drill	6, 25	A drilling fixture may be used.
B-4	Drill switch hole	Hand drill or power drill, 7/16" twist drill	6	
B-5	Drill wire hole	Hand drill or power drill, 1/4" twist drill	6	
B-6	Sand	Belt sander or finishing sander	15	
B-7	Apply self-stick vinyl	Scissors, knife		Be sure to cut holes completely through vinyl.

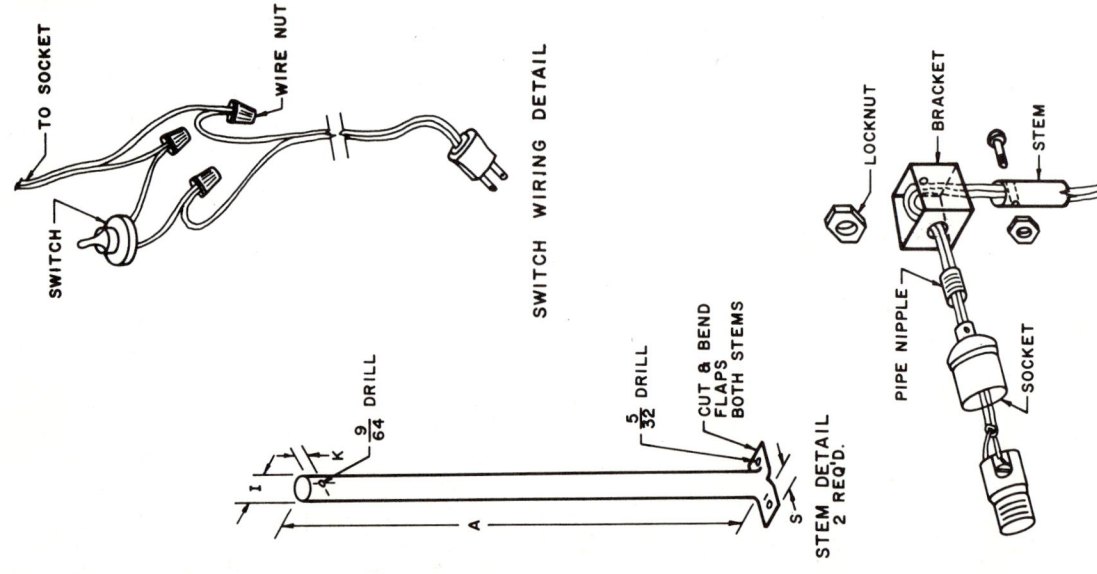

Desk Lamp (cont.)

STEMS				
ST-1	Cut to length	Tubing cutter or hacksaw	4 or 5	
ST-2	Drill holes	Drill press or portable electric drill, 9/64" twist drill	6	A drilling fixture may be used.
ST-3	Cut tabs	Hacksaw, tin snips	4, 5	Make the saw cut first. Then trim to size.
ST-4	Bend tabs	Pliers	3	
ST-5	Deburr	File	15	
ST-6	Polish			Use steel wool to give a satin finish.
BRACKET				
BR-1	Cut to size	Squaring shears, tin snips	4	
BR-2	Punch socket hole	3/8" hand punch	4 or 6	Could be drilled.
BR-3	Punch mounting holes	1/8" hand punch	4 or 6	Could be drilled.
BR-4	Bend to shape	Hand seamer	19	
BR-5	Assemble socket to bracket	Wrench, screwdriver	3, 9	
ASSEMBLY				
DL-1	Install stems	Screwdriver	3, 9	Use two #6×3/8" sheet metal screws PH for each stem.
DL-2	Install electrical wire in stem			Thread 12" piece of wire through the stem that will support light socket.
DL-3	Install switch	Pliers	3	
DL-4	Install electrical wire and plug			
DL-5	Complete wiring to switch and socket	Wire strippers	3	Strip wire ends. Do no wiring. Use wire nuts. No bare wire should show.
DL-6	Apply felt to bottom of base		11	Use glue. Four small felt circles may be used.
DL-7	Assemble shade and bracket to stems	Screwdriver	3, 9	

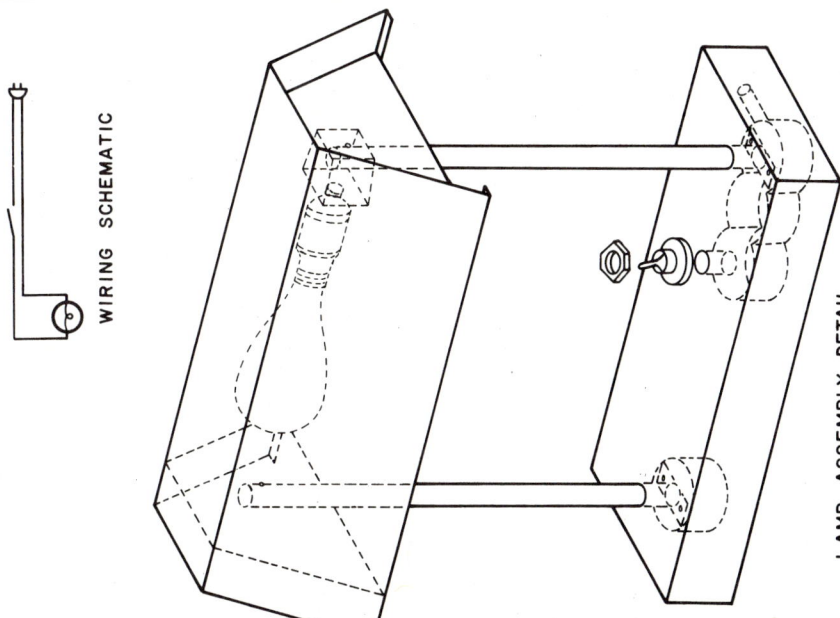

WIRING SCHEMATIC

LAMP ASSEMBLY DETAIL

Note: Teacher must inspect wiring before lamp is operated.

Calculator Stand

PARTS AND MATERIALS

Qty.	Part	Size	Material
1	Stand	1/8" × B × A	Acrylic Plastic

DIMENSIONS

Dimension Symbol	Metric mm	Cust. in.
A	230	9
B	100	4
C	25	1
D	50	2

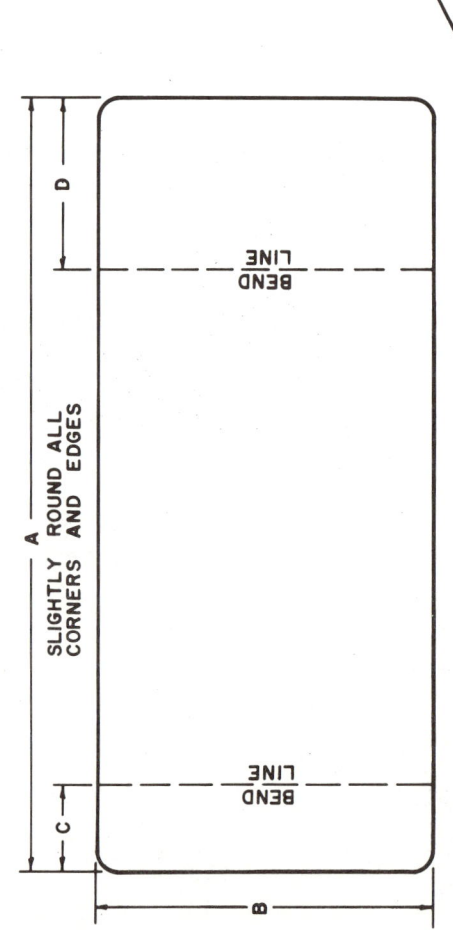

LAYOUT

SLIGHTLY ROUND ALL CORNERS AND EDGES

BEND LINE

Calculator Stand

PROCEDURE

CALCULATOR STAND

Operation Number	Operation	Tools & Equipment	Topics	Notes
CS-1	Cut to size	Table saw or band saw	5	Use a fine-tooth blade.
CS-2	Round corners	File or belt sander or disk sander	15	
CS-3	Smooth and buff edges	Abrasive paper and buffer	15	
CS-4	Bend ends	Strip heater or oven	20, 25	A bending fixture may be used.

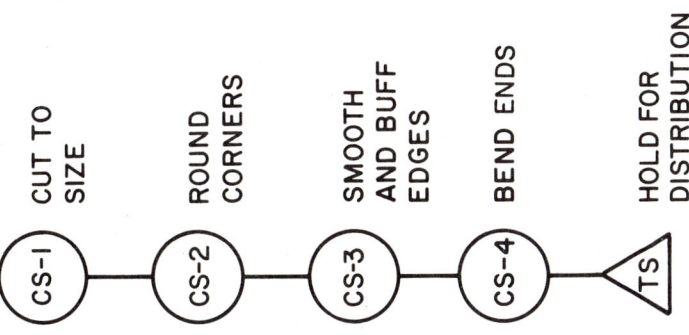

CUT TO SIZE — ROUND CORNERS — SMOOTH AND BUFF EDGES — BEND ENDS — HOLD FOR DISTRIBUTION

CS-1 — CS-2 — CS-3 — CS-4 — TS

Letter Opener

DIMENSIONS

Dimension Symbol	Metric mm	Cust. in.
A	150	6
B	75	3
C	12	1/2
D	24	1

PARTS AND MATERIALS

Qty.	Part	Size	Material
1	Blade	1/8" x A x D	Acrylic Plastic

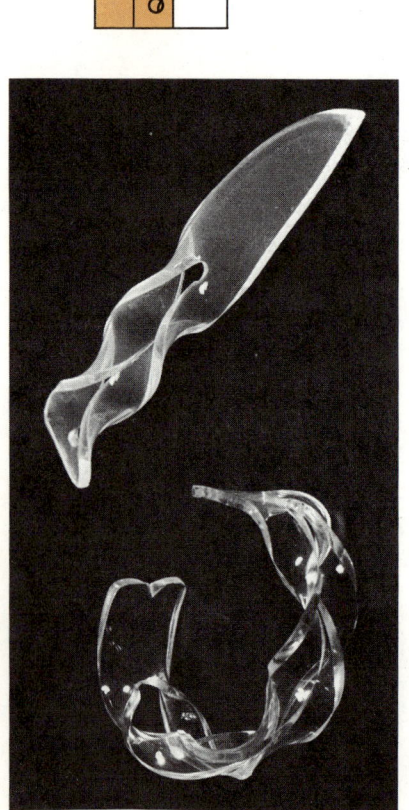

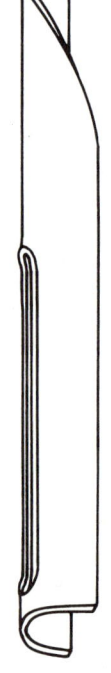

OPENER FORMED SO SLOT EDGES MAY BE BUFFED

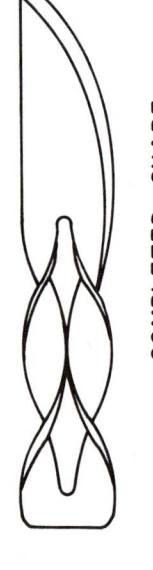

COMPLETED SHAPE

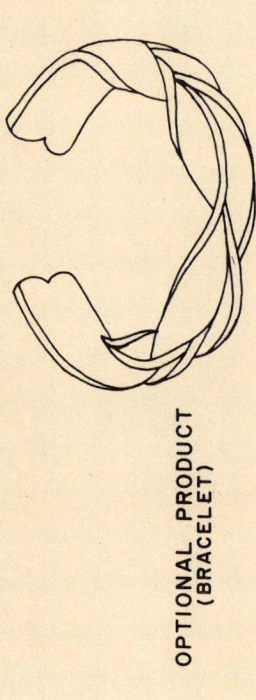

OPTIONAL PRODUCT (BRACELET)

DESIRED BLADE SHAPE

$\frac{3}{16}$ DRILL, SAW LINES TANGENT TO HOLES

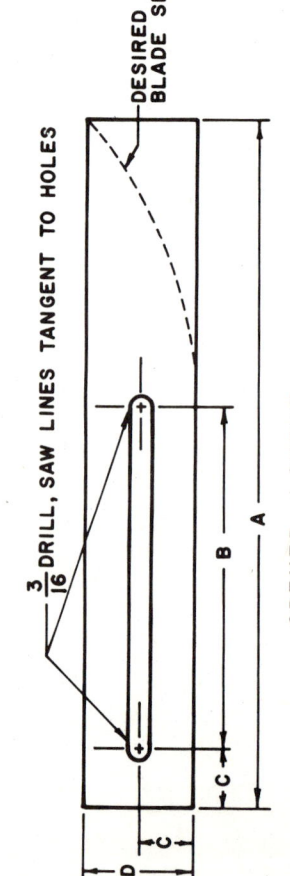

OPENER LAYOUT

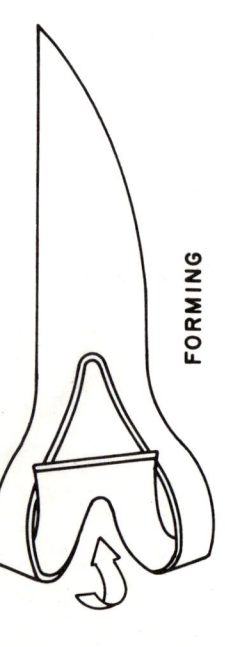

FORMING

Letter Opener

PROCEDURE

Operation Number	Operation	Tools & Equipment	Topics	Notes
		LETTER OPENER		
LO-1	Cut to size and shape	Coping saw or scroll jaw or band saw	5	
LO-2	Sharpen cutting edge	File or belt sander	15	Make sure cutting edge is smooth.
LO-3	Smooth and buff edges	File or abrasive paper, buffer	15	
LO-4	Drill holes	Hand drill or portable electric drill or drill press, 3/16" twist drill	6	Drill at low speed.
LO-5	Cut slot	Coping saw or scroll saw	5	
LO-6	Heat plastic and open slot	Oven	20	Heat and bend as shown in drawing.
LO-7	Smooth and buff slot edges	File or abrasive paper, buffer	15	
LO-8	Reheat and bend to shape	Oven		Heat, open slot, and pass end through slot. Hold flat until cool.

Gizmo

DIMENSIONS

Dimension Symbol	Metric mm	Cust. in.
A	90	3-1/2
B	20	3/4
C	16	5/8
D	8	5/16
E	35	1-3/8
F	45	1-3/4
G	25	1
H	127	5
I	178	7
J	10	3/8
K	27	1-1/16
L	6	1/4

PARTS AND MATERIALS

Qty.	Part	Size	Material
1	Base	B×A×F	Wood
1	Frame	1/8"×A×I	Plastic
1	Clip	G×H	Sheet Metal (22 gage)
1	Backing	A×H	Cardboard
3	Screws PH	#6 x 1/2"	
			Finish

Note: Can also be used as a calculator stand or a picture holder.

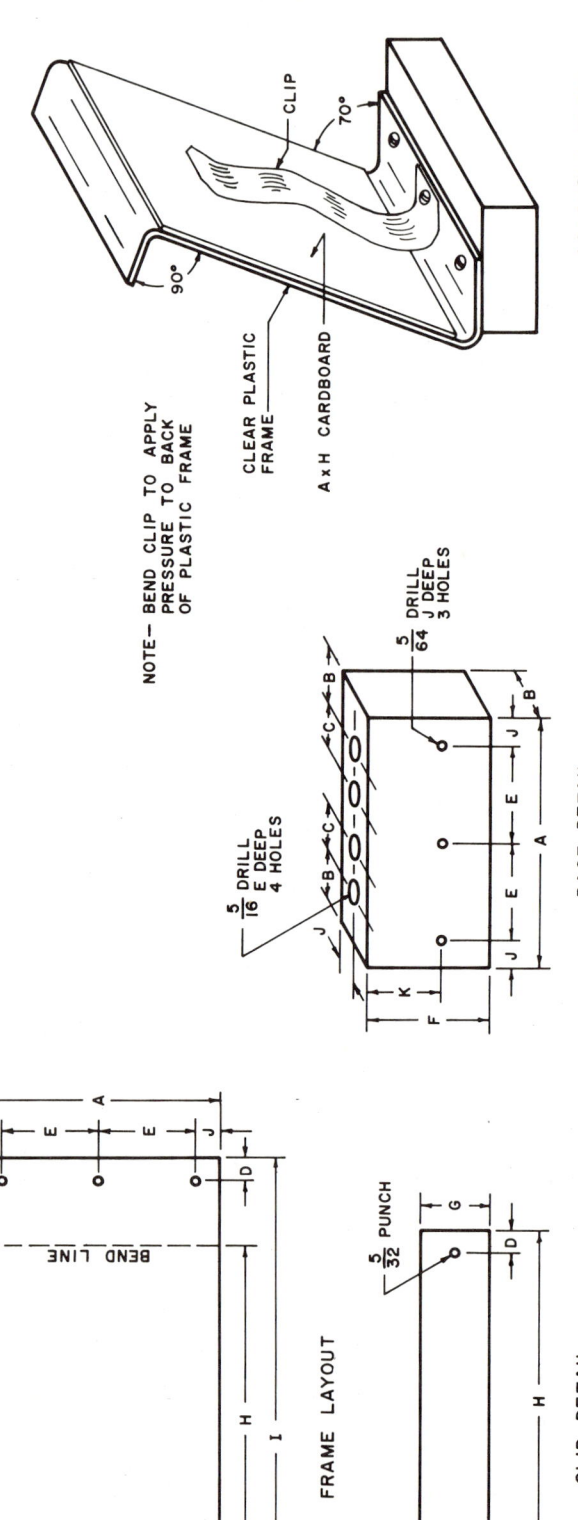

Gizmo

PROCEDURE

Operation Number	Operation	Tools & Equipment	Topics	Notes
BASE				
B-1	Cut to size	Handsaw or power saw	5	
B-2	Drill 3 pilot holes	Hand drill or portable electric drill or drill press, 5/64" twist drill	6, 25	A drilling fixture may be used.
B-3	Drill 4 pencil holes	Portable electric drill or drill press, 5/16" twist drill	6, 25	A drilling fixture may be used.
B-4	Sand	Abrasive paper	15	
B-5	Finish	Spray or brush	23	
FRAME				
F-1	Cut to size	Handsaw or power saw	5	Use a fine-tooth blade.
F-2	Drill 3 holes	Hand drill or portable electric drill or drill press, 9/64" twist drill	6, 25	A drilling fixture or a template may be used for marking hole locations. Run the drill at slow speed.
F-3	Smooth edges	File or abrasive paper	15	
F-4	Polish edges	Buffer	15	
F-5	Bend ends	Strip heater or oven	20	
CLIP				
C-1	Cut to size and shape	Squaring shears or tin snips	4	
C-2	Punch hole	9/64" hand punch	4	May be drilled.
C-3	Bend to shape	Vise, mallet	3, 19	A bending jig could be made.
ASSEMBLY				
G-1	Assemble base, frame, and clip	Screwdriver	3, 9	
G-2	Insert cardboard backing			

BASE: B-1 CUT TO SIZE → B-2 DRILL 3 PILOT HOLES → B-3 DRILL 4 PENCIL HOLES → B-4 SAND → B-5 FINISH

FRAME: F-1 CUT TO SIZE → F-2 DRILL 3 HOLES → F-3 SMOOTH EDGES → F-4 POLISH EDGES → F-5 BEND ENDS

CLIP: C-1 CUT TO SIZE AND SHAPE → C-2 PUNCH HOLE → C-3 BEND TO SHAPE

→ G-1 ASSEMBLE BASE, FRAME, AND CLIP → G-2 INSERT CARDBOARD BACKING → TS HOLD FOR DISTRIBUTION

Funnel

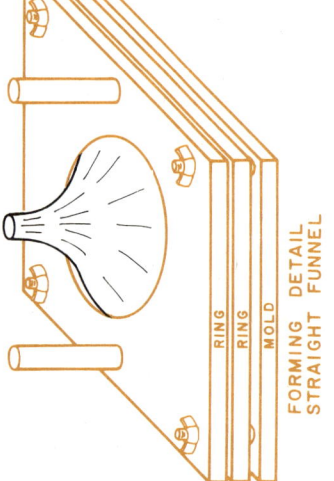

FORMING DETAIL
STRAIGHT FUNNEL

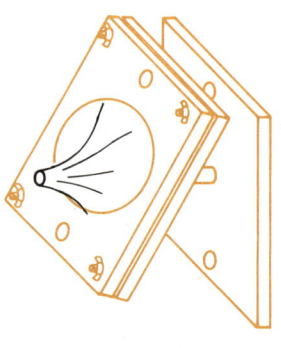

FORMING DETAIL
OFFSET FUNNEL
(GUIDEPINS REMOVED)

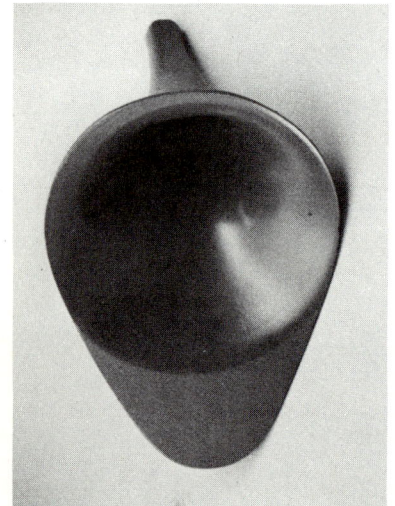

PARTS AND MATERIALS

Qty.	Part	Size	Material
1	Funnel	.060" × B × B	ABS Plastic
1	Base	C × D × E	Wood
2	Rings	C × D × E	Wood
1	Post	3/8" dia. × H	Wood Dowel
2	Pins	1/2" × H	Wood Dowel
4	Stove Bolts	1/4"-20 × 1-1/2"	
4	Wing Nuts	1/4"-20	

DIMENSIONS

Dimension Symbol	Metric mm	Cust. in.
A	75	3
B	150	6
C	12	1/2
D	230	9
E	190	7-1/2
F	20	3/4
G	95	3-3/4
H	130	5
I	114	4-1/2

Funnel

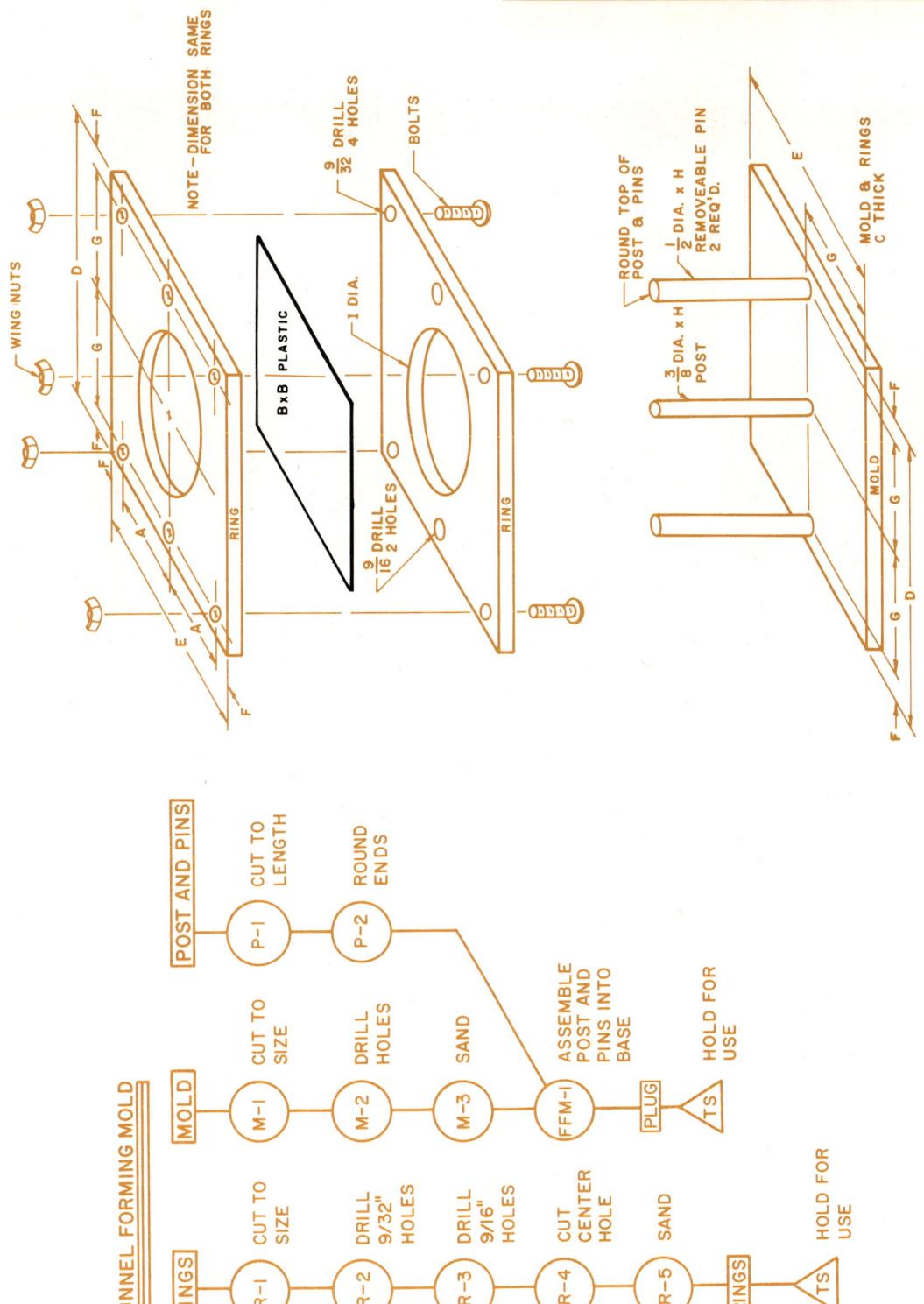

(continued next page)

Funnel (cont.)

PROCEDURE

Operation Number	Operation	Tools & Equipment	Topics	Notes
FUNNEL FORMING MOLD				
	RINGS			
R-1	Cut to size	Handsaw or power saw	5	
R-2	Drill 9/32" holes	Portable electric drill or drill press, 9/32" twist drill	6	
R-3	Drill 9/16" holes	Portable electric drill or drill press, 9/16" twist drill	6	
R-4	Cut center hole	Coping saw or scroll saw or drill press with flycutter	5 or 6	
R-5	Sand	Abrasive paper	15	
	MOLD			
M-1	Cut to size	Handsaw or power saw	5	
M-2	Drill holes	Drill press, 3/8" twist drill	6	
M-3	Sand	Abrasive paper, sanding block	15	
	POST AND PINS			
P-1	Cut to length	Backsaw or coping saw or miter saw	5	
P-2	Round ends	Abrasive paper	15	Slightly round one end of post and each pin.
	ASSEMBLY			
FFM-1	Assemble post and pins into base	Hammer	3, 11	Glue post in place. Pins are removable.

Funnel (cont.)

FUNNEL				
F-1	Cut plastic to size	Handsaw or band saw or tin snips or aviation snips	5 or 4	
F-2	Clamp plastic in rings			Smooth side of plastic should be down.
F-3	Heat plastic and rings	Oven	20	Wear gloves when handling hot materials.*
F-4	Form to shape			Align rings on plug and push downward. Use an even pressure. Hold until cooled. Then remove from rings.**
F-5	Cut to shape	Coping saw, scroll saw, band saw or tin snips	5 or 4	Lay out shape with felt tip pen.
F-6	Cut funnel end	Coping saw or belt sander or disk sander	5 or 15	
F-7	Sand edges	Abrasive paper	15	

* **Note**: Preheat oven to 375°F (190°C). Plastic is ready to form when it begins to sag in the middle.

** **Note**: To form an offset funnel, remove the guide pins. Center the plastic over the post. Then push down until one end of the ring touches the base. Hold the other end up about 1 inch (25 mm).

FUNNEL
- F-1 CUT PLASTIC TO SIZE
- F-2 CLAMP PLASTIC IN RINGS
- F-3 HEAT PLASTIC AND RINGS
- F-4 FORM TO SHAPE
- F-5 CUT TO SHAPE
- F-6 CUT FUNNEL END
- F-7 SAND EDGES
- TS HOLD FOR DISTRIBUTION

Trivet

DIMENSIONS

Dimension Symbol	Metric mm	Cust. in.
A	140	5-1/2
B	12	1/2
C	25	1
D	204	8

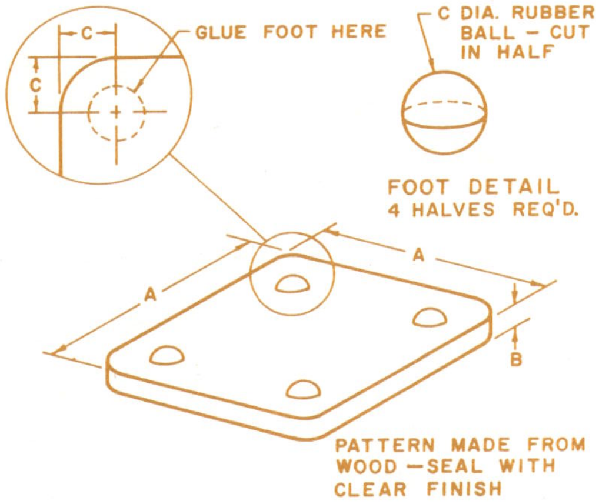

PARTS AND MATERIALS

Qty.	Part	Size	Material
8 oz.	Trivet		Plastic (Casting Resin with Catalyst)
1	Mold	.060" × D × D	High Density Polyethylene Plastic
1	Board	B × A × A	Wood
4	Feet	1" dia.	2 Rubber Balls

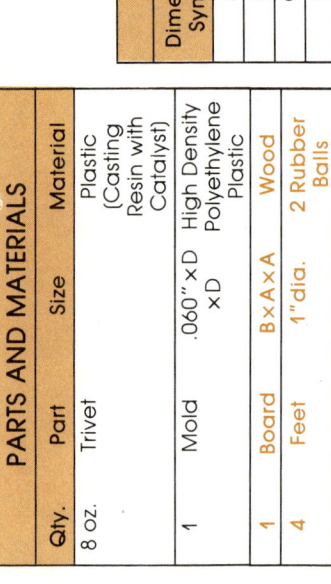

.060 HIGH DENSITY POLYETHYLENE MOLD BLANK

WOOD PATTERN

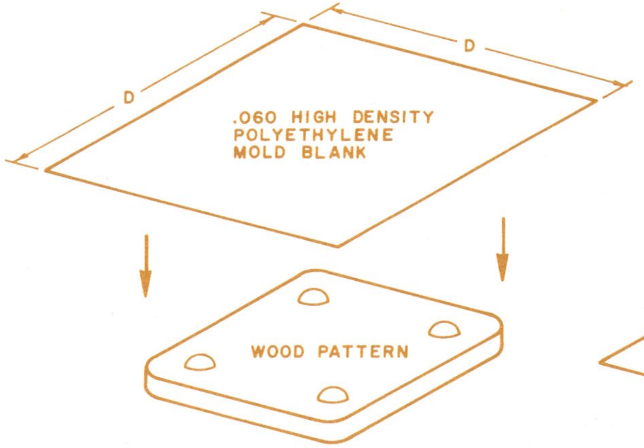

MOLD IS VACUUM-FORMED OVER PATTERN

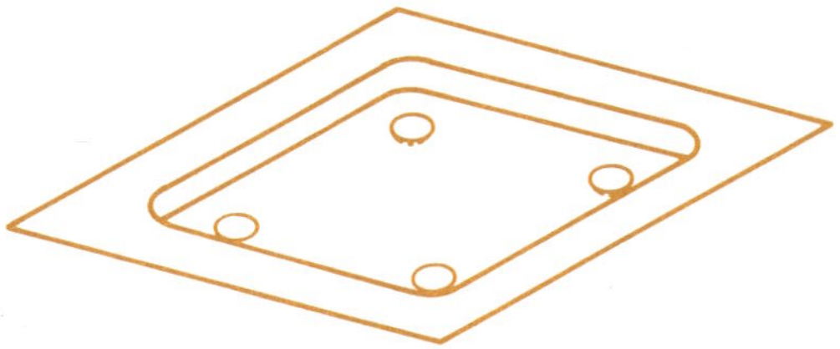

COMPLETED MOLD

TRIVET PATTERN

PATTERN MADE FROM WOOD—SEAL WITH CLEAR FINISH

FOOT DETAIL 4 HALVES REQ'D.

C DIA. RUBBER BALL — CUT IN HALF

GLUE FOOT HERE

POURING PLASTIC IN MOLD

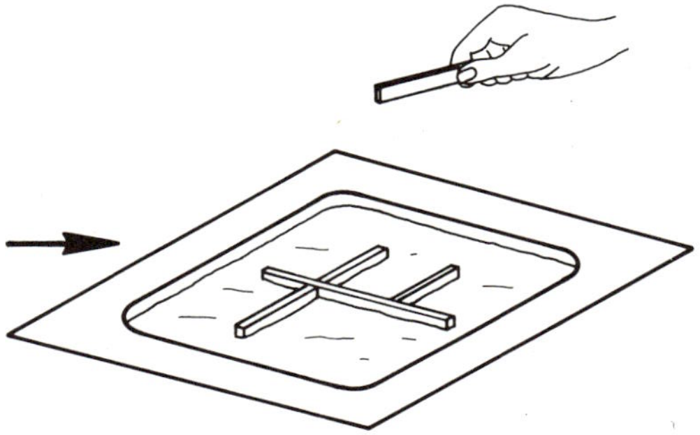

PLACING EMBEDMENTS

FOR EMBEDMENTS USE SEEDS, DRIED FLOWERS, PENNIES, SPICES ETC. USE WOOD STRIPS FOR DIVIDERS

FINISHED TRIVET

Trivet

PROCEDURE

Operation Number	Operation	Tools & Equipment	Topics	Notes
TRIVET PATTERN AND MOLD				
MOLD				
M-1	Cut to size	Scissors or squaring shears	4	
M-2	Form mold over pattern	Vacuum former	20	Heat plastic first. Follow the directions for your machine.
FEET				
F-1	Cut in half	Knife	3	
BOARD				
B-1	Cut to size and shape	Handsaw or power saw	5	
B-2	Sand	Belt sander or finishing sander, abrasive paper	15	Slightly round all corners and edges.
B-3	Assemble feet to board	Clamps	11, 24	Glue feet to board.
TRIVET				
T-1	Mix and pour first layer	Paper cup, stirring stick, mold	20	Mix and pour only enough to cover bottom of mold. Allow to harden.
T-2	Place embedments (items to be enclosed in plastic)			Could enclose seeds, flowers, coins, spices, or other items in plastic. Dividers may be used.
T-3	Mix and pour final layer	Paper cup, stirring stick, mold	20	Mix and pour to top of mold. Allow to harden. Remove from mold.
T-4	Clean and polish	File, abrasive paper, buffer	15	

Note: Plastic must harden just enough that embedments do not sink when placed on it.

TRIVET PATTERN AND MOLD

MOLD → M-1 CUT TO SIZE

BOARD → B-1 CUT TO SIZE AND SHAPE → B-2 SAND → B-3 GLUE FEET TO BOARD

FEET → F-1 CUT IN HALF → (to B-3)

(B-3) → PATTERN → M-2 FORM MOLD OVER PATTERN → TS HOLD FOR USE

TRIVET

T-1 MIX AND POUR FIRST LAYER → T-2 PLACE EMBEDMENTS → T-3 MIX AND POUR FINAL LAYER → T-4 CLEAN AND POLISH → TS HOLD FOR DISTRIBUTION

Skateboard

PARTS AND MATERIALS

Qty.	Part	Size	Material
1	Mylar Film	10 Mill I×J	Plastic
1	Surface Mat	1-1/2 oz. I×J	Fiber glass
2	Decorative Cloth	I×J	Cotton
3	Woven Cloth	1-1/2 oz. I×J	Fiber glass
6	Mat	1-1/2 oz. I×J	Fiber glass
1	Polyester Resin (with catalyst)	1 qt.	Plastic
1	Mold	E×D	Sheet Metal (20-24 Gage)
2	Skateboard Trucks		
	Plastic Gloves		
	Paper Cups	8 oz.	
	Stirring Sticks		

DIMENSIONS

Dimension Symbol	Metric mm	Cust. in.
A	610	24
B	150	6
C	25	1
D	660	26
E	406	16
F	585	23
G	100	4
H	200	8
I	635	25
J	178	7

Skateboard

PROCEDURE

Operation Number	Operation	Tools & Equipment	Topics	Notes
SKATEBOARD MOLD				
SM-1	Cut to size and shape	Squaring shears or tin snips or aviation snips	4	
SM-2	Bend sides to shape	Box and pan brake	19	
SM-3	Form curve	Stake, mallet	3, 19	Could be formed over a pipe clamped in vise.
SKATEBOARD				
SB-1	Cut layers to size	Scissors	4	
SB-2	Laminate layers	Plastic gloves, paper cups, stirring sticks, brush	20	*See drawings. Allow resin to cure before going to next procedure.
SB-3	Cut skateboard to size	Band saw	5	
SB-4	Smooth edges	Disk sander or belt sander	15	
SB-5	Drill truck mounting holes	Portable electric drill or drill press, twist drill	6	
SB-6	Mount trucks	Screwdriver, adjustable wrench	3, 9	

* **Note**: Mix 3 or 4 oz. of resin. Discard cup and stirring stick. Mix a new batch for every layer using a new cup and stick.

(continued next page)

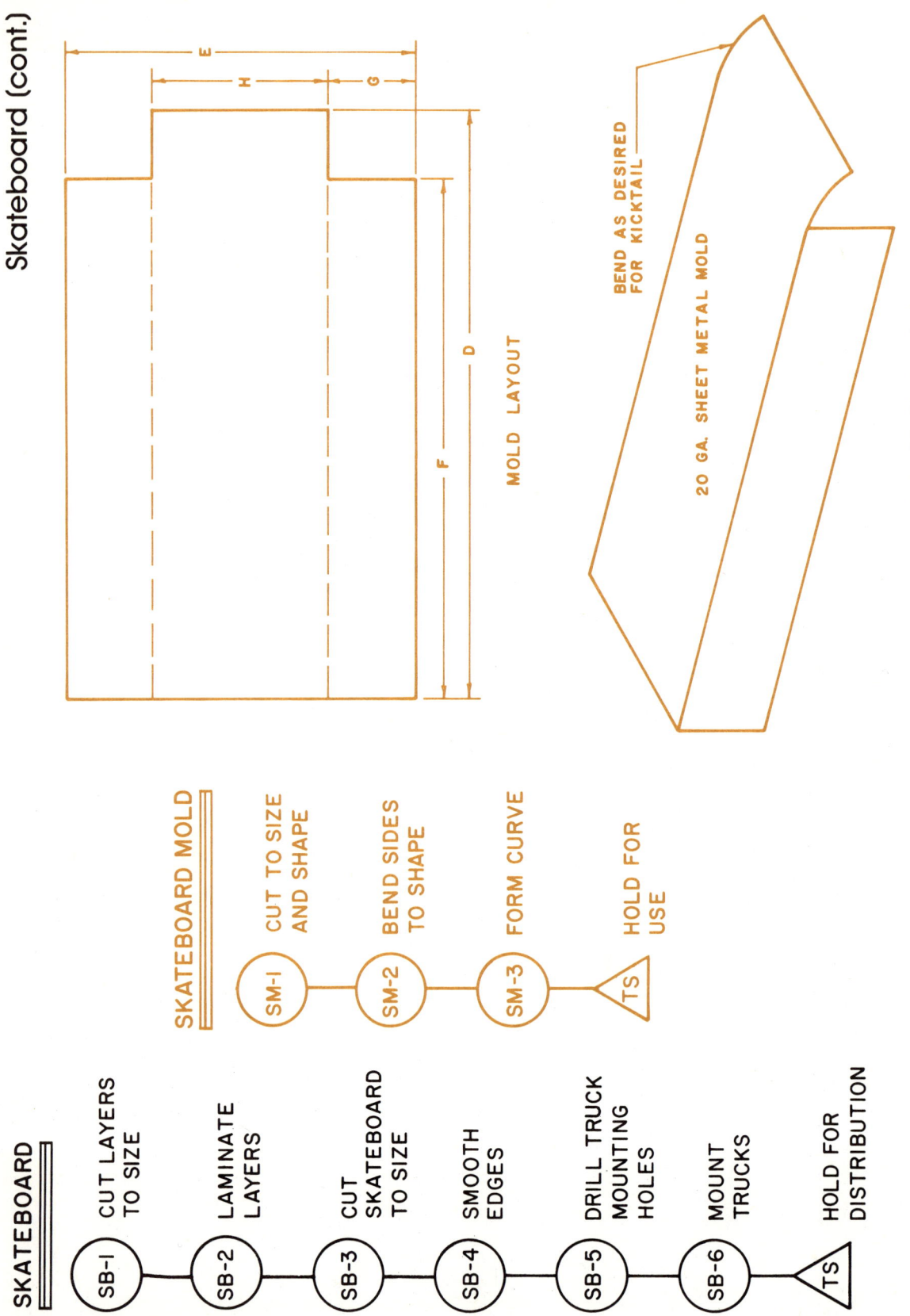

Skateboard (cont.)

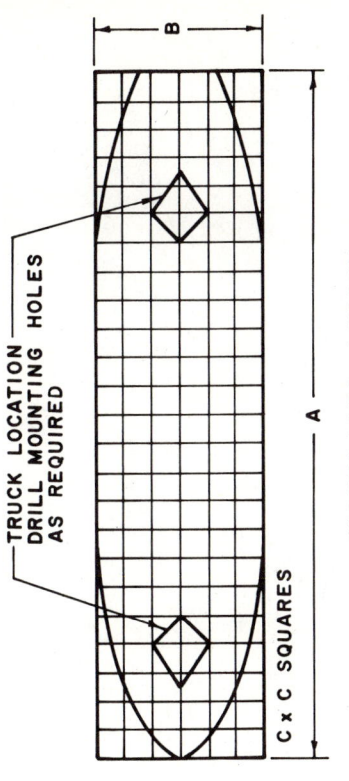

SKATEBOARD LAYOUT
- TRUCK LOCATION DRILL MOUNTING HOLES AS REQUIRED
- C x C SQUARES

MAT & CLOTH LAYOUT
- CUT ALL PIECES OF FIBER GLASS MAT, CLOTH AND DECORATIVE CLOTH THIS SIZE

MAT & CLOTH APPLICATION SEQUENCE

- TWELFTH LAYER — DECORATIVE COTTON CLOTH RESIN
- ELEVENTH LAYER — WOVEN FIBER GLASS CLOTH RESIN
- TENTH LAYER — FIBER GLASS MAT RESIN
- NINTH LAYER — FIBER GLASS MAT RESIN
- EIGHTH LAYER — FIBER GLASS MAT RESIN
- SEVENTH LAYER — WOVEN FIBER GLASS CLOTH RESIN
- SIXTH LAYER — FIBER GLASS MAT RESIN
- FIFTH LAYER — FIBER GLASS MAT RESIN
- FOURTH LAYER — FIBER GLASS MAT RESIN
- THIRD LAYER — WOVEN FIBER GLASS CLOTH RESIN
- SECOND LAYER — DECORATIVE COTTON CLOTH RESIN
- FIRST LAYER — FIBER GLASS SURFACE MAT RESIN
- MOLD RELEASE — MYLAR FILM
- MOLD

Tote Bag

DIMENSIONS

Dimension Symbol	Metric mm	Cust. in.
A	915	36
B	380	15
C	20	3/4
D	10	3/8
E	100	4

PARTS AND MATERIALS

Qty.	Part	Size	Material
1	Bag	A×B	Heavy Cloth
1	Cord	1/8" dia. × 36"	
6	Grommets	1/4" dia.	
1	Fusible Tape	1/2" × 72"	

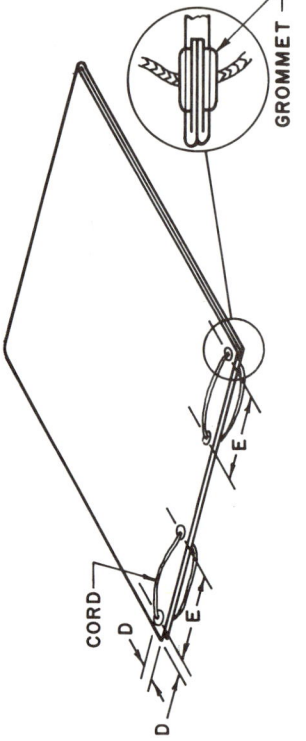

ASSEMBLY DETAIL

HEM DETAIL

SEAM DETAIL

Tote bag

PROCEDURE

Operation Number	Operation	Tools & Equipment	Topics	Notes
BAG				
B-1	Cut to size	Scissors	4	
B-2	Fold and fuse hems	Iron	11	Fold hems wrong sides together
B-3	Fold and fuse seams	Iron	11	Fold with right side of cloth on the inside. Seam. Then turn bag right side out.
B-4	Punch holes	Hammer, punch	3	
B-5	Install grommets	Hammer, grommet setting tool	3, 10	Only middle grommets go through hem. Outer grommets go through both sides for added reinforcement.
CORD				
C-1	Cut to length	Knife	3	
ASSEMBLY				
TB-1	Lace cord and tie ends			

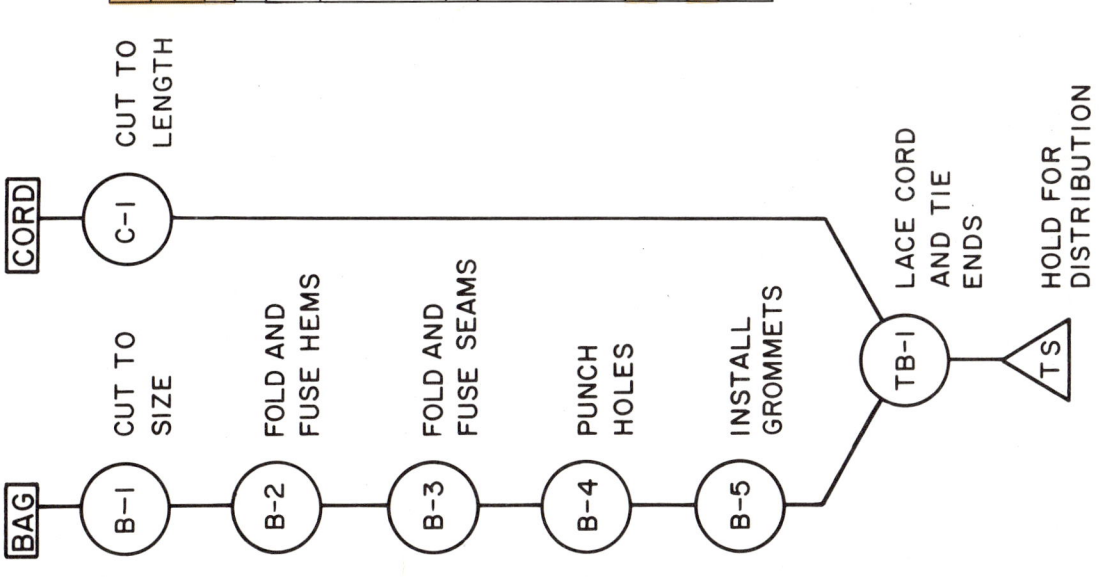

Shop Apron

DIMENSIONS

Dimension Symbol	Metric mm	Cust. in.
A	915	36
B	630	23-3/4
C	254	10
D	570	22-1/2
E	20	3/4
F	10	3/8

PARTS AND MATERIALS

Qty.	Part	Size	Material
1	Apron	A×D	Heavy Cloth
4	Grommets	1/4" dia.	
2	Ropes	1/8" dia. ×36"	
1	Fusible Tape	1/2" ×9'	

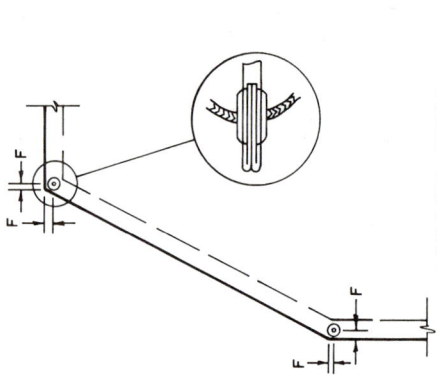

LACING DETAIL

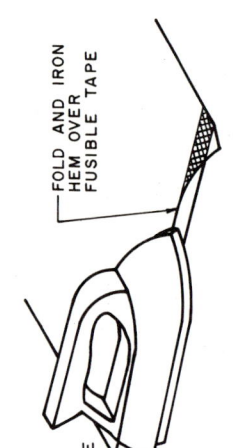

GROMMET DETAIL

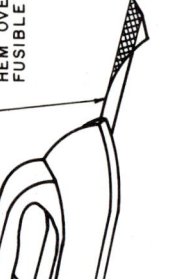

HEM DETAIL

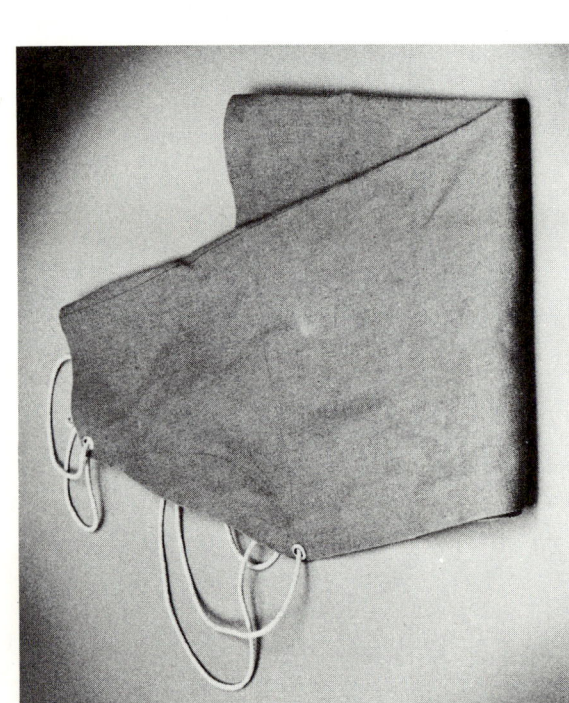

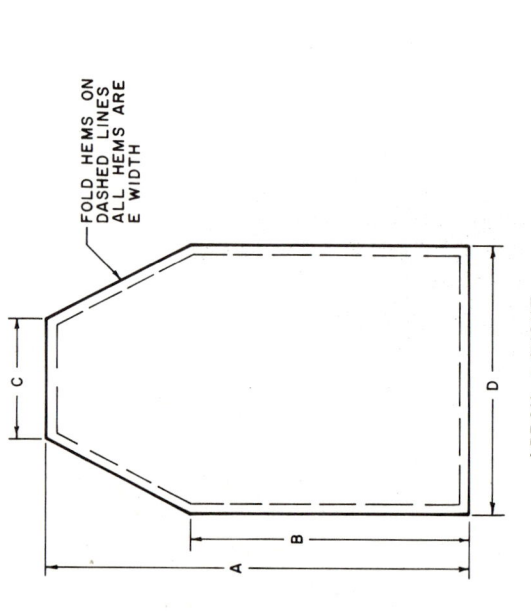

APRON LAYOUT

Shop Apron

PROCEDURE

Operation Number	Operation	Tools & Equipment	Topics	Notes
APRON				
A-1	Cut to size and shape	Scissors	4	
A-2	Fold and fuse hems	Iron, fusible tape	11	Fold hems wrong sides together.
A-3	Punch grommet holes	Hammer, punch	3, 4	
A-4	Install grommets	Hammer, grommet setting tool	3, 10	
A-5	Screen stencil design	Screen stenciling equipment	23	(Optional) School name and emblem could be printed.
ROPES				
R-1	Cut to length	Knife	4	
ASSEMBLY				
SA-1	Install rope and tie knots			See drawings.

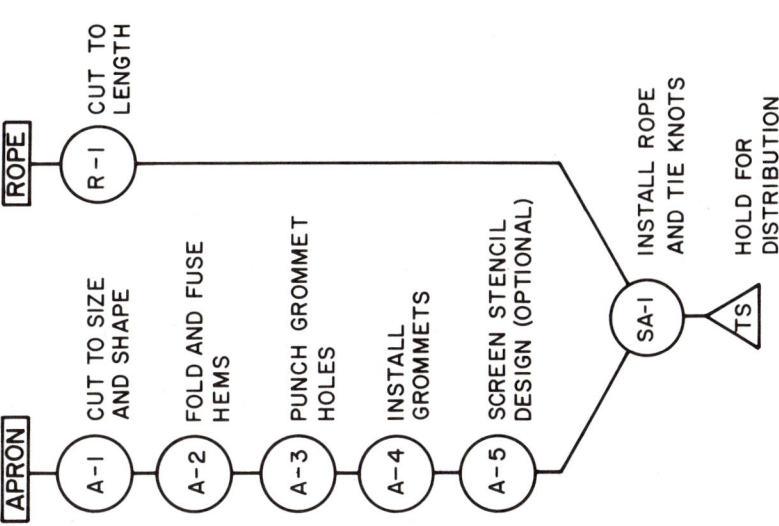

Plant Holder

PARTS AND MATERIALS

Qty.	Part	Size	Material
1	Disk	F dia.	8-oz. Leather
2	Thongs	4' long	Leather
			Dye

DIMENSIONS

Dimension Symbol	Metric mm	Cust. in.
A	92	3-1/2
B	80	3
C	65	2-1/2
D	50	2
E	6	1/4
F	184	7-1/2

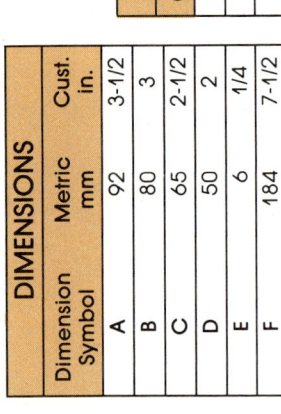

FORMING THE SHAPE

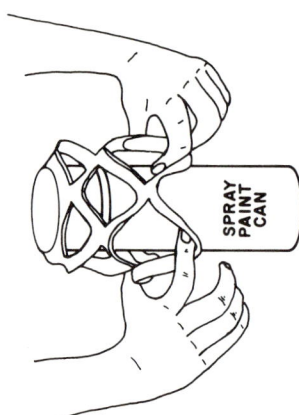

ASSEMBLY DETAIL

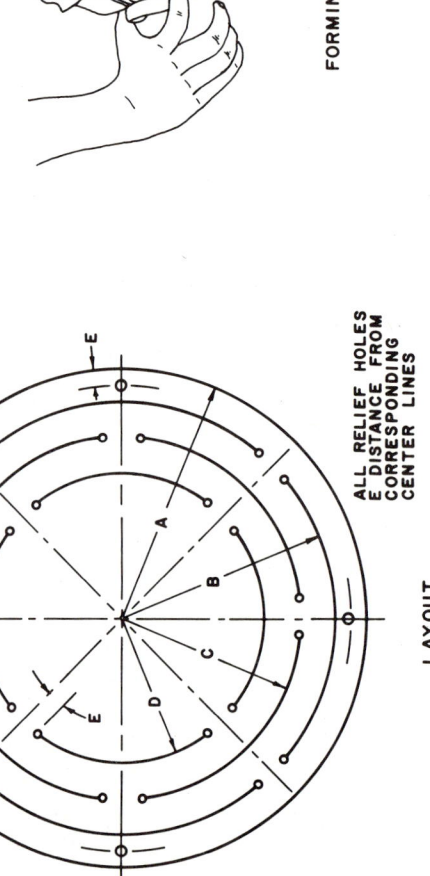

LAYOUT

Plant Holder

PROCEDURE

Operation Number	Operation	Tools & Equipment	Topics	Notes
DISK				
D-1	Cut to shape	Knife or shears	4	
D-2	Punch thong holes	Mallet, No. 3 drive punch	4, 10	
D-3	Punch relief holes	Mallet, No. 1 drive punch	4, 10	
D-4	Cut slits	Knife	4	
D-5	Wet and form to shape			Wet rough side. Form over can or other cylindrical shape. Let dry. See drawing.
THONGS				
T-1	Cut to length	Knife	4	
ASSEMBLY				
PH-1	Assemble thongs to disk			Tie knots in thongs. Tie loop in thongs.
PH-2	Dye or color			(optional)

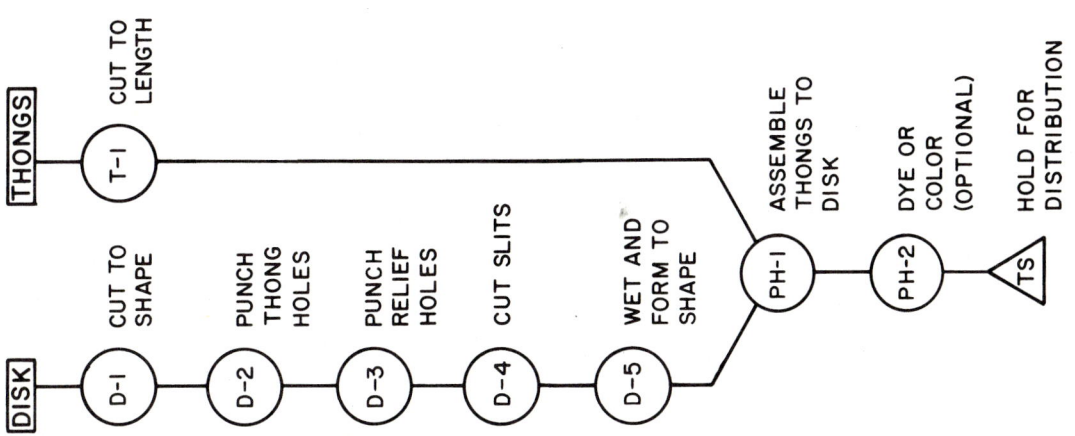

Camera Case

DIMENSIONS

Dimension Symbol	Metric mm	Cust. in.
A*	120	4-3/4
B*	60	2-3/8
C*	25	1
D	10	3/8
E*	128	5-1/8
F*	75	3
G	30	1-3/16
H	3	1/8
I	68	2-3/4
J	6	1/4
K	203	8-1/8
L	78	3-1/8
M	153	6-1/8
N	12	1/2

***Note**: Measure your camera. (If you wish to make a calculator case, measure your calculator.) Be sure to include any buttons or knobs. Use these measurements to make a wood block. The front will be formed over this block.

PARTS AND MATERIALS

Qty.	Part	Size	Material
1	Front	E×M	3- to 4-ounce Leather
1	Back	L×K	3- to 4-ounce Leather
1	Belt Loop	I×C	3- to 4-ounce Leather
2	Rapid Rivets		
45"	Lacing		3/32" Leather or Plastic
			Dye (optional)

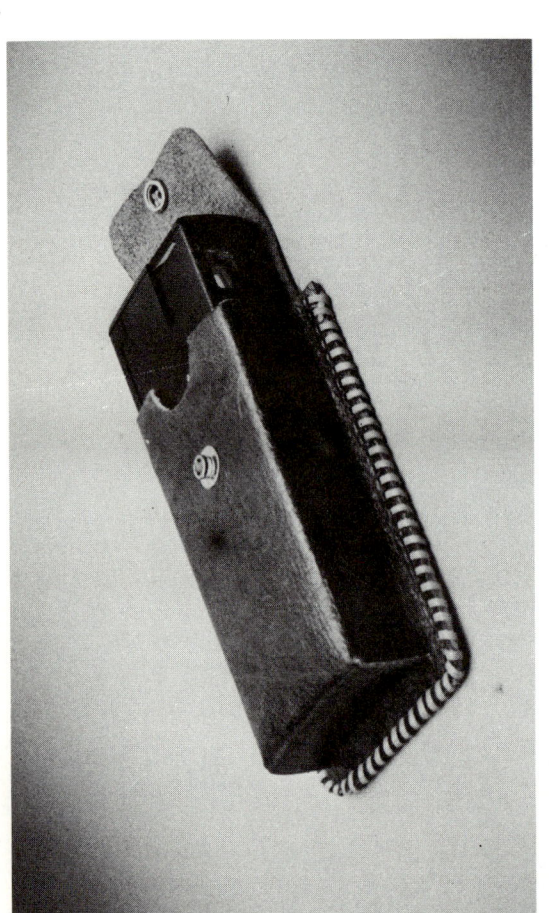

Camera Case

PROCEDURE

Operation Number	Operation	Tools & Equipment	Topics	Notes
FRONT				
F-1	Cut to overall size	Knife or shears	4	
F-2	Punch holes	Leather punch	4	Use a small punch. Relief holes make cutting and bending corners easier.
F-3	Cut to shape	Knife or shears	4	
F-4	Groove fold lines	Leather gouge or swivel knife	4	Cut groove on flesh or rough side of leather. Do not cut deeper than 1/2 the thickness of the leather.
F-5	Wet and form to shape			Moisten leather on rough side. Form around block and clamp until dry.
BACK				
B-1	Cut to size and shape	Knife or shears	4	
ASSEMBLY				
CC-1	Assemble front and back		11	Use rubber cement to cement edges together.
CC-2	Cut lacing holes	3/32" thonging chisel leather mallet	10	Lacing holes should be 1/8" from edge and evenly spaced.
CC-3	Install lacing	Lacing needle	10	Use whipstitch.
CC-4	Install snap	Leather punch, snap anvil and setter, and mallet	10	
CC-5	Tool design on front	Stamping tools	21	(Optional)
CC-6	Attach belt loop with rivets	Leather punch and mallet	10	
CC-7	Dye or color			(Optional)

(continued next page)

Camera Case (cont.)

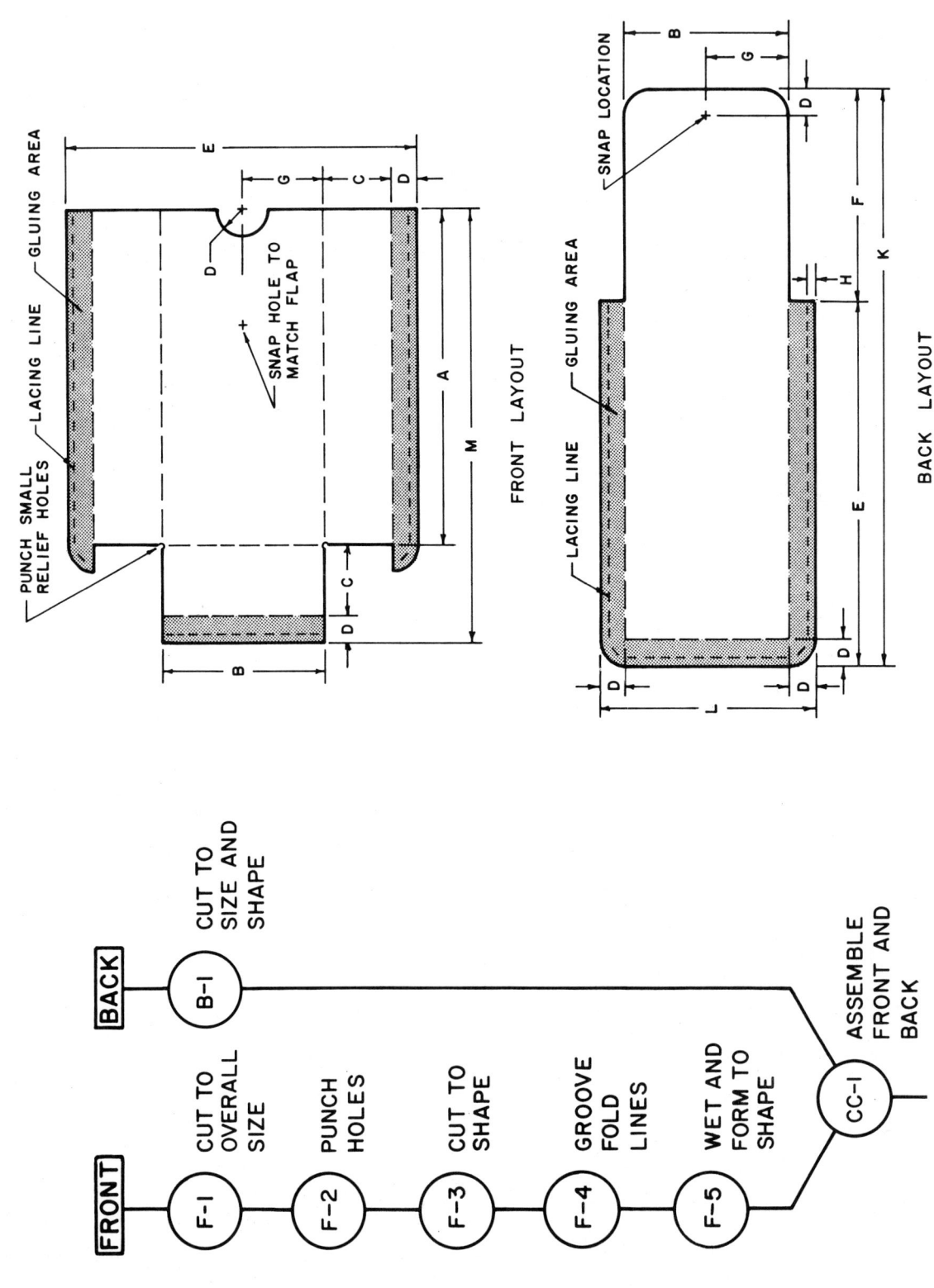

Camera Case (cont.)

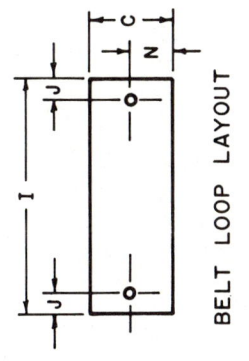

BELT LOOP LAYOUT

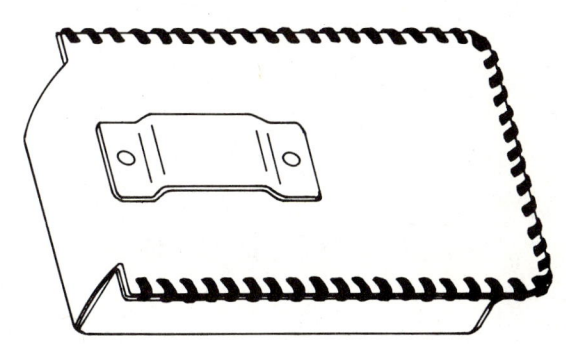

ASSEMBLY DETAIL

- CC-2 CUT LACING HOLES
- CC-3 INSTALL LACING
- CC-4 INSTALL SNAP
- CC-5 TOOL DESIGN ON FRONT (OPTIONAL)
- CC-6 ATTACH BELT LOOP
- CC-7 DYE OR COLOR (OPTIONAL)
- TS HOLD FOR DISTRIBUTION

Rocket

DIMENSIONS

Dimension Symbol	Metric mm	Cust. in.
A	292	11
B	108	4-1/4
C	75	3
D	457	18
E	89	3-1/2
F	25	1
G	64	2-1/2
H	44	1-3/4
I	12	1/2
J	70	2-3/4
K	6	1/4
L	57	2-1/4
M	20	3/4
N	100	4

PARTS AND MATERIALS

Qty.	Part	Size	Material
BODY			
1	Engine Retainer	18ga. × N	Wire
3	Fins	1/16" × E × F	Cardboard
1	Launch Lug	3/16" dia. × E	Soda Straw
1	Paper	B × A	Paper
1	Tape	C × A	Gummed Paper Tape
1	Tape	C × D	Gummed Paper Tape
NOSE CONE			
1	Cork	#4	Cork
1	Cork	#8	Cork
1	Dowel	3/4" dia. × I	Wood
STREAMER			
1	Streamer	F × A	Crepe Paper
2	Strings	D	Cotton
1	Rubber Band		Rubber
	Glue		
	Staples		
	Paint		

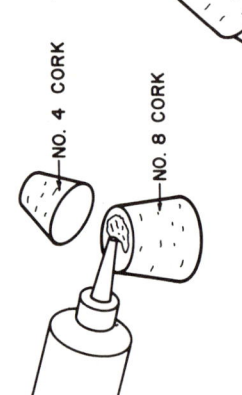

NOSE CONE ASSEMBLY DETAIL

NO. 4 CORK
NO. 8 CORK

LAUNCH LUG DETAIL
SODA STRAW

ENGINE RETAINER DETAIL
USE PAPER CLIP OR 18 GA. WIRE N LENGTH

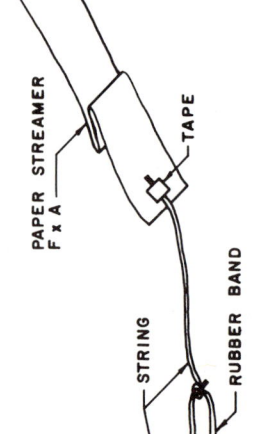

STREAMER DETAIL
PAPER STREAMER F × A
TAPE
STRING
RUBBER BAND

FIN DETAIL 3 REQ'D.
NOTE – NOTCHED SIDE TOWARD ROCKET BODY

Rocket

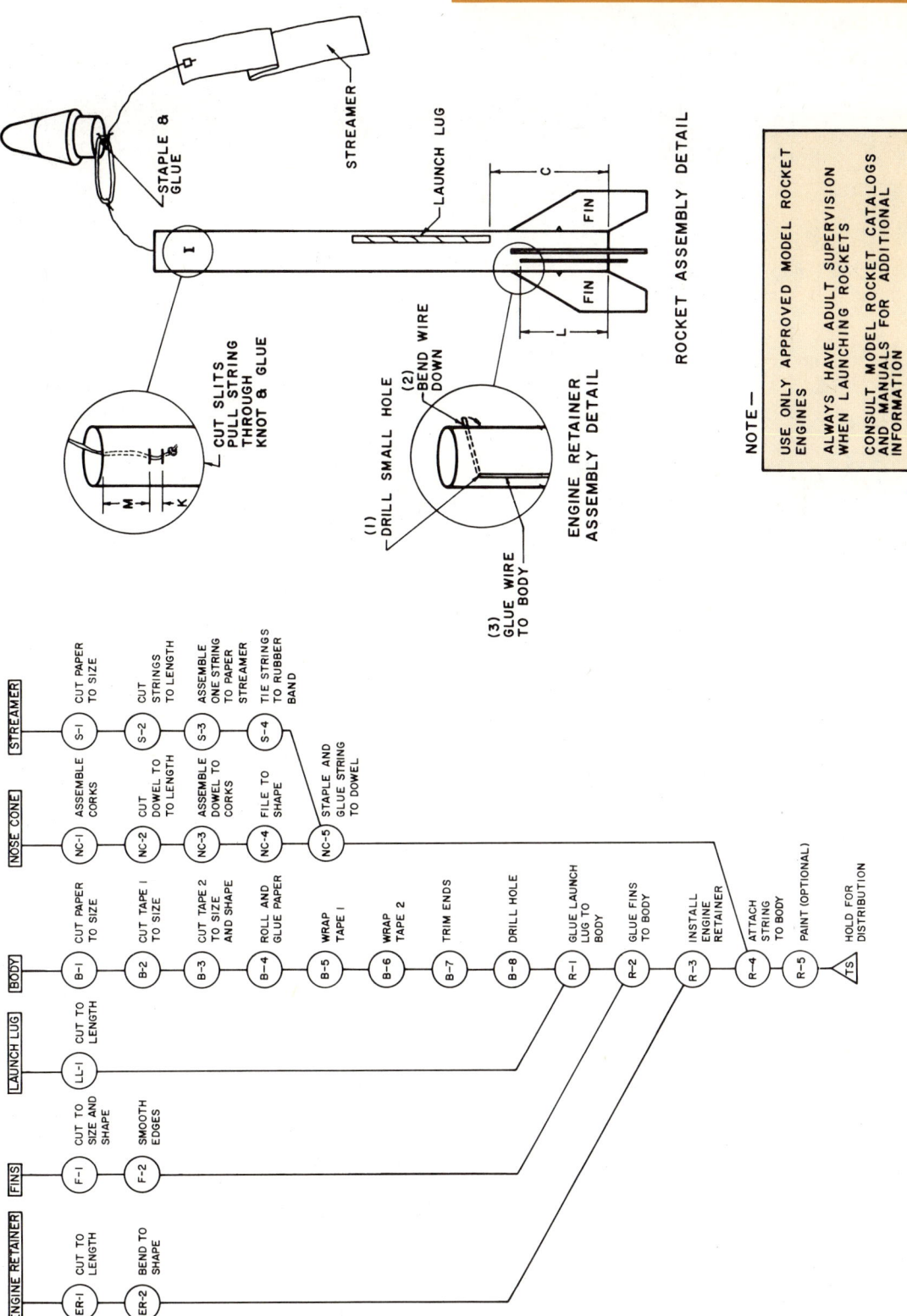

(continued next page)

Rocket (cont.)

PROCEDURE

Operation Number	Operation	Tools & Equipment	Topics	Notes
ENGINE RETAINER				
ER-1	Cut to length	Wire cutters	3	A straightened paper clip may be used.
ER-2	Bend to shape	Pliers	3, 19	A bending jig may be used.
FINS				
F-1	Cut to size and shape	Knife or scissors or paper cutter	4	A template may be used.
F-2	Smooth edges	Abrasive paper	15	Make edges smooth for less wind resistance.
LAUNCH LUG				
LL-1	Cut to length	Knife	4	
BODY				
B-1	Cut paper to size	Scissors or paper cutter	4	
B-2	Cut tape 1 to size	Scissors	4	
B-3	Cut tape 2 to size and shape	Scissors	4	Be sure to cut angled end.
B-4	Roll and glue paper		11	See drawings. Roll paper tightly. Use a small amount of glue.
B-5	Wrap tape 1			See drawings. Moisten tape and roll tightly over paper.
B-6	Wrap tape 2			See drawings. Moisten tape and roll tightly over tape 1. Note angle.
B-7	Trim ends	Scissors		See drawings. Allow to dry before trimming ends.
B-8	Drill hole	Hand drill, 1/16" twist drill	6	Hole should go through both sides.

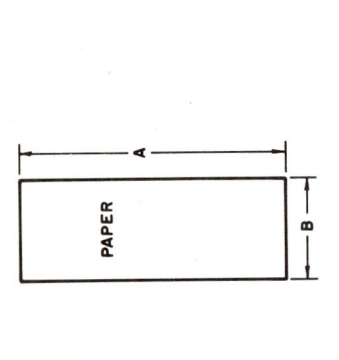
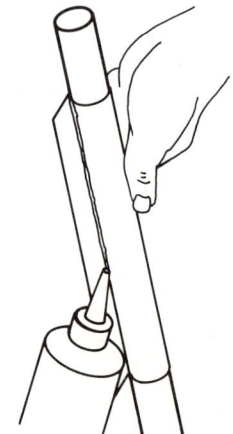
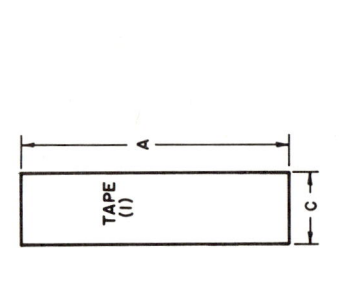
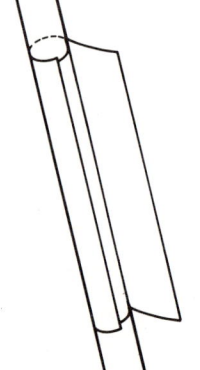

Rocket (cont.)

NOSE CONE				
NC-1	Assemble corks		11	See drawings. Use glue.
NC-2	Cut dowel to length	Coping saw or back-saw or miter saw	5	
NC-3	Assemble dowel to corks		11	See drawings. Use glue.
NC-4	File to shape	File or abrasive paper	15	Make nose smooth for less wind resistance.
NC-5	Staple and glue string to dowel	Stapler	10, 11	Staple first, and then glue.
STREAMER				
S-1	Cut paper to size	Scissors	4	
S-2	Cut strings to length	Scissors or knife	4	Both strings are D length.
S-3	Assemble one string to paper streamer			Use masking tape.
S-4	Tie strings to rubber band			See drawing.
ASSEMBLY				
R-1	Glue launch lug to body		11	Make sure lug is properly spaced and straight on the body.
R-2	Glue fins to body		11	A fin assembly fixture will help space and align fins while glue dries.
R-3	Install engine retainer		11	See drawings.
R-4	Attach string to body		11	See drawing.
R-5	Paint			(Optional)

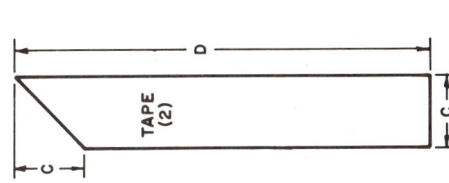

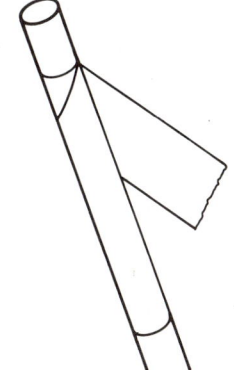

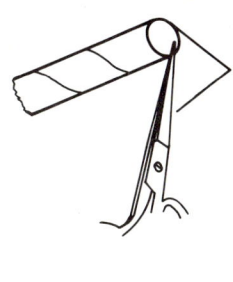

BODY TUBE ASSEMBLY

RACE CAR

CAN YOU RESEARCH & DEVELOP A RACE CAR?

HERE'S THE IDEA:

A model race car can be powered by a CO_2 cartridge. The car can be guided by a piece of fishing line stretched along the floor. Each car may be raced and timed, or cars may be raced against each other.

RESEARCH & DEVELOPMENT DECISIONS:

Design information is given for the car body, wheels, axles, and bearings. You will need to R&D the **BODY SHAPE**, the **WHEELBASE** (distance between the front and rear axles), and the **LENGTH OF EACH AXLE**.

- WHAT IS THE BEST SHAPE FOR A RACE CAR?

- HOW CLOSE OR HOW FAR APART SHOULD THE FRONT WHEELS BE? THE REAR WHEELS?

- HOW FAR APART SHOULD THE AXLES BE SPACED?

- HOW MUCH SHOULD THE CAR WEIGH?

- WHAT SHOULD BE THE SAME ON EVERYONE'S CARS? WHAT COULD BE DIFFERENT?

- WHAT RULES SHOULD BE FOLLOWED WHEN RACING THE CARS?

- WHAT OTHER DECISIONS MIGHT BE NECESSARY?

287 RACE CAR

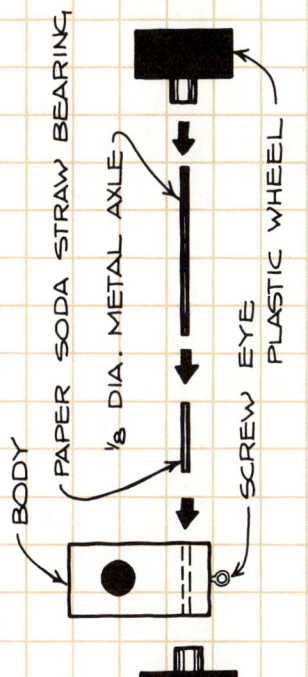

WHEEL ASSEMBLY

PUSH BEARING INTO BODY. PUSH AXLE THRU BEARING, AND PUSH ON BOTH WHEELS.

DIMENSIONS		
DIMENSION SYMBOL	METRIC MM.	CUST. IN.
A	295	12
B	42	$1\frac{1}{2}$
C	65	$2\frac{3}{4}$
D	33	$1\frac{1}{4}$
E	8	$\frac{3}{8}$
F	21	$\frac{3}{4}$
G	50	2

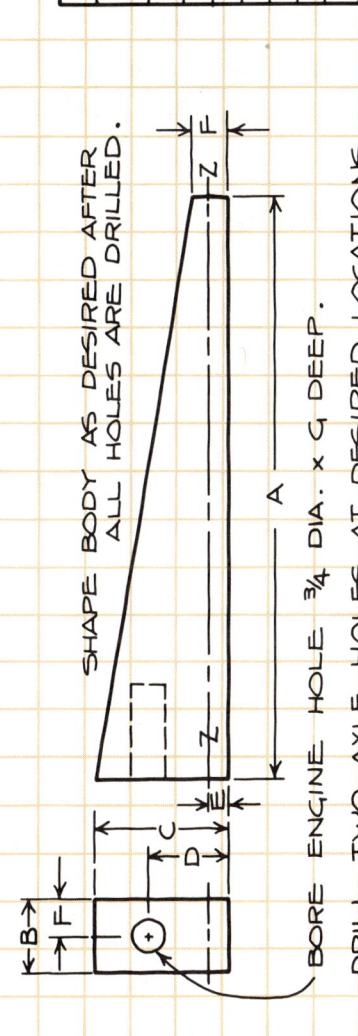

RACE CAR ASSEMBLY

BODY DETAIL
(HOLE LOCATIONS AND MAXIMUM SIZE LIMITS)

SHAPE BODY AS DESIRED AFTER ALL HOLES ARE DRILLED.

BORE ENGINE HOLE $\frac{3}{4}$ DIA. × G DEEP.

DRILL TWO AXLE HOLES AT DESIRED LOCATIONS ON LINE ZZ. CHOOSE DRILL SIZE EQUAL TO THE OUTSIDE DIAMETER OF THE BEARINGS. THE BEARINGS SHOULD FIT SNUGLY INTO AXLE HOLES.

PINBALL

CAN YOU RESEARCH & DEVELOP A PINBALL GAME?

HERE'S THE IDEA:

IF YOU STRETCH RUBBER BANDS AROUND NAILS DRIVEN INTO A BOARD, A STEEL BALL BEARING WILL BOUNCE OFF OF THEM. THIS ACTION IS MUCH LIKE THAT IN A PINBALL MACHINE.

RESEARCH & DEVELOPMENT DECISIONS:

DESIGN INFORMATION IS GIVEN FOR A BASIC BOARD, A SHOOTER, AND FLIPPERS. YOU WILL NEED TO R&D THE GAME. <u>LAYOUT</u> AND THE <u>COURSE</u> FOR THE BALL TO FOLLOW. MAKE THE COURSE CHALLENGING!

- HOW TIGHTLY SHOULD THE RUBBER BANDS BE STRETCHED?
- WHAT GAME LAYOUT OR DESIGN WOULD BE FUN AS WELL AS CHALLENGING?
- IN WHAT WAYS COULD POINTS BE SCORED?
- HOW MIGHT THE BOARD BE DECORATED?
- CAN YOU IMPROVE ON THE DESIGN FOR THE SHOOTER OR FLIPPERS?
- WHAT ADDITIONAL PARTS COULD YOU ADD? SPINNERS? BELLS? MOVING PARTS?
- WHAT OTHER DECISIONS MIGHT BE NECESSARY?

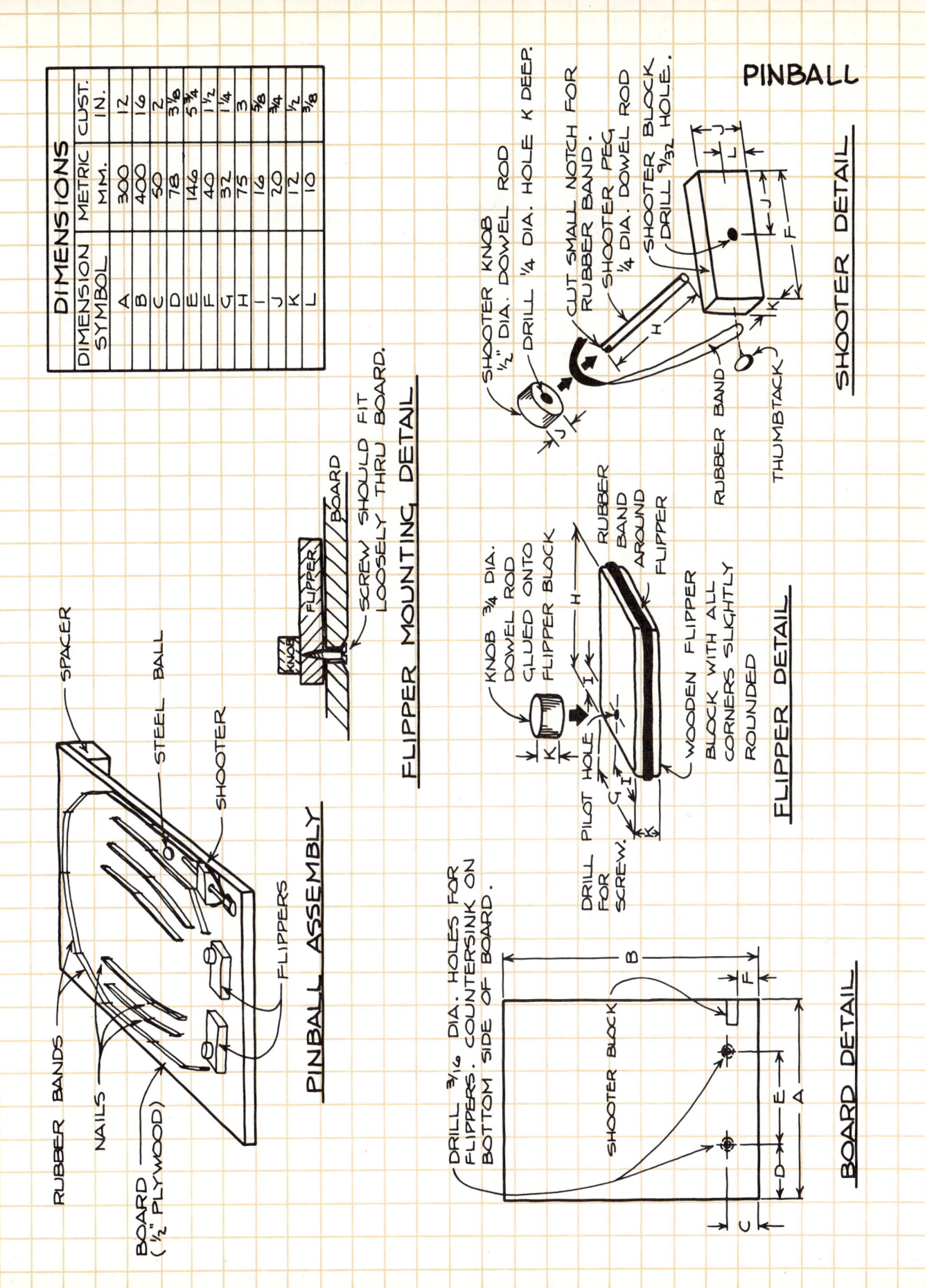

SPORTS EQUIPMENT RACK

CAN YOU RESEARCH & DEVELOP A SPORTS EQUIPMENT RACK?

HERE'S THE IDEA:

FOOTBALLS, TENNIS RACQUETS, FRISBEES, AND OTHER SPORTS ITEMS CAN BE DIFFICULT TO STORE. A SPECIAL RACK CAN HELP YOU TO STORE AND TO TAKE BETTER CARE OF YOUR EQUIPMENT.

RESEARCH & DEVELOPMENT DECISIONS:

DESIGN INFORMATION IS GIVEN FOR ONE BASIC TYPE OF RACK. YOU WILL HAVE TO R&D THE **SIZE** AND THE **LAYOUT** OF THE RACK. DESIGN THE RACK TO HOLD YOUR OWN SPORTS EQUIPMENT.

- WHAT SPORTS EQUIPMENT DO YOU HAVE?

- WHAT SIZE SHOULD YOUR SPORTS EQUIPMENT RACK BE?

- WHAT SIZE HOLES, SLOTS, OR PEGS ARE NEEDED TO HOLD YOUR EQUIPMENT?

- WHAT MATERIALS WOULD MAKE THE BEST RACK?

- WHERE MIGHT YOU PUT A SPORTS EQUIPMENT RACK? HOW WOULD IT BE MOUNTED?

- COULD THE RACK BE DESIGNED TO HOLD SOMETHING OTHER THAN SPORTS EQUIPMENT? PHOTO EQUIPMENT? HOBBY MATERIALS? TROPHIES AND AWARDS?

- WHAT OTHER DECISIONS MIGHT BE NECESSARY?

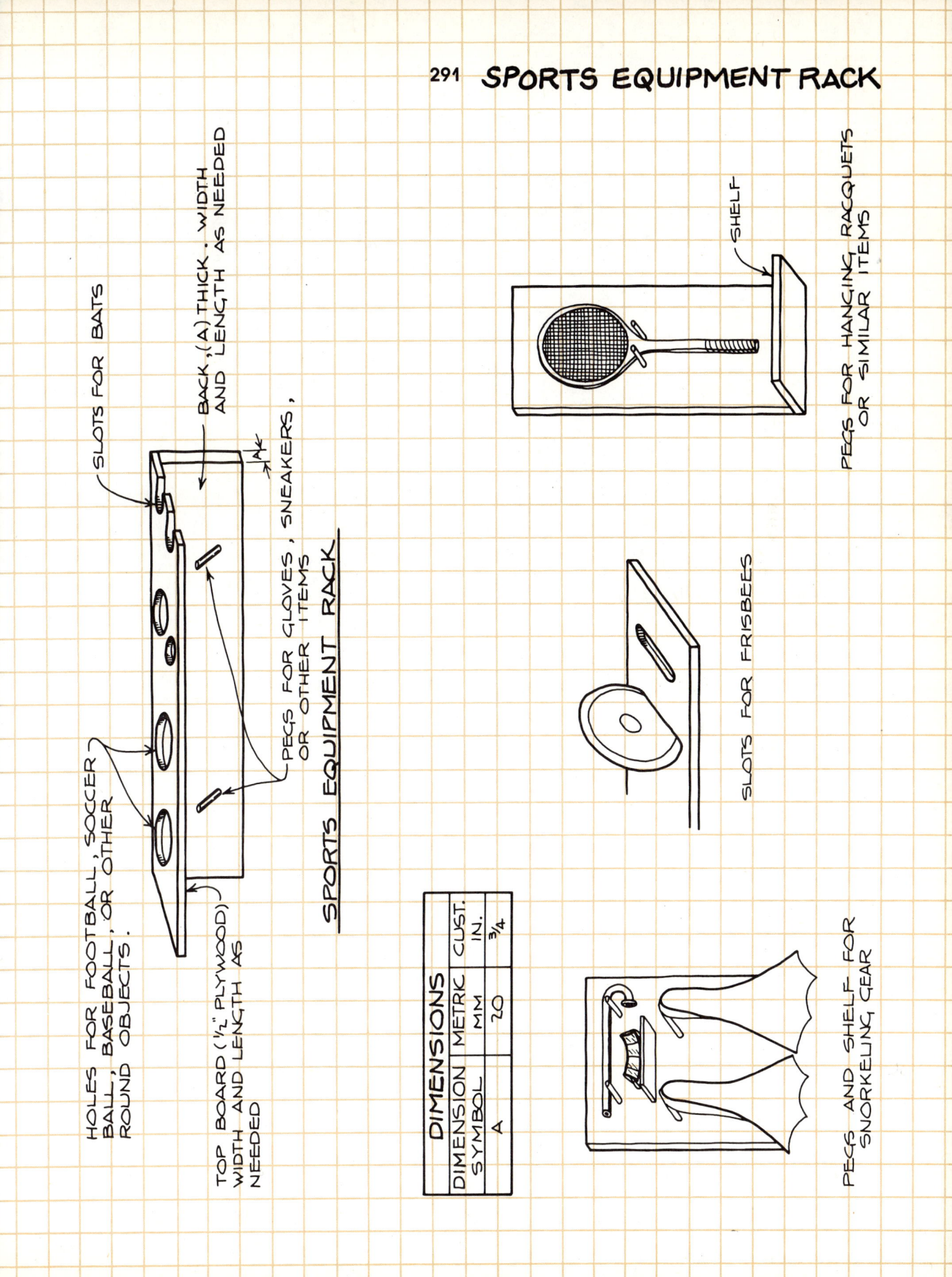

PENNY SPORTS

CAN YOU RESEARCH & DEVELOP A PENNY SPORTS GAME?

HERE'S THE IDEA:

USE A PENNY AS A BALL OR PUCK IN A MINIATURE VERSION OF YOUR FAVORITE SPORT. A FLAT BOARD CAN BE THE PLAYING FIELD. A FENCE IS NEEDED AROUND THE SIDES.

RESEARCH & DEVELOPMENT DECISIONS:

DESIGN INFORMATION IS GIVEN FOR A BASIC GAME BOARD. YOU WILL NEED TO R&D THE <u>TYPE OF GAME</u>, THE <u>LAYOUT</u>, AND THE NECESSARY <u>ACCESSORIES</u> (HELPFUL ADDITIONS) FOR THE GAME.

- HOW COULD YOU MAKE A BASKETBALL GAME? FOOTBALL? SOCCER? HOCKEY?

- CAN YOU MAKE UP YOUR OWN GAME?

- HOW CAN YOU SHOOT, "KICK", OR HIT THE PENNY?

- WHAT OBSTACLES COULD YOU ADD TO MAKE THE GAME MORE CHALLENGING?

- CAN YOU MAKE A DIFFERENT GAME ON THE OTHER SIDE OF THE BOARD?

- WHAT ACCESSORIES MIGHT YOU MAKE? GOAL POSTS? BASKETS? A GOALIE? HITTING STICKS?

- WHAT OTHER DECISIONS MIGHT BE NECESSARY?

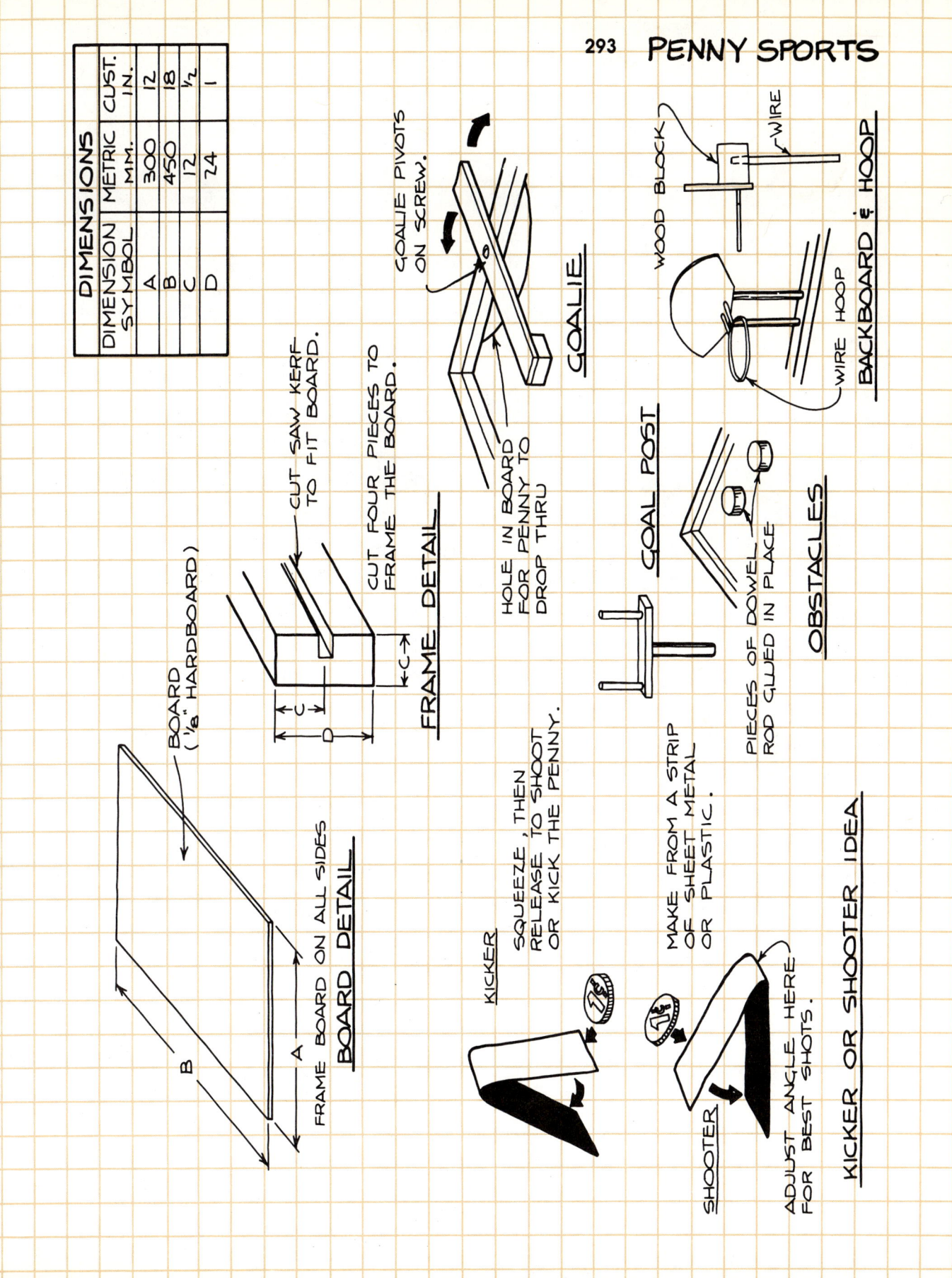

LOCKER ORGANIZER

CAN YOU RESEARCH & DEVELOP A LOCKER ORGANIZER?

HERE'S THE IDEA:

SOME THINGS GET LOST EASILY IN A LOCKER. A SPECIAL HOLDER FOR SMALL ITEMS CAN HELP YOU TO ORGANIZE YOUR LOCKER.

RESEARCH & DEVELOPMENT DECISIONS:

ONE IDEA FOR A LOCKER ORGANIZER IS SHOWN IN THE PHOTOGRAPH. USE YOUR IMAGINATION TO VISUALIZE (SEE IN YOUR MIND) OTHER IDEAS. YOU WILL NEED TO R&D THE <u>ENTIRE</u> PRODUCT.

- WHAT SMALL OBJECTS DO YOU KEEP IN YOUR LOCKER? PENCILS? COMB? GUM?
- HOW MIGHT A LOCKER ORGANIZER BE MADE? WHAT SHAPE? WHAT SIZE? HOW MANY SPACES?
- WHICH MATERIAL WOULD BE BEST? WOOD? PLASTIC? METAL? A COMBINATION OF MATERIALS?
- WHAT FINISH MIGHT BE USED?
- WHAT EXTRAS COULD BE INCLUDED? A MIRROR? A CALENDAR?
- HOW WILL IT BE FASTENED TO YOUR LOCKER?
- WHAT OTHER DECISIONS MIGHT BE NECESSARY?

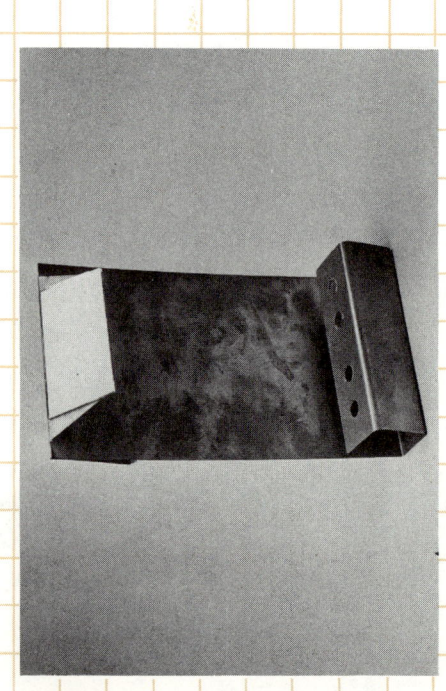

UNIT V

TOOLS, MATERIALS, PROCESSES

Topic 1 Metrics

Systems of Measurements

There is more than one system of measurement. In the United States we are mostly concerned with two different systems: metric and customary.

The customary system is also called the **U.S. Customary** system of measurement. People in the United States have been using this system for many, many years. It is **not** an **easy** system. But people have learned it and are used to it. As time passes, however, the customary system will be used less and less. The metric system is being adopted by the United States.

The metric system is also known as the **SI Metric** system. SI is an abbreviation of International System of Units. This system is used almost all over the world.

The chart given in Figure T1-1 shows the units of measure commonly used in the customary and metric systems. Also shown are the kinds of measurements they describe. Examine especially the metric terms used in describing length, weight, volume, and speed. Compare these with the customary terms for the same measurements. These metric terms illustrate one factor that makes the metric system easier than the customary system. Do you know what it is? Can you figure it out?

The Metric System

The metric system uses **base units** and **prefixes.** Look again at the metric terms for length. The third word on the list is the base unit for measuring length, **meter.** The prefixes of other terms describe lengths greater or less than one meter. In metrics, prefixes are used with base units to describe measurements.

The chart presented in Figure T1-2 lists the **prefixes** most used and what they mean. Notice that the metric system is a **decimal** system just as our money system is a decimal system. Units are related by a power of 10. Power refers to a number times itself, in this case 10. In metrics, units are powers of 10 in terms of the base unit. For example, meter is a base unit. A kilometer is 1000 meters. The number 1000 is a power of 10. It is 10 to the third power (10^3) or 10 × 10 × 10. Other units are related in a similar way. Refer again to Figure T1-2. In figure T1-3, the **base units** commonly used are given.

Figure T1-1

Comparison of Metric and Customary Units of Measurements

Measurement	Metric Units	Customary Units
Length	millimeter centimeter meter kilometer	inch foot yard mile
Weight	gram kilogram metric ton	ounce pound ton
Volume	milliliter liter	ounce cup pint quart gallon
Area	square millimeters square meters	square inches square feet square yards
Temperature	degree Celsius	degree Fahrenheit
Speed	kilometer per hour	miles per hour

Consider the terms **centimeter** and **kilometer.** The prefix **centi-** means 1/100 or 0.01. What then does centimeter mean? (Remember, one **cent** is $0.01 of a dollar.) A **centimeter** is 1/100 or 0.01 of a meter. One hundred cents make up a dollar. One hundred centimeters make up a meter. The prefix **kilo-** means 1000. A kilometer is 1000 meters.

■ Is a milligram greater than or less than a kilogram? Less, of course. In metrics, the prefixes tell you **comparative** (greater or less) sizes or amounts. Milli- is a fraction of one base unit, kilo- is 1000 base units.

■ What is being measured? Weight — **gram** is the base unit for measuring weight.

Understanding the sizes or amounts of base units, and knowing the meanings of the prefixes will help you to learn the measurements used in metrics. See Figure T1-4. These are

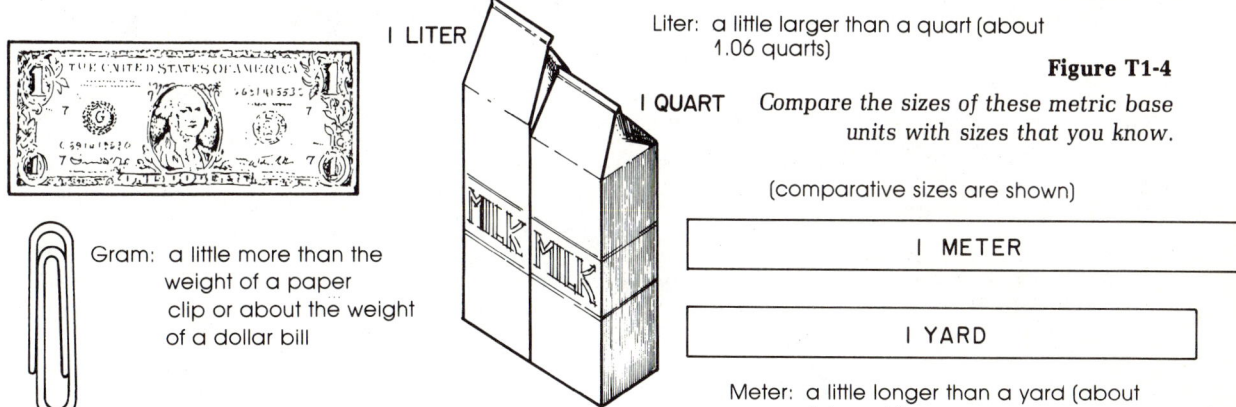

Figure T1-2
Metric Prefixes

Prefix	Symbol	Meaning
milli-	m	0.001 or $\frac{1}{1000}$ or $\frac{1}{10 \times 10 \times 10}$ $\frac{1}{(10^3)}$
centi-	c	0.01 or $\frac{1}{100}$ or $\frac{1}{10 \times 10}$ $\frac{1}{(10^2)}$
kilo-	k	1000. or $10 \times 10 \times 10$ (10^3)

Figure T1-3
Metric Base Units

Base Unit	Symbol	Measurement
meter	m	length
gram	g	weight
liter	L	volume

Figure T1-4
Compare the sizes of these metric base units with sizes that you know.

Gram: a little more than the weight of a paper clip or about the weight of a dollar bill

Liter: a little larger than a quart (about 1.06 quarts)

Meter: a little longer than a yard (about 1.1 yards)

comparisons that will help you to understand and to remember the sizes or amounts of base units.

Reading and Writing Metrics

Look again at the charts in Figures T1-2 and T1-3. Notice the column marked "Symbol." Units of measure may be represented by symbols. This makes reading and writing metrics much easier. Just as metric **prefixes** are put before the **term** for the base unit, so the **symbol for the prefix** is placed before the **symbol for the base unit.** For example, **mL** represents milliliter, **cm** centimeter, and **kg** kilogram. What does the symbol **mm** represent?

Notice that nearly all symbols are lowercase letters. The exception shown here is **L** for liter. A capital letter is used in this case because the lowercase letter **l** can be mistaken for the numeral one (1). Another exception will be shown later when we discuss temperature measurement.

Metric numbers and symbols should be written as follows:

- Whenever possible, use decimals instead of fractions. Write 0.1 mm, **not** 1/10 mm.
- Leave a space between the number and the symbol: 5 kg, **not** 5kg.
- If a number contains many digits, count three places from the decimal point and leave a space, count three more places and leave a space. In other words, keep digits in groups of three beginning at the decimal point: 5 678 910.012 345 67 mm. If the number has only four digits from the decimal point, do not leave a space: 1234.1234 mm. If the four-digit number is part of a list, space the digits as the other numbers on the list are spaced.
- Write a very small number as follows: 0.1 g, **not** .1 g. Place a zero before the decimal point.
- Area is measured most often in square meters. The symbol for this is m². A **superscript** is used. A small "2" is placed to the right and partly above the "m." Small areas are measured in square millimeters (mm²).

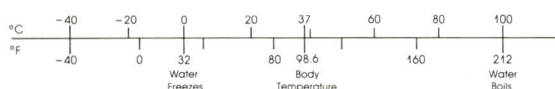

Figure T1-5

Compare the Celsius temperature scale with the Fahrenheit temperature scale.

Measuring Temperature

In metrics, temperatures are measured in **degrees Celsius** (°C). Notice that the symbol used for this measurement is a capital letter. This is because the unit of measure was named after the man who invented the system, **Anders Celsius**. This scale is easy to use. Water freezes at 0°C and boils at 100°C. See Figure T1-5. Compare the Celsius scale to the Fahrenheit scale used in the customary system.

■ Are you ill if your body temperature is 40°C?
■ Would you wear a coat if the temperature outside is 25°C?

Notice above that no space is left between the symbol for degrees and the symbol for Celsius.

Metrics and Manufacturing

As you know, people in the United States have been using the U.S. Customary system of measurement. This includes people who make products and parts for products — manufacturers. Measuring is **very** important in making a product. Changing the system of measurement will change the ways that products are made. New tools and machines are often necessary, and workers must learn to use metric measurements. See Figure T1-6.

Different sizes of tools and machine parts will probably be needed. This is because a customary measurement does not usually convert (change) to a simple metric measurement. For example, the distance from . _____ to _____ . is 1-1/4 inches in customary terms. It is still exactly the same distance if we measure it in millimeters, but we call it 31.75 mm. This is known as **soft conversion**. Metric terms are substituted for customary terms but the size remains the same.

Figure T1-6

Manufacturers encourage their workers to learn metrics.

There is another type of conversion. For example, the head of a 1/4-inch bolt is 7/16 inch. When these measurements are converted into the **equivalent** metric measurements, complex numbers result: 1/4 inch = 6.35 mm and 7/16 inch = 11.1125 mm. A wrench must fit the head of a bolt. Can you imagine buying an 11.1125 mm wrench?

Manufacturers want to keep things simple for consumers. Therefore the **sizes** of the bolts and wrenches are changed when metric measurements are used. Instead of 1/4-inch bolts with 7/16-inch heads, 6-mm bolts with 10-mm heads are made. Instead of 7/16-inch wrenches, 10-mm wrenches are made. The consumer who buys metric bolts must also buy metric wrenches. The customary wrenches will not fit the new bolts.

Changing actual sizes when changing the system of measurement is known as **hard conversion**. Remember also that when sizes are changed (bolts), related measurements (wrenches to fit the bolts) are also affected.

In manufacturing, **length** is the most common measurement needed. Length is generally measured in **millimeters**. To use metric measurements when making the products presented in this book, you will need to measure length in millimeters. Be certain that your rule or meter stick is divided into millimeters. See Figure T1-7.

Figure T1-7

This rule can be used to measure length in millimeters.

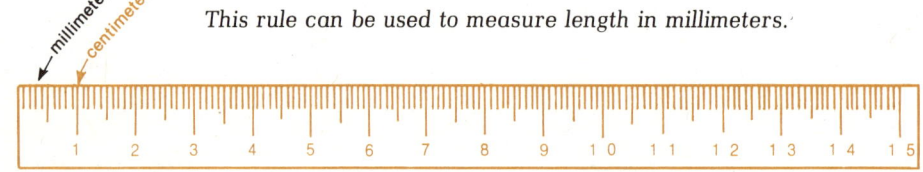

Figure T1-8   Metrics   **299**

Conversion Tables

Measure	To change from	To	Multiply by
Length	millimeters	inches	0.04
	inches	millimeters	25.4
	centimeters	inches	0.4
	inches	centimeters	2.54
	meters	feet	3.3
	feet	meters	0.3
	meters	yards	1.1
	yards	meters	0.9
	kilometers	miles	0.6
	miles	kilometers	1.6
Weight (mass)	grams	ounces	0.035
	ounces	grams	28.3
	kilograms	pounds	2.2
	pounds	kilograms	0.45
Volume	milliliters	ounces	0.03
	ounces	milliliters	29.6
	liters	pints	2.1
	pints	liters	0.47
	liters	quarts	1.06
	quarts	liters	0.95
	liters	gallons	0.26
	gallons	liters	3.8
Area	square millimeters	square inches	0.002
	square inches	square millimeters	645.2
	square meters	square yards	1.2
	square yards	square meter	0.8
Temperature	°C	°F	(°C × 1.8) + 32
	°F	°C	(°F − 32) × .555
Speed	miles per hour	kilometers per hour	1.6
	kilometers per hour	miles per hour	0.6

Kilometers	Hectometers*	Decameters*	Meters	Decimeters*	Centimeters	Millimeters
1 km =	10 hm =	100 dam =	1000 m =	10 000 dm =	100 000 cm =	1 000 000 mm
0.1 km =	1 hm =	10 dam =	100 m =	1 000 dm =	10 000 cm =	100 000 mm
0.01 km =	0.1 hm =	1 dam =	10 m =	100 dm =	1 000 cm =	10 000 mm
0.001 km =	0.01 hm =	0.1 dam =	1 m =	10 dm =	100 cm =	1 000 mm
0.0001 km =	0.001 hm =	0.01 dam =	0.1 m =	1 dm =	10 cm =	100 mm
0.00001 km =	0.0 001 hm =	0.001 dam =	0.01 m =	0.1 dm =	1 cm =	10 mm
0.000001 km =	0.00 001 hm =	0.0001 dam =	0.001 m =	0.01 dm =	0.1 cm =	1 mm

* Units of measure having hecto-, deca-, and deci- prefixes are seldom used. They are included here to help you to understand how metric units are related.

Figure T1-9

It is possible in the metric system to change from one metric measure to another by moving the decimal point. Remember though that the **base unit** must be the same. What is the base unit for the measurements shown in this chart?

Making Conversions

It is better to use **either** customary measurements **or** metric measurements when making products. Mixing terms can become confusing, and mistakes can be made. Products in this book may be made according to the metric **or** the customary system of measurement. You should choose one system or the other because the actual sizes may be different. With plans from another source, however, or because of machine limitations or other reasons, you may someday need to change measurements from one system to the other. Figure T1-8 is a table which will help you.

Because metric units of measure are related by powers of 10, **metric measurements** can be changed from one metric unit to another by simply moving the decimal point. See Figure T1-9.

To change from a smaller unit of measure to a larger one, move the decimal point to the **left**. To change from a larger unit of measure to a smaller one, move the decimal point to the **right**. For example, to change millimeters to centimeters (smaller to larger), move the decimal point one place to the left. (1 mm = .1 cm). To change kilometers to meters (larger to smaller), move the decimal point three places to the right. (9.1234 km = 9123.4 m). To change millimeters to meters, in which direction would you move the decimal point? How many places to the **left** would you have to move it? Three? Correct! (9321 mm = 9.321 m)

Only measurements having the **same** base unit can be changed in this manner. For example, you **cannot** change from grams to kilometers by moving the decimal point. These units do not measure the same thing. Grams measure weight, and kilometers measure length or distance. But you **can** change from **grams** to kilo**grams** and from kilo**meters** to milli**meters** by correctly moving the decimal point because the same **kinds** of measurements are being made.

Metrics in the United States

As the United States converts to the metric system of measurement, changes occur. Certainly the changeover is influencing American manufacturing. And it can be a good influence. Metric measures are more precise and easier to work with. Exporting products is easier when both countries use and understand the units of measure.

Already 95 percent of the world's population is using metrics. We in the United States are joining the rest of the world by adopting this logical and useful system of measurement — metrics.

Topic 2 Layout

Layout is measuring and marking materials to size. Tools used for layout include rules, squares, dividers, center punches, and many others.

Layout is done by transferring dimensions from the drawings to the material on which you are working. For example, if a dimension of a part is given as 3" on the drawings, then measure and mark 3" on the material. Practice in layout will help you to make products accurately. See Figure T2-1.

Measuring Tools

Using a Rule

A rule is used for measuring **lengths**. Common rules have markings in inches or millimeters, and some rules have both. Use a rule carefully. Your measurements must be exact. See Figure T2-2.

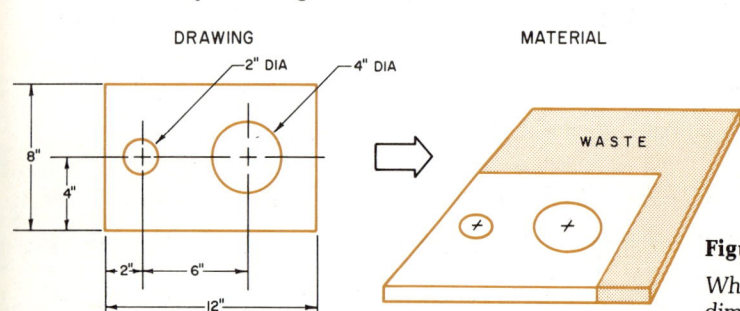

Figure T2-1

When making a layout, the dimensions shown on the drawing are measured and marked on the materials.

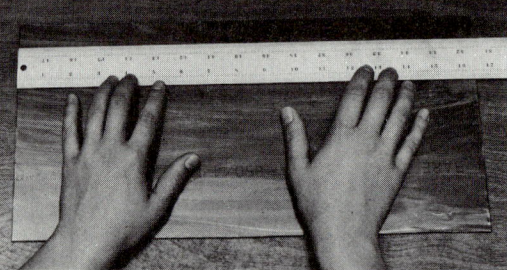

Figure T2-2

Measuring length with a bench rule.

Using Calipers

Calipers are used for measuring **diameters.** An outside caliper is used for measuring outside diameters. Dowel rods may be measured by using an outside caliper. An inside caliper is used for measuring the diameter of holes. See Figure T2-3.

Use a rule to set the correct distance on the outside caliper. Then try to slip the caliper over the item you are measuring. The diameter is correct when the tips of the caliper touch the surface very lightly.

To measure the distance across a hole with an inside caliper, slip the legs of the caliper inside the hole. Adjust the caliper so that the tips of the legs are touching the inside surfaces with a very light pressure. Then measure the distance between the tips.

Using a Micrometer

A micrometer is used to accurately measure **thickness.** A micrometer may measure in hundredths of a millimeter (.01 mm) or thousandths of an inch (.001″).

To adjust the micrometer, turn the thimble until the anvil and the spindle lightly touch opposite sides of the stock. Then add the number on the thimble to the reading on the sleeve to determine the measurement. A micrometer is used for precise measurements. Using this tool requires practice. See Figure T2-4.

Figure T2-3

Measuring diameters with calipers.

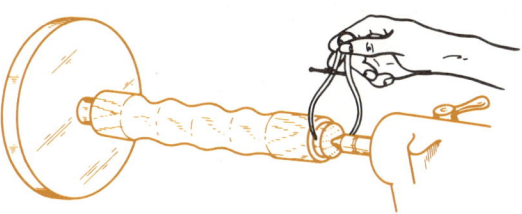

Checking a diameter using an outside caliper.

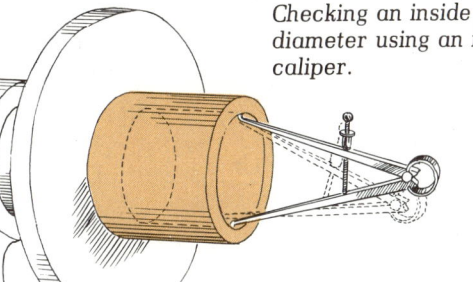

Checking an inside diameter using an inside caliper.

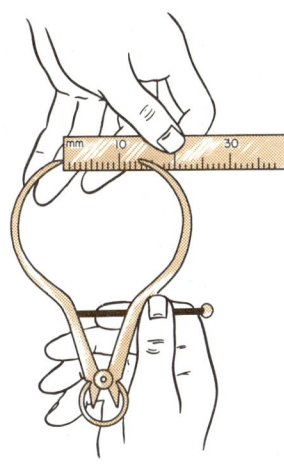

Setting an outside caliper to the proper dimension.

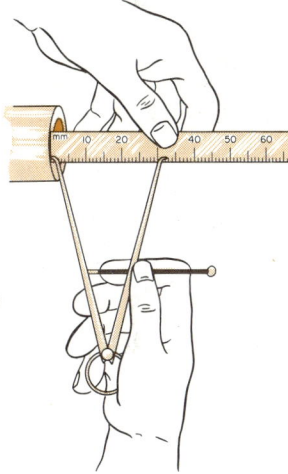

Setting an inside caliper to the proper dimension.

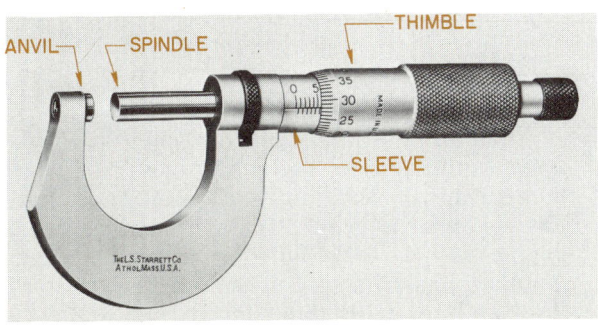

Figure T2-4

A micrometer.

The parts of a micrometer.

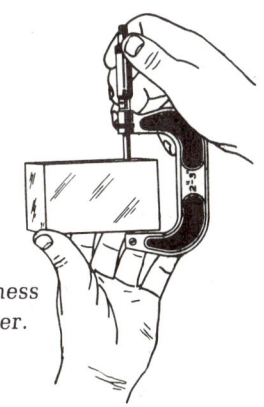

Measuring thickness with a micrometer.

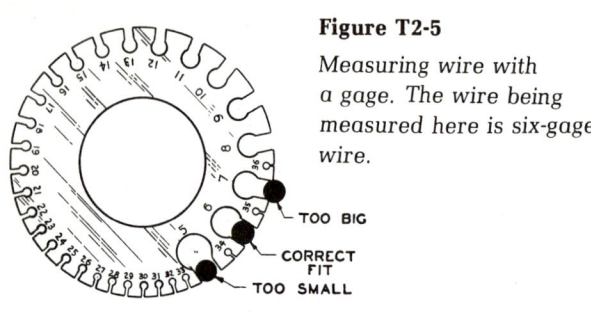

Figure T2-5

Measuring wire with a gage. The wire being measured here is six-gage wire.

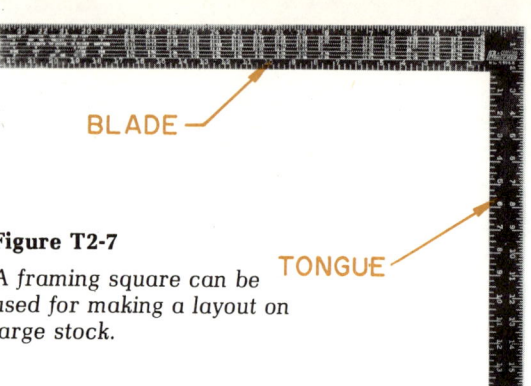

Figure T2-7

A framing square can be used for making a layout on large stock.

Figure T2-6

Using a try square.

Making a layout.

Checking squareness.

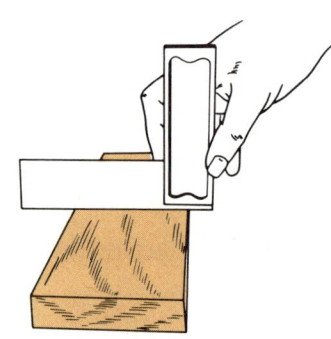

Checking the flatness of a surface.

Using a Wire and Sheet Metal Gage

Wire and sheet metal are too **thin** to be measured with a rule. Therefore, a gage with numbered slots is used to measure the different sizes. Slip the material into a slot that fits. The number marked beside the slot is the size. See Figure T2-5.

Do not force material into an opening that is too small. Remove all burrs (ridges or rough spots), and hammer all dents straight before measuring.

Using a Try Square

Use a try square to measure and mark straight **lines** across wood, metal, and plastic materials (stock) square to the edge of the stock (at a 90-degree angle). The handle of the try square is held against the edge. A line is drawn along the blade of the try square. See Figure T2-6.

You can also use a try square to check **squareness** of stock. Hold the handle against one surface and the blade against the other surface. If the two surfaces are square, no light will show between the try square and the material.

The blade of the try square may be used to see if the surface of material is **flat**. Place the blade in different locations on the surface. Check to see if light shows between the blade of the try square and the surface of the wood. If the surface is flat, no light will show.

Using a Framing Square

If the stock is too **large** for laying out lines with a try square, you can use a framing square. Use it in the same way that you would use a try square. See Figure T2-7.

Using a Combination Square

A combination square is used to **lay out** 90-degree and 45-degree angles. Since the combination square is adjustable, it can be used as a **depth gage** also. See Figure T2-8.

Using a Center Square

You can find the center of a **round** piece using a center square. The center head can be adjusted anywhere along the blade. Make sure the center head is tight against the stock. Mark a line on the stock. Turn the stock and mark another line. The center is the point at which the two lines cross. See Figure T2-9.

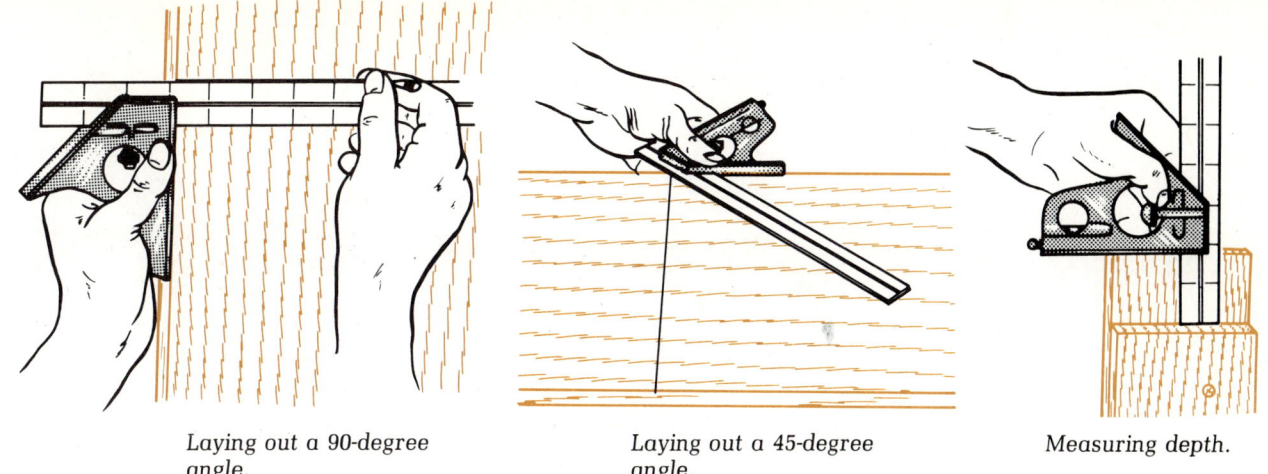

| Laying out a 90-degree angle. | Laying out a 45-degree angle. | Measuring depth. |

Figure T2-8
Using a combination square.

Using a T-Bevel

A sliding T-bevel is used for laying out **angles** other than 90 degrees. Using a protractor as a guide, set the blade at the angle you want. Then place the T-bevel against the material and draw a line along the blade. You can also duplicate (make an exact copy of) an angle using the T-bevel. See Figure T2-10.

Using Templates

Templates (patterns) are special layout tools. Some templates can be traced around to mark **shapes** on materials, such as wood, metal, or plastics. See Figure T2-11. Other templates may be used to locate **holes.** Sometimes the **lines** along which metal or plastic will be bent (bend lines) are found by using a template. (See also Topic 25, **Production Tooling,** for the use of templates.)

Marking Tools

Using a Scratch Awl

A scratch awl is used to **scribe** (scratch) lines on surfaces such as sheet metal. It is used with a guide, such as a straightedge or a template. (See Topic 25, **Production Tooling.**) Hold the straightedge or template firmly on the material. It must not slip while the material is being marked. See Figure T2-11.

Using a Center Punch

A center punch is used to mark metal to show where a **hole** is to be drilled. Place the pointed end of the punch at that location. Hit the top of the center punch with a hammer. This makes a

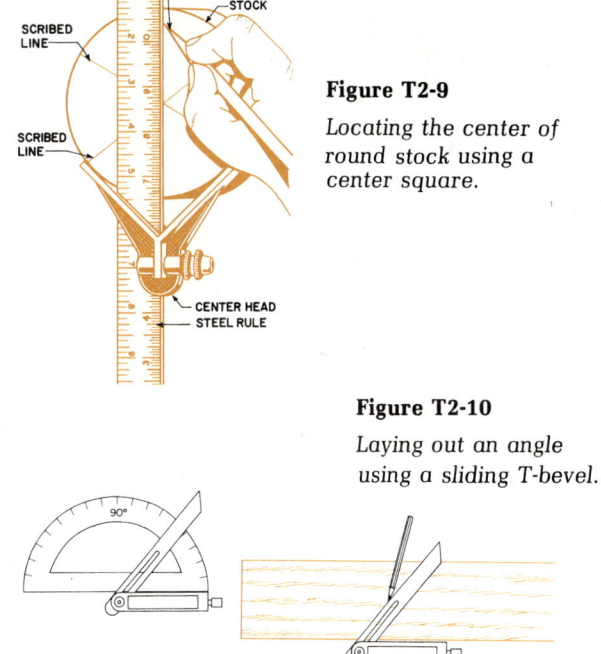

Figure T2-9
Locating the center of round stock using a center square.

Figure T2-10
Laying out an angle using a sliding T-bevel.

Figure T2-11
Marking a shape on stock. A scratch awl is being used to scribe a line around a template.

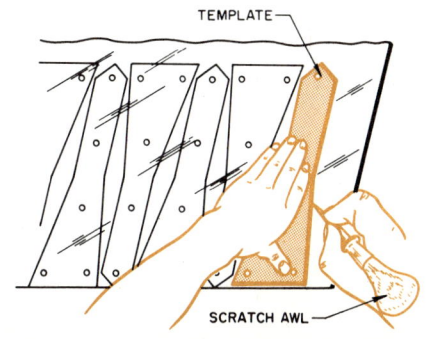

dent in the surface of the metal. The dent centers the drill in the exact location for the hole. It also keeps the drill from wandering out of position. See Figure T2-12.

Using a Marking Gage

You can scribe a line **parallel*** to the edge of wood or plastic with a marking gage. An adjustable head acts as a guide when the gage is moved. Set the gage to the correct dimension. Push or pull the marking gage to make a line on the stock. See Figure T2-13.

Using Dividers

Dividers can be used for **marking circles and arcs** (parts of a circle) on wood, metal, or plastic much as you use a compass to mark circles and arcs on paper. Dividers have two sharp points. Set the points for the correct radius. The radius is one-half of the distance across the center of a circle (half the diameter). See Figure T2-14.

Dividers can be used to **transfer measurements** from rules to materials. This tool is especially good for making repeated measurements. Using a rule, set the dividers at the desired measurement. Then use the dividers to measure and mark. Scratch or punch a mark on the material. With one point of the dividers at this location, make a mark where the **other** point touches the material. Repeat this procedure as often as necessary.

* Parallel lines extend in the same direction and are always the same distance apart.

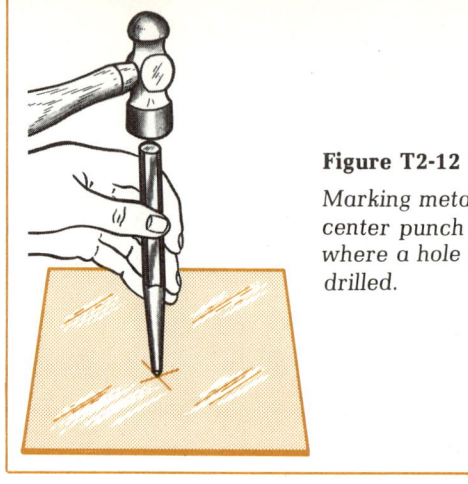

Figure T2-12
Marking metal with a center punch to show where a hole should be drilled.

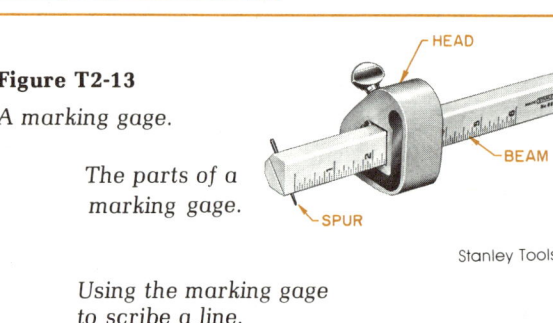

Figure T2-13
A marking gage.

The parts of a marking gage.

Stanley Tools

Using the marking gage to scribe a line.

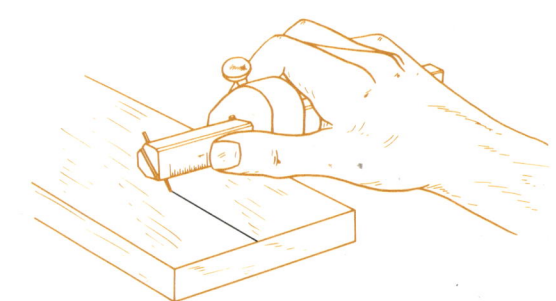

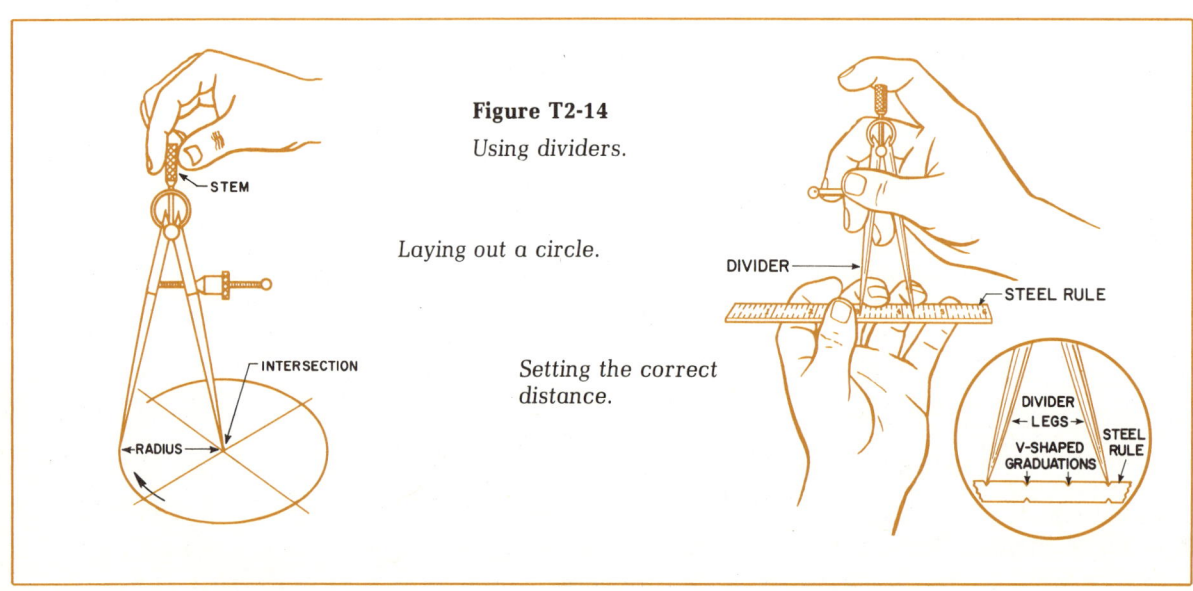

Figure T2-14
Using dividers.

Laying out a circle.

Setting the correct distance.

Topic 3 Bench Tools

Hammers, screwdrivers, pliers, and wrenches are common bench tools. There are several kinds of each of these tools to meet special needs. Bench tools are hand tools that are generally used on a workbench.

Hammers

Hammers are used to **drive** (move) **or shape objects** with a striking force. There are different hammers for different jobs. Use the proper hammer for the task you have to do. See Figure T3-1.

Using a Claw Hammer

A claw hammer is used to **drive nails** into wood. It can also be used to **pry** (pull) **nails** from wood. A claw hammer is classified in size by the weight of its head. Common weights are 13 ounces and 16 ounces. Use a claw hammer **only** for driving and pulling nails.

Using a Mallet

A mallet is used most often to **strike other tools.** You may also use a mallet to pound on a **material** whose surface you do not want to damage. Mallets are commonly made of wood, plastic, rawhide, or rubber. Mallets range in size from just a few ounces to several pounds.

Using a Ball Peen Hammer

Use a ball peen hammer for hammering **metal.** You can use it to pound metal parts together or to bend metal in a vise. It can also be used to round the end of a rivet or to hammer sheet metal into shape. Using the ball end (peen) of the head to round a rivet or to hammer metal into a bowl shape is called **peening.** Ball peen hammers range in weight from 6 ounces to 40 ounces.

Using a Setting Hammer

A setting hammer is used for **shaping sheet metal.** The wedge-shaped end is used for striking narrow strips of metal. The square end of the hammer is used for flattening hems and for bending the edges of sheet metal.

Hammers

1. Do not use a hammer that has a cracked or broken handle.
2. Make sure that the hammerhead is firmly attached to the handle.

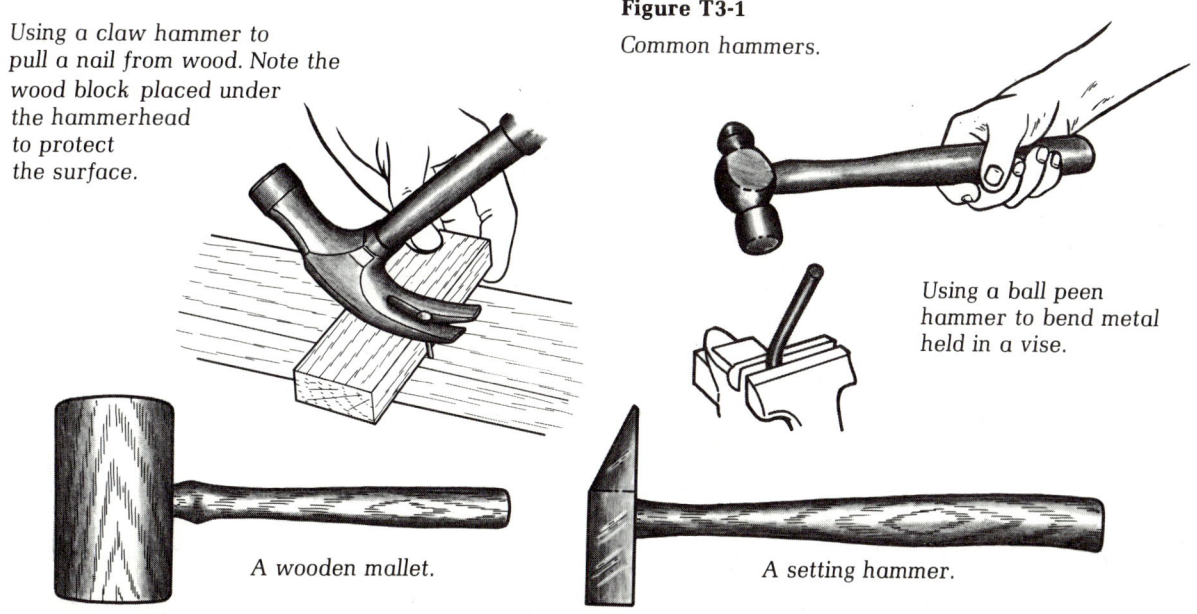

Using a claw hammer to pull a nail from wood. Note the wood block placed under the hammerhead to protect the surface.

Figure T3-1
Common hammers.

Using a ball peen hammer to bend metal held in a vise.

A wooden mallet.

A setting hammer.

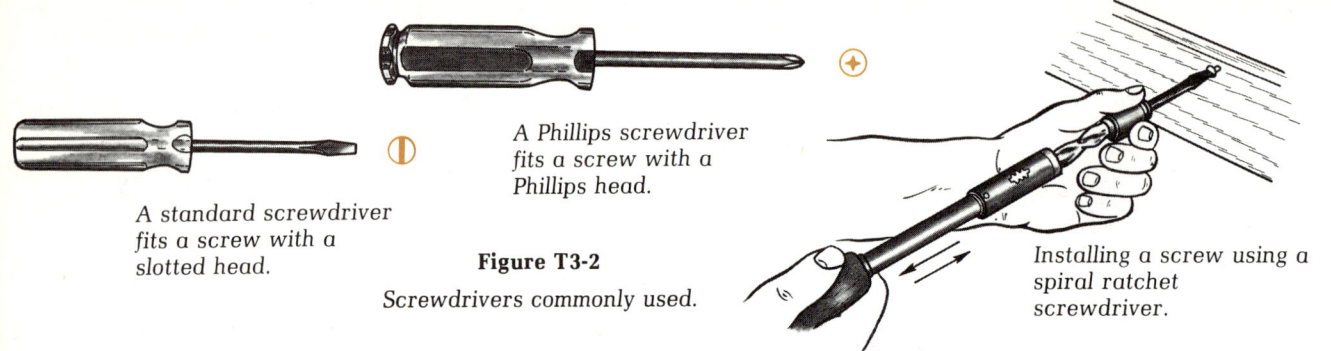

A standard screwdriver fits a screw with a slotted head.

A Phillips screwdriver fits a screw with a Phillips head.

Figure T3-2
Screwdrivers commonly used.

Installing a screw using a spiral ratchet screwdriver.

Screwdrivers

A screwdriver is a tool for **installing and removing screws.** (See Topic 9, **Threaded Fasteners and Taps & Dies.**) There are many kinds of screwdrivers. Each kind has a special purpose. See Figure T3-2.

Using Common Screwdrivers

Select a screwdriver with the kind of blade needed for your purpose. Most screws are either slotted or Phillips head. The blade tip for common screwdrivers is either straight across or shaped for a Phillips head screw. Select a screwdriver with a tip that fits snugly into the screwhead.

Using a Spiral Ratchet Screwdriver

A spiral ratchet screwdriver is useful when installing or removing screws. When you push down on the handle, the blade turns. This does the work much faster than turning a regular screwdriver. Either a regular or a Phillips blade can be used in the spiral ratchet screwdriver.

Pliers

Pliers are used for **holding and cutting.** Some pliers may also be used for twisting wire. There are many kinds of pliers. Each has a special purpose. Common pliers include slip-joint, adjustable, long-nose, cutting, and locking pliers. See Figure T3-3.

Using Slip-Joint Pliers

Slip-joint pliers are the most common type of pliers. They are used mostly for **holding** stock. The slip joint allows the jaws to be opened extra wide if needed. Some slip-joint pliers can also be used to **cut** wire.

Using Adjustable Pliers

Adjustable pliers have long handles. The jaws can be adjusted to several opening sizes. The long handles increase your **holding** power.

Using Long-Nose Pliers

Long-nose pliers are used to **hold** small objects. They are good for bending wire and reaching into small places.

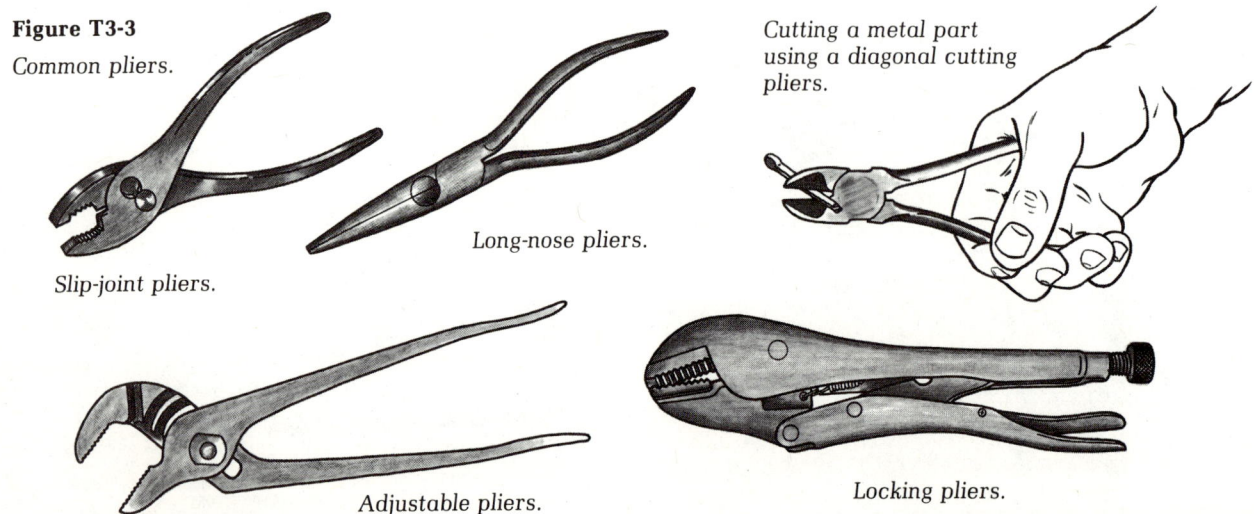

Figure T3-3
Common pliers.

Slip-joint pliers.

Long-nose pliers.

Cutting a metal part using a diagonal cutting pliers.

Adjustable pliers.

Locking pliers.

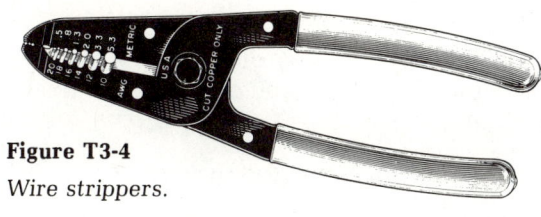

Figure T3-4
Wire strippers.

The Cooper Group

Using Wire Strippers

Wire strippers are used to **strip insulation** from electrical wires. The blades cut through the insulation but not necessarily through the wire. See Figure T3-4.

To use wire strippers, select the proper setting for the size of wire you are stripping. Squeeze the blades around the wire and pull toward the end of the wire. Wire strippers can also be used for **cutting** wire to length.

Using Diagonal Cutting Pliers

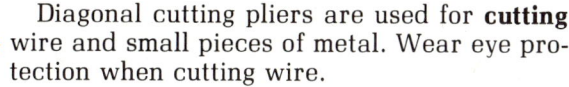 Diagonal cutting pliers are used for **cutting** wire and small pieces of metal. Wear eye protection when cutting wire.

Using Locking Pliers

A pair of locking pliers is useful when **holding** stock. The jaws can be adjusted to the right size and then locked onto the stock. You do not have to keep squeezing the pliers to hold the jaws on the work. Some locking pliers have straight jaws for holding flat or square stock. Others have curved jaws for holding round stock.

Wrenches

Wrenches are used for **holding** and **turning** nuts and bolts. (See Topic 9, **Threaded Fasteners and Taps & Dies**.) Many different kinds and sizes of wrenches are available. You should use the proper wrench for each job. See Figure T3-5.

Using an Adjustable Wrench

An adjustable wrench has one fixed jaw and one jaw that is movable. You can adjust this wrench to fit any size nut or bolt head. When using this wrench, pull the wrench with the force against the fixed jaw. Make sure that the wrench is tight on the nut before applying force.

Using an Open-End Wrench

An open-end wrench has a fixed size opening at each end. Sets of open-end wrenches are made to fit standard-sized nuts and bolts. Select a wrench that fits snugly on the **sides** of the nut or bolt head.

Using a Box Wrench

A box wrench is used on hexagonal-shaped nuts and bolts. Each of these nuts and bolts has six sides and six corners. A box wrench grips (holds) the **corners** of the nut or bolt rather than the sides. Like an open-end wrench, each end of a box wrench fits one size nut or bolt. An advantage over the open-end wrench is that the box wrench can be moved a shorter distance before changing its position on the nut. This is useful when working in a place where there is little room to move the wrench.

Using a Combination Wrench

A combination wrench has an open-end wrench on one end and the same size box wrench on the other end. The **grip** on a nut or bolt can be changed quickly without changing tools.

Using a Hex Wrench

A hex wrench grips a bolt from the **inside** rather than from the outside. A setscrew or headless bolt has a hex-shaped hole in the top. Use the hex wrench which fits the setscrew.

Figure T3-5
Common wrenches.

An adjustable wrench can be tightened to fit your needs. Pull in the direction of the arrow.

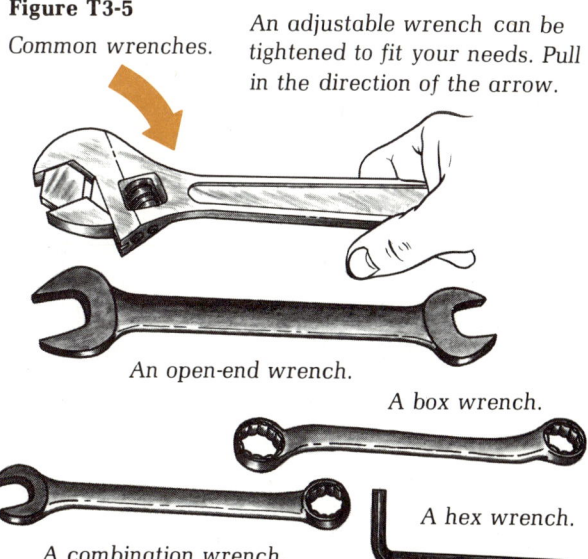

An open-end wrench.

A box wrench.

A combination wrench.

A hex wrench.

Topic 4 Shearing

Thin materials, such as paper, sheet metal, and some plastics are often cut by shearing. Among the many shearing tools are scissors, tin snips, cold chisels, squaring shears, and hand punches.

Using a Scissors

A scissors is used to cut many different kinds of **thin materials,** such as cloth and paper. It can be used to shear either **straight or curved** lines. See Figure T4-1.

Using a Utility Knife

A utility knife is a sharp, single-edged tool. It is used to cut **soft materials,** such as cardboard or leather. The blades are disposable. This means that they are thrown away after they become dull. A straightedge or a template is usually needed as a guide for the blade. The blade will cut materials that are too thick to be cut by a scissors. The utility knife, is **very** sharp. Be careful that you do not cut yourself. See Figure T4-2.

Using Tin Snips and Aviation Snips

Tin snips and aviation snips are used to shear sheet metal and, sometimes, leather and plastics. These tools are more powerful than scissors and can cut straight or curved lines efficiently in **harder materials.** See Figures T4-3 and T4-4.

Using a Notcher

A notcher is a special shear used to **cut notches** in sheet metal. A hand-held notcher is small and portable. Larger notchers are stationary and mounted on a stand or bench. By using a notcher, two cuts can be made at the same time. A notcher usually cuts 45- or 90-degree notches. See Figure T4-5.

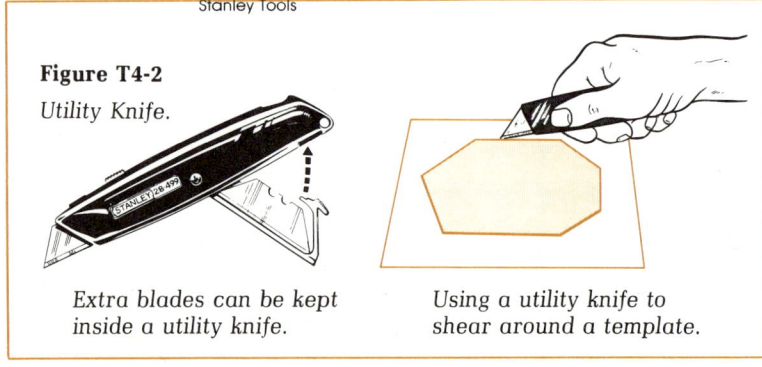

Figure T4-2
Utility Knife.

Extra blades can be kept inside a utility knife.

Using a utility knife to shear around a template.

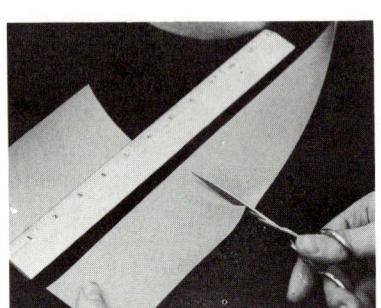

Figure T4-1
Using common scissors.

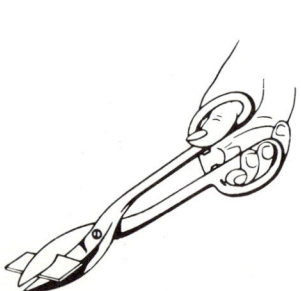

Figure T4-3
Using tin snips to cut metal.

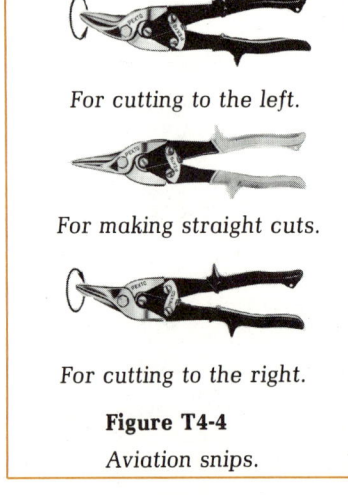

For cutting to the left.

For making straight cuts.

For cutting to the right.

Figure T4-4
Aviation snips.

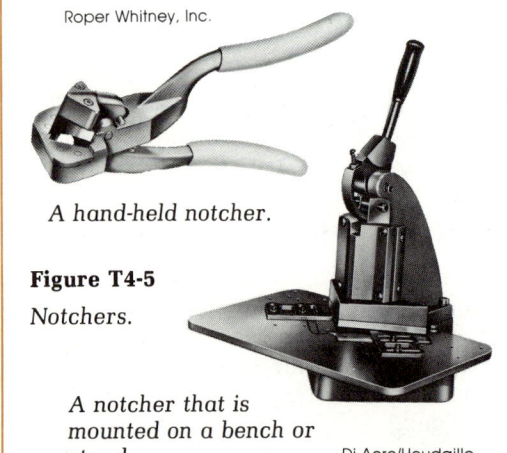

A hand-held notcher.

Figure T4-5
Notchers.

A notcher that is mounted on a bench or stand.

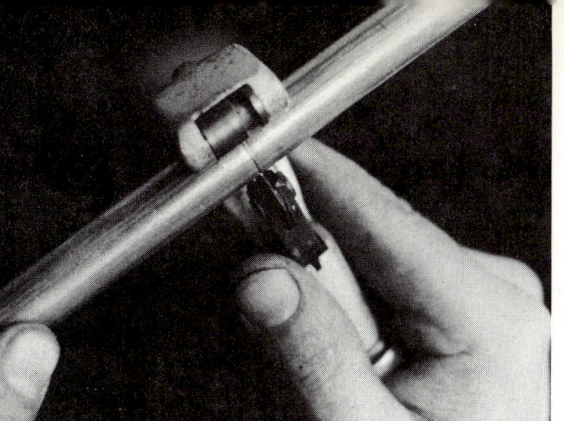

Figure T4-6
Using a tubing cutter to cut copper tubing.

Figure T4-7
Using a throatless shear to cut sheet metal.

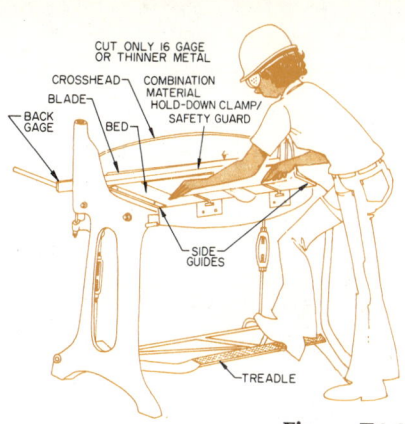

Figure T4-8
The parts of a squaring shears.

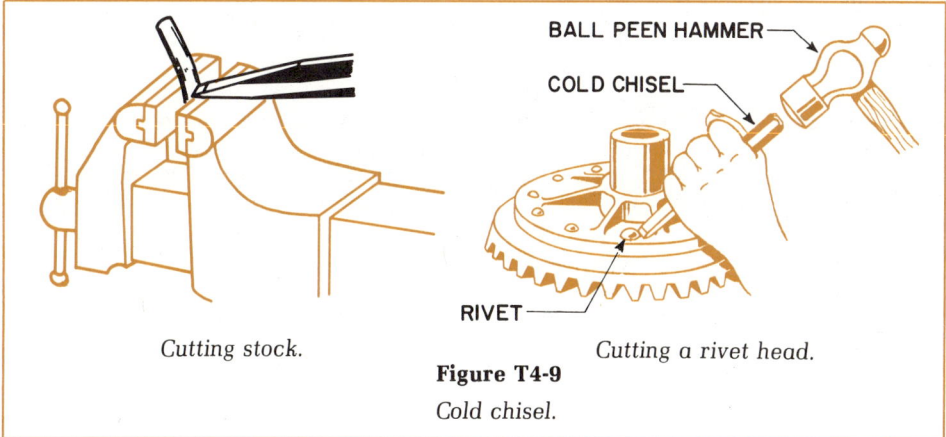

Figure T4-9
Cold chisel.

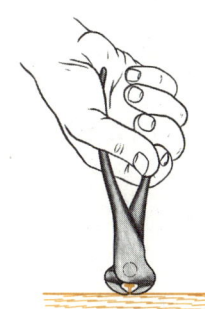

Figure T4-10
Using an end cutting nipper to cut a nail.

Using a Tubing Cutter

A tubing cutter is a special tool made for cutting **soft metal tubing,** such as copper and aluminum. A tubing cutter is fast and accurate. A sharp cutter wheel shears the tubing as the cutter is rotated around the tubing. See Figure T4-6. Pressure to make the cut is provided by tightening the adjusting screw.

Using a Throatless Shear

A throatless shear cuts sheet metal on **straight or curved** lines without bending or deforming (changing the shape) the metal. This is a bench-mounted tool with a long handle. The long handle provides good leverage (increase of force.) Because of this, heavier sheet metal can be sheared. See Figure T4-7.

Using a Squaring Shears

A squaring shears is used to cut sheet metal in **straight** lines. Large sheets of metal can be cut accurately on this machine. The size of each squaring shears is determined by the widest material that it will shear.

A **back gage** can be adjusted behind the blade to cut several pieces the same size. Guides may also be bolted to the table in front of the **blade** for use when cutting angles. To make square cuts, hold the edge of the metal against the **guide** at the side of the table. Do not place your foot under the **treadle.** See Figure T4-8.

Using a Cold Chisel

A cold chisel is most often used to cut **metal rods.** Sometimes it is used to cut sheet metal that is thicker than 20 gage. (The lower the gage number, the thicker the metal.) Use a ball peen hammer to strike the chisel. Make certain that no one is in line with chips that fly from the metal that is being sheared. See Figure T4-9.

Using an End Cutting Nipper

An end cutting nipper is designed with one edge tapered (narrowed gradually) to a sharp cutting edge. This tool is used to shear wire, rivets, and nails. A nipper has good leverage and heavy jaws. It can efficiently shear these items flush (even) with the surface of a material. See Figure T4-10.

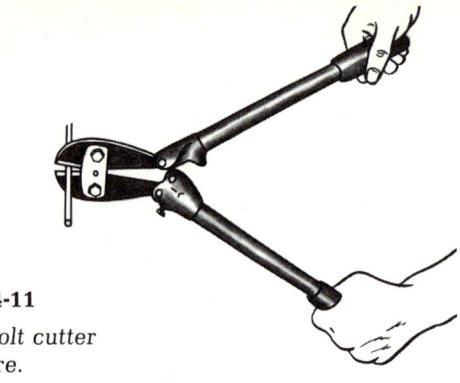

Figure T4-11
Using a bolt cutter to cut wire.

Using a Bolt Cutter

A bolt cutter is used to quickly shear bolts, metal rods, and thick wire. The long handles produce powerful leverage. This tool can shear stock quickly. See Figure T4-11.

Using Hand and Bench Punches

Hand and bench punches are used to punch holes in **sheet metal**. Each tool has a metal punch and die. The hole in the die matches the diameter of the punch. The punch and die can be changed with different size sets for larger or smaller holes. Punches make clean holes quickly and safely. See Figure T4-12.

A hand punch is shaped much like a pair of pliers. The handles are squeezed together to punch the hole.

A bench punch is used by positioning metal under the punch and pulling a lever down to punch the hole.

Using Leather Punches

Two kinds of leather punches shear holes in **leather or cloth**. See Figure T4-13. These punches are tubes. Each tube has one end tapered to a sharp cutting edge. One kind of leather punch is the **rotary punch**. Rotate the wheel to the tube which will cut the size hole you wish to make.

The other kind of punch is the **round drive punch**. Round drive punches come in different sizes. Hit the punch with a mallet. Be sure to use a block of wood under the work that you are punching. (See Topic 3, **Bench Tools**.)

Using a Swivel Knife

A swivel knife is used to make decorative and useful cuts in leather. You can make **straight or curved** cuts. The knife blade swivels (turns) to help you make curved cuts more neatly.

Place the leather on a hard surface. Hold the swivel knife as shown in Figure T4-14. Push down with your index finger. Use your thumb and other fingers to hold the blade straight or to make the blade turn as you pull the knife toward you. Be sure to keep the knife perpendicular (at a 90-degree angle) to the leather.

Using a Leather Gouge

A leather gouge shears a **small groove** in leather. This small groove makes folding or bending leather easier. Normal gouging depth is about 1/2 the thickness of the leather. Gouging is done on the flesh side (opposite the good side) of the leather. Gouging is usually easier if you moisten the leather first. See Figure T4-15.

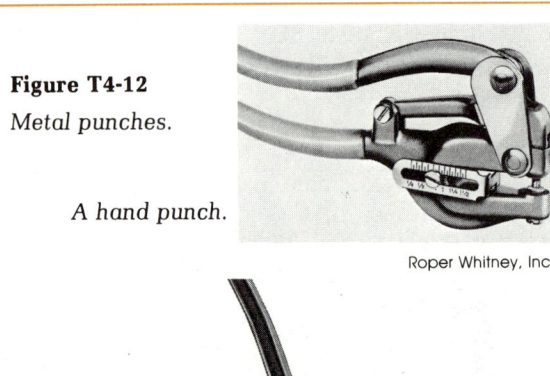

Figure T4-12
Metal punches.

A hand punch.

Roper Whitney, Inc.

A bench punch.   Roper Whitney, Inc.

Figure T4-13
Leather punches.

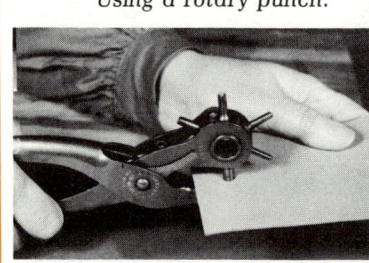

Using a rotary punch.

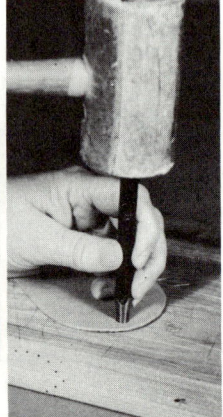

Using a drive punch.

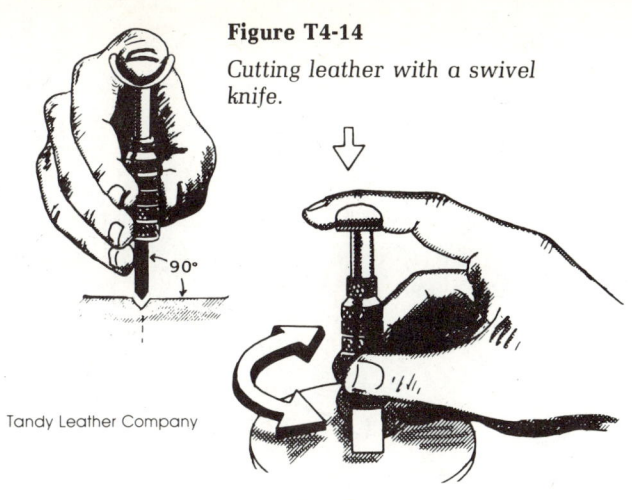

Figure T4-14
Cutting leather with a swivel knife.

Tandy Leather Company

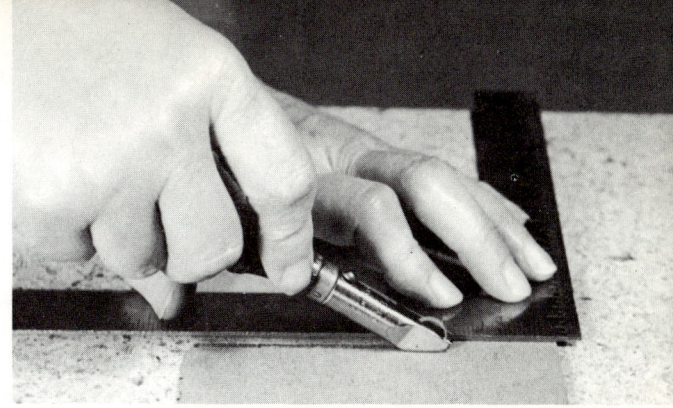

Figure T4-15
Using a leather gouge to cut a groove in leather.

Topic 5 Sawing

Sawing is a common way of cutting material to size or shape. Wood, metal, and plastics can be cut easily by sawing.

Always plan your cut before you saw. First lay out the line to be cut. Use this line as a guide. When you saw, a small channel of material is removed from the stock. This is called **kerf.** Always cut on the waste side of the line (the side not planned for use) to allow for the kerf.

 There are many kinds of saws. Each kind is useful in its own way. Use all saws carefully!

Handsaws

Handsaws are used most often to **cut wood.** All but the coping saw cut on the forward stroke. Start the cut by pulling the saw toward you. Use a long stroke. Push the saw forward, and then pull it back toward you. Keep the saw straight while you cut. When the cut is nearly complete, take short, easy strokes to avoid splintering.

Using a Crosscut Saw

A crosscut saw has small, sharp teeth. It is used to cut a board **across the grain.** A crosscut saw may be used to cut a board to length. See Figure T5-1.

Using a Ripsaw

Ripsaws have larger teeth than crosscut saws. The teeth are chisel-shaped. Use a ripsaw to cut wood **with the grain** (rip). A ripsaw is used to cut a board to width. See Figure T5-2.

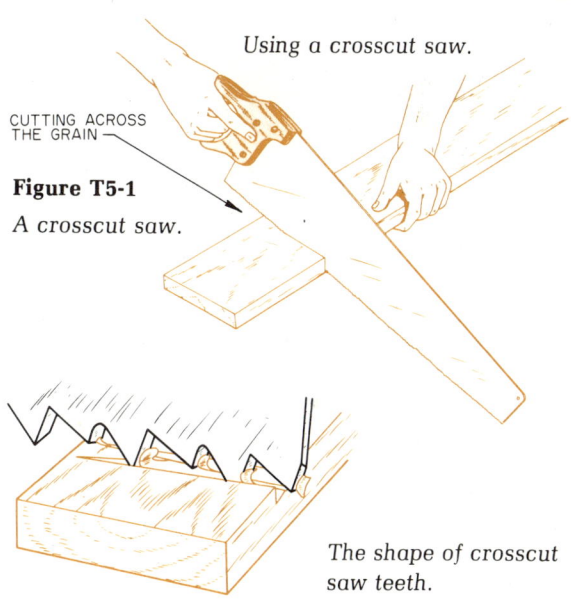

Using a crosscut saw.

CUTTING ACROSS THE GRAIN

Figure T5-1
A crosscut saw.

The shape of crosscut saw teeth.

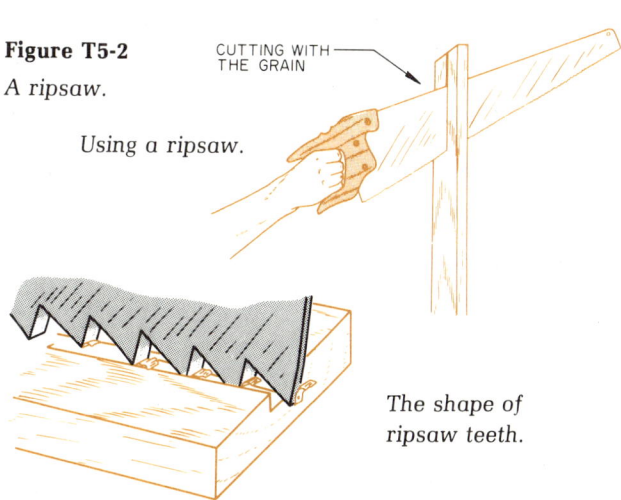

Figure T5-2
A ripsaw.

CUTTING WITH THE GRAIN

Using a ripsaw.

The shape of ripsaw teeth.

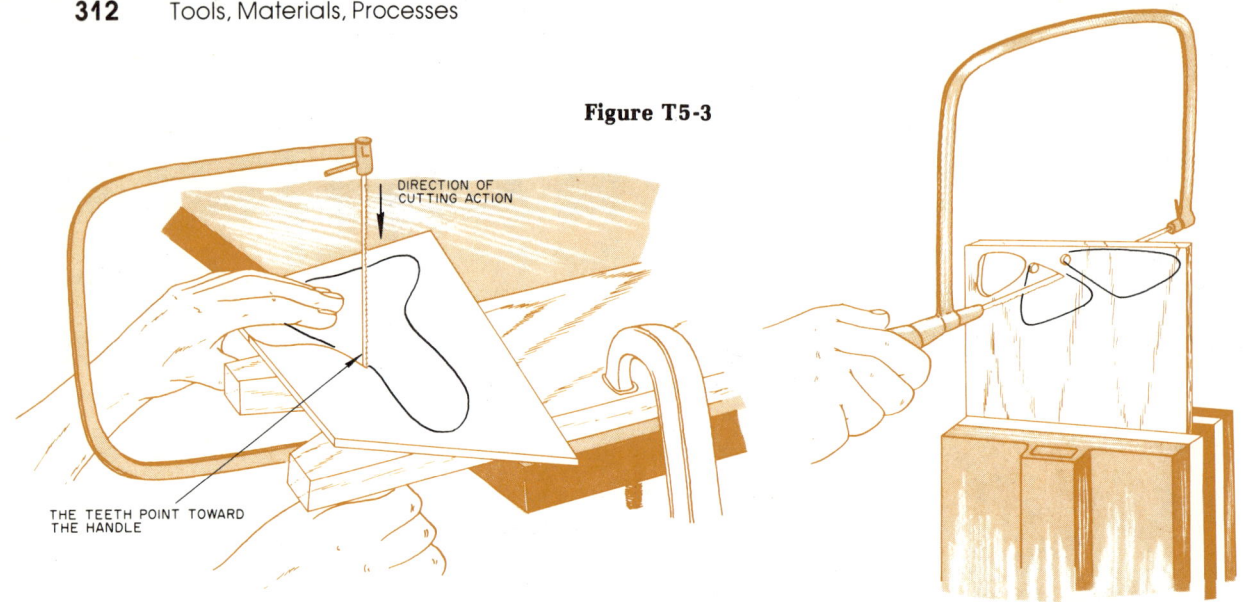

Cutting outside curves using a coping saw.

Cutting inside curves using a coping saw.

Using a Coping Saw

A coping saw is used to make **curved and irregular cuts** through wood or plastics. A coping saw has replaceable blades. The blade may be taken out of the metal frame and a new one put in. You can also turn the blade in the frame. This will help you to saw in tight places. The blades have fine teeth that point toward the handle. The saw cuts as you pull it. See Figure T5-3.

Inside cuts can be made by first drilling a hole. Then disconnect one end of the blade, and insert it through the hole. Reattach the blade to the saw, and cut the hole.

Using the Backsaw

The backsaw is used for **crosscutting wood.** It has a stiff back and fine teeth. The backsaw is used to make a smooth, straight cut. A guide block clamped onto the board will help you to make an accurate cut. See Figure T5-4.

Using a Miter Box

A miter box and saw are used to accurately **cut angles in wood.** The saw is held in the guides. Adjust the guides to the desired angle. Hold the stock firmly against the fence, and make the cut.

Figure T5-4
Using a backsaw.

Miter boxes that cannot be adjusted can be bought or made easily. A backsaw is used in this type of miter box. See Figure T5-5.

Using a Hacksaw

A hacksaw is used to **cut metal.** A replaceable blade is held in a metal frame. See Figure T5-6.

The teeth should point away from the handle. This means that the saw will cut when you push forward. Use the proper blade for the material that you are cutting. The finer the material, the more teeth the blade should have. Use a blade with 18 teeth per inch for cutting drill rod and metal thicker than 1/4″. Use a blade with 24 teeth per inch for cutting angle iron and other

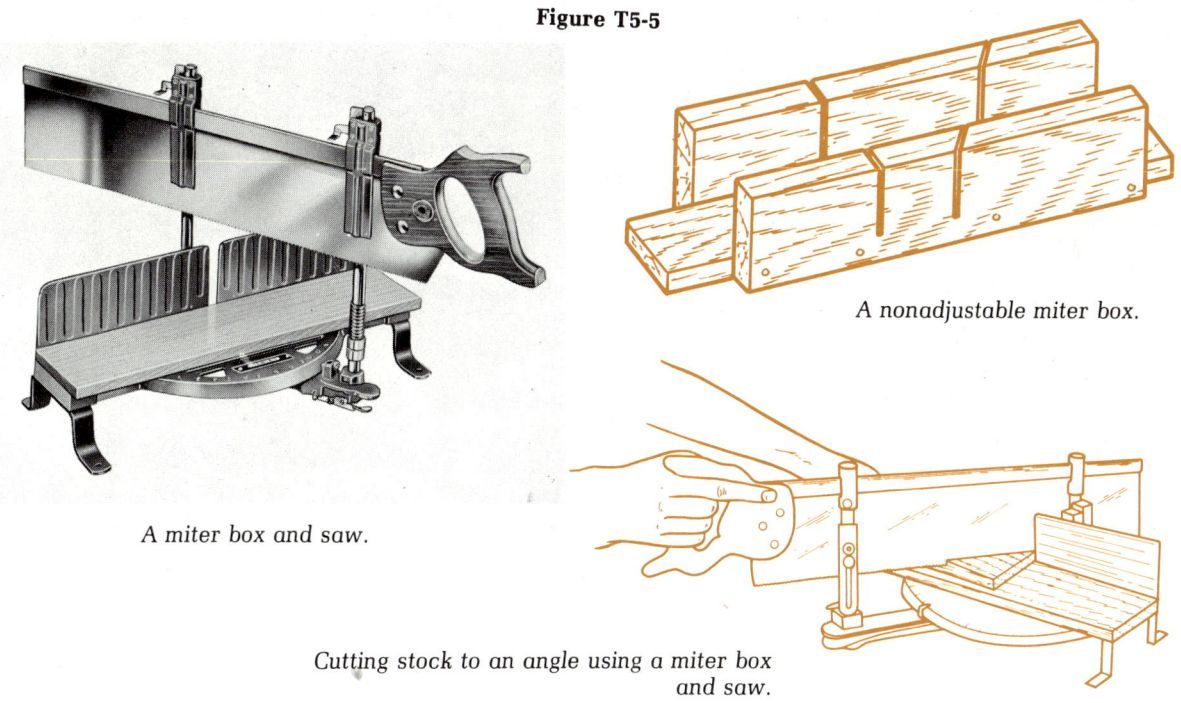

Figure T5-5

A nonadjustable miter box.

A miter box and saw.

Cutting stock to an angle using a miter box and saw.

The parts of a hacksaw.

Using a hacksaw.

Figure T5-6

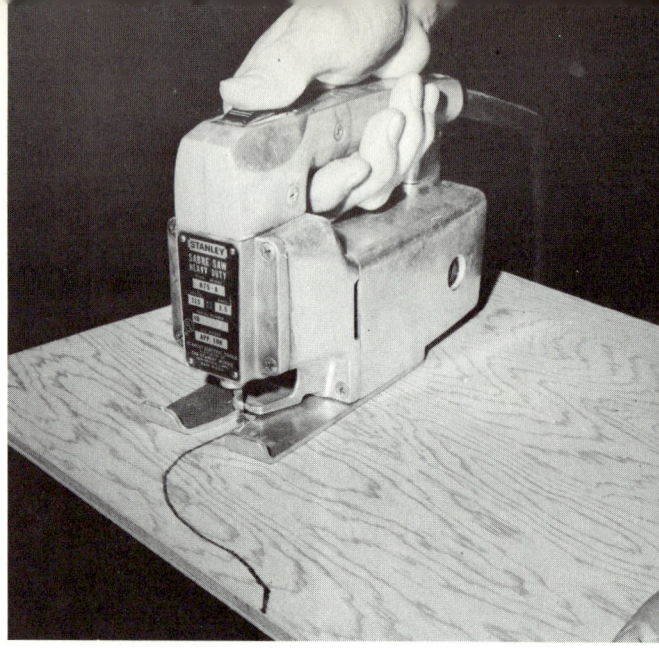

Figure T5-7

Using a saber saw to cut a curve.

metals with similar shapes. A fine-tooth blade, 32 teeth per inch, should be used for cutting sheet metal, tubing, or other thin metals.

Power Saws

Power saws are useful tools. Used properly, they will help you to make many interesting products. But used improperly, they are **dangerous!** Use extreme care when doing work with a power saw. Be sure you know **exactly** how to use it. Never use it on your own. First get permission from your teacher. **Review, remember, and follow all safety rules!**

Using a Saber Saw

A saber saw is a portable electric tool. It is used to cut wood, metal, and plastics in **curved** or **straight** lines. See Figure T5-7.

The blade moves up and down rapidly. It cuts on the upstroke. A saw blade with small, fine teeth will make a smoother cut than a blade with large teeth. The material being sawed with a saber saw should be clamped or held firmly to prevent vibration (shaking).

Using the Portable Circular Saw

The portable circular saw may be used to **crosscut** and **rip** boards and wood composition materials, such as particle board, when the table saw or the radial arm saw is not available. See Figure T5-8. This high-speed tool is used to cut straight lines on materials. It is most accurate when used with guides, but it can also be moved **freehand**.

Adjust the depth of cut carefully. The blade should extend slightly below the bottom side of the material being cut. Make all adjustments before connecting the power. Never lay down a portable saw that is still running.

Figure T5-8

Using a portable circular saw.

Crosscutting.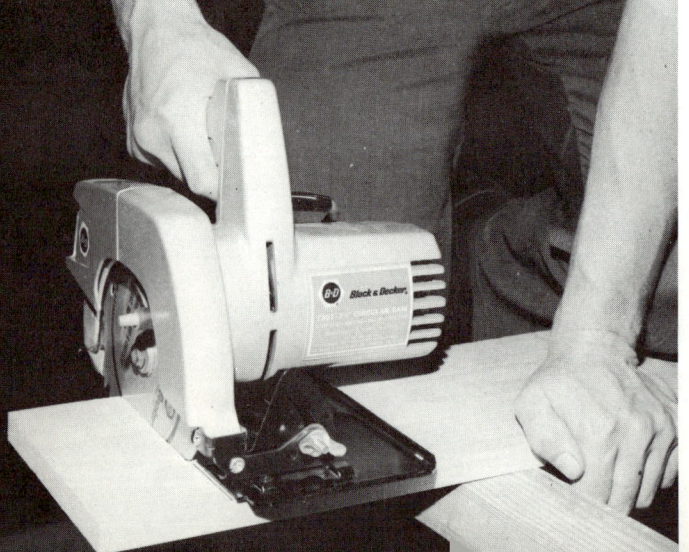

Ripping.

Using a Scroll Saw

A scroll saw (jigsaw) is used to cut **curves** and **irregular shapes** in wood and other materials. It is safer and easier to use than some other saws. See Figure T5-9.

The blade of a scroll saw can be changed, and the speed can be adjusted. Select the best blade and speed for the material being cut. Use slow speeds when cutting metal, thick wood, and other hard materials. Use a fast speed when sawing thin wood and cutting fine details. Also, the thinner the stock, the finer the teeth on the blade should be.

In addition to outside cuts, the scroll saw can be used for inside cuts. Drill a hole. Insert the blade through the hole, and make the inside cut.

Using a Band Saw

A band saw is most often used for cutting **curves** and **irregular shapes** in wood and plastics. The blade is a flexible steel band which has teeth on one edge. It runs on two large wheels. The band saw cuts much more rapidly than the scroll saw. Ripping and crosscutting can be done on a band saw, but not accurately. Some band saws can be used to cut metal. These saws run more slowly and use a fine-tooth metal-cutting blade.

Certain safety measures should be taken before operating the band saw. Before cutting sharp curves, **relief cuts** should be made so that the saw blade will not bind or twist. **Keep your fingers away from the blade.** Keep the upper guide slightly above the top of the material being cut. See Figure T5-10.

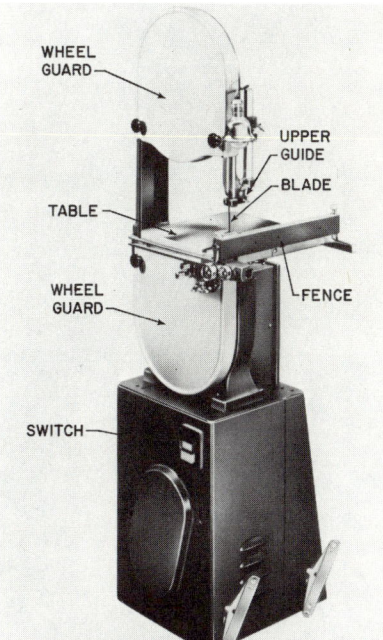

The parts of a band saw.

The parts of a scroll saw.

Using a scroll saw.

Figure T5-9
A scroll saw.

Figure T5-10
A band saw.

Cutting a curve on a band saw. Note the relief cuts.

The parts of a table saw.

Figure T5-11

A table saw. The guard has been removed in these illustrations to show how the wood is cut. **Always use the guard when using the table saw.**

Ripping can be done on a table saw.

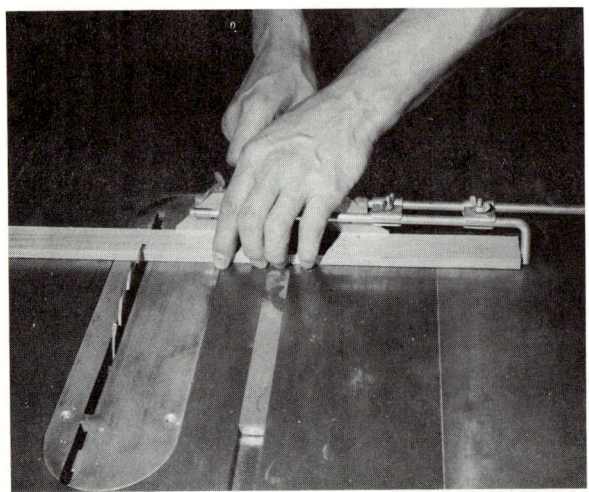

Crosscutting can be done on a table saw.

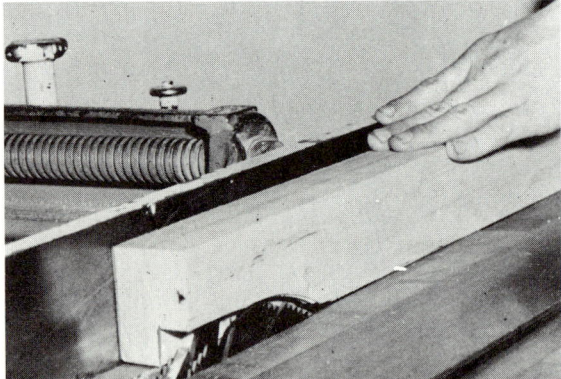

Using an adjustable dado head to cut a rabbet on a table saw.

Using a Table Saw

A table saw is used to quickly and accurately crosscut and **rip** wood materials. See Figure T5-11. A circular blade sticks up through the table. It is adjusted to the proper height by turning a handwheel. The blade should be slightly higher than the material being cut. You must push the wood as it is being cut.

During ripping operations, one edge of the wood moves against the fence. The fence can be adjusted for ripping different widths. Check with your teacher to see if a push stick should be used to push the wood through the saw.

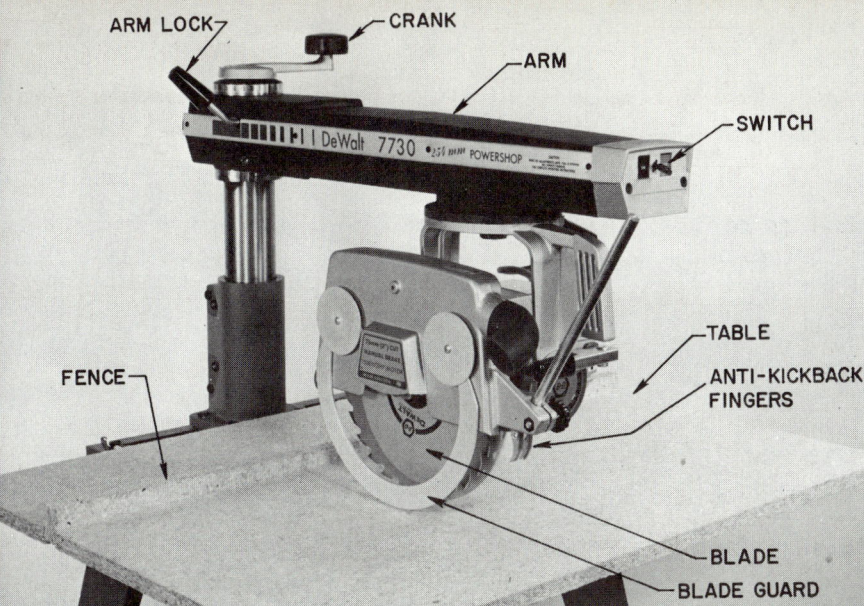

The parts of a radial arm saw.

Figure T5-12
A radial arm saw.

Using a radial arm saw for crosscutting.

When crosscutting, use the miter gage to hold and guide the wood. The miter gage can be adjusted to cut any angle from 30 to 90 degrees.

Different blades for different purposes are used on a table saw. A combination blade can be used for most operations. However, a special fine-tooth blade gives a smooth, splinter-free edge when sawing plywood. And acrylic plastic can be cut with a special blade.

In addition to crosscutting and ripping, other operations can be done easily and accurately on the table saw. A dado (groove) or rabbet can be cut using a dado head instead of a saw blade. The dado head can be adjusted to cut the desired width. The depth is set by raising or lowering the blade. (See Topic 25, **Production Tooling,** for ideas on cutting multiple (several) parts using a table saw.)

Eye protection is especially important. Never make freehand cuts on the table saw. Set the blade at the proper height. Always use the guards.

Using a Radial Arm Saw

A radial arm saw is most commonly used to **crosscut long pieces** of wood to length. See Figure T5-12. Since the upper arm pivots (turns), angles can be cut across a board by moving the arm to the desired angle. A crank raises and lowers the blade. You must hold the board and pull the saw carefully toward you.

The radial arm saw may be used to quickly and accurately cut a number of pieces of wood to the same length. Always make sure that your hands are out of the way.

Saws

1. Always disconnect the power before changing any parts on a power saw.
2. Be sure that all power tools are grounded before using them.
3. Roll up sleeves, and tuck in loose clothing that could get caught in the machinery.
4. Wear safety goggles whenever power machinery is being used.
5. Never place your fingers in front of a blade that is moving.
6. Allow any power saw to reach top running speed before using the saw.
7. Check all materials to be sawed for nails and other items.
8. Use only sharp blades.
9. Clear all scrap materials from the machines and from the floor.
10. Only one person at a time should operate a machine.
11. Ask permission before using a power tool.
12. Do not leave a machine that is running. Be sure that the blade has completely stopped moving before you leave it.

Topic 6 Drilling and Boring

Holes are drilled or bored in wood, metal, and plastics. Originally, holes were **bored** in wood and **drilled** in metal. Today these terms are generally used interchangeably. The drill or bit makes holes by removing chips of the material. You must select the proper type of drill or bit for the material being drilled or bored.

Boring is done using a bit brace. This tool is hand-operated. Drilling is done using a hand drill (hand-operated) or using power drills, such as a portable electric drill or a drill press.

Drilling and Boring Tools

Using a Bit Brace

A bit brace is used for boring holes in **wood**. The crank-type handle of the bit brace makes it possible to apply a lot of force. All bits used in a bit brace have a special square-tapered **tang** (end). The tang fits into a special **chuck** on the brace. See Figure T6-1.

Using a Hand Drill

A hand drill is used for drilling small holes in wood, metal, and plastic materials. A twist drill is fastened in the chuck and tightened by hand. You must supply the drilling power. However, the **gears** increase the force that you apply. Crank the handle clockwise when drilling. See Figure T6-2.

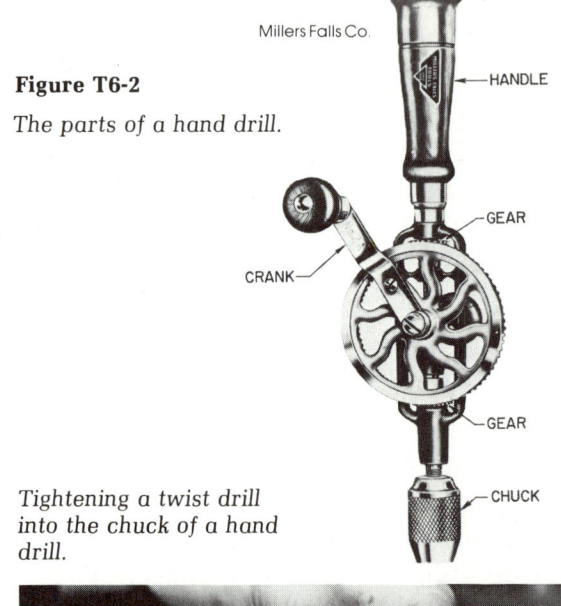

Figure T6-2

The parts of a hand drill.

Tightening a twist drill into the chuck of a hand drill.

Figure T6-1

The parts of a bit brace.

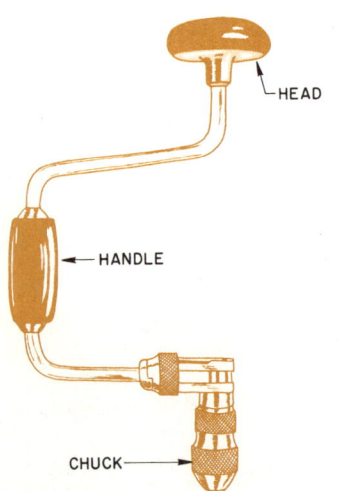

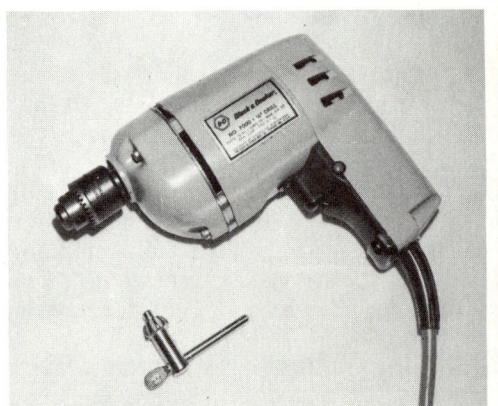

A. A portable electric drill and chuck key.

B. Using a portable electric drill to drill a hole.

Figure T6-3

Drilling and Boring 319

Using a Portable Electric Drill

A portable electric drill is a commonly used tool. This drill operates at a high speed. Drilling can be done quickly and easily. The size of a portable electric drill is determined by the largest-diameter **twist drill** that will fit into the **chuck.** Common sizes of drills are 1/4" and 3/8".

The chuck has three jaws. These jaws grip the smooth, round end of the twist drill when the chuck is tightened with the **chuck key.** When starting a hole, use both hands to guide and operate the drill. See Figure T6-3.

Using a Drill Press

The drill press is powered by an electric motor. The drill press is not only fast and easy to operate, but is much more **accurate** than other methods of drilling. Place the material to be drilled in position on the table. As you pull on the **feed lever,** the rotating twist drill or bit cuts into the material. See Figure T6-4.

The speed of a drill press can easily be varied. This is done by changing the **belt's** position on the **cone pulleys.** The speed can be slowed down for drilling metal and increased for drilling wood. Generally speaking, small drills and bits should be operated at high speeds and larger drills and bits at slow speeds. All drills or bits used in the drill press must have a shank that has a round end (no tang). Use the chuck key to tighten the jaws of the chuck around the twist drill or bit.

Always remember to remove the chuck key before starting the drill press. Sheet metal and small objects should be securely **clamped** in position for drilling. (See Topic 24, **Clamping**.) This is done to prevent injury to the operator or damage to the materials. Cylindrical (round) stock can be held in a **V-block** for accurate and safe drilling.

A drill press vise is also very useful for holding round and flat stock. Reduce pressure as the twist drill starts to come through the bottom surface of the work.

Figure T6-4

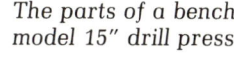

The parts of a bench model 15" drill press.
Rockwell International

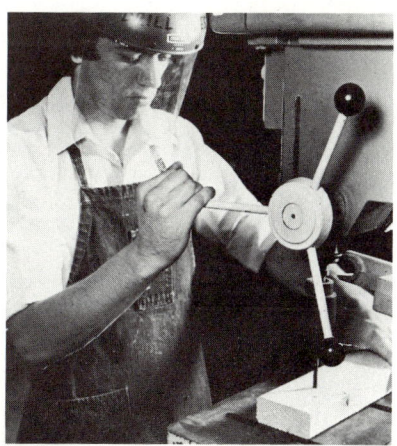

Using a drill press. Guide the drill into the material by pulling on the feed lever.

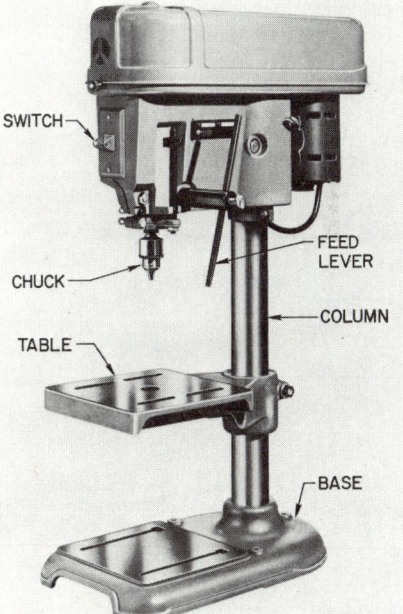

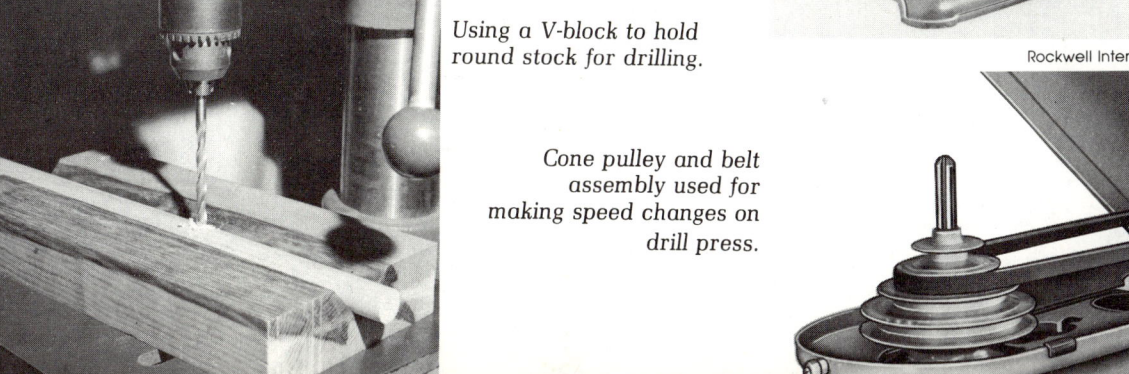

Using a V-block to hold round stock for drilling.

Cone pulley and belt assembly used for making speed changes on drill press.

Rockwell International

Drills and Bits for Making Holes

There are many drills and bits used for making holes in wood, metal, and plastic materials. Each has a special purpose. Remember: a bit that has a square-tapered tang should only be used in a bit brace. All drills and bits that have round shanks can be used in hand drills, portable electric drills, or drill presses.

Wood Bits

Several different kinds of bits are used for boring holes in wood materials. The kinds most commonly used are auger bits, expansive bits, spade bits, Forstner bits, and multispur bits.

Auger Bits. Auger bits are used for drilling holes in wood. This kind of bit has a screw-like point that pulls the bit into the wood. An auger bit is always used in a bit brace. It is never used in a power drill. See Figure T6-5.

Auger bits are available in sizes from 4 to 16. These numbers refer to the size in sixteenths of an inch. Thus, a number 4 auger bit is 4/16" (or 1/4") in diameter. Number 8 is 8/16" (or 1/2") in diameter, and so on.

Expansive Bits. The expansive bit can be adjusted to bore a hole any size between one 1" and 3" in diameter. The size is changed by adjusting the distance between the outside edge of the cutter and the center of the point. This distance is the desired radius of the hole. Expansive bits are designed only for boring wood and are used in a bit brace. These bits have a screw-like point that pulls the bit through the wood as it is turned. See Figure T6-6.

Spade Bits. Spade bits are commonly used for boring larger holes through wood materials. They are generally used in portable electric drills or drill presses. Spade bits range from 1/4" to 1-1/2" in diameter. The round shank is small enough to fit into any drill chuck. See Figure T6-7. Using a spade bit is a fast way to bore a hole, but it may split or chip the surface of the wood.

Forstner Bits. A Forstner bit is used to make a flat-bottomed hole with very smooth edges and sides. It is an excellent tool for boring a **counterbored** hole in wood. See Figure T6-8. Because the Forstner bit has no sharp point, you must be very accurate when starting the hole. A Forstner bit with a square-tapered tang may only be used in a bit brace. A Forstner bit with a round shank is generally used in a drill press.

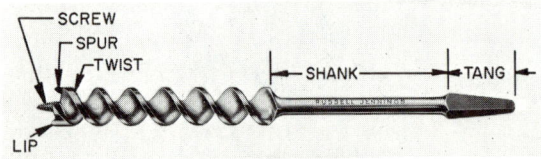

Stanley Tools

Figure T6-5

The parts of an auger bit. Note the square-tapered tang.

Figure T6-6

An expansive bit.

Figure T6-7

A spade bit. Compare the round shank with the square-tapered tangs of the auger bit and the expansive bit.

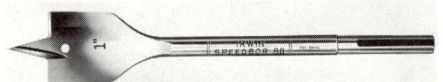

The Irwin Auger Bit Co.

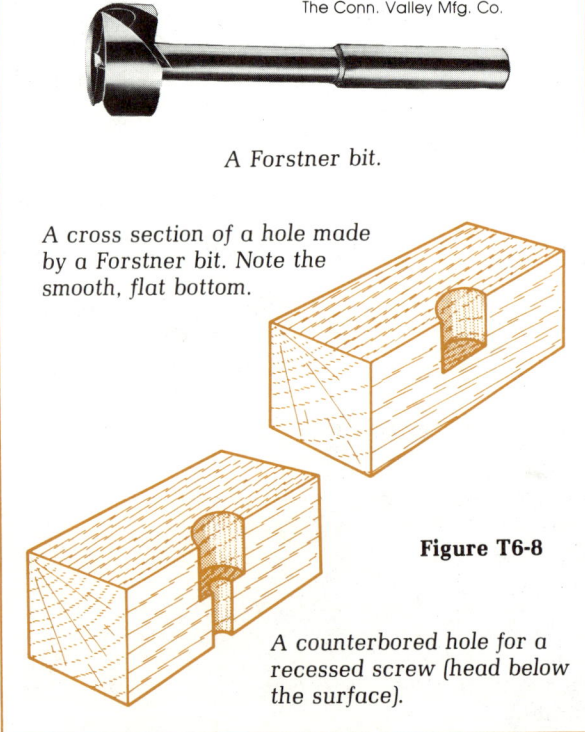

The Conn. Valley Mfg. Co.

A Forstner bit.

A cross section of a hole made by a Forstner bit. Note the smooth, flat bottom.

Figure T6-8

A counterbored hole for a recessed screw (head below the surface).

Drilling and Boring 321

Figure T6-9
A multispur bit.

Multispur Bits. A multispur bit is used in a drill press for boring large diameter holes. These bits are called "multispur" because of the spurs around the outer edges. These spurs make this kind of bit ideal for boring holes into plywood. There is very little splitting or chipping. Multispur bits are available in several sizes, ranging from 1/2" to 4" in diameter. See Figure T6-9.

Twist Drills

Twist drills are the most commonly used tools for drilling holes in wood, metal, and plastics. They are used in hand drills, portable electric drills, and drill presses. Twist drills are available in many sizes. Sizes from 1/16" to 1/2" in diameter are used most often. See Figure T6-10.

Special Bits

Several kinds of special drilling tools are available. A few examples of these are countersinks, flycutters, and center drills.

Countersinks. A countersink is used to make a tapered hole in wood, metal, or plastic so that flat-head screws can be set flush (even) with the surface of the material. See Figure T6-11. The angle on the cutting edges of the countersink matches the angle of the screwhead. When using a countersink, drill deeply enough that the screwhead will fit flush or just below the surface.

A countersink with a square-tapered tang should only be used in a bit brace. Round-shank countersinks can be used in hand drills, portable electric drills, and drill presses. See Figure T6-12.

A combination drill and countersink can be used to drill the pilot hole, body hole, and countersink hole all at the same time. See Figure T6-13.

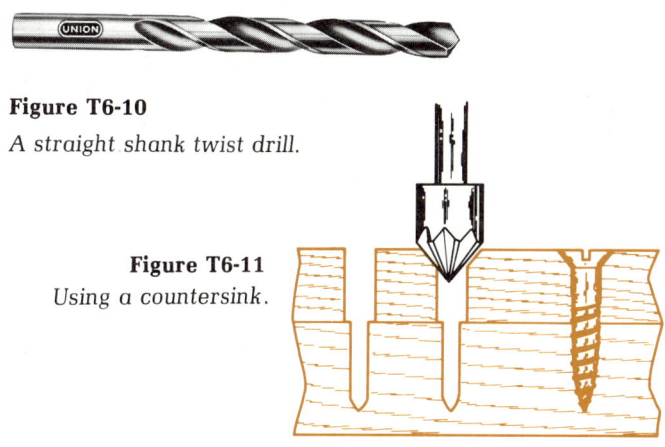

Figure T6-10
A straight shank twist drill.

Figure T6-11
Using a countersink.

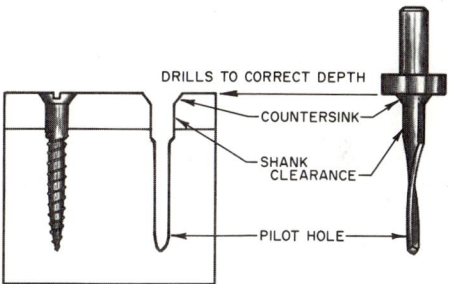

Figure T6-13

A combination drill and countersink can be used to prepare a hole for countersinking a screw.

A countersink for use in a bit brace. Note the square-tapered tang.

Figure T6-12

A countersink for use in a hand drill, portable electric drill, or a drill press. This countersink is used to drill wood.

A high-speed countersink to use when drilling metal.

Flycutters. A flycutter is used for cutting very large holes in wood. This type of bit can be adjusted to cut holes up to 8" (200 mm) in diameter. Flycutters should be used only in a drill press. Be very careful when using a flycutter. Make sure that the wood is clamped in place and that your hands are out of the way of the bit. Use a very slow speed. See Figure T6-14.

Center drills. A center drill is used to drill an accurately located hole for further drilling operations. For example, if the holes on two parts of a product must match, they must be drilled accurately. A center drill would be used to drill them. Another important use is to prepare the end of round (cylindrical) metal stock for turning on the metal lathe. Center drills can be used in the drill press or in the drill chuck of a lathe. See Figure T6-15.

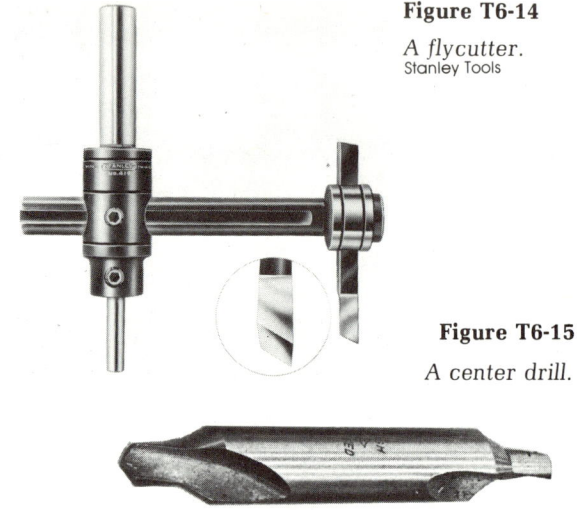

Figure T6-14
A flycutter.
Stanley Tools

Figure T6-15
A center drill.

Drilling

1. Wear eye protection when using any drilling or boring tools.
2. Disconnect the power to the power tool before making any adjustments or repairs.
3. Do not allow loose clothing to get near rotating parts of power tools. Roll up your sleeves, and wear an apron.
4. If a piece of material works its way loose from a clamped position, step back and turn off the machine.
5. Use the correct speed for the size of drill or bit being used and the kind of material to be drilled. Generally, adjust the drill speed to be faster for drilling smaller holes and slower for drilling larger holes.

Topic 7 Wood Planing, Shaping, and Turning

Chisels, routers, jointers, planers, and wood lathes are used to shape wood. As the tools are used, chips of wood are removed. All of these tools are sharp or have sharp parts. Read and follow all safety directions carefully.

Wood Chisels

Wood chisels are used to cut grooves and slots, or to smooth small areas of wood. The size of a chisel is determined by the width of the cutting blade. Sizes range from 1/4" to 2" wide. Select a chisel according to the size of the work area that is to be chiseled. Use a narrow chisel to smooth a small area and a wide chisel to smooth a larger area.

Using A Wood Chisel

Before chiseling, fasten the wood securely to the bench or in a vise. For shallow cuts, hold the chisel in both hands and direct your cut away from you. For deeper cuts, hold the chisel with one hand and strike the top of the chisel with the palm of your other hand or a mallet. See Figure T7-1.

Figure T7-1
Using a wood chisel.

Figure T7-2

Common planes.

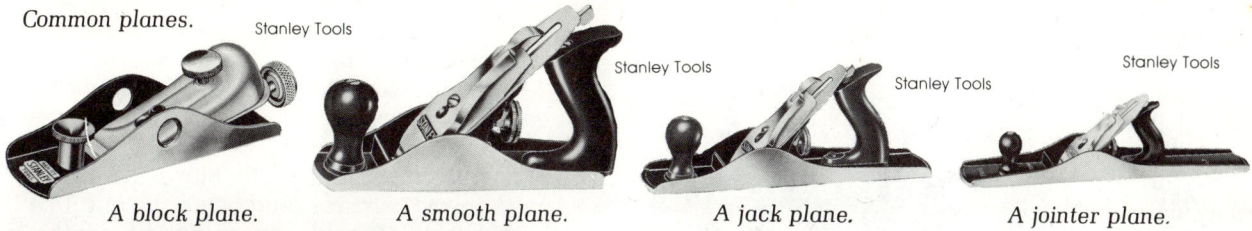

A block plane.   A smooth plane.   A jack plane.   A jointer plane.

Figure T7-3

Planing with the grain of the wood.

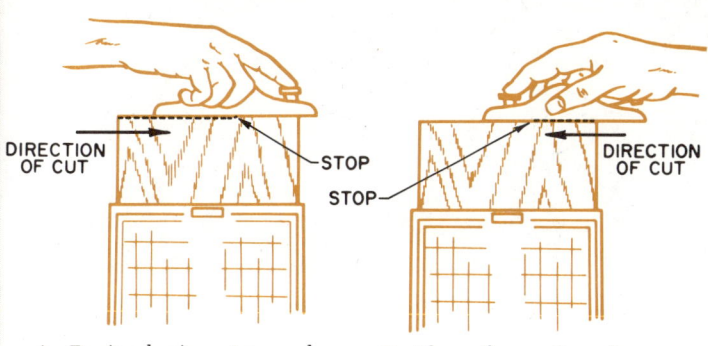

A. Begin planing at one edge. Note the point at which you should stop.

B. Plane the surface from the opposite edge to the point where the cut was stopped.

Figure T7-4

Planing end grain using a block plane.

SAFETY DIRECTIONS

Wood Chisels

1. Always direct a chisel away from you as you cut.
2. Do not touch the sharp edge of a chisel.
3. Use a chisel carefully.
4. Use only sharp chisels.

Hand Planes

Hand planes are used to smooth surfaces of wood. You must push the plane to make it cut. The four common types of hand planes are block planes, smooth planes, jack planes, and jointer planes. See Figure T7-2.

- **Block planes** are small. Generally, they are 6″ or 7″ long. They are used to remove material from the ends of a board (end grain).
- **Smooth planes** are used for smoothing small surfaces. These planes are usually about 9″ long.
- **Jack planes** are all-purpose planes. They are 14″ or 15″ long.
- **Jointer planes** are used for working on large areas. These planes are 22″ or 24″ long.

Using a Hand Plane

When planing the face or edge of a board, always plane **with the grain.** This means that you should move the plane along the grain (lines) of the wood, not across it. This will produce a smooth surface. See Figure T7-3.

When planing end grain, plane first from one direction and then from the other so that you will not splinter the wood. See Figure T7-4. When you set the plane down, place it on its side. Resting the plane on its cutting edge may dull it.

Routers

Routers are fast and accurate tools. They are commonly used for cutting decorative edges on wood and for cutting dadoes, grooves, and rabbets when making wood joints. They are also used for making internal cuts, such as the lettering on a sign. See Figure T7-5.

The type of cut you make depends mostly on the **bit** you select. There are many kinds of router bits. Each will cut a different shape. The **depth of cut** can also be adjusted to change the type of cut. The depth of cut is set by raising or lowering the motor.

Most internal cuts are done freehand. A guide may be used to make straight cuts with the router.

Figure T7-5

Making an internal cut using the router.

A bit that is used to rout an edge has a **pilot** (smooth, round piece of metal) in the center. This pilot guides the router bit along the edge of the wood. Care and practice are needed to keep the pilot against the wood to make an even cut. The pilot rests against the edge of the stock as the cut is made. Because of this, the routed edge will be only as straight and even as the edge of the wood. If this edge was sawed unevenly, then the routed edge will be uneven. See Figure T7-6.

Using a Router

1. Select a bit for the shape you wish to cut, and lock it into the router.
2. Adjust the base so that the router will cut to the correct depth.
3. Secure the wood to be routed to the workbench. It must not move when you apply pressure to the router.
4. Start the router. Carefully place the cutter on the work surface, and begin the cut. Use both hands to hold and guide the router. Remember, the bit will cut wherever the router is moved. Keep the router base flat against the work surface at all times.
5. Move the router along the cut. The movement must be slow enough not to bog (slow) the motor, but fast enough that the wood is not burned.
6. When the cut is completed, turn off the router. Do not lift the router from the work until the bit has stopped turning.
7. Disconnect the power.

Router

1. Always wear a face shield when using the router. Wood chips may fly back at you.
2. Hold and guide the router with both hands.
3. Disconnect the power before changing bits.
4. Always lay the router on its side or top. Never set it down on the bit.

Jointers

Jointers are stationary (not movable) power tools. They have knives that rotate to plane the

Figure T7-6

Cuts made by different router bits. Notice the pilots on bits used to rout edges.

Stanley Tools

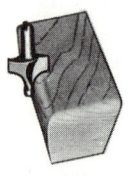 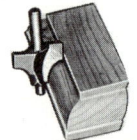   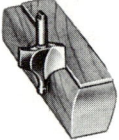  

ROUNDING BEADING OGEE CORE BOX ROUNDING RABBET DOVETAIL

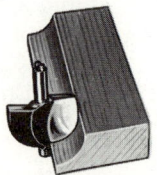 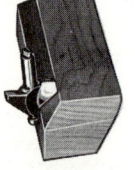  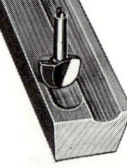  

COVE CHAMFER STRAIGHT CORE BOX ROUNDING CHAMFERING

wood. The knives are part of the **cutterhead**. See Figure T7-7. They cut quickly and accurately.

Jointers are used to plane the edges of boards. They are also used to plane (smooth) the faces of boards before the boards are planed to thickness with a thickness planer. Jointers may also be used to cut rabbets, tapers, and bevels on stock.

The size of a jointer is determined by the width of the widest cut the jointer can make. Eight-inch jointers are used in most school shops. The **infeed table** is usually lowered 1/16″ below the **outfeed table** to remove 1/16″ of material. The **fence** can be tilted to make bevels and chamfers. See Figure T7-8.

Using a Jointer

1. For **straight jointing** (planing with a jointer) of stock, set the fence at a 90-degree angle to the infeed table. See Figure T7-9. Use a try square to measure the angle.
2. If the stock is to be **beveled,** set the fence at the desired angle. See Figure T7-10. The fence should also be adjusted for the width of the stock. The stock to be cut should cover all but a small portion of the cutterhead.
3. By turning the adjusting handwheel, lower the infeed table 1/16″ below the outfeed table. The cut made will be 1/16″ deep.
4. Turn on the machine.
5. Press the stock against the fence and the infeed table.
6. Push the stock into the rotating cutterhead to make the cut. Use a **push shoe** when planing the surface of stock. Use a **push stick** when planing narrow stock. See Figure T7-11. When pushing stock over the jointer, never place your hands directly over the cutterhead.
7. When the cut is complete, turn off the machine. Stay at the machine until the cutterhead has come to a complete stop.

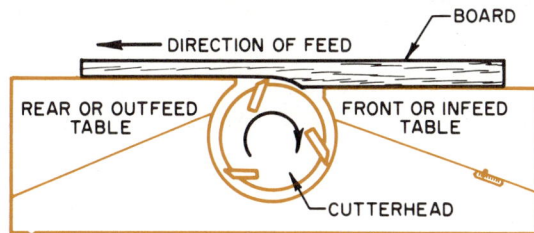

Figure T7-7

On a jointer, stock is cut by a rotating cutterhead.

Figure T7-8

The parts of a jointer.

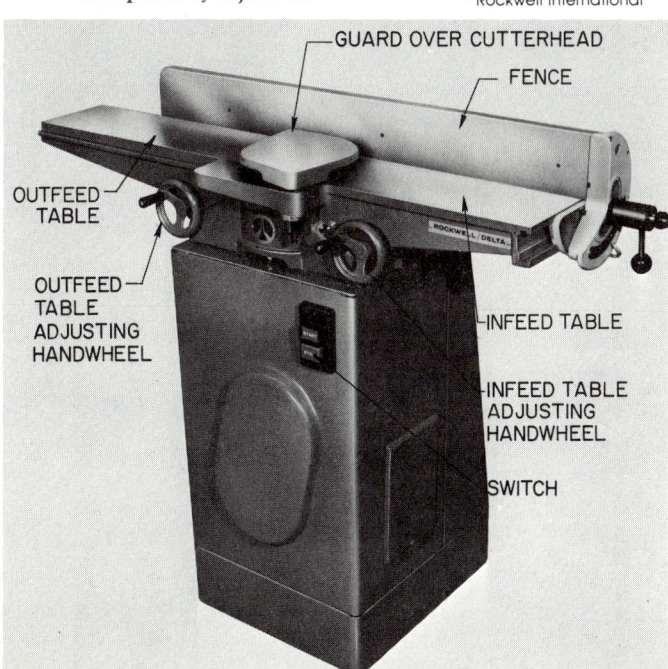

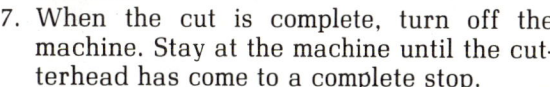

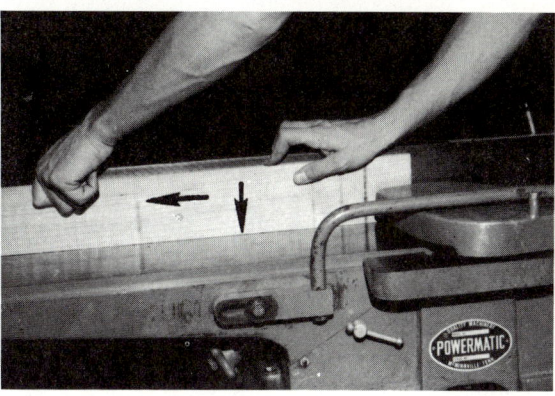

Figure T7-9

Jointing the edge of a board. Push forward while pressing the stock downward and against the fence.

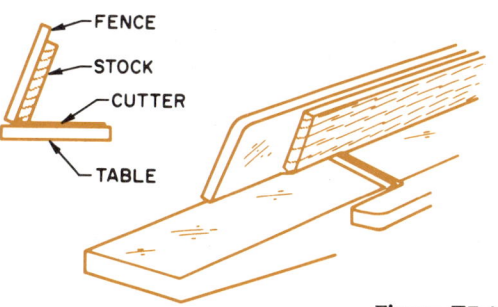

Figure T7-10

Planing a chamfer on the jointer. Note that the fence has been tilted toward the table.

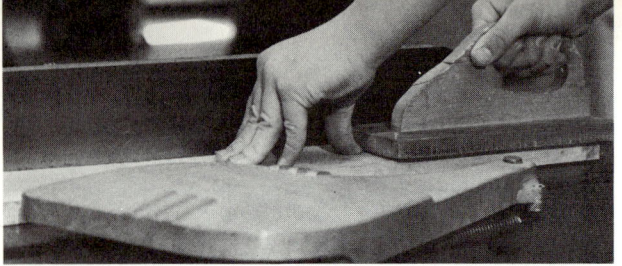

Using a push shoe.   **Figure T7-11**   Using a push stick.

Jointer

1. Always wear eye protection.
2. Keep the guard **in place** over the cutterhead.
3. Do not use the jointer to smooth a piece of wood less than 12" long or 1/4" thick.
4. Make two or more light cuts rather than one deep cut.
5. Use a push shoe when planing the face of stock with a jointer.
6. Use a push stick when planing narrow stock with a jointer.
7. Always disconnect power before making adjustments.

The Uniplane

The Uniplane® is a stationary power tool. It is used in much the same way as a jointer. The edges, ends, and faces of boards can be smoothed by using a jointer. The special advantage of the Uniplane® is that it can be used to smooth small surfaces and end grain. The Uniplane® uses eight small cutters (cutter bits) that rotate downward. There is no **kickback***. The disadvantage of the Uniplane® is that it cuts across the grain of the board, leaving a somewhat rough cut.

The Uniplane® is very useful for cutting bevels and chamfers on wood materials. A miter gage can be used to help you plane an end square or to an angle. The angle of the table is adjustable. See Figure T7-12.

Using a Uniplane

1. If you want to make a square cut, set the table 90 degrees to the fence. Check the angle with a try square. (See Topic 2, **Layout**.)
2. Set the depth of cut by rotating the **control knob**. Do not set the depth more than 1/16".

*Kickback is when material is moved suddenly and forcefully back toward you.

Figure T7-12

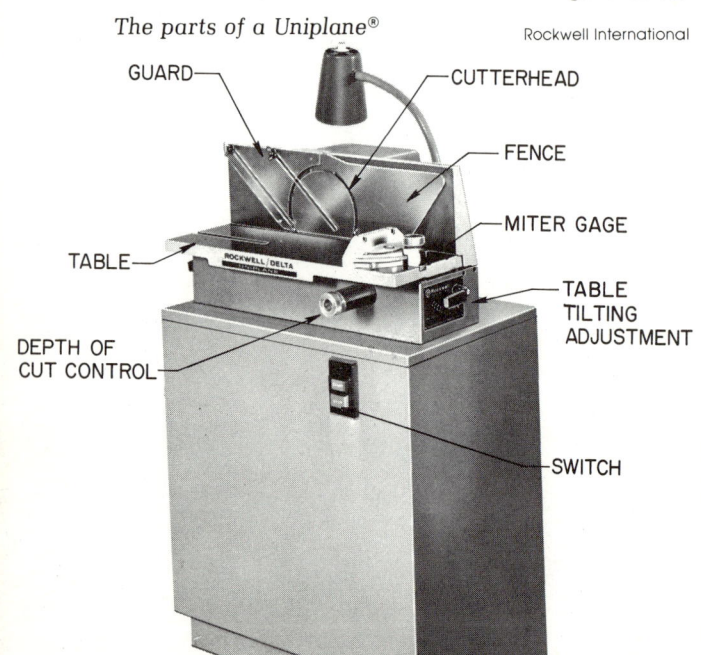

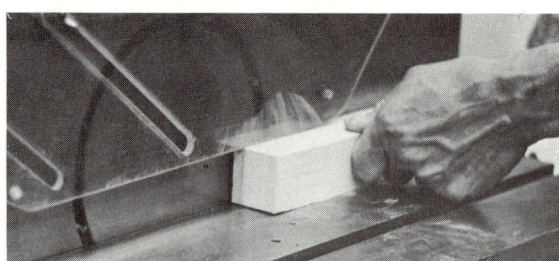

Smoothing a board on a Uniplane®.

Using the miter gage when chamfering small stock on a Uniplane®.

3. Press the wood against the fence and the table, and slide it across the cutting blades.
4. After the cut is made, turn off the machine. Stay at the machine until the cutting blades have come to a complete stop.

Uniplane

1. Always wear eye protection.
2. Keep the cutterhead covered by the guard.
3. Keep your fingers away from the cutter bits.
4. Use push sticks for small materials.
5. Unplug the Uniplane® before making any adjustments.

Thickness Planers

Planers are used to reduce the thickness of a board and smooth its surface. See Figure T7-13. Work done on the surface of a board is called **surfacing.** Most planers surface only one side of a board at a time.

The size of a planer is determined by the width of the widest board that you can surface on the machine. The depth of cut is determined by the height of the table. If the board is very rough or uneven, smooth one side on the jointer before putting it through the planer.

Using a Thickness Planer

1. Adjust the machine for the proper depth of cut. This is done by rotating the handwheel until the **thickness gage** shows 1/16" less than the thickness of the stock.
2. Turn on the machine, and slide the stock on the table until it is caught by the **feed rolls.** The feed rolls automatically feed the stock through the cutterhead. See Figure T7-14.
3. Remove the stock from the other side of the machine as the stock clears the feed rolls.
4. Adjust the machine for each cut. One complete turn of the handwheel is the greatest adjustment you should make at one time. Run the stock through the planer as many times as are needed.
5. When surfacing thin stock, place a thicker board under the stock for support. Adjust the machine for both thicknesses.
6. When surfacing is complete, turn off the machine. Do not leave the machine until it has come to a complete stop.

Thickness Planer

1. Wear eye and ear protection.
2. The wood should be longer than the distance between the rollers. This makes it possible for the machine to control the stock. Generally, stock no less than 12" in length should be run through a planer.
3. Keep your hands away from the infeed rolls.
4. Stand to one side of the machine when stock is being fed into the infeed table. A machine that is not used properly can throw material back with great force (kickback).
5. Make several light cuts rather than one deep cut.
6. Turn off the thickness planer before making adjustments.

Powermatic/Houdaille

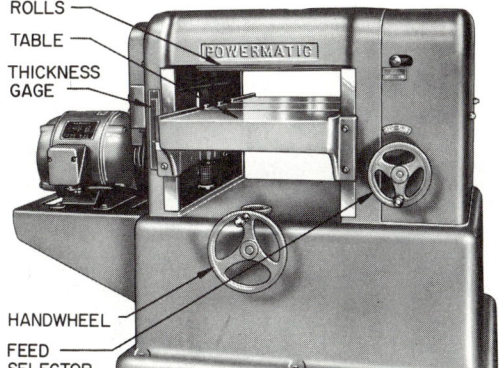

Figure T7-13

The parts of a thickness planer.

Figure T7-14

Inserting a board into a thickness planer.

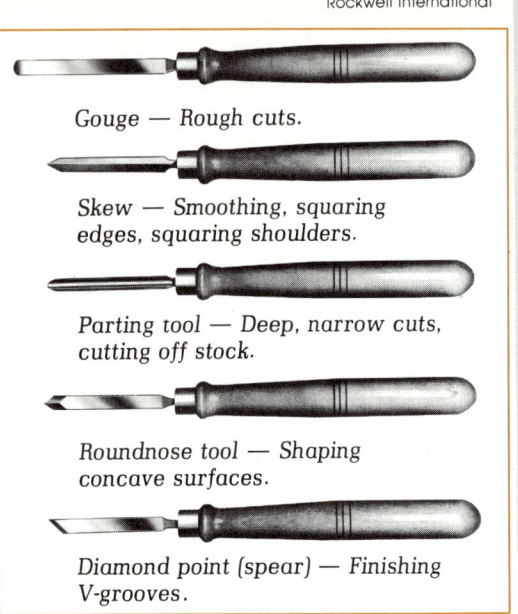

Figure T7-15
Lathe tools and their uses.

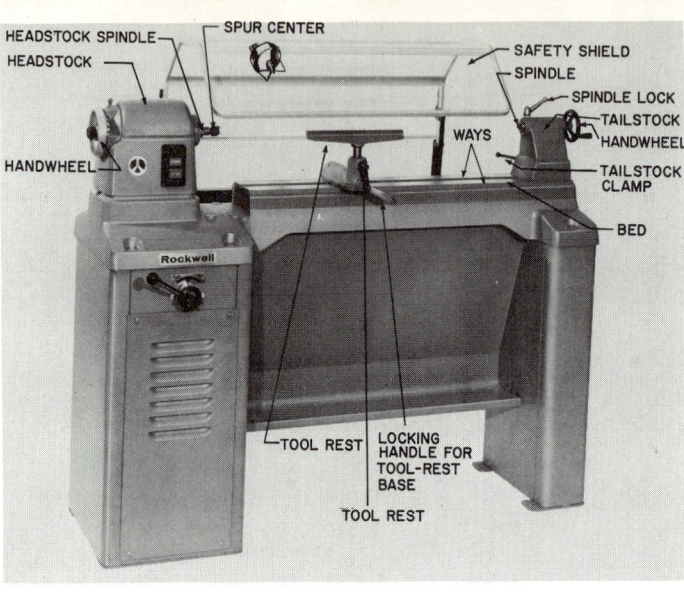

Figure T7-16
The parts of a wood lathe.

Wood Lathes

Wood lathes shape wood by rotating it against **lathe tools** (turning chisels). Lathe tools are used to cut or scrape wood to produce various shapes. Different tools are used to make different kinds of cuts. For example, some make sharp cuts while others cut curves. See Figure T7-15.

The size of a lathe is determined by the **swing**. This is the distance from the **live center** (the center that turns) to the **bed** and the distance between centers. See Figure T7-16.

Turning Stock between Centers (Spindle Turning)

Mounting the stock:
1. Select a piece of stock about 1" longer and 1/4" larger in diameter than the final size of your product.
2. Square the ends and find the center of each (See Topic 2, **Layout**).
3. On one end, make two diagonal saw cuts about 1/8" deep across the center of the stock. From corner to corner works well on square stock.
4. Center punch the center of the other end. (See Topic 2, **Layout**.) Put a small amount of lubricant (such as wax) in the hole.
5. Remove the spur center from the headstock of the lathe. See again Figure T7-16.
6. With a wooden mallet, tap the spur center into the center of the end that you have cut with a saw. The spurs of the center should fit into the cuts.
7. Mount the stock and the spur center on the lathe. The spur center should fit snugly into the headstock spindle.
8. While supporting the other end of the stock, move the tailstock to about 1" from the end of the stock. Tighten the tailstock clamp.

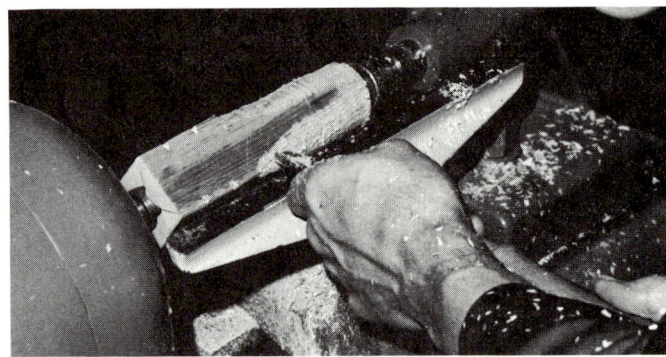

Figure T7-17
Turning between centers. Rough cuts are made first.

Figure T7-18
Making finish cuts on round stock. Note the pencil marks showing where cuts should be made.

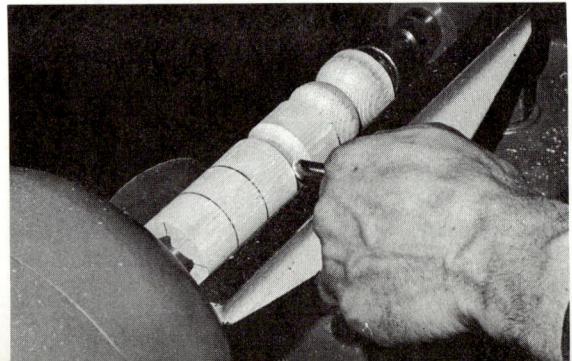

Figure T7-19
Outboard faceplate turning.

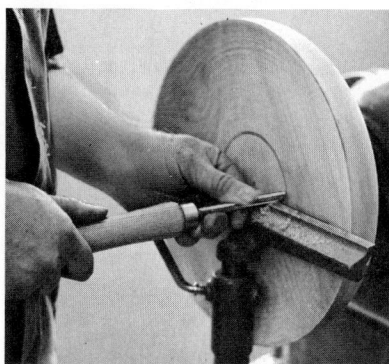

Figure T7-20
Inboard faceplate turning.

Figure T7-21
Preparing for faceplate turning.

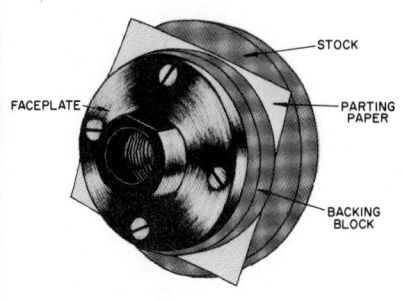

9. Turn the tailstock handwheel to engage the dead center (the center that does not turn) in the hole made by the center punch. Apply moderate (medium) pressure.
10. Lock the tailstock spindle.
11. Adjust the tool rest so that it just clears the stock (1/16″ to 1/8″). Rotate the stock by hand to make sure.
12. Start the machine at its slowest speed.

Turning the stock:
1. Set an outside caliper to the desired diameter. (See Topic 2, **Layout**.)
2. Using a gouge, make rough cuts until the stock is round and almost to the finished size. See Figure T7-17.
3. Make light finish cuts until the stock is the desired size. Use a gouge, roundnose chisel, or a skew chisel to make the finish cuts. See Figure T7-18. Measure the stock with the outside caliper.
4. If curves and grooves are to be cut in the stock, use a ruler and pencil to lay them out. Do this while the stock is rotating.
5. Use a parting tool or a skew chisel to make sharp cuts. Use a roundnose chisel or a gouge to make rounded cuts.
6. When final shaping is complete, remove the tool rest, and sand the workpiece with abrasive paper while it is still rotating.
7. When finished sanding, turn off the lathe.
8. Remove the workpiece (turned stock) by releasing the tailstock spindle and turning the tailstock handwheel. Pull the workpiece from the spur center.

Faceplate Turning

Curved objects, such as bowls, are turned on a faceplate instead of between centers. This method allows you to hollow out the object.

Stock used for faceplate turning is usually very thick. Often, two or more pieces of wood are glued together. Many designs can be created by gluing different types and colors of wood together.

For very large turnings, the faceplate and stock may be mounted on the **outboard** (handwheel) side of the headstock. See Figure T7-19. This is done so the stock will not hit the bed. Most turnings, however, are done on the **inboard** side (over the bed). See Figure T7-20.

Turning on a faceplate:
1. Lay out (mark) the shape of the turning on the stock.
2. Cut the stock to a round shape on the band saw. The cut should be about 1/4″ larger than the finished size.
3. Center and mount the stock (workpiece) on the faceplate. Drive screws through the faceplate into the workpiece. If you do not want screw holes in the finished product, or if there is a chance that the lathe tool will hit the screws, use a backing block. Glue this block to the base of the workpiece. Place a piece of heavy paper between the block and the work. By doing this, you will be able to remove the block when the product is finished. When the workpiece is mounted to the faceplate, the screws extend into the backing block and not into the workpiece. See Figure T7-21.
4. Attach the faceplate and workpiece to the lathe. Then screw the faceplate onto the spindle.
5. Adjust the tool rest so that it just clears the work. Rotate the workpiece by hand to be sure.
6. Set the lathe at its slowest speed before starting the lathe.
7. Turn the workpiece to shape. Set the tool

rest for the area from which you are removing material.
8. When shaping is complete, remove the tool rest and sand the workpiece with abrasive paper while it is rotating.
9. Remove the workpiece by locking the spindle and unscrewing the faceplate from the lathe.
10. Unscrew the faceplate from the workpiece. If a backing block was used, separate the workpiece from it by tapping a flat chisel between the block and the workpiece.

— SAFETY DIRECTIONS —

Wood Lathe

Although the lathe does not have a moving cutting tool, it must be used carefully.
1. Have your teacher check the setup of the lathe before you begin.
2. Wear both safety glasses and a face shield at all times when using the lathe.
3. Do not wear loose clothing.
4. Remove all jewelry before you begin work.
5. Securely lock the tailstock, tailstock spindle, and tool rest and tool support base before starting the machine.
6. Always start a new turning on the lathe's slowest speed. If the speed is increased, always return the setting to the slow speed. This adjustment may be made before or after turning off the lathe depending on the type of lathe.
7. Remove the tool rest before sanding on the lathe.

Topic 8 Metal Turning

Round metal parts can be shaped on a metalworking lathe. The lathe rotates (spins) the work. Special cutting tools are moved against the work to remove unwanted material. Material can be removed from either the outside or the inside of the rotating stock.

The Metalworking Lathe

There are many types and sizes of metalworking lathes. They can be large and complicated, such as those found in industry. Or they may be smaller and easier to use, such as those found in school shops. See Figure T8-1. Whether large or small, all lathes work in basically the same way.

Using a Metalworking Lathe

A metalworking lathe is used for many operations. It is used to remove material from the end of stock. This is called **facing**. It is used to remove material along the length of the stock. This is usually done when stock is **turned between centers**. A lathe may also be used to **taper or bevel** round metal parts. A special tool may be used in a lathe for **knurling** (creating a rough surface). Both internal (inside) or external (outside) **threads** can be cut on a lathe. (See Topic 9, **Threaded Fasteners and Taps & Dies**.)

There are many different types of cutting tools which can be used in a lathe. Each tool has a specific purpose. Your teacher can help you select the right tool for the job. See Figure T8-2.

Facing

Facing is the process of squaring the end of stock. The stock is usually held in a **three-jaw chuck.**
1. Place the stock in the chuck. Allow about 1″ (25 mm) to stick out. (Large stock may have to stick out farther.) Tighten jaws with the **chuck key.** See Figure T8-3.
2. Mount the **cutting tool** in the **tool holder.** Fasten it to the **tool post** on the **compound rest.** The tip of the cutting tool should line up with the center of the stock. Make sure that only the cutting edge will touch the work.
3. Lock the **carriage** in place with the **carriage lock screw.** To face the work you will need to use only the **compound-rest knob** and the **cross-feed knob.**
4. Adjust the spindle speed. This is usually done by moving the belts on the **spindle** and motor pulleys. Your teacher can help you to determine the proper speed for the diameter of the stock.
5. Do not start the machine until your teacher checks your setup.

Metal Turning   **331**

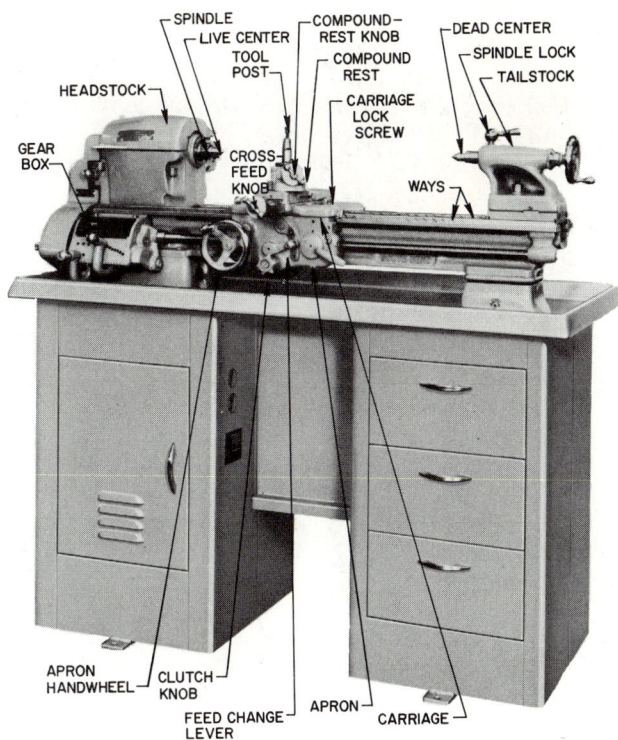

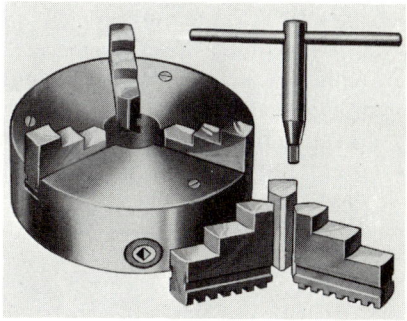

Figure T8-3

A three-jaw chuck and chuck key.

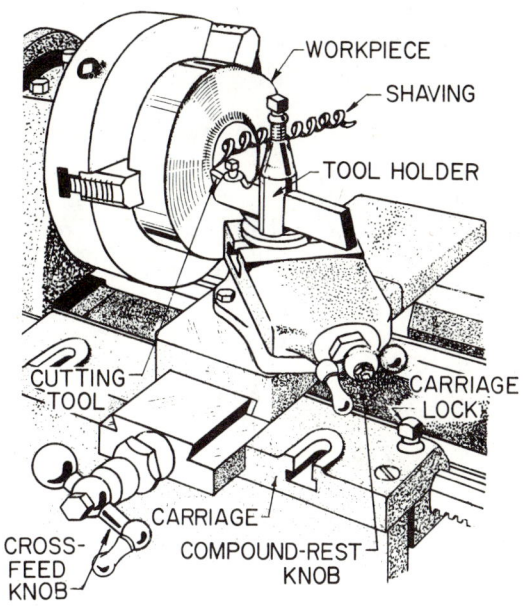

Figure T8-1

The parts of a metalworking lathe.

Figure T8-2

Tools for making various cuts when turning metal stock.

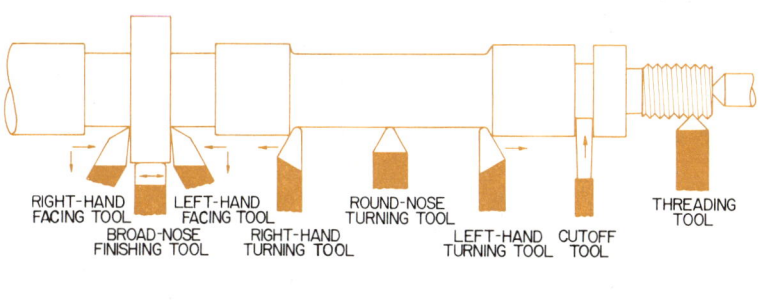

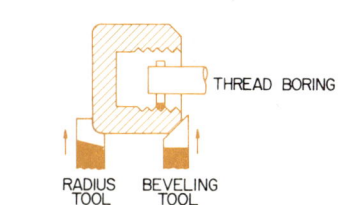

Figure T8-4

Facing is done to make stock shorter and to make the ends even and square.

6. Face the stock by turning the **cross-feed knob.** Always work from the center of the work to the outside. This will help you to avoid cutting too deeply. See Figure T8-4.
7. Be sure that the cutting tool cuts completely along the face of the stock. If you must make more than one cut, turn the compound-rest knob to adjust the depth of the cutting tool.

Center-Drilling

Center-drilling is done to prepare for between-center (straight) turning. Begin with stock that is 1" (25 mm) longer than the planned final size of your product.
1. Face the stock. Leave the stock mounted in the three-jaw chuck.
2. Insert a **drill chuck** into the **tailstock.**
3. Tighten a **center drill** into the drill chuck.
4. Use the center drill to cut a small, tapered hole in each end of the stock. See Figure T8-5.

Turning Between Centers

1. Face and center-drill both ends of the stock.
2. Install a live center and **faceplate** on the **headstock.** Install a dead center in the tailstock. (A live center will turn. A dead center will not.)
3. On one end of the center-drilled stock, mount a lathe dog. A **lathe dog** connects the stock to the faceplate. This makes the stock rotate. See Figure T8-6. A soft piece

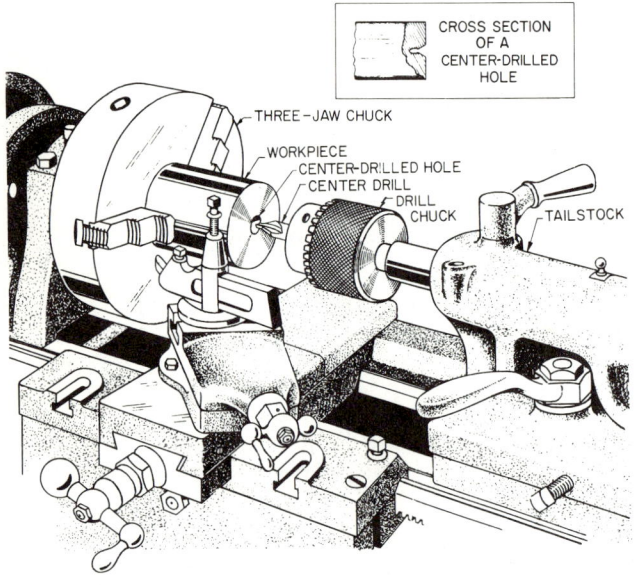

Figure T8-5

Centers hold the stock in the lathe. A hole must be made in each end of the stock for the centers. This is center-drilling.

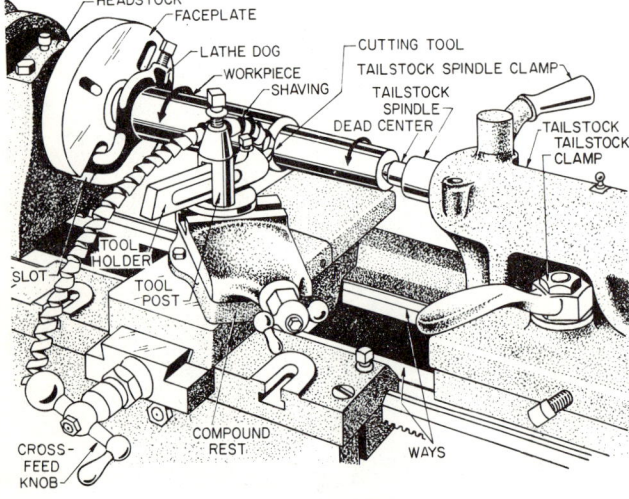

Armstrong Brothers Tool Company

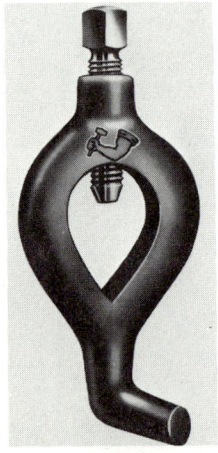 
South Bend Lathe

Figure T8-6

Straight turning is done down the length of the stock. A lathe dog is used to connect the work to the faceplate.

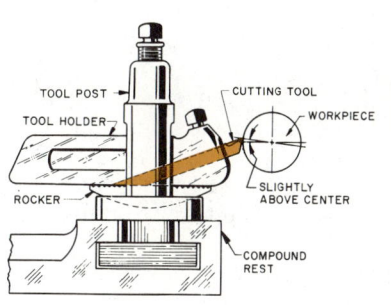

Figure T8-7

A cutting tool in position for cutting.

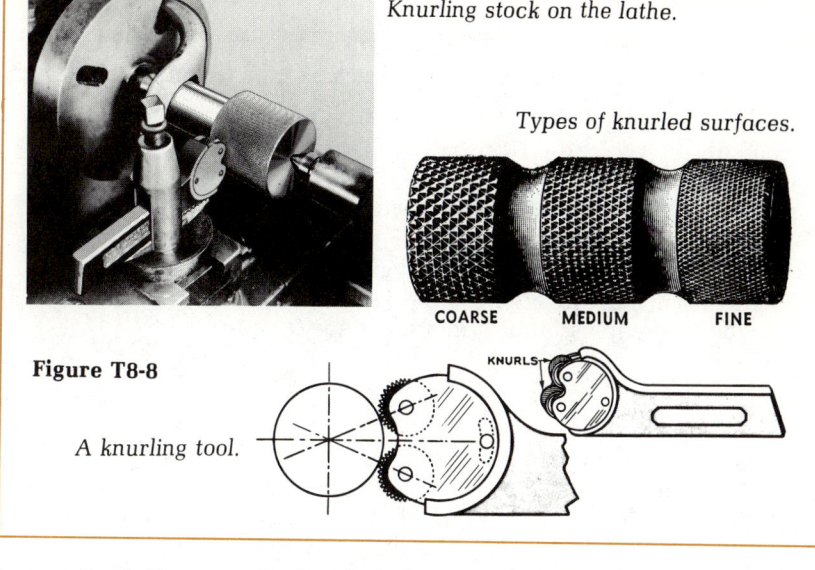

Figure T8-8

A knurling tool.

of metal may be placed between the lathe dog and the stock to protect the finish.

4. Mount the end of the stock with the lathe dog on the live center. Slip the bent leg of the lathe dog into the slot on the faceplate.
5. Put a small amount of center lubricant on the dead center.
6. While supporting the other end of the stock with your hand, move the tailstock up to the end and engage the dead center.
7. Lock the tailstock to the **ways.** Refer again to Figure T8-1.
8. Tighten the **tailstock handwheel** until the dead center is just lightly snug. Then tighten the **spindle lock.**
9. Mount a cutting tool in the tool holder. Fasten it to the tool post on the compound rest. See Figure T8-7.
10. The lathe should now be ready to turn. Check with your teacher to make sure that the lathe is properly set up and that the lathe speed is correct.
11. Turn the stock. Always turn from the tailstock to the headstock. See again, Figure T8-6. Use the cross-feed knob to adjust the depth of cut. Use the **apron handwheel** to control the cutting motion.
12. After turning is complete, remove the center-drilled ends by using a cutoff tool or a facing operation.

Knurling

Knurling puts a shallow pattern of diamond shaped grooves in a piece of metal. See Figure T8-8. This is done to make the metal easier to grip. A metal screwdriver, for example, is much easier to use if the handle has been knurled.

Before knurling, make sure that the stock is firmly supported between centers.

1. Clearly mark the area to be knurled.
2. Mount the knurling tool in the tool post.
3. Adjust it so that it contacts (touches) the stock at an exact right angle (90 degrees).
4. Center the tool vertically (up and down).
5. Make certain that it contacts the work equally on each side of the centerline.
6. Set the lathe to run very slowly.
7. The carriage must be set to move automatically. Do **not** set this yourself. Your teacher will help you to make this adjustment.
8. Begin knurling. Feed the knurling tool back and forth along the stock. Continue to do this until the knurl is made up of clean diamond shapes with slightly flat tops.

Metalworking Lathe

1. Wear eye protection when using the metalworking lathe.
2. Always remove the chuck key from the chuck after tightening the jaws of the chuck.
3. Do not wear loose clothing.
4. If your hair is long, tie it back.
5. Do not run the cutting tool into the chuck.
6. Turn off the lathe before cleaning it.
7. Keep arms and hands away from headstock.

Topic 9 Threaded Fasteners and Taps & Dies

Threaded Fasteners

Threaded fasteners are used to hold parts together. See Figure T9-1. Nuts, bolts, and screws are examples of threaded fasteners. Parts that are put together with these fasteners may be taken apart later without damage to the pieces.

Bolts and Nuts

Bolts are available in many shapes and sizes. See Figure T9-2. Bolts are classified by the diameter of the thread, the number of threads per inch (see Figure T9-3), the length (not including the head), and the shape of the head. For instance, a bolt may be classified as 1/2 - 13 UNC × 2-1/2 Hex HD. This means that the bolt is 1/2" in diameter, has 13 threads per inch, is 2-1/2" long, and has a hexagon-shaped (six-sided) head. Bolts are inserted into threaded holes or used with nuts. They are tightened with wrenches.

Nuts are made to be threaded onto bolts. See Figure T9-4. For a proper fit, use a nut that is the same size and has the same number of threads per inch as the bolt. Some common nuts include hexagon nuts, wing nuts, and square nuts. See Figure T9-5.

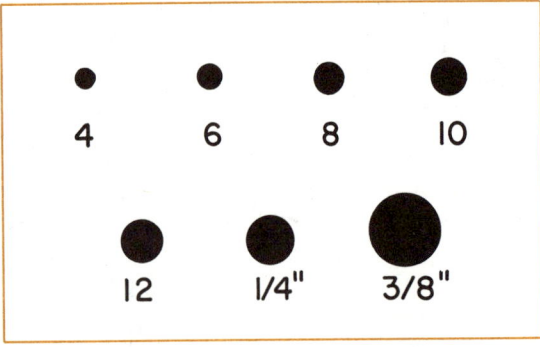

Figure T9-2

Common diameters of screws and bolts given by gage and inch.

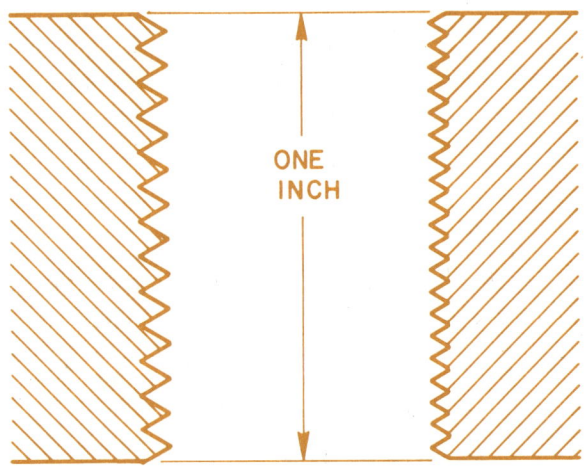

Figure T9-3

A cross section of a fastener with National Fine threads is shown on the right. Compare these with the cross section showing National Coarse threads on the left. How many threads in one inch for each type of fastener?

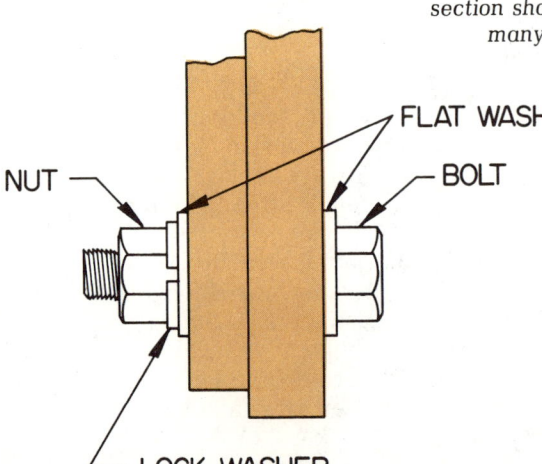

Figure T9-4

Typical nut and bolt assembly with washers.

Figure T9-1

Common Fasteners

Name	Head Shapes	Sizes & Lengths	Used On	Comments
WOOD SCREWS	 round head oval head flat head	Diameters: no. 4 thru no. 12 Lengths 1/2" thru 4"	Wood	Looks nice, fast to use. Decorative, requires countersink. Fits flush with surface or below surface, can be covered, requires countersink.
SELF-TAPPING SCREW	 pan head	Diameters: no. 4 thru no. 12 Lengths: 1/4"-2"	Sheet Metal, Sheet Plastic, Wood	Cuts its own threads, has coarse threads, threaded from head to point.
MACHINE SCREWS	flat head round head	Diameters: no.4 through 1/4" Lengths: 1/2" thru 4"	Wood, Metal, Plastic	Usually requires a nut, available in fine threads or coarse threads.
SET SCREWS	 socket head	Diameters: no. 4 thru 3/8" Lengths: 1/4" thru 1"	Metal	Can be screwed in below surface, requires a tapped hole, small, easily hidden.
CAP SCREWS	 hex head	Diameters: 1/4" thru 3/4" Lengths: 1/2" thru 4"	Metal	Available in fine and coarse threads, generally used in a threaded hole.
MACHINE BOLTS	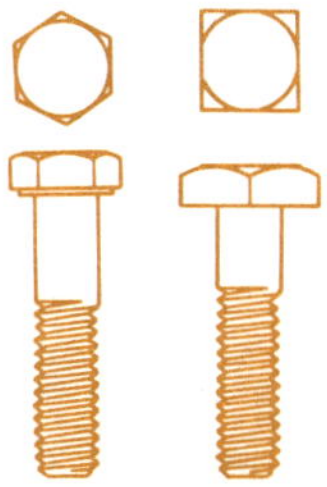 hex head square head	Diameters: 1/4" though 1" Lengths: 1/2" through 2"	Metal	Requires a nut, generally used with washers. Available in fine or coarse thread.

Screws

Screws are classified in much the same way as are bolts. **Machine screws,** for instance, are used in ways similar to bolts. However, they are very small and have slotted heads. See Figure T9-6. They are tightened with a screwdriver. Machine screws are classified by diameter (wire gage), length, head type, and number of threads per inch.

Wood screws and **self-tapping screws,** are different from bolts and machine screws. The threaded portions of these screws taper (narrow) to points. They need no threaded holes. These screws cut their own threads as they are driven into soft materials. Wood screws and self-tapping screws are classified by wire gage, length, and type of head. For example, #8 × 1-1/2″ RH is the classification of a screw. This screw is number 8 gage in diameter, is 1-1/2″ long, and has a round head. Common head types are flat head, round head, and oval head. Some self-tapping screws are made with pan-shaped heads.

Washers

Washers are flat metal disks that are used with threaded fasteners. There are two basic types: flat washers and lock washers. See Figure T9-7. **Flat washers** serve two purposes. They help to distribute pressure. Then, if too much force is applied to the joined parts, the washer will break, not the parts or the bolts. Washers also protect the surfaces of parts being fastened. **Lock washers** lock nuts into place. When they are used, they are generally used with flat washers. See again Figure T9-4.

Taps and Dies

Taps and dies are used to cut threads. A tap is used to cut internal (inside) threads, such as in a hole in a metal part or pipe. A die is used to cut external (outside) threads, such as on bar stock.

Using a Tap

A tap is used to cut internal threads for a bolt, machine screw, or threaded rod. A tap is turned with a **tap wrench.** See Figure T9-8. To use a tap, first drill a hole to the correct size. The hole size is determined by the bolt diameter and the number of threads per inch. On a **tap and drill chart,** find the correct size twist drill for drilling the hole. See Figure T9-9

After the hole is made, use a tap wrench to turn the correct tap into the hole. See Figure T9-10. There are three types of taps: taper, plug, and bottoming. See Figure T9-11.

When tapping a hole that does not extend completely through the workpiece, all three taps must be used. The **taper tap** is used first. It has a tapered end and is easiest to start in the hole. The **plug tap** is less tapered and is used next. The **bottoming tap** has no taper. It is used to tap to the bottom of the hole.

When tapping a hole completely through the workpiece, you will need to use only a taper tap.

Make sure the tap is started **straight** in the hole. Turn the tap very slowly, and keep the cutting surface lubricated with oil. Turn the tap backward one turn for every two turns forward. This will remove waste material from the cutting surface. See Figure T9-12.

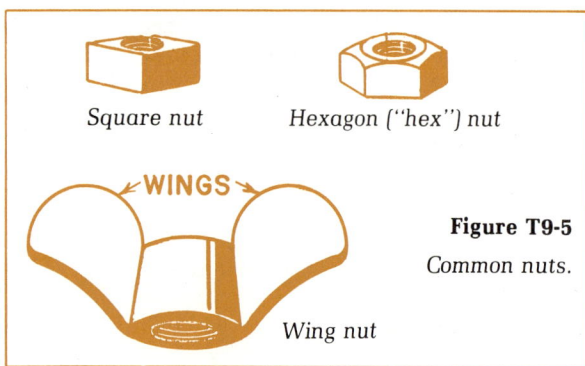

Figure T9-5
Common nuts.

Slotted head   Phillips head

Figure T9-6
Common types of screws.

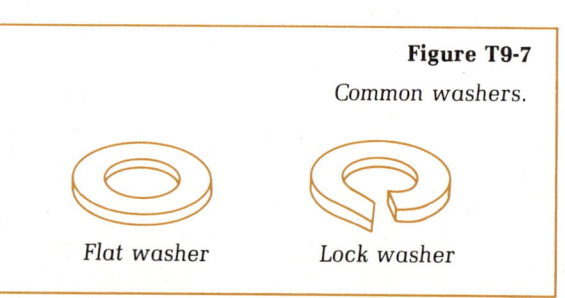

Figure T9-7
Common washers.

Flat washer   Lock washer

Threaded Fasteners and Taps & Dies 337

A T-handle tap wrench.

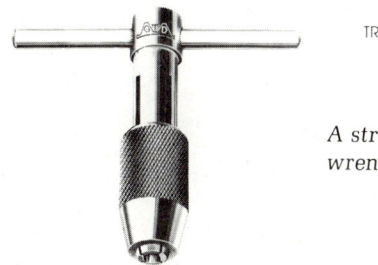

A straight-handle tap wrench.

Figure T9-8

Tap wrenches

Figure T9-9

Sizes of taps and tap drills.

National Fine		National Coarse	
Size and Thread	Tap Drill (Inch)	Size and Thread	Tap Drill (Inch)
Gage		Gage	
4 — 48	3/32	4 — 40	3/32
5 — 44	7/64	5 — 40	7/64
6 — 40	1/8	6 — 32	7/64
8 — 36	9/64	8 — 32	9/64
10 — 32	5/32	10 — 24	5/32
Inch		Inch	
1/4 — 28	7/32	1/4 — 20	13/64
5/16 — 24	9/32	5/16 — 18	17/64
3/8 — 24	11/32	3/8 — 16	5/16
7/16 — 20	25/64	7/16 — 14	3/8
1/2 — 20	29/64	1/2 — 13	27/64
9/16 — 18	33/64	9/16 — 12	31/64
5/8 — 18	37/64	5/8 — 11	17/32
3/4 — 16	11/16	3/4 — 10	21/32
7/8 — 14	13/16	7/8 — 9	49/64
1 — 14	15/16	1 — 8	7/8

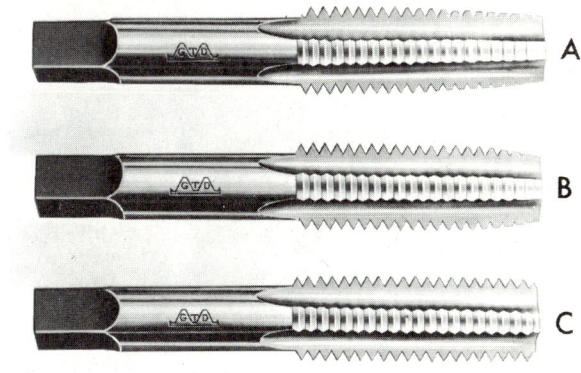

Figure T9-11

The three basic kinds of taps.
A. Taper tap
B. Plug tap
C. Bottoming tap

Figure T9-12

Checking the tap for squareness.

Figure T9-10

Hold the tap wrench securely when starting threads.

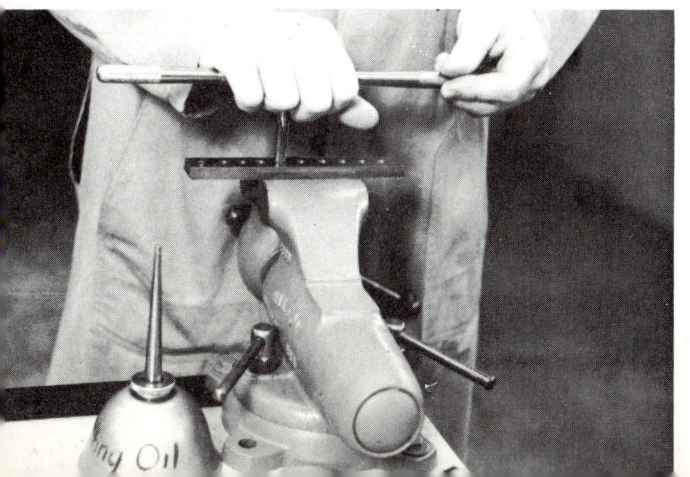

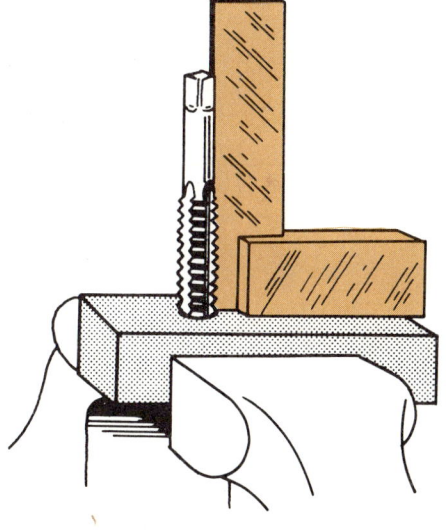

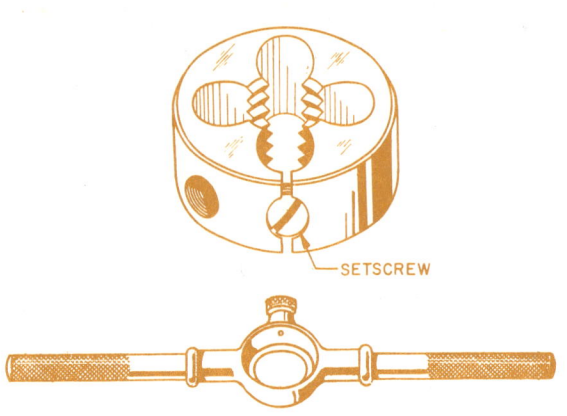

Figure T9-13

A threading die fits into a diestock. A rod to be threaded is placed into the die.

Using a Die

A threading die is used to cut **external threads** on round metal. The die is held and turned by a **diestock.** See Figure T9-13. Dies are classified by diameter and number of threads per inch.

Chamfer (angle) the end of the rod to be threaded. Chamfering the end will make starting the cut easier. Use a grinder or a file to make the chamfer. See Figure T9-14.

Hold the diestock by the handles, and slowly turn the die onto the rod. Keep the diestock square (at a right angle) to the rod being threaded. See Figure T9-15. Apply firm, even pressure as you turn the diestock onto the work. Turn the diestock one turn backward for every two turns forward. Apply oil to the cutting surface to make cutting easier. Oil will also help to remove waste material.

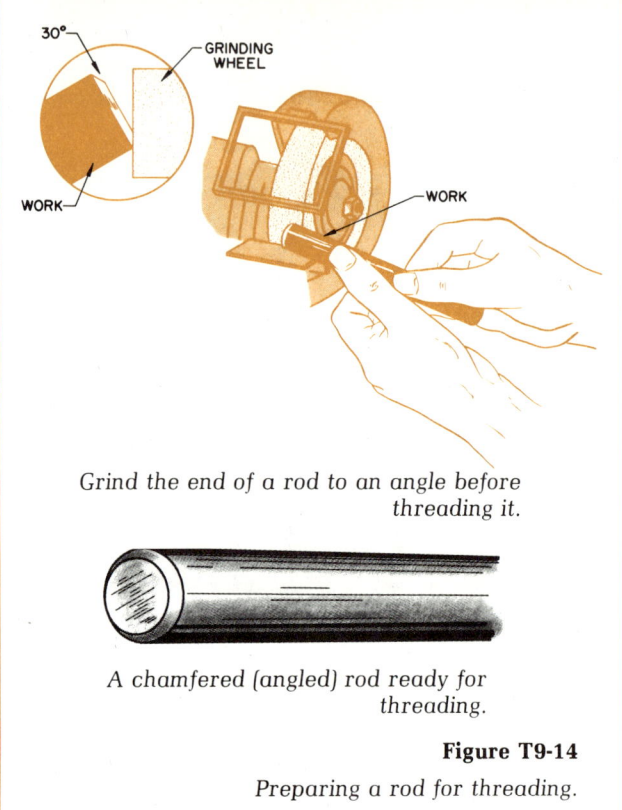

Grind the end of a rod to an angle before threading it.

A chamfered (angled) rod ready for threading.

Figure T9-14

Preparing a rod for threading.

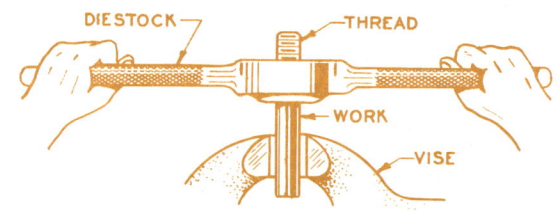

Figure T9-15

Turning the diestock to cut the threads.

Topic 10 — Non-Threaded Fasteners and Fastening Tools

Non-threaded fasteners are used to join materials together. Some non-threaded fasteners are rivets, nails, and staples. They are used to join parts together quickly and permanently. If a product does not have to be taken apart after it is put together, this type of fastener may be used. Some other non-threaded fasteners, such as snap fasteners, are used for joining parts that may need to be taken apart again later.

Rivets

Rivets are made from several metals, including steel, brass, copper, and aluminum. Leather, canvas, cloth, plastics, and many kinds of metal are materials that can be joined by rivets. Each rivet has a head and shank. Its size is determined by **diameter** and **body length.** The head is not included in the body length unless the rivet is a countersink rivet.

Non-Threaded Fasteners and Fastening Tools

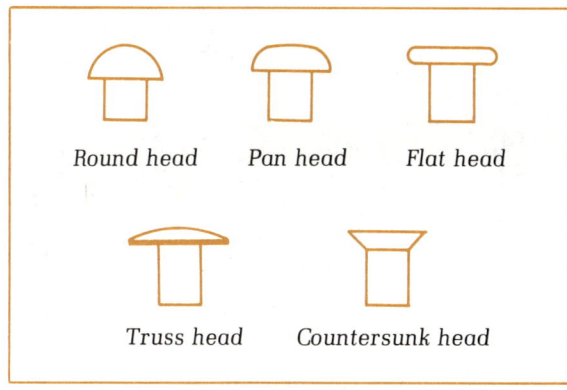

Figure T10-1

Common shapes of rivets.

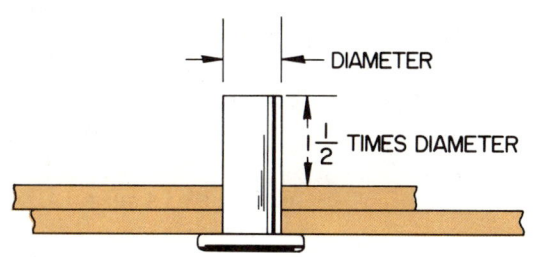

Figure T10-2

Determining how long the rivet should be.

To use a rivet, first select the proper kind and size. See Figures T10-1 and T10-2. Drill or punch a hole through the parts to be riveted. The hole should be slightly **larger** than the diameter of the rivet.

A rivet that is 1/8″ in diameter should extend 3/16″ above the work surface so that a head can be formed. A **rivet set** is used to squeeze materials together for riveting. The rivet set is also used to smooth and round (shape) the head of the rivet. See Figure T10-3.

Two-part Rivets

Two-part rivets are sometimes called "**rapid rivets**." They can be used on such materials as leather, canvas, plastics, and textiles. Each rivet consists of a **cap** and a **post**.

To use a rapid rivet, first punch a hole through both parts. Then, insert the post into the hole from underneath. Place the cap over the post, and hit the top of the cap with a mallet. See Figure T10-4.

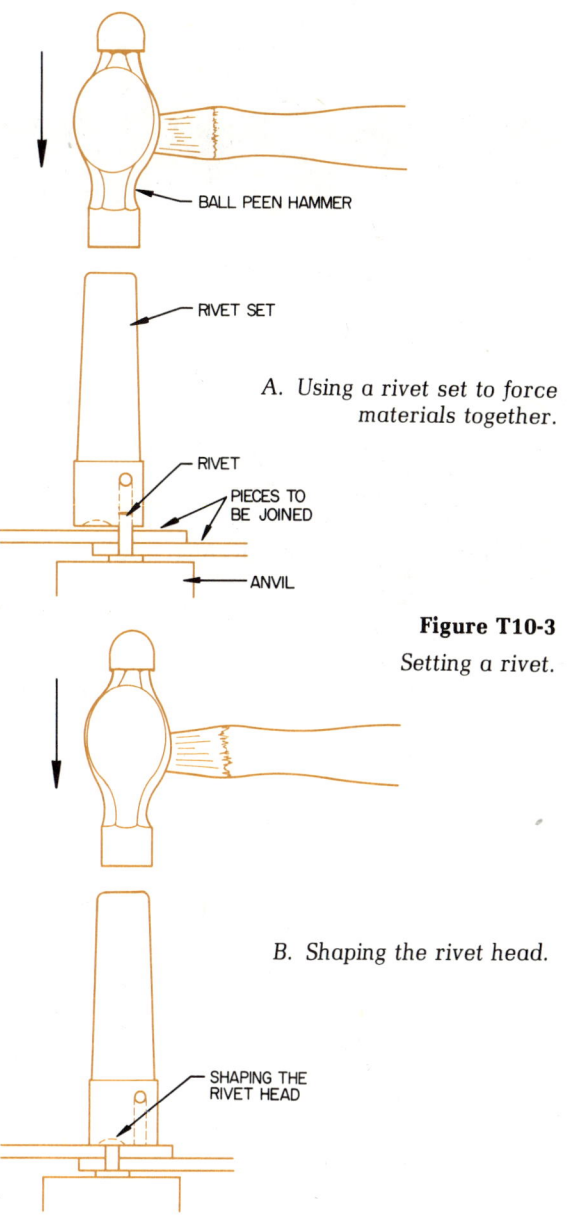

A. *Using a rivet set to force materials together.*

Figure T10-3

Setting a rivet.

B. *Shaping the rivet head.*

Figure T10-4

Installing a two-part rivet.

340 Tools, Materials, Processes

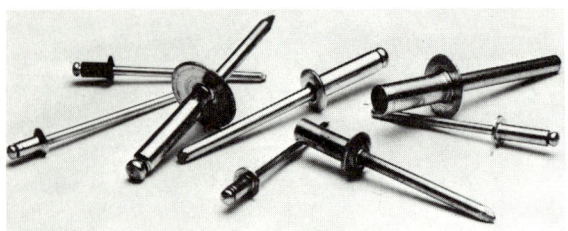

Figure T10-5 USM Corporation
Blind rivets.

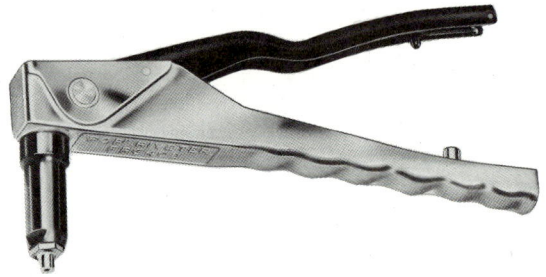

A riveting tool. USM Corporation

Blind Rivets

Blind rivets are often called "Pop"® rivets. They are used on metals or plastics. They are called "blind" rivets because one side of each rivet enters the "blind" (or unseen) side of the metal or plastic. See Figure T10-5.

A blind rivet is made up of two parts. One part has a **stem** (mandrel). This stem extends through the other part of the rivet, the **head**. The head is hollow.

To use this fastener, insert the rivet stem into the riveting tool. Then, insert the blind rivet through the material to be riveted. Squeeze the handles of the riveting tool to pull the stem into the hollow part of the rivet. This produces a bulge on the "blind" side. Keep squeezing until the stem of the rivet breaks. See Figure T10-6.

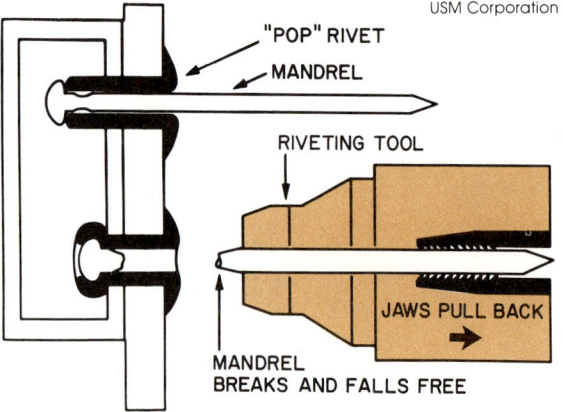

Figure T10-6
A cross section of materials being joined by a blind rivet.

Fasteners for Leather or Cloth

Snap Fasteners

Snap fasteners are used to hold two pieces of leather or cloth together. When snap fasteners are used, parts can be snapped together and taken apart whenever necessary. Each snap fastener has four parts. The **cap** and **socket** form one half of the fastener. The **eyelet** and **stud** make up the other half. See Figure T10-7.

The snap fastener is installed in much the same way as a rivet. Mark and punch the location of the two holes. Position the socket on a hard, flat surface, and place the material over the socket. Place the cap on the material over the socket. Using a setting tool and mallet, **set** this assembly.

In the second piece of material, insert the eyelet in the hole, and **place** it on a hard, flat surface. Position the stud over the eyelet and **set** this assembly with a setting tool. See Figure T10-8. Adjust the stud for proper tightness.

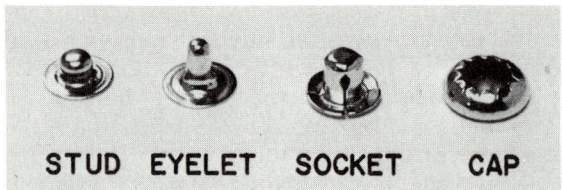

Figure T10-7
The parts of a snap fastener.

Figure T10-8
A special setting tool for setting a snap fastener.

Non-Threaded Fasteners and Fastening Tools   341

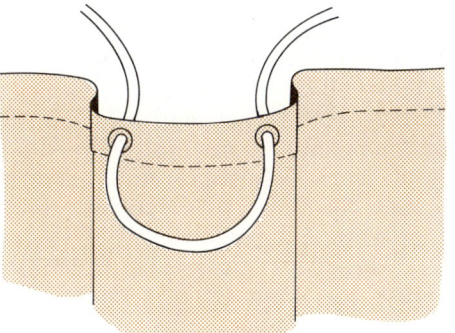

Figure T10-9
Using grommets to reinforce holes in materials.

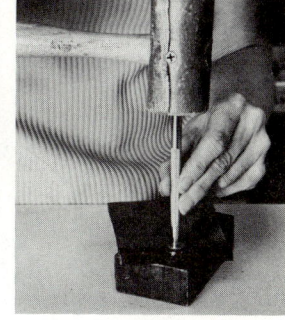

Figure T10-10
Using a special tool to join the two parts of a grommet.

Grommets

Grommets are metal eyelets. They are used to reinforce holes in cloth, leather, or plastics. See Figure T10-9. Drawstrings are often threaded through grommets. Aprons, bags, and many other leather and textile products contain grommets. Grommets can also be used to fasten pieces of material together. See Figure T10-10.

Lacing

Flexible materials, such as pieces of leather, are often laced together. Evenly punched holes must be made along the edge of the work to be laced. A **thonging chisel** can be used to punch the holes for lacing. See Figure T10-11.

The **single whipstitch** requires lacing three and one-half times as long as the distance being laced. The **cross whipstitch** needs lacing about six times the lacing distance. The **single buttonhole stitch** conceals the edge of materials being joined. Lacing needed for the single buttonhole stitch is also about six times the distance to be laced. See Figures T10-12 and T10-13.

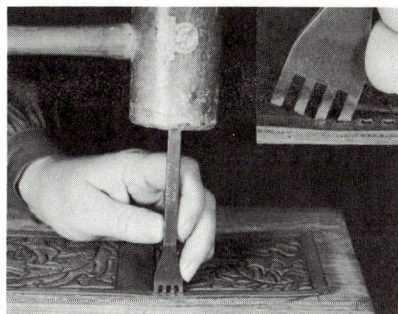

Figure T10-11
Using a thonging chisel to make lacing holes in leather.

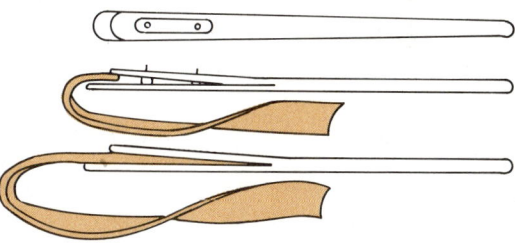

Figure T10-12
Attaching lacing leather to the lacing needle.

Figure T10-13
Patterns for lacing.

Single whipstitch.   Cross whipstitch.   Single buttonhole stitch.

Nails

Nails are general-purpose fasteners to hold pieces of wood together. Nails are driven into the wood with a hammer. The holding power of nails is less than that of threaded fasteners, such as screws.

The sizes of common, box, and finish nails are numbered according to a system called the **penny** system. The small letter "d" is used as a symbol for the term "penny." Nails that are commonly used range from 2d (1") to 16d (3-1/2") in size. Small wire nails and brads are measured by their length (in inches or fractions) and their gage (wire diameter). Brads are usually 1/2" to 1" long. The higher the gage number, the smaller the diameter.

Kinds of Nails

Five kinds of nails are used most often. These are common nails, box nails, finishing nails, casing nails, and brads. See Figure T10-14.

- **Common nails** have flat heads. They are strong and have good holding power.
- **Box nails** have flat heads. These nails are smaller in diameter than common nails. Box nails are less likely to split the wood.
- **Finishing nails** have small heads that can be sunk below the work surface. Use these nails when you do not want the head of the nail to show. Finishing nails are used especially for finish work, such as attaching trim or doing cabinetwork.
- **Casing nails** are much like finish nails. However, the diameter of a casing nail is larger than that of a finish nail, and the head has greater holding power.
- **Brads** have small diameters and small heads. Brads are very useful for fastening narrow strips of wood.

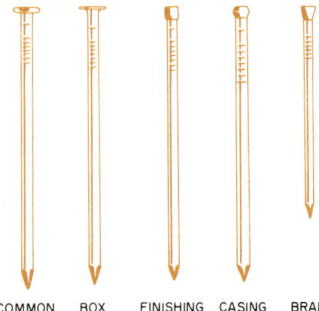

Figure T10-14
Kinds of nails.

COMMON BOX FINISHING CASING BRAD

Nailing

1. Wear safety goggles when using nails.
2. Hit the nail carefully. Nails can fly out in any direction when they are hit the wrong way.
3. When you are using a hammer to strike a nail, move your fingers out of the way.
4. Never hold nails in your mouth.

Special Wood Fasteners

Tacks and Corrugated Fasteners

Tacks are small flat-head fasteners. They are used to attach materials such as carpet to wood surfaces. See Figure T10-15.

Corrugated fasteners are often used to join mitered joints such as picture-frame corners. One edge of the fastener is sharp. The other edge of the corrugated fastener is struck with a hammer to drive the fastener into wood. See again Figure T10-15.

Staples

Staples are designed to fasten thin materials, such as cloth, screen, fencing, or electric wire to fairly soft materials, such as wood or cardboard. The lengths of staple legs range from 1/4" to 9/16". Staple guns are used to insert staples into a surface. See Figure T10-16. Never point the staple gun at anyone. Keep your fingers away from the area being stapled.

Figure T10-15
Common fasteners for wood.

A tack.

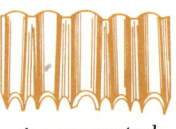
A corrugated fastener.

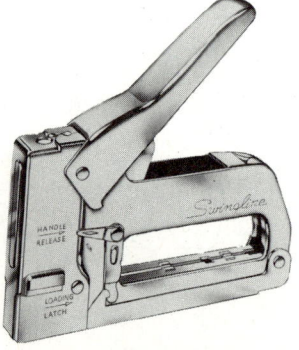

Figure T10-16
A hand-held staple gun and staple.

Swingline Incorporated

Topic 11 Glues, Cements, and other Adhesives

Bonding agents such as adhesives are used to make materials stick together. Some adhesives cure (set) as water evaporates from them. Some adhesives become liquid when heated and stick to surfaces as they cool. Other bonding agents cure by chemical reaction.

Bonding

In **adhesion** bonding, materials (can be the same kind or different kinds) are stuck together with another material called an **adhesive**. Adhesives all require time to cure. However, they do not all cure at the same rate. In determining the best type of adhesive to use for a particular job, keep in mind the **curing time** of each kind.

The **curing method** is also important. White glue, for example, cures as it dries. It will not cure if the glued area is sealed from air, because air allows moisture to evaporate. Epoxy glues, on the other hand, cure by chemical action. A sealed area does not prevent these glues from curing.

In **fusion** bonding, two similar materials are fused (melted together) by means of heat or chemical action. Chlorinated solvents* and plastic cement chemically melt (soften) some plastic materials together, joining the two surfaces into one.

Adhesives

Selecting an Adhesive

Choose the best adhesive for the job that you are doing. Some jobs require a glue that has a lot of **holding power.** A part that is to receive a great deal of strain, such as a tool handle, will need a very strong glue.

Some products must stay flexible. For example, when gluing two pieces of cloth together, you would **not** use a glue that dries very stiff.

Sometimes parts must be clamped together in order for the glue to cure properly. Using this type of glue would not be practical if the object is very large or has an odd shape. Clamping would be very difficult.

Keep in mind where the object will be used. An item that will be exposed to heat or moisture, such as a toy to be used outside or in the water, must be glued with a **heat-resistant or moisture-resistant** adhesive.

Another factor to consider is whether to use a **fast-setting** or **slow-setting** adhesive. Generally, when gluing an object that has many parts that connect, you will probably use a slow-setting glue. This will give you time to clamp all the parts together before the glue dries. Use a fast-setting glue for an item that cannot be clamped or that you will need to use right away.

Some adhesives **chemically affect** (change) the materials being bonded. Some household cements dissolve polystyrene foam (such as Styrofoam®), for example, but white glue does not. White glue, on the other hand, tends to curl paper and may stain felt if applied heavily.

White Glue

White glue (polyvinyl acetate) is one of the most commonly used glues. In liquid form, this glue is white, but it dries clear. See Figure T11-1. White glue dries quickly, but requires clamping. White glue works well on **wood, paper,** and **foamed plastic.** It is not waterproof and will soften at high temperatures.

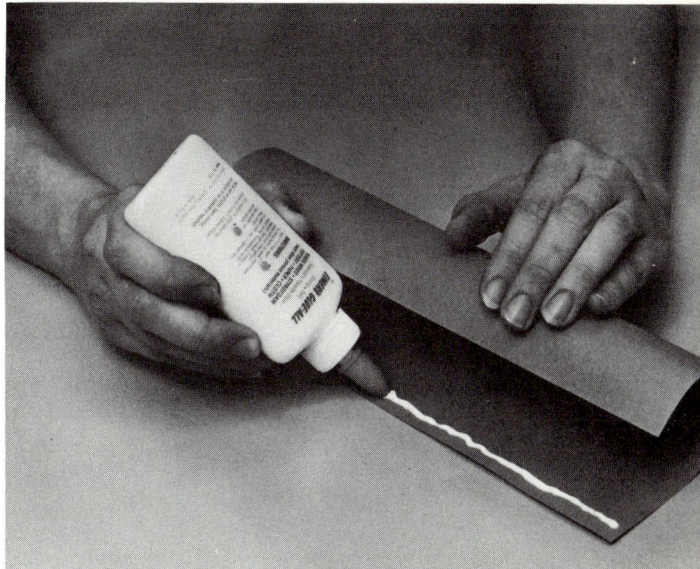

Figure T11-1

Using white glue to make a paper mock-up of a product.

---

* A solvent is a liquid in which other substances can be dissolved.

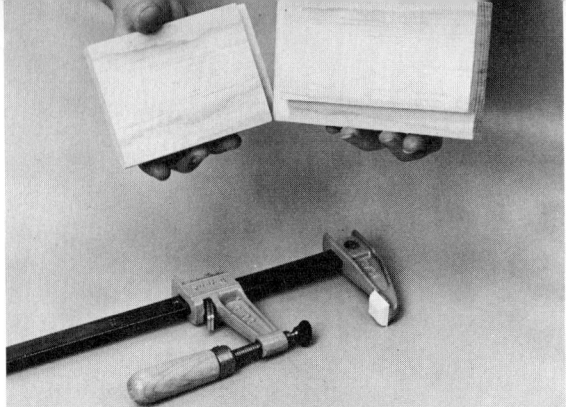

Figure T11-2

Using aliphatic resin glue to join the two parts of a wood joint.

Figure T11-3

Mixing water with plastic resin glue powder.

Figure T11-4 The Terrell Corporation

A hot melt glue gun and the sticks of glue used in it.

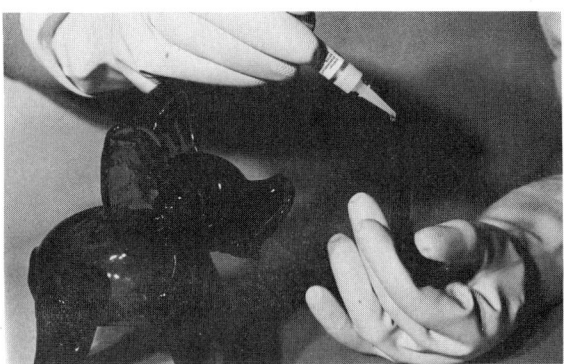

Figure T11-5

Using "super" glue to repair a broken glass figurine.

Aliphatic Resin Glue

Aliphatic resin glue is a yellowish-colored liquid. It is excellent for gluing **wood**. It sets quickly, dries clear, and makes a very strong bond. This glue also resists heat and solvents. However, it is not moisture resistant. Aliphatic resin glue is used on wood, cloth, paper, and other porous (moisture-absorbing) materials. See Figure T11-2.

Plastic Resin Glue

Plastic resin glue is a powdered adhesive. See Figure T11-3. It is used for gluing **wood**. It is moisture resistant and very strong. Plastic resin glue must be mixed with water to a creamy thickness. Mix according to the directions given. This glue cures by chemical action. You should mix only the amount that you will need. Once mixed, plastic resin glue must be used within one hour.

Hot Melt Glue

Hot melt glue is most often sold in sticks. It is applied with a glue gun. See Figure T11-4. Glue guns heat the sticks of glue and melt them into a quick-setting, usable form. Hot melt glue is not a strong adhesive, but it works well for gluing **small parts.** It is not strong enough to hold together parts that carry a load or that have force applied to them.

The glue is ready to use when it flows freely out of the tip of the gun as you squeeze the trigger. Remove all excess (extra) glue from the material with a cloth while the glue is still hot. Apply a small bead of glue and press the materials together a little at a time. Do not touch the tip of the gun. It may be hot. Make sure all glued surfaces have cooled before touching them with your hands.

Cyanoacrylate Glue ("Super Glue")

Cyanoacrylate glue, (such as Super Glue®) is fast-setting. Almost anything can be bonded with this kind of glue. It is powerful! Be extremely careful not to get any on your skin. This kind of glue is used for **small jobs** (fixing a broken dish, for example). You will need to use only a drop or two for most jobs. See Figure T11-5.

Epoxy Glue

Epoxy glue consists of a resin and a catalyst. A catalyst helps a chemical reaction to take place. In this case, the glue hardens. To use epoxy glue, mix equal quantities of the catalyst and the resin. See Figure T11-6. Epoxy is very strong, waterproof, and heat resistant. Epoxy can be used to glue **almost any material.**

Rubber Cement

Rubber cement is used to bond porous (moisture-absorbing) materials like **leather, textiles, and paper.** See Figure T11-7. It is easy to apply. Rubber cement is excellent for use with thin paper. There is no water evaporation to stain or wrinkle the paper. This cement is used to adhere abrasive paper to the disk of a disk sander. Use the cement as directed on the container.

To make a bond that lasts, apply rubber cement to each of the two surfaces. Allow the rubber cement to dry for a few minutes before joining the surfaces.

Contact Cement

Contact cement bonds two surfaces on contact. See Figure T11-8. Once the two glued surfaces touch one another, it is nearly impossible to separate them. This cement is used to join **veneers to wood** materials like countertops or tabletops. Clamping is not necessary because contact cement bonds as soon as it touches something. Use contact cement as directed on the container. Paper placed between cement-coated surfaces prevents the materials from coming into contact before they are properly positioned.

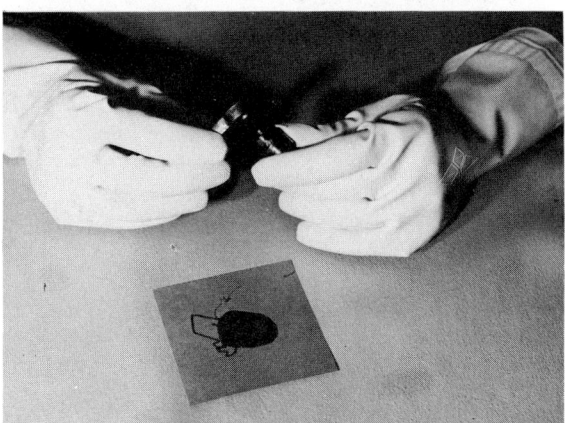

Figure T11-6
Using epoxy glue to join two metal pieces.

Figure T11-7
Applying rubber cement to the back of tooled leather before joining it with another piece of leather.

Figure T11-8
Using contact cement to apply veneer.

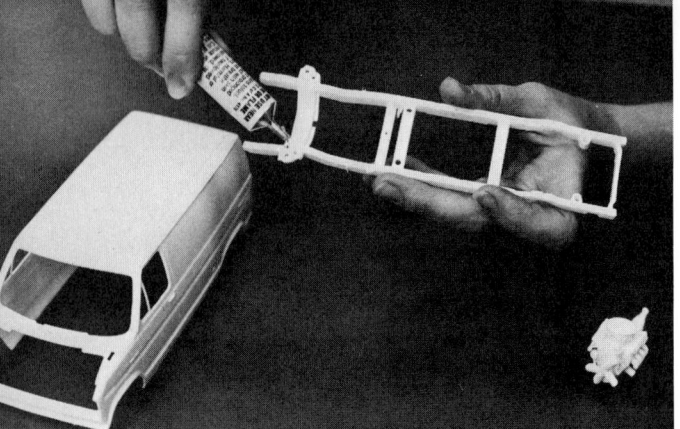

Figure T11-9

Using plastic cement to assemble a plastic model.

Figure T11-10

Using chlorinated solvent to join two pieces of plastic.

Plastic Cement

Plastic cement is used to bond **styrene plastic.** This cement cures quickly. Use only a small amount of plastic cement. Wipe off any excess immediately. Plastic cement is sometimes called **model glue** because it is used in model-building. See Figure T11-9.

Chlorinated Solvents

A chlorinated solvent is a mixture of methylene chloride and trichlorethylene. This mixture is a clear liquid. It is used to bond **acrylic plastics.** Like plastic cement, this liquid melts (softens) the acrylic by chemical action. The two pieces become welded into one. A chlorinated solvent can be applied with a small brush or with a solvent applicator. See Figure T11-10.

Fusible Tape

Fusible tape or webbing is a special adhesive used to hold pieces of **cloth** together. When heat is applied, it melts, bonding the cloth. It is a fast and easy way to make hems and seams in cloth without stitching. See Figure T11-11.

To use fusible tape follow this procedure:
1. Cut a piece of the tape to the same length as the hem or seam. Place the tape between the layers of fabric (cloth) that you are bonding.
2. Cover the fabric with a damp cloth.
3. Using a steam iron set on the hot ("wool") setting, press for ten seconds. Do **not** slide the iron back and forth.
4. Turn the fabric over, and repeat the process. Use the iron carefully. Remember, it is hot.
5. Allow the fabric to cool. Then pull on it gently to see if the bond is tight. If it isn't, press again for five to ten seconds on each side.

Figure T11-11

A. Positioning fusible tape in the hem of a fabric.

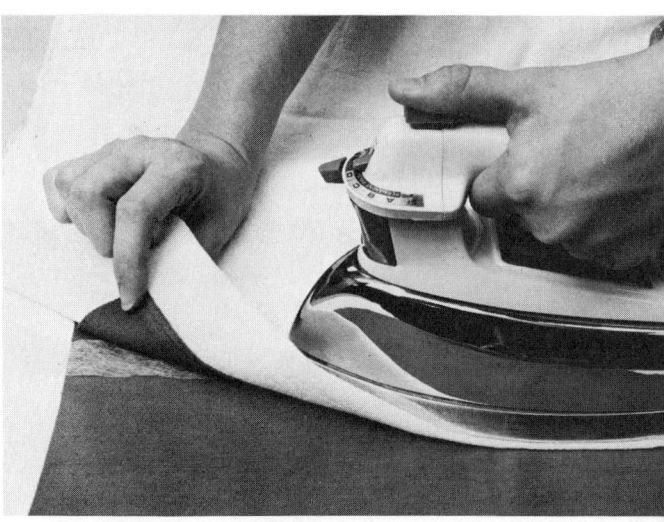

B. Completing the bond by pressing with a steam iron. The fabric is covered by a pressing cloth.

Topic 12 Soldering

Soldering is a common process used to join metal surfaces together. A special metal called **solder** is melted. It adheres (sticks) to the metals being joined. The base metals must be hot. Then the melted solder will flow and adhere properly. Sheet metal, copper pipe, and electric wires are materials that are commonly soldered.

Tools Used in Soldering

Four tools are commonly used in soldering. See Figure T12-1.

- A **soldering copper** is usually needed for sheet-metal work. A gas-heated bench-type furnace is used to heat the soldering copper.
- An **electric soldering copper** is used for general-purpose soldering. It can be used to solder electrical connections or for light sheet-metal work.
- An **electric soldering gun** is used for soldering light metalwork. It is especially useful for soldering electrical connections.
- A **propane torch** is used when a lot of heat is needed. For example, it is used when soldering copper pipe.

A bench furnace.

A propane torch.

Figure T12-1
Tools used for soldering.

A soldering copper.

An electric soldering gun.

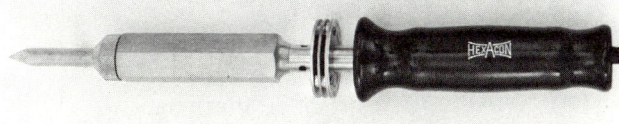

An electric soldering copper.

348 Tools, Materials, Processes

Figure T12-2

A. Solder is melted to join metal surfaces.

B. Flux cleans the metals to be joined and helps the melted solder to flow.

Solder and Flux

Tin and lead are usually alloyed (mixed) in equal amounts to make solder. This solder melts at about 400° F (204° C).

A substance called **flux** is used in soldering. Flux cleans the metal. It also helps the hot solder to flow and to stick. See Figure T12-2.

There are two basic types of flux, corrosive and noncorrosive. A **corrosive** flux may be used with galvanized (zinc-coated) sheet metal. Zinc chloride is a corrosive flux. It contains an acid. Because a corrosive flux eats away at a material it cannot be used for electrical work.

Noncorrosive flux is used on electrical work. Resin flux and rosin flux are noncorrosive. They are often sold in paste form.

A special hollow solder called **fluxcored solder** may be either corrosive or noncorrosive. Fluxcored solder may be called acid core or rosin core depending on the type. Flux is also contained in the core. Fluxcored solder is often used for small jobs.

Soldering

1. Clean the material to be soldered by rubbing the surface with steel wool or emery cloth. (See Topic 15, **Filing and Sanding**.)
2. Apply flux to further clean and protect the work surface.
3. Clean the soldering copper with a file if the surface is rough. See Figure T12-3.
4. Prepare the soldering copper by coating its surface with solder. See Figure T12-4.
5. Apply solder to the work surface. See Figure T12-5. Press the entire face of the soldering copper on the work to heat it. Do this until the solder melts on the surface. When soldering with the propane torch, apply heat to the side **opposite** the soldered surface. The solder will flow toward the heat source when it melts.
6. Hold the work surfaces together with another object until the solder cools.

Soldering

1. Wear eye protection when soldering.
2. Use extreme care in handling hot metals and soldering tools.
3. If flux gets on your skin, wash it off immediately.
4. Warn others about metal that might be hot.

Soldering   **349**

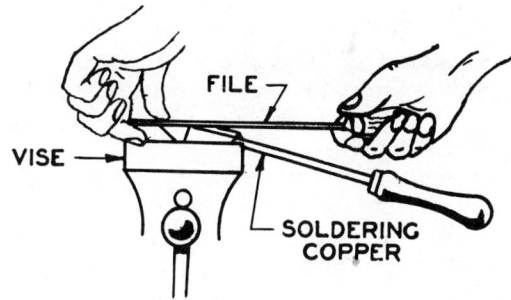

Figure T12-3
Cleaning a soldering copper with a file.

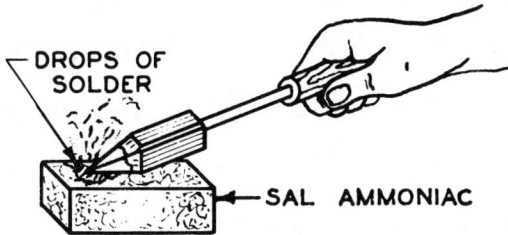

Figure T12-4
Coating the soldering copper with solder on a sal ammoniac block (cleaning block).

Figure T12-5
Soldering sheet metal.

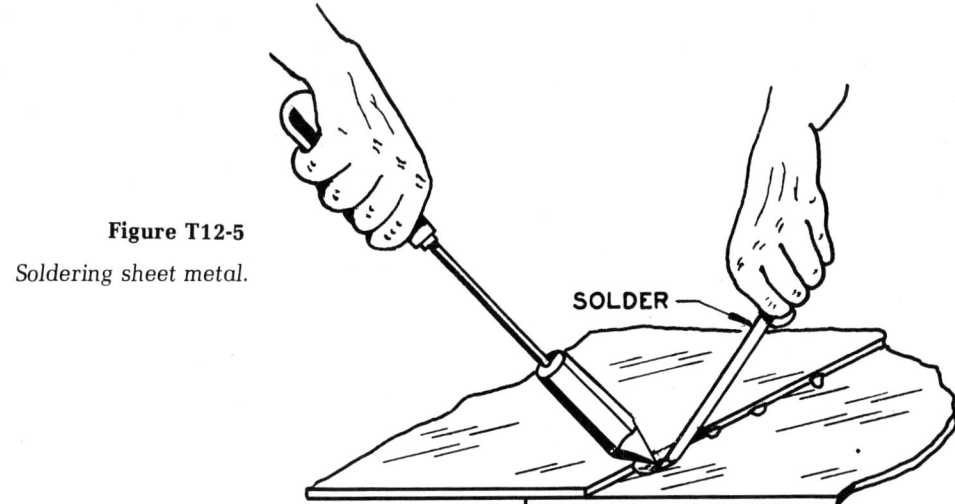

Topic 13 Oxyacetylene Brazing, Welding, and Cutting

Many metals, such as steel, can be joined or separated using heat from an oxyacetylene torch. The torch mixes oxygen and acetylene gases. This mixture of gases burns easily. The flame produced may be used for brazing, welding, or cutting metal.

Using Oxyacetylene Equipment

An oxyacetylene welding outfit is basically made up of the following: cylinders of oxygen and acetylene, regulators, hoses, and a torch. See Figure T13-1.

Oxygen and acetylene are stored in separate **cylinders.** A great deal of gas is forced into each cylinder. This results in very high pressures. The high pressures are controlled and adjusted by the **regulators.**

Hoses carry the gases from the cylinders to the torch. The green hose carries oxygen. The red hose carries acetylene.

The gases are mixed inside the **torch.** When burned, they produce a very hot flame. The flame is regulated (adjusted and controlled) by interchangeable torch **tips** and by the torch **valves.** See Figure T13-2.

Never operate an oxyacetylene torch without safety glasses and a tinted face shield or welding goggles. The colored lenses protect your eyes from the bright light. You can also see the work better. See Figure T13-3.

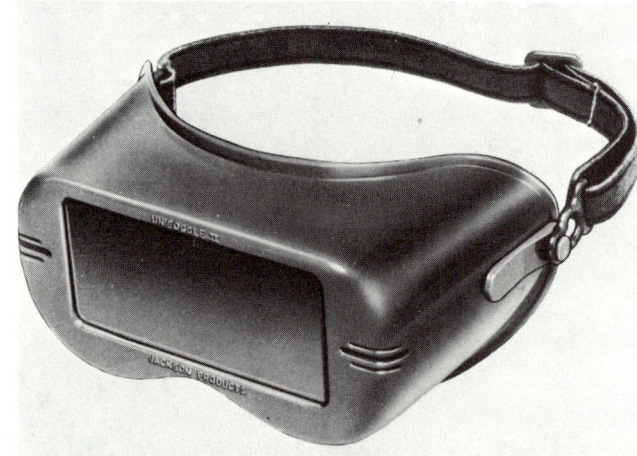

Figure T13-3
Welding goggles have special dark lenses to protect your eyes. Always wear the proper eye protection when welding.

Airco, Inc.

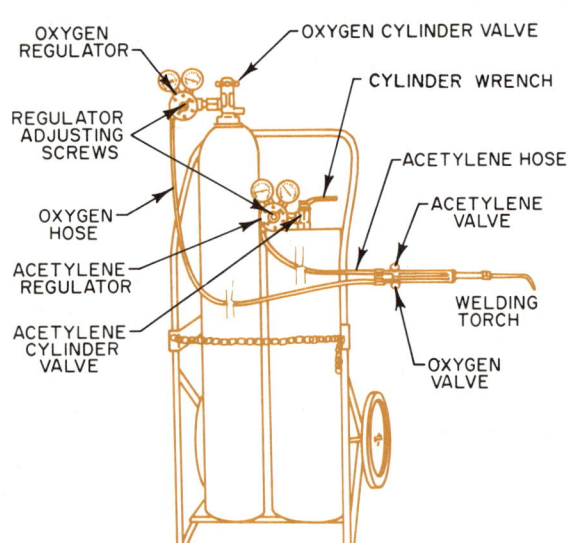

Figure T13-1
Oxyacetylene welding equipment.

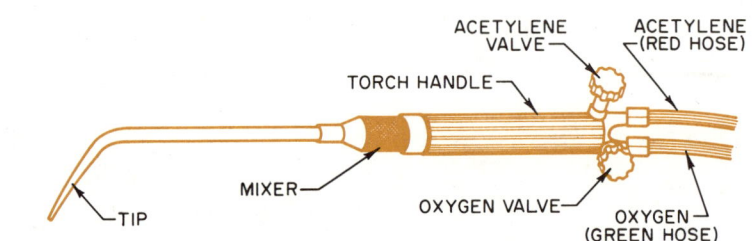

Figure T13-2
The parts of an oxyacetylene welding torch.

Lighting the Torch

1. Select the correct size tip for the thickness of material to be welded. Thin metal will require a small tip with a small opening. Thick metal will require a larger tip with a larger opening. (Larger openings produce larger flames.) Numbers stamped on the tip show its size. The larger the number, the larger the tip. See Figure T13-4.

    Once the tip size is determined, insert the tip into the torch. Tighten the knurled nut, hand tight. **Do not tighten with a wrench.**
2. Turn the regulator adjusting knobs (screws) counterclockwise until they turn freely.
3. Slowly open the oxygen cylinder. Turn the valve as far as it will go.
4. Slowly open the acetylene valve 1/4 to 1/2 turn. Do not remove the cylinder wrench from the cylinder.
5. Check the pressure for both cylinders. If either cylinder shows less than 25 psi (pounds per square inch), **do not use the cylinder.**
6. Open the oxygen adjustment valve on the torch one full turn. (Remember, the oxygen valve is connected to the green hose.) With this valve open, turn the regulator knob (screw) clockwise until the correct operating pressure is reached. The information in Figure T13-4 may be used as a guide, but always follow your teacher's directions. When the regulator is adjusted, turn **off** the oxygen valve.
7. Adjust the acetylene operating pressure. This is done in the same manner as for the oxygen (step 6). The acetylene valve is connected to the red hose. Adjust the acetylene to the correct operating pressure.
8. After both regulators are properly adjusted, the torch is ready to light. Point the tip of the torch away from the cylinders, yourself, and others. Slightly open the acetylene valve. Light the acetylene with a **spark lighter.** See Figure T13-5. Continue to turn the acetylene valve until the flame burns with very little smoke. Then slowly turn on the oxygen valve until a small blue cone forms in the center of the flame. Adjust the torch for the type of flame that you need. See Fig. 13-6.

Figure T13-4

Select the correct size tip to use for the thickness of material to be welded.

Tip Size	Oxygen Pressure (psi)*		Acetylene Pressure (psi)		Metal Thickness
	Min.	Max.	Min.	Max.	
00	1	2	1	2	1/64" — 3/64"
0	1	3	1	3	1/32" — 5/64"
1	1	4	1	4	3/64" — 3/32"
2	2	5	2	5	1/16" — 1/8"
3	3	7	3	7	1/8" — 3/16"
4	4	10	4	10	3/16" — 1/4"
5	5	12	5	15	1/4" — 1/2"

* Pounds per square inch

NEUTRAL FLAME

REDUCING FLAME

OXIDIZING FLAME

Figure T13-6

Types of flames.

Neutral Flame — Equal volumes of oxygen and acetylene

Reducing Flame — Too much acetylene or not enough oxygen

Oxidizing Flame — Too much oxygen or not enough acetylene

Figure T13-5

After the correct regulator pressures have been set, use the spark lighter to light the torch.

Union Carbide

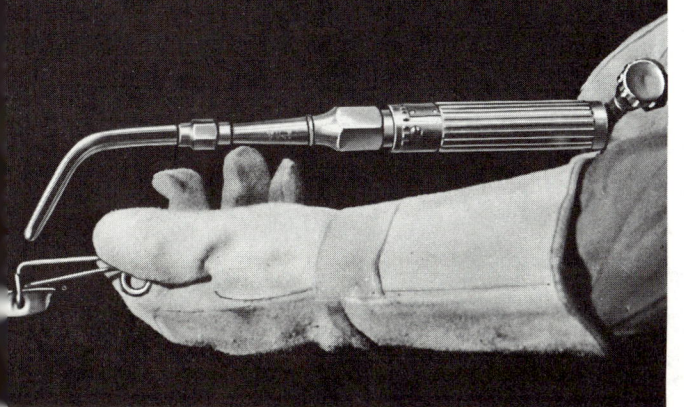

Oxyacetylene Welding

1. Position steel or iron parts for welding. Place them so they will not move while being welded. The joint should be easy to reach.
2. Select the welding rod. The rod should be made of the same metal as the parts being welded. Its diameter should be the same as the thickness of the metal.
3. Put on welding goggles and gloves.
4. Light the torch. Adjust it to a neutral flame.
5. Tack weld the joint at each end so that the metal will not warp or twist when it is heated. Long pieces may need to be tacked in the middle also. If one piece of material is thicker than the other, direct more heat at the thicker piece. This will cause the two edges to melt at the same rate.
6. Position the torch and rod at a 45-degree angle to the joint. See Figure T13-7. The inner blue cone of the flame should be just above the surface of the metal.
7. Melt one edge of the joint. When a puddle forms, dip the end of the welding rod into its center. This produces a mound-shaped bead. See T13-18.
8. Slowly move the torch in a tight U-shaped pattern as you move the puddle along the length of the joint. Keep the welding rod near the flame. Add metal to the puddle when it is needed.
9. As you weld, the metal parts become hotter. Keep the parts at the proper temperature for welding by tilting the torch slightly away from the welding area.

Problems in Oxyacetylene Welding

If the weld is pitted and bubbly you may be using an oxidized flame (too much oxygen). If the weld is too thin and did not penetrate (enter the joint) well, you probably did not heat the metal enough. The tip size may be too small, or you may have moved the torch along the joint too rapidly. See Figure T13-8.

Shutting Off the Welding Unit

1. First close the acetylene valve, and then the oxygen valve.
2. Tightly close the valves on both cylinders.
3. Open the valve on either the oxygen or acetylene at the torch. Allow time for the gas to drain from the hose and regulator. Repeat the procedure for the other gas line.
4. When the high-pressure and low-pressure gages on both cylinders read zero, turn the regulator screws counterclockwise as far as they will go.
5. Close the valves on the torch.

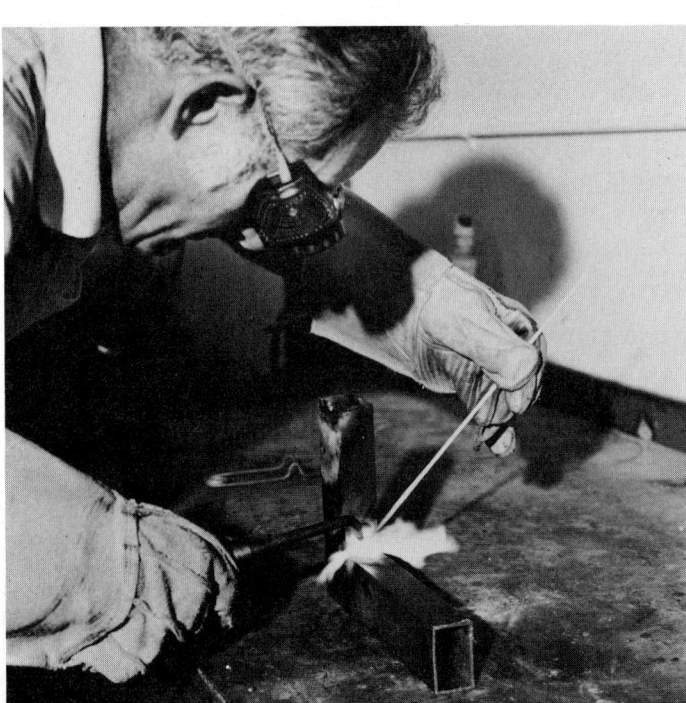

Figure T13-7
Joining metal parts by welding.

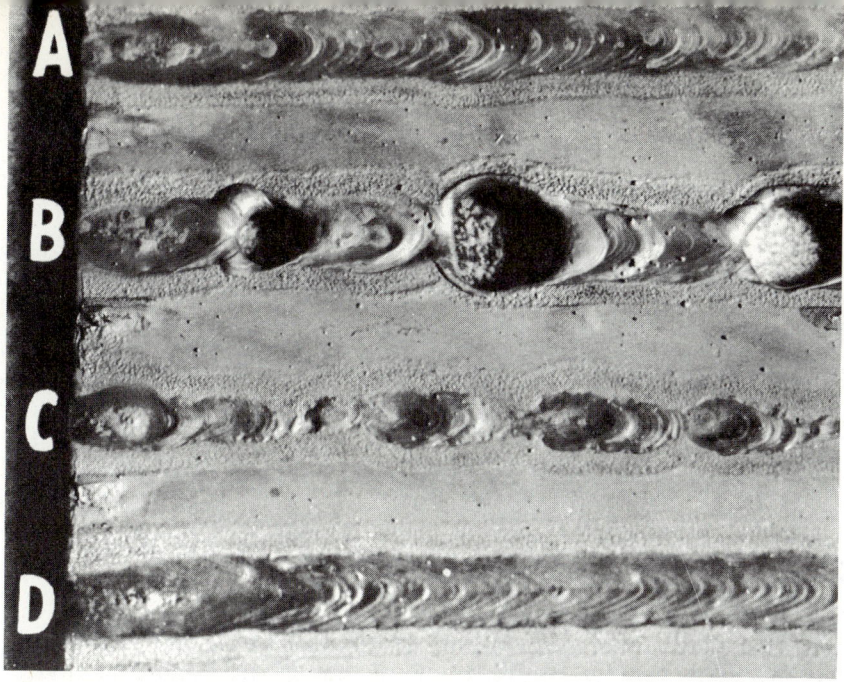

Figure T13-8
Appearance of oxyacetylene welds.

Union Carbide

A. Satisfactory weld.
B. Too much heat. Torch moved too slowly.
C. Not enough heat. Torch moved too fast.
D. Satisfactory weld.

Oxyacetylene Brazing

Brazing is much like soldering. (See Topic 12, **Soldering**.) See Figure T13-9. Steel, copper, and brass parts are bonded together with molten (melted) metal usually bronze. Pieces of metal are heated with an oxyacetylene torch, and bronze is applied to the joint. Metals joined by brazing must have a higher melting point than bronze.

Apply flux when brazing. Flux cleans the metal further, and makes the bronze flow better.

Some brazing rods are already coated with flux when you buy them. This flux melts as the rod is heated. If your brazing rod is not coated, you can coat it yourself. Heat the tip of the bronze rod with the torch. Then dip the hot rod into a can of flux. The heat will bond the flux to the surface of the rod. Repeat this as often as necessary when you are brazing.

Follow this procedure:

1. Clean the metal surface with a wire brush, steel wool, or emery cloth. Bronze will not adhere (stick) to a dirty surface.
2. Position the clean metal.
3. Put on welding goggles and gloves.
4. Adjust the torch to a neutral or slightly reducing (carbon) flame. The reducing flame will have a greenish feather extending from the tip of the blue inner cone.
5. Apply heat evenly to both surfaces to be brazed.
6. When the metal becomes a dull red, apply a thin layer of bronze to the surface.
7. Go back and reheat the joint. This time, apply enough bronze to form a dome-shaped build-up along the joint. Do this by creating a small puddle of molten bronze with the torch. Add bronze to this puddle as you move the torch in small arcs down the length of the joint.

Use enough heat to make the bronze flow easily. Do not allow the bronze to melt away from the joint.

Figure T13-9
Bonding metal parts by brazing.

Handy and Harman

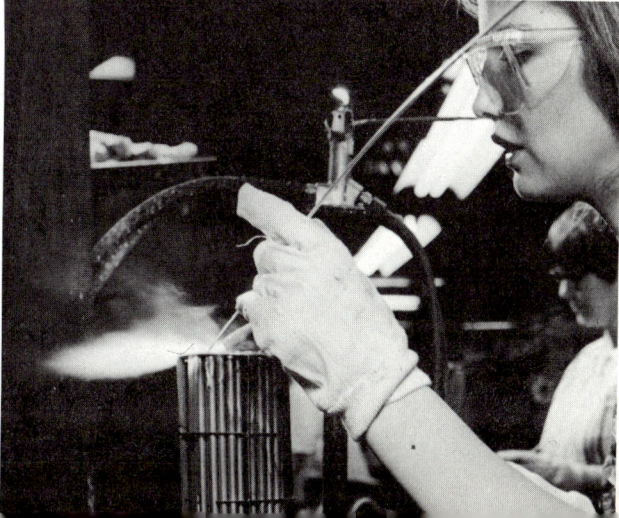

Oxyacetylene Cutting

To cut steel using an oxyacetylene outfit, use a cutting torch instead of a welding torch. See Figure T13-10. Adjust the regulators for the torch tip and the thickness of the metal to be cut. Your instructor will help you set up this equipment.

1. Light the torch, and adjust it to a neutral flame.
2. Locate the inner core of the flame just above the work surface.
3. Tilt the tip slightly, and preheat the metal in a single spot. Keep heating until it is cherry red.
4. Push down on the oxygen cutting lever. A jet stream of oxygen burns through the metal.
5. Position the torch at a 90-degree angle to the work surface, and continue cutting along the desired line. Move the torch smoothly. See Figure T13-11.

Oxyacetylene Welding

1. Make sure all welding equipment is in good order.
2. **Never** operate any cylinder with less than 25 psi pressure. Gas can be drawn into the wrong hose by unequal pressure, and a fire or an explosion can result.
3. Wear welding goggles when you are welding or watching someone else weld.
4. **Never** point the tip of the torch at cylinders, other people, or yourself.
5. Remove all flammable materials from the welding area before starting to work.
6. Follow your teacher's instructions carefully.

Figure T13-10

The parts of a torch for oxyacetylene cutting.

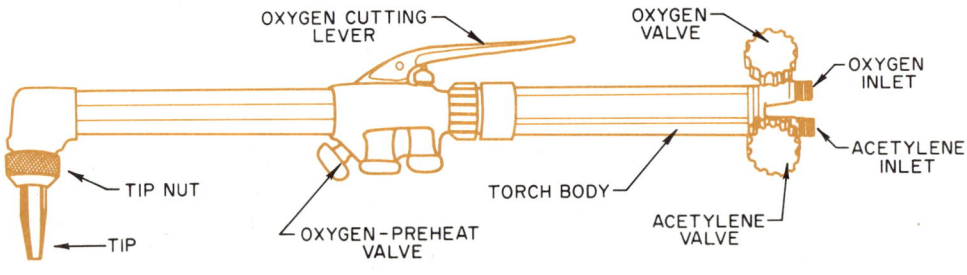

Figure T13-11

Separating metal by oxyacetylene cutting.

Union Carbide

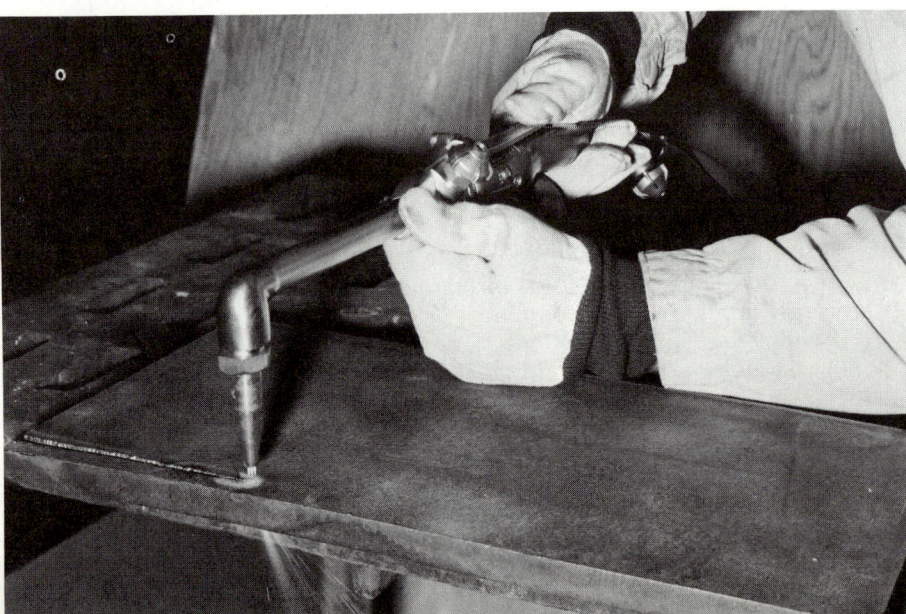

Topic 14 Electric Welding

Two methods of electric welding are generally used in the school shop. Resistance welding is also known as **spot welding.** Shielded metal-arc welding is commonly called **arc welding.**

Spot Welding

Spot welding joins two pieces of sheet metal together by pressure and heat. The heat is produced by resistance to the flow of electricity. A flow of high current is passed from one tong of a spot welder to the other. See Figure T14-1. Heat is built up in the metal pieces. They melt and fuse (join) together.

Using a Spot Welder

1. Clean all surfaces to be welded.
2. Adjust the tongs to the correct pressure. Use more pressure for thick metals than for thin ones.
3. If the spot welder has a timer, set it to the correct number of **cycles** of electric current. (A cycle is equal to a given number of seconds. The number varies.) The number of cycles you must use will depend on the kind and thickness of the metals to be joined. More cycles are needed to melt and fuse thick pieces of metal than are needed for thin sheet metal.
4. Clamp the metal between the tips of the tongs.
5. Apply pressure by lowering the operating lever. Do not press down on the lever until you are sure the weld will be in the right place.
6. Turn on the welder. Usually this is done by pressing a button or a trigger switch.
7. Continue to apply pressure to the materials for two or three seconds after the current has stopped.
8. Handle the metal carefully. The welded area will be hot.

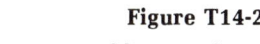

Spot Welding

1. Wear eye protection.
2. Wear gloves.
3. Allow the metal to cool before you touch it.

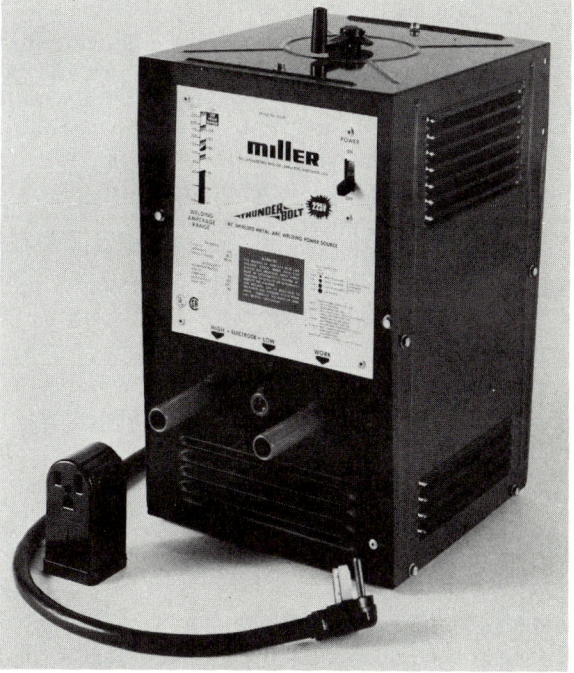

Figure T14-2
An arc-welding machine.

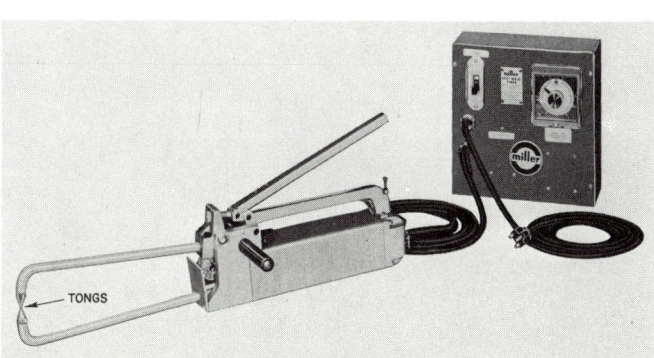

Figure T14-1

Portable spot welder. The tongs open, and the metal parts to be spot-welded are placed between them.

Miller Electric Manufacturing Company   Miller Electric Manufacturing Company

Arc Welding

Arc welding uses the heat from an electric arc to melt and fuse base metals and welding-rod metal into one piece. The heat from the arc melts the flux (cleaner) coating on the rod.

There are two basic kinds of arc welding machines. Some provide **direct current (DC)**. Others provide **alternating current (AC)**. Many arc welding machines found in the school shop can provide either AC or DC. You select the amount of electric current (**amperes**). This amount will depend on the size and kind of materials. Thick metals require more amps than thin metals. See Figure T14-2.

Using an Arc Welder

To use an arc welder, make sure that one **cable** of the arc welder is connected to the **electrode holder,** and the other to the **grounding clamp.** The grounding clamp should be attached to the welding table or the work itself. An **electrode** (welding rod) is placed in the electrode holder. See Figure T14-3.

1. Wear protective clothing and a welding helmet with a tinted face shield or welding goggles. See Figure T14-4.
2. Clean all surfaces to be welded.
3. Clamp the metal onto the welding table.
4. Attach the grounding clamp to the table or the work.
5. Select and insert the proper electrode into the electrode holder. Set the machine on the amount of current needed. The size of the electrode will determine this amount. Your teacher will help you to select the right electrode and machine setting for the job.
6. Hold the electrode just above (but not touching) the area to be welded.
7. Turn on the arc welder. Lower the protective helmet over your face.

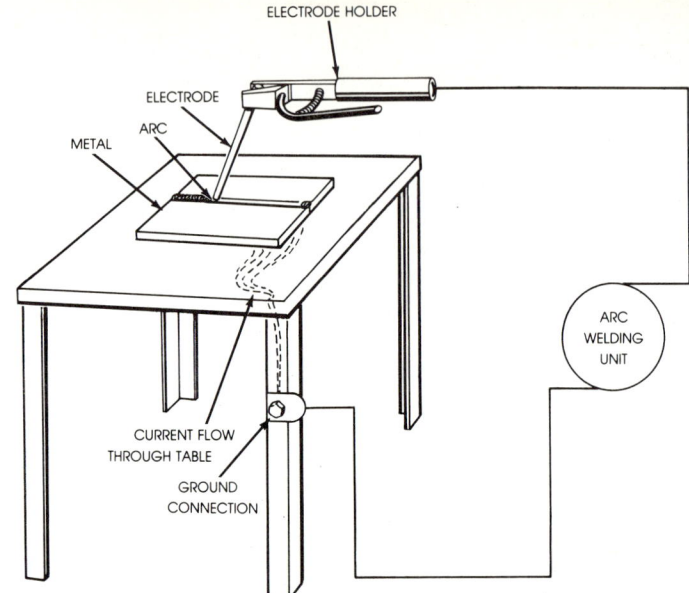

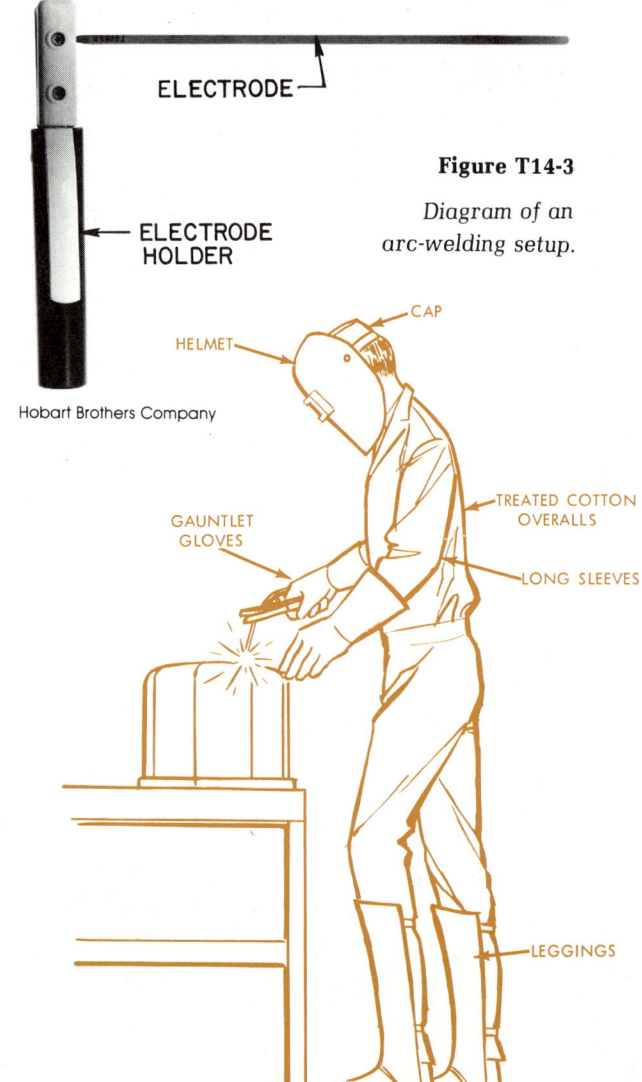

Figure T14-3

Diagram of an arc-welding setup.

Figure T14-4

Wear protective clothing when welding.

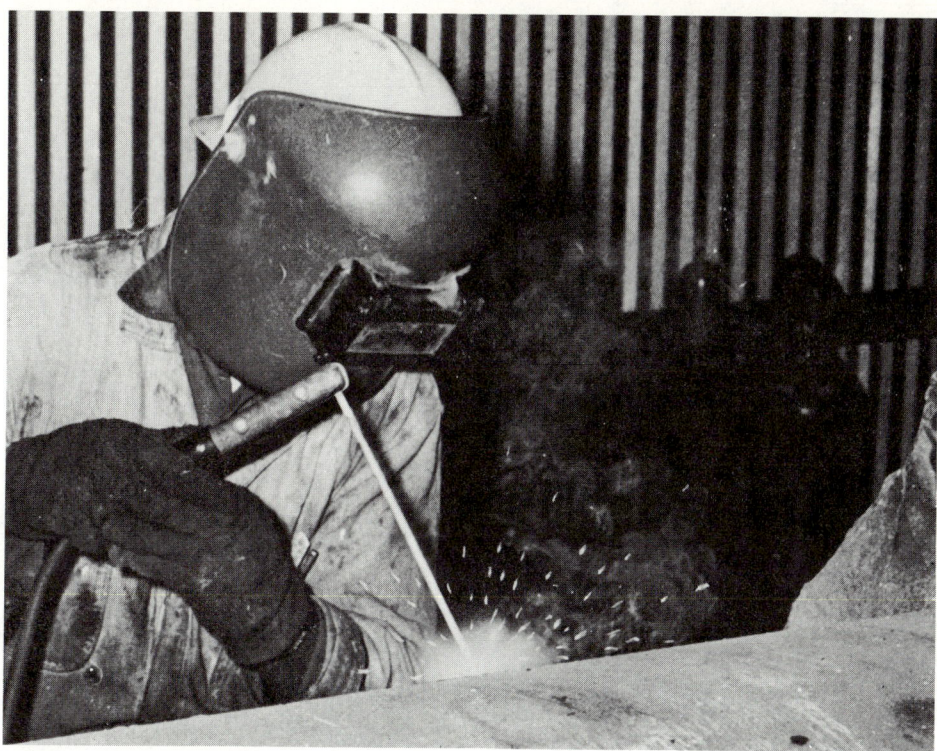

Miller Electric Manufacturing Company

Figure T14-5

Joining metal parts using the arc-welding process.

8. Start the arc by one of two methods. One method uses a "pecking" motion. This is called the **down-up method.** The other way to start the arc is much like striking a match. This is called the **scratch method.**
9. When joining two pieces of metal, tack-weld them, (weld them at the ends). This will keep them from moving while being welded.
10. When starting the actual weld, position the tip just above the base metal. Lean the electrode at a slight angle toward the direction of the weld. See Figure T14-5.
11. Once you have started the weld, be sure to maintain a steady welding speed. Watch the puddle (pool of molten metal). Move the arc slowly enough to maintain a good puddle, but quickly enough that the puddle does not melt through the base metal.
12. After the weld has cooled, chip the **slag** (layer of waste metal) from the surface to inspect the weld. When chipping, direct the chips away from yourself and away from others. Be sure that you are wearing eye protection.

Arc Welding

1. Wear a welding helmet that has a filtering lens to protect your eyes. The rays from the arc can damage unprotected eyes.
2. Wear gloves, shoes, a lab coat, and long sleeves (tucked into gloves) for protection from heat and harmful rays.
3. Be careful around hot metals. Treat all metal pieces as though they are hot until you are sure that the metal has cooled.
4. Use pliers or tongs to pick up hot metals.
5. Take care not to accidentally strike the arc on the welding table.
6. Be careful when chipping slag from the metal. Never chip in an area where people are likely to be hit with the chips.

Topic 15 Filing and Sanding

Filing, sanding, and polishing are processes used to remove excess (unneeded) material and to prepare surfaces for finishing.

Filing

Filing is a process used to shape and finish materials, such as wood, metal, and plastics. See Figure T15-1. Files and rasps are tools used in filing. Files have rows of cutting edges. The shape of the cutting edges (**teeth**) is called **cut**. There are two basic kinds of cuts: **single-cut** and **double-cut**. **Single-cut files** have rows of straight cutting edges. **Double-cut files** have rows of diamond-shaped cutting edges. See Figure T15-2.

Rasps have rows of wedges. These make deep cuts into the surface of wood. Rasps remove large amounts of wood at one time. They are used mostly on unfinished wood.

Files come in a great variety of shapes and sizes. Three of the most common shapes are **flat, half-round,** and **round**. See Figure T15-3. Eight-inch, ten-inch, and twelve-inch files are used most often in school shops. Always use a file that has a wooden or plastic handle.

Using a File

1. For most filing operations, place the workpiece in a **vise**. Be sure that the area you wish to file can be reached easily. Protect the work from the **jaws** of the vise. Make sure that the area to be filed is close to the top of the jaws.

2. Press down on the **forward** stroke. Push the file straight ahead or at a slight angle to the work.

3. Lift the file on the return stroke.

4. Continue to file using an even, steady pressure. Too little pressure allows the file to slide over the work surface without cutting. Too much pressure overloads the file and chips or clogs the teeth.

Figure T15-1

Using a file. Always use a file that has a handle.

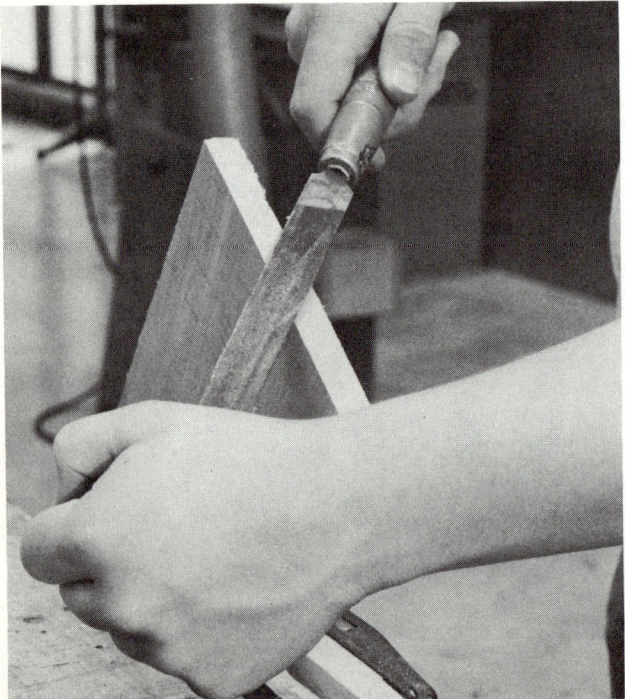

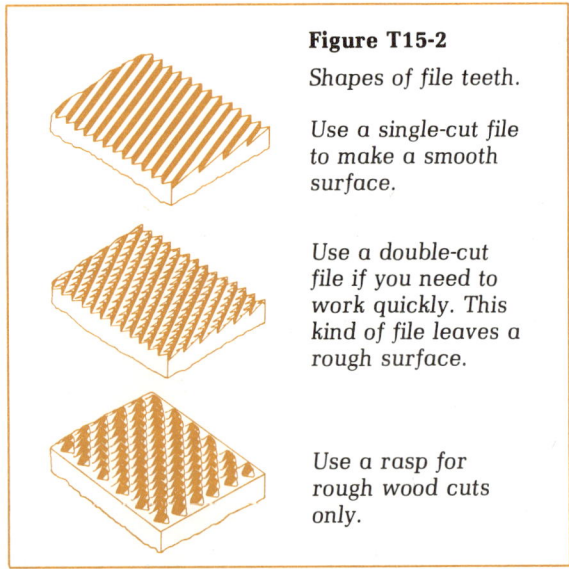

Figure T15-2

Shapes of file teeth.

Use a single-cut file to make a smooth surface.

Use a double-cut file if you need to work quickly. This kind of file leaves a rough surface.

Use a rasp for rough wood cuts only.

Figure T15-3

Shapes of files.

Flat

Round

Half-round

Filing and Sanding   359

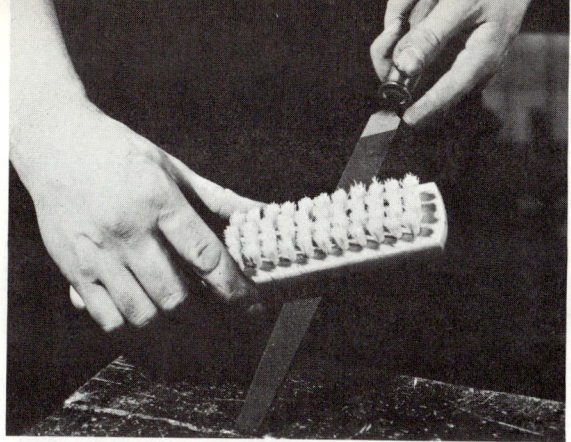

Figure T15-4 Nicholson File Company

Cleaning a file with a file card.

After you have finished filing, or when the file becomes clogged, use a file card (special wire brush) to clean filings from the file teeth. See Figure T15-4. Store files separately. Files may become dull by touching other metal objects.

Filing

1. Wear eye protection.
2. Use a file that has a handle. The handle is separate and must be attached to the file.
3. Use a file card to clean a file. Do not try to knock filings out of the teeth. Files break easily.

Sanding

Sanding is a process of smoothing the surface of wood, metal, or plastic. Sanding is usually done with abrasive paper or cloth. Abrasive paper is often called **sandpaper**. **Emery cloth** is an abrasive cloth. Abrasive paper and cloth is made into sheets, disks, and rolls. These are made to fit different sanding equipment. See Figure T15-5.

Abrasives with large **grains** can be used to remove materials quickly. However, these abrasives leave a coarse, rough surface. Abrasives with small grains remove less material. They are used to produce a smooth surface.

The grains of abrasive are often called **grit**. A number on the paper tells the size of the grain. Grades for abrasives commonly range from 50 grit (coarse) to 600 grit (very fine). Generally a coarse-grit abrasive is used first for rough sanding. Then a finer-grit abrasive is used for finish (final) sanding.

Hand Sanding

The best way to hand sand a flat surface is to use a **sanding block**. See Figure T15-6. Simply fold a piece of abrasive paper or cloth around a wood block. Hold or fasten it securely in place. If you are sanding wood, rub the sanding block with (in the same direction as) the grain of the wood. The block provides a flat sanding area which helps to level the surface being sanded.

For sanding irregular, round, or convex (curved outward) shapes, use abrasive paper or cloth with **finger pressure**.

Another method would be to wrap a piece of abrasive paper around a short piece of **dowel rod**. The dowel rod will allow you to hold and control the abrasive easier.

Figure T15-5

Abrasive papers are made in a variety of grits, as well as many shapes for different uses.

Norton Company

Figure T15-6

Using abrasive paper. Note the sanding block.

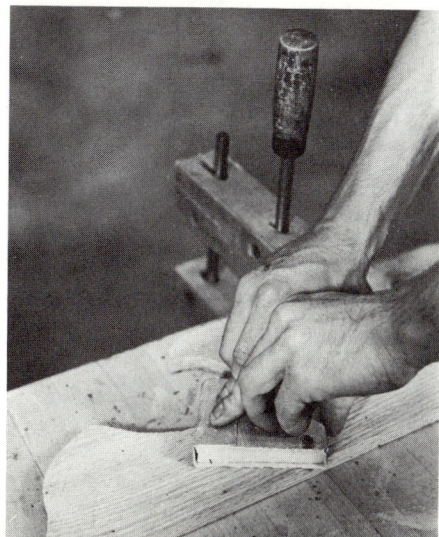

Figure T15-7
Using a portable finishing sander.

Black and Decker (U.S.) Inc.

Power Sanding

Electric or air-powered sanding tools reduce the time and energy required for hand sanding. Some power sanders are small and portable. Others are large, stationary models.

Portable Finishing Sanders

Portable finishing sanders use electricity or air power (compressed air) to move abrasive paper across wood materials. See Figure T15-7. There are two basic kinds of finishing sanders: orbital and in-line. An **orbital finishing sander** cuts faster than an in-line finishing sander. However, when coarse-grit abrasive paper is used, it leaves swirl marks on the work surface. These marks must be removed before applying finish. Use a fine-grit abrasive paper in the sander for the final sanding. The **in-line finishing sander** moves back and forth, in much the same way as hand sanding. It leaves a surface that is ready to receive finish.

Using a Portable Finishing Sander

1. Install abrasive paper or cloth on the sanding pad.
2. Clamp the wood to the bench.
3. Start the sander, and then gently lower it onto the wood.
4. Begin sanding. Move the sander with the grain of the wood. Hold the sander lightly. It is heavy enough by itself to provide pressure for sanding.
5. When you have finished sanding, lift the sander and turn it off. Never set the sander down on the bench when the motor is still operating.
6. If a smoother surface is needed, change to a finer-grit abrasive paper and repeat the operation.

Finishing Sanders

1. Wear eye protection.
2. If your hair is long, tie it back.
3. Use both hands to control the sander.
4. Disconnect the sander before making any adjustments or repairs.

Portable Belt Sanders

Portable belt sanders are used to remove material quickly from flat surfaces, such as cutting boards. See Figure T15-8. Belt sanders are used in many shops to rough-sand products after they have been glued together. Remove excess (extra) glue before sanding.

The size of a belt sander is measured by the width and length of the abrasive belt. See Figure T15-9. Belt widths usually range from 3″ to 4″. Many grit sizes are available.

Using a Portable Belt Sander

1. Release the tension on the movable roller and install the proper-grit abrasive paper. Be sure that the arrow on the inside of the belt faces the direction of the sander's rotation.
2. Clamp the wood to the bench.
3. Check the position of the belt by turning the rollers one complete turn by hand.
4. Plug the sander cord into an electrical outlet.
5. Start the sander **before** placing it on the wood. Be sure that the belt is tracking properly (staying in position as it moves).
6. Slowly lower the sander onto the work surface, and begin sanding. Sand in the direction of the wood grain. Do not let the sander remain in one area too long or a low spot will be made.
7. When the surface has been sanded smooth, lift the sander from the work, and turn off the motor. Never set the sander down on the bench top when the motor is still operating.
8. If smoother sanding is needed, replace the belt with a finer-grit abrasive and repeat the sanding operation.

— SAFETY DIRECTIONS —

Portable Belt Sanders

1. Wear eye protection.
2. If your hair is long, tie it back.
3. Keep the sander flat on the work surface as you work.
4. Keep the power cord away from the moving belt.
5. Be sure that the starter switch is turned **off** before you plug in the sander.

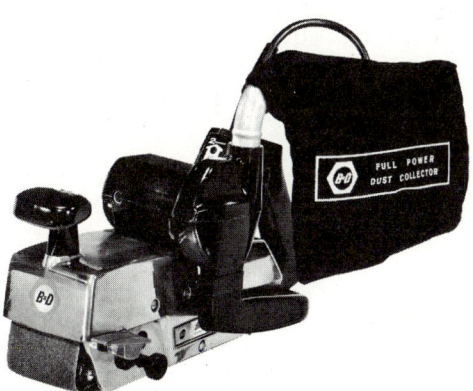

Black and Decker (U.S.) Inc.

Figure T15-8
A portable belt sander.

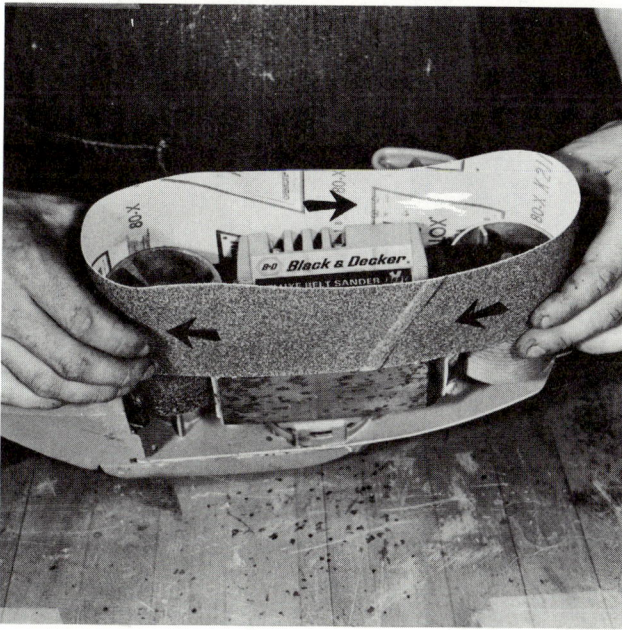

Figure T15-9
Installing a new belt on the belt sander. The arrows show the direction in which the belt will move when the sander is used.

Stationary Disk Sanders

The stationary disk sander smooths and shapes materials by means of a spinning abrasive disk. See Figure T15-10. Both straight and convex (outside curve) surfaces can be sanded with a disk sander.

The size of a disk sander is measured by the diameter of the disk. Abrasive paper is cemented to the disk. A table is located in front of the disk. The table supports the workpiece as it is sanded. See Figure T15-11. This table can be tilted for sanding angles, such as bevels and chamfers.

Using a Stationary Disk Sander

1. Make sure that the abrasive paper or cloth is cemented tightly to the metal disk.
2. If necessary, the table may be tilted to an angle. A miter gage may be used to guide the wood when sanding an exact angle.
3. Mark the wood with a pencil line to show how much material should be removed.
4. Turn on the sander.
5. Place the wood flat against the table. Slowly and firmly press it against the moving disk. Make sure that the wood touches the side of the disk that is moving **downward**. Move the stock back and forth slightly to avoid wear in one area of the disk. When sanding convex (outside) curves, do not allow the stock to rest in one spot. A flat area could result.
6. When you have finished sanding, remove the wood from the sander, and turn off the machine. Do not leave the machine until it has come to a complete stop.

Stationary Disk Sanders

1. Wear eye protection.
2. Keep your fingers away from the sanding disk.
3. Always support the work securely on the table when sanding.

Stationary Belt Sanders

Stationary belt sanders are used to smooth and shape flat or curved surfaces on wood, metal, and plastics. Sanding is done as an abrasive belt moves across the work. Ends, sides, and faces of stock can be evenly sanded with a stationary belt sander. See Figure T15-12.

The size of a stationary belt sander is determined by the width and length of the **belt**. Many grit sizes are available for both rough and

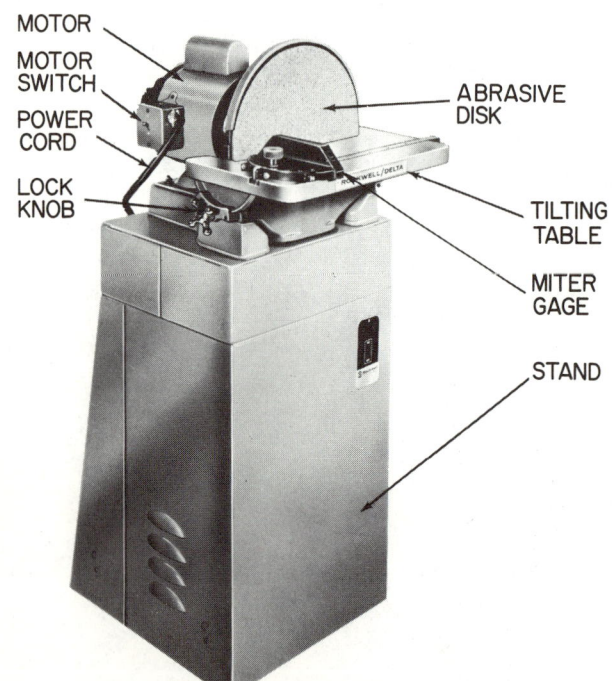

Figure T15-10

The parts of a stationary disk sander.

Figure T15-11

Using a stationary disk sander. Note that the wood is touching the side of the disk that is moving downward.

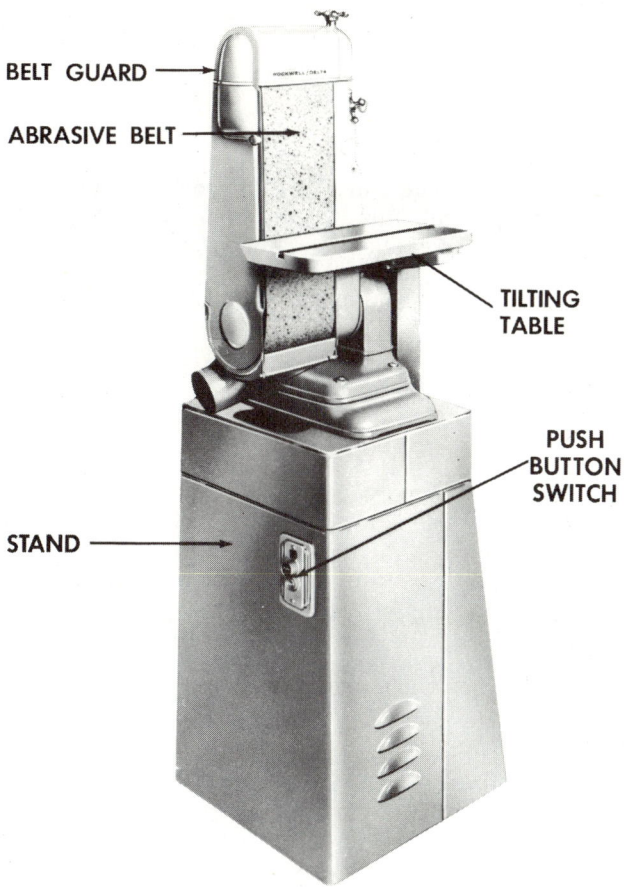

Using a stationary belt sander set in the horizontal position.

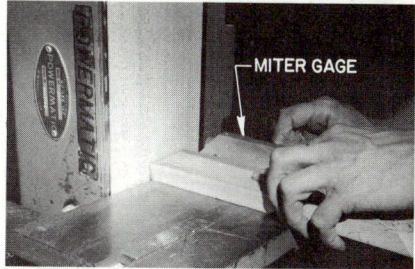

Using a stationary belt sander set in the vertical position. The miter gage helps you to keep the board square when sanding end grain.

Rockwell International

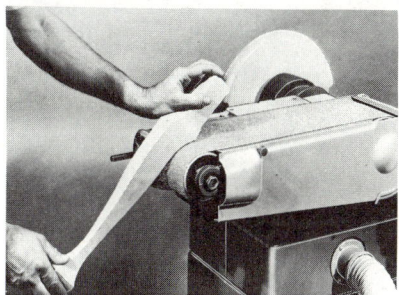

Using a stationary belt sander in the horizontal position to sand a concave curve.

Figure T15-12   Rockwell International

The parts of a stationary belt sander.

finish sanding. The **table** on this sander can be tilted. This table has a slot for a miter gage.

The sander can be positioned horizontally (lying flat) or vertically (up and down). See Figure T15-13. Convex edges (outside curves) can be sanded only while the belt is in a vertical position. Concave surfaces (inside curves) can be sanded only with the belt in a horizontal position. In the horizontal position, the table acts as a fence, preventing work from moving off the sander.

Using a Stationary Belt Sander

1. Inspect the belt for tension (tightness) and tracking. Make sure that the belt used has the proper abrasive grit.
2. Turn on the machine.
3. When the sander is in a vertical position, place the wood flat on the table. When the sander is in a horizontal position, place the wood on the belt.
4. Press the wood against the abrasive belt.

Figure T15-13

5. Move the wood back and forth across the belt so the work will wear evenly.
6. When you have finished sanding, remove the wood from the table or belt, and turn off the machine. Do not leave the machine until it has come to a complete stop.

— SAFETY ⬍ DIRECTIONS —

Stationary Belt Sanders

1. Wear eye protection.
2. If your hair is long, tie it back.
3. Before turning on the machine, be sure that all parts of the belt sander are clamped tightly in place.
4. Do not place your fingers between the work and the table.
5. Keep the workpiece against the table or fence as you sand.

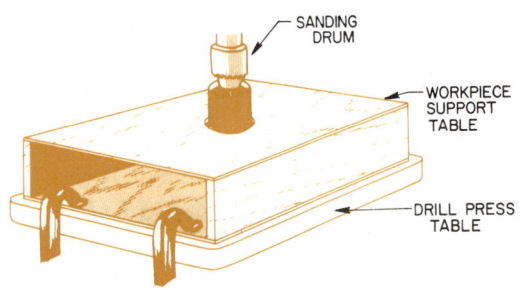

Figure T15-14

Using a sanding drum attachment on a drill press. Note the special table which has been attached to the drill press table.

Using a Sanding Drum

A sanding drum is a hard rubber cylinder covered with a sleeve (tube) of abrasive paper or cloth. See Figure T15-14. It is used to smooth the edges of irregular curves. The sanding drum is used in a drill press. Install the sanding drum by inserting it into the chuck of the drill press. A special wooden table should be placed over the drill press table. The special table supports the stock being sanded. It has a recess for the drum. The drum can be raised or lowered through this recess so that all parts of the drum surface can be used.

Figure T15-15

Using steel wool to smooth a wood surface.

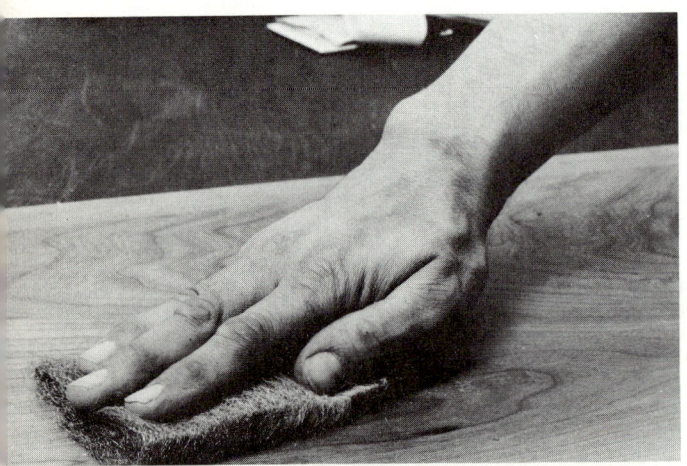

Polishing

Using Steel Wool

Steel wool is used for polishing **metal** surfaces and for smoothing **wood** surfaces. See Figure T15-15. Wood is often rubbed with steel wool between finish coats. This results in a smoother final finish.

Steel wool is made of steel shavings. These steel shavings have small, sharp edges. The coarseness of steel wool varies from 0000 (very fine) to 3 (very coarse).

Buffers

Buffers are used for polishing **metal** or **plastics.** See Figure T15-16. Buffing wheels are made of many layers of cloth. The cloth is usually cotton muslin. Some buffing wheels have layers of cloth that are sewn together. Use these for buffing a hard material such as metal. Unsewn buffing wheels work best for plastics. Never buff metal on a buffing wheel used for plastics. And never buff plastics on a buffing wheel used for metal.

An abrasive polishing compound must be applied to the wheels during buffing. The abrasive in the buffing compound removes small amounts of material to give the work surface a smooth, shiny finish. The abrasive compound chosen, for the most part, determines the smoothness of the finished surface. The compounds are usually in stick form and come in various degrees of coarseness. Tripoli com-

Figure T15-16

A bench-top buffer with two buffing wheels attached.

pound has a medium abrasive. It is often used for general-purpose work. Emery compound is coarse. Rouge compound and jeweler's rouge compound are fine abrasives.

A wheel-shaped wire brush is sometimes used on the buffing machine. It is used to produce a satin finish on materials such as aluminum. The metal workpiece is held against the rotating wire brush.

Using a Buffer

1. Sand the surface to be buffed with fine-grit abrasive paper.
2. Select the buffing compound best suited for the kind of material to be buffed and for the type of finish you wish to produce. If you are going to produce a highly polished surface on an unbuffed surface, start with a coarse-grit compound and finish with a fine-grit compound.
3. Start the buffer. Apply the buffing compound to the revolving wheel. This is done by pressing the compound stick against the **lower half** of the wheel. Only a small amount is needed.
4. Hold the workpiece securely. Press it lightly against the **lower portion** of the wheel. Take special care to use the lower portion. Otherwise, the buffer can easily "catch" an edge and throw your work to the floor.
5. Apply even pressure to the work. Heat caused by too much pressure may melt the edge of plastic.
6. Keep turning the work as you buff. Make sure that all areas are polished. If too much buffing compound is left on the work, wash it off with soap and water.

―SAFETY  DIRECTIONS―

Buffers

1. Wear eye protection.
2. If your hair is long, tie it back.
3. Never buff metal on a buffing wheel used for plastics.
4. Never buff plastics on a buffing wheel used for metal.
5. Always buff on the lower half of the wheel.

Topic 16 Grinding

Grinding is the process of removing small chips of material from steel by pressing it against a rotating abrasive wheel. Grinding machines used in school shops are usually **bench** or **pedestal grinders.** See Figure T16-1. A bench grinder is mounted on a bench. The pedestal grinder is supported by a floor-mounted pedestal.

Grinders

A **grinder** (grinding machine) is an electric motor with an abrasive grinding wheel at each end. See Figure T16-2.

Wheels used on grinding machines are made of abrasive grains held together with a bonding material. The many sharp edges of the abrasive grains are the cutting edges that remove chips as the wheel rotates.

A **tool rest** is positioned in front of each wheel. It provides support for the material being ground. The tool rest should be clamped in place very close to the wheel.

Figure T16-1

The parts of a pedestal grinder.
Baldor Electric Company

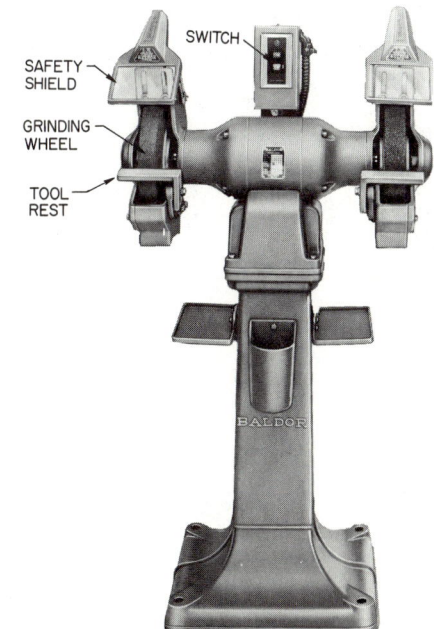

Make sure **safety shields** are in place when you are using the grinder. Also, keep a container of water near the grinder. The water is used to cool the metal as it becomes heated from grinding.

Nonferrous metals (metals that are not iron), such as aluminum and brass, should not be used on a grinder. They clog the pores of the wheel. If this happens, the wheel must be resurfaced with a **wheel dresser**. See Figure T16-3.

Grinding wheels may become grooved through use. A wheel dresser may be used to correct this situation also. The wheel dresser removes abrasive material from the wheel to form a flat grinding surface.

Using a Grinder

1. Inspect the grinding wheel. Make sure that it has no chips or cracks.
2. Check the position of the safety shield and the tool rest.
3. Turn on the grinder. Always stand to one side when starting a grinder.
4. Place the metal on the tool rest, and slowly push the work onto the working surface of the grinding wheel.
5. Move the metal back and forth across the wheel. Use even pressure so that the wheel will wear evenly. If you are grinding a round piece of metal, keep turning it so that you do not produce a flat spot.

SAFETY DIRECTIONS

Grinding

1. Wear eye protection, such as goggles or a face shield.
2. Do not use the side of the grinding wheel for grinding.
3. Keep your attention on the workpiece while it is on the tool rest.
4. Always stand to one side when starting a grinding wheel.
5. Do not grind thin materials, such as sheet metal, on the grinder.

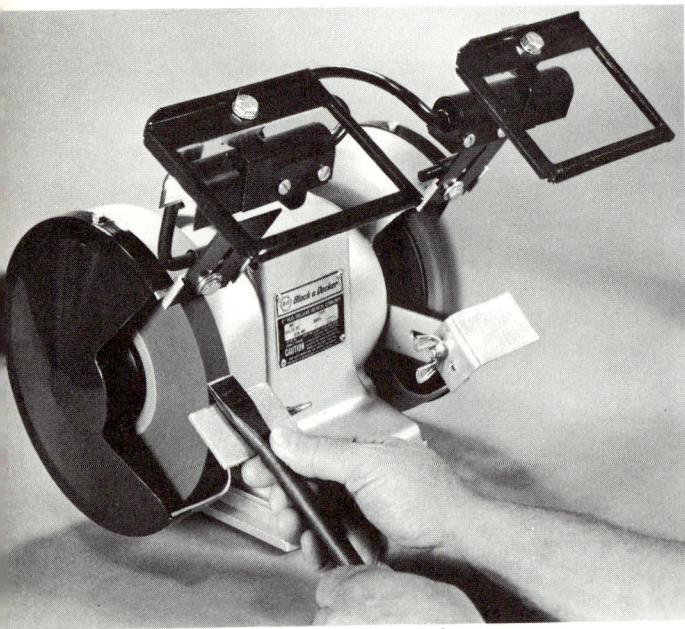

Black and Decker (U.S.) Inc.

Figure T16-2

Using a grinder to sharpen a cold chisel.

Norton Company

Figure T16-3

Resurfacing a grinding wheel with a wheel dresser.

Topic 17 Induced Fracture and Etching

Induced Fracture

Induced fracture means breaking a hard, brittle material along a line of weakness. Glass and acrylic plastics are two materials that are commonly separated by induced fracture.

Fracturing Glass

Glass is scored (scratched) with a glass cutter to produce a line of weakness. See Figure T17-1. Pressure on each side of the scored line will cause the glass to fracture (break) along the line.

To separate (cut) glass:
1. Lay the glass on a clean, flat surface.
2. Place a straightedge (rule) along the line that you intend to cut.
3. Dip the glass-cutter wheel into kerosene. This will make it cut more easily.
4. Start the score by placing the glass cutter on the far end of the glass.
5. Pull the glass cutter toward you. See Figure T17-2. Use **one** firm, steady stroke. Do not make a second cut.
6. Complete the fracture by snapping the scored line over an object with a straight edge. See Figure T17-3.

If long, thin slivers of glass are to be removed, first score a line in the usual manner. Then, use square-nose pliers or the square notches of the glass cutter to break off the piece of glass.

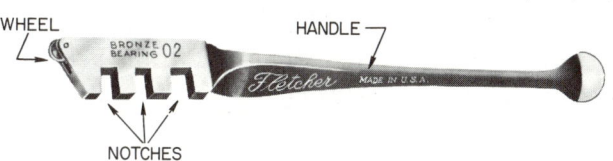

The Fletcher-Terry Company

Figure T17-1
The parts of a glass cutter.

Fracturing Glass

1. Always wear eye protection.
2. Be careful when working with glass. It can cut!

Figure T17-2
Scoring a line using a glass cutter.

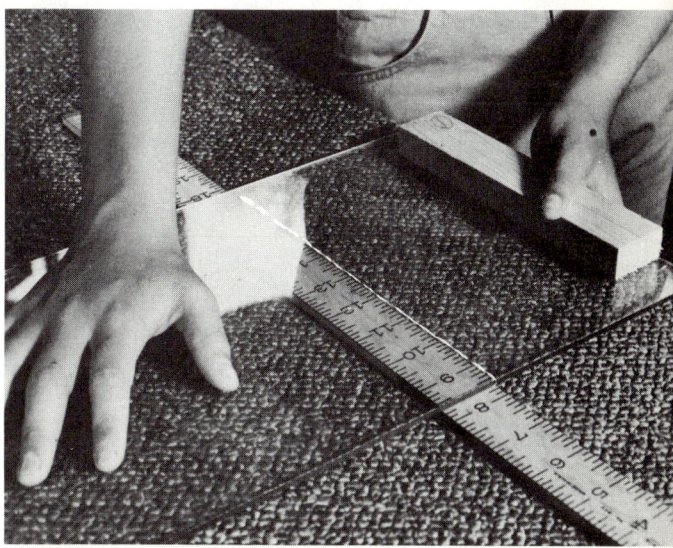

Figure T17-3
Snapping the glass along the scored line.

Fracturing Plastic

Acrylic plastic sheet stock can also be fractured by scoring a line on the surface. A straightedge and sharp scratch awl can be used to scratch a deep line into the surface of this material. (See Topic 2, **Bench Tools**.) Make several passes of the tool. If the score is made properly, the plastic will fracture in the right place. Complete the fracture by the same procedure that is used in fracturing glass.

Chemical Etching

Chemical etching is used to **mark** or **decorate** materials such as metal. Use this process to make pleasing designs on the surfaces of materials that are difficult to mark by other methods. An acid is used to eat into material in or around design areas. Areas that are not to be etched are coated with beeswax or asphaltum, a black liquid. See Figure T17-4. This protective coating is called a **resist** (meaning **resistant to acid**).

Aluminum is the metal most often etched in school shops. It is etched with hydrochloric acid. Add one part acid to one part water. **Always** pour acid into water. **Never** pour water into acid! It could explode!

Copper and brass are etched with nitric acid. Add one part acid to one part water. Non-acid etching solutions are also available.

Etching Metal

1. Make sure the work area is well-ventilated. Do **not** inhale acid fumes!
2. Clean and polish the metal to be etched. Be careful not to get fingerprints on the surface of the metal. The oil in the fingerprints acts as a resist and causes the image to be blurry.
3. Dab or brush the resist onto the metal and allow it to dry for 24 hours. If the **background** is to be etched, apply the resist to the **design area.** If the **design** is to be etched, coat the **entire surface** of the metal with resist and scratch the design through the resist using a sharp, pointed tool. See Figures T17-4 and T17-5.
4. Select the proper acid for the metal to be etched.
5. Pour the water into a glass container.

Figure T17-4

Applying resist to metal.

Figure T17-5

Scratching design areas through the resist.

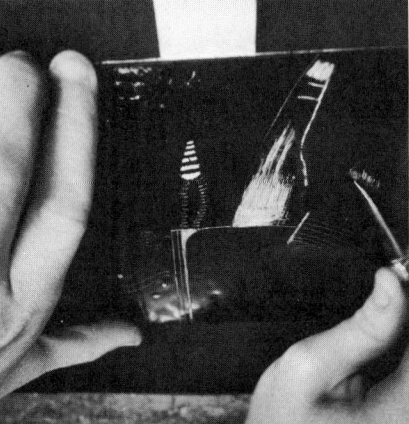

Figure T17-6

Completed etched design after resist has been removed.

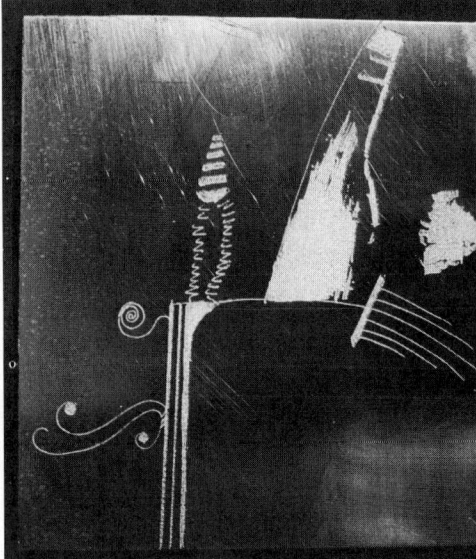

6. Add the acid to the water. **Always add acid to water.**
7. Completely submerge (cover) the metal in the acid mixture. Hold the workpiece as instructed by your teacher.
8. Allow time for the acid to work. It takes about one-half hour for hydrochloric acid to work, and about one and one-half hours for nitric acid.
9. Using hot water, wash all acid off the metal.
10. Remove the resist with lacquer thinner. Beeswax can be removed with hot water.

Your etching is finished. See Figure T17-6.

Etching

1. Wear eye protection.
2. Always add **acid to water,** not the reverse. Add the acid carefully to avoid splashing.
3. Work in a well-ventilated area. Do not inhale acid fumes.
4. When etching metal or other materials, wear rubber gloves, an apron, and a long-sleeved shirt.
5. Use glass containers. Metal or plastics may react chemically.
6. Wipe up any spilled acid immediately so that others will not touch it.
7. Dispose of used acid as your teacher tells you. Do not pour it down a sink unless told to do so. Some acids will damage the metal pipes of a sink.

Topic 18 Metal Casting

Metal casting is the process of making parts or products by pouring molten (melted) metal into a mold. As the metal cools, it becomes hard and takes the shape of the mold. The part or product made is called a **casting.** Even complicated parts can be made accurately by using this method. For this reason, metal casting is a very important manufacturing process. One of the most common metal-casting procedures is **sand casting.**

Figure T18-1
A one-piece pattern.

Sand Casting

Sand casting is done by pressing a special type of moist sand, called **molding sand,** around a pattern to make a mold. Sand is removed from certain areas, leaving a series of gates (passageways) leading to and from the pattern. The pattern is removed, leaving a cavity (hole) in the sand. This cavity is the mold that will be filled with the liquid metal. The result will be a casting that is identical in shape to the original pattern.

Patterns

One-piece patterns are commonly used in school shops. See Figure T18-1. They are easy to make and to use. The back of the one-piece

Figure T18-2

Draft helps you to remove a pattern from a sand mold without damaging the mold.

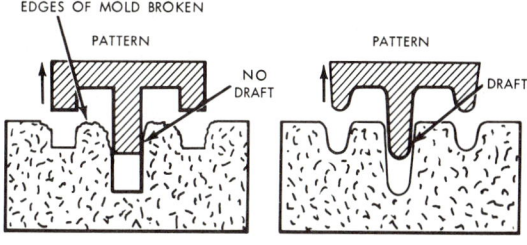

pattern is flat. The rest of the pattern has tapered surfaces. The slight angle of the taper is called **draft**. Draft helps in removing the pattern without destroying the edges of the mold. See Figure T18-2.

Complicated patterns are usually made in two pieces. These are called **split patterns**. See Figure T18-3.

A pattern that is easy to build and to use is the **full-mold foamed polystyrene plastic pattern**. See Figure T18-4. This pattern is left in the mold. You pack sand around the pattern (into the cavities in the pattern). You do not have to remove any part of the pattern or gating system. The heat from the molten metal dissolves the polystyrene foam, and the metal forms against the sand. This type of mold is good to use when patterns are complicated in shape. Of course, you must make a new mold for each casting.

Preparing a Sand Mold

Sand casting is done in a **foundry.** For this reason, sand-casting equipment is sometimes called foundry equipment. The sand is kept in a **molding bench.** See Figure T18-5. The mold is prepared on a molding board on top of the molding bench. The box in which the pattern and sand are held is called a **flask.** The flask has two parts: the **cope** and the **drag.** They are held together by **pins** and **sockets.** See Figure T18-6.

To prepare a sand mold:
1. Place the flat side of the pattern on a molding board.
2. Place the drag upside down around the pattern. The pins will be down.
3. Sprinkle parting compound (a powdery material) on the pattern to keep the sand from sticking to it. See Figure T18-7.

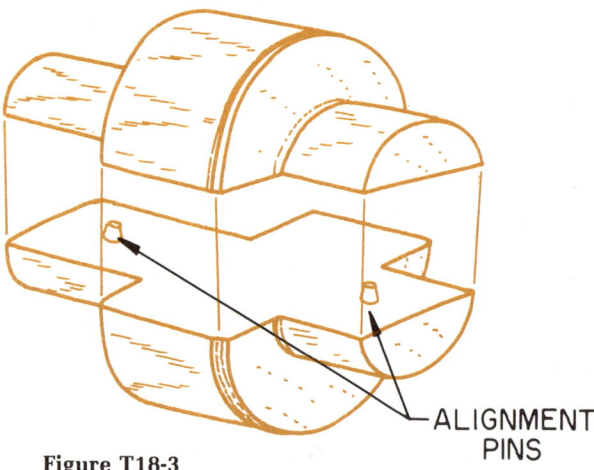

Figure T18-3

A split pattern.

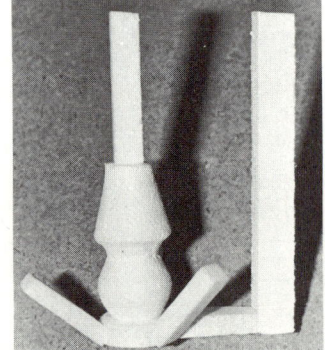

Figure T18-4

A foamed styrene pattern.

McEnglevan Heat Treating and Manufacturing Co.

Figure T18-5

A. A molding bench with storage space for tools and sand.

Metal Casting **371**

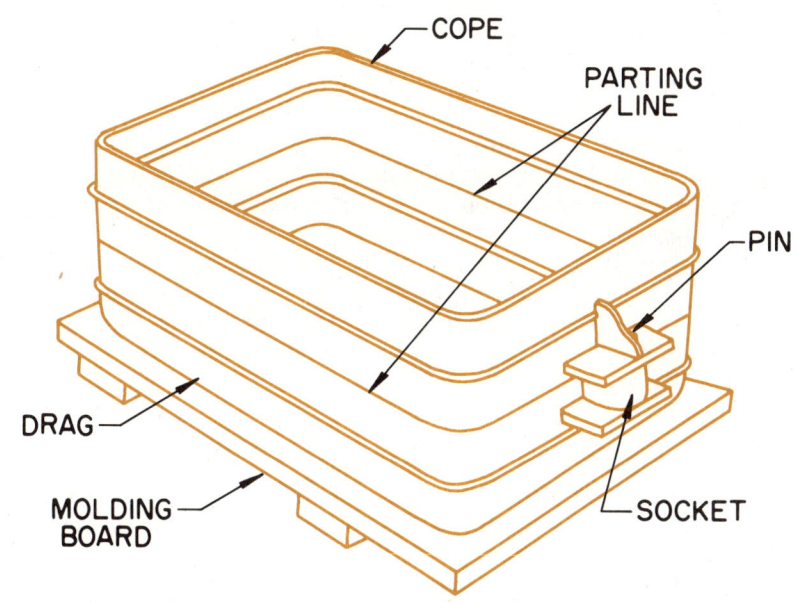

Figure T18-6
Flask for sand casting.

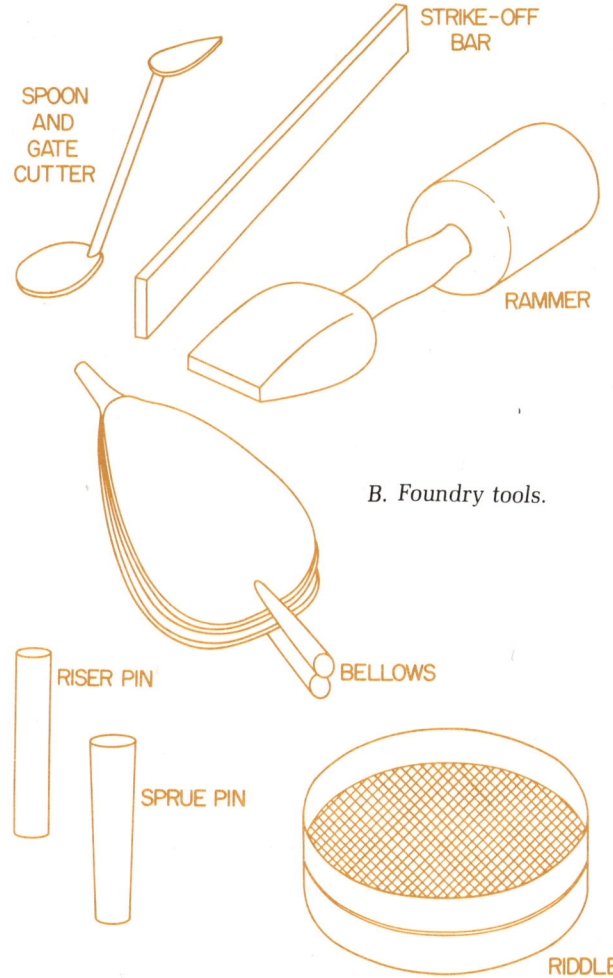

B. Foundry tools.

4. Fill a riddle (screen) with sand. Sift 2" of sand on the pattern. See Figure T18-8. Press the sand down.
5. Dump (do not sift) the remaining sand into the drag. Pack it down with a **rammer**. See Figure T18-9. Do not hit the pattern or the sides of the flask.
6. Level the top of the sand with a strike-off bar (straightedge). See Figure T18-10.
7. Sprinkle molding sand over the top. Rest the **bottom** board on the drag.
8. Hold the molding board, drag, and bottom board together at the sides. Carefully turn them all over at the same time. The pins in the drag will now be pointing **up**.
9. Remove the molding board (now on top). Smooth the surface with the **spoon** end of the spoon and gate cutter or a **trowel**. See Figure T18-11. Make sure that the sand is packed to the edges.
10. Blow off the loose sand with **bellows**. See Figure T18-12.
11. Sprinkle some parting compound over the pattern and mold to keep the halves from sticking together.
12. Set the cope on the drag, and set the sprue and riser pins in place. See Figure T18-13. The sprue pin makes a hole called the **sprue**. Molten (melted) metal will be poured through the sprue into the mold. The riser pin makes a hole called the **riser**.

Metal that is poured into the mold will come up into the riser when the mold is full.

13. Riddle sand into the cope, and ram it with a rammer. Do not pack the sand as tightly on this side. The sand must be loose enough to allow gases to escape.
14. Strike off the top of sand, and remove the pins. Round off the top of the sprue hole into the shape of a funnel. Use your fingers to do this. See Figure T18-14.
15. Make tiny holes (vents) with wire. Insert the wire into the sand. Stop about 1″ from the pattern. See Figure T18-15.
16. Lift off the cope, and lay it carefully on its side.
17. Slightly dampen the edges of the mold near the pattern using a **bulb sponge.** See Figure T18-16. This makes the sand firmer so that pattern can be lifted out without the mold breaking.
18. Lift the pattern out with a **draw pin.** See Figure T18-17.
19. Cut a **gate** (small channel) in the drag sand between the sprue and the hole made by the pattern. See Figure T18-18. Make the gate a little less than 1/4″ deep and 1″ wide. Use the gate-cutter end of the spoon and gate cutter.
20. Cut a gate between the riser and the hole made by pattern.
21. Patch all breaks, and blow off any loose sand with bellows.
22. Put the cope back on the drag, and place a **flask weight** on top to keep the molten metal from lifting up the cope during pouring.
23. Clamp all parts together, and position the mold on a sandy surface for pouring.

Melting and Pouring Aluminum

Alloy (mixed) aluminum is most often used for casting in school shops. The metal is placed into a **crucible** (bucket) made of graphite. The crucible is then placed into a **crucible furnace.** See Figure T18-19. This equipment is used to heat the metal until it is melted and at the proper temperature for pouring.

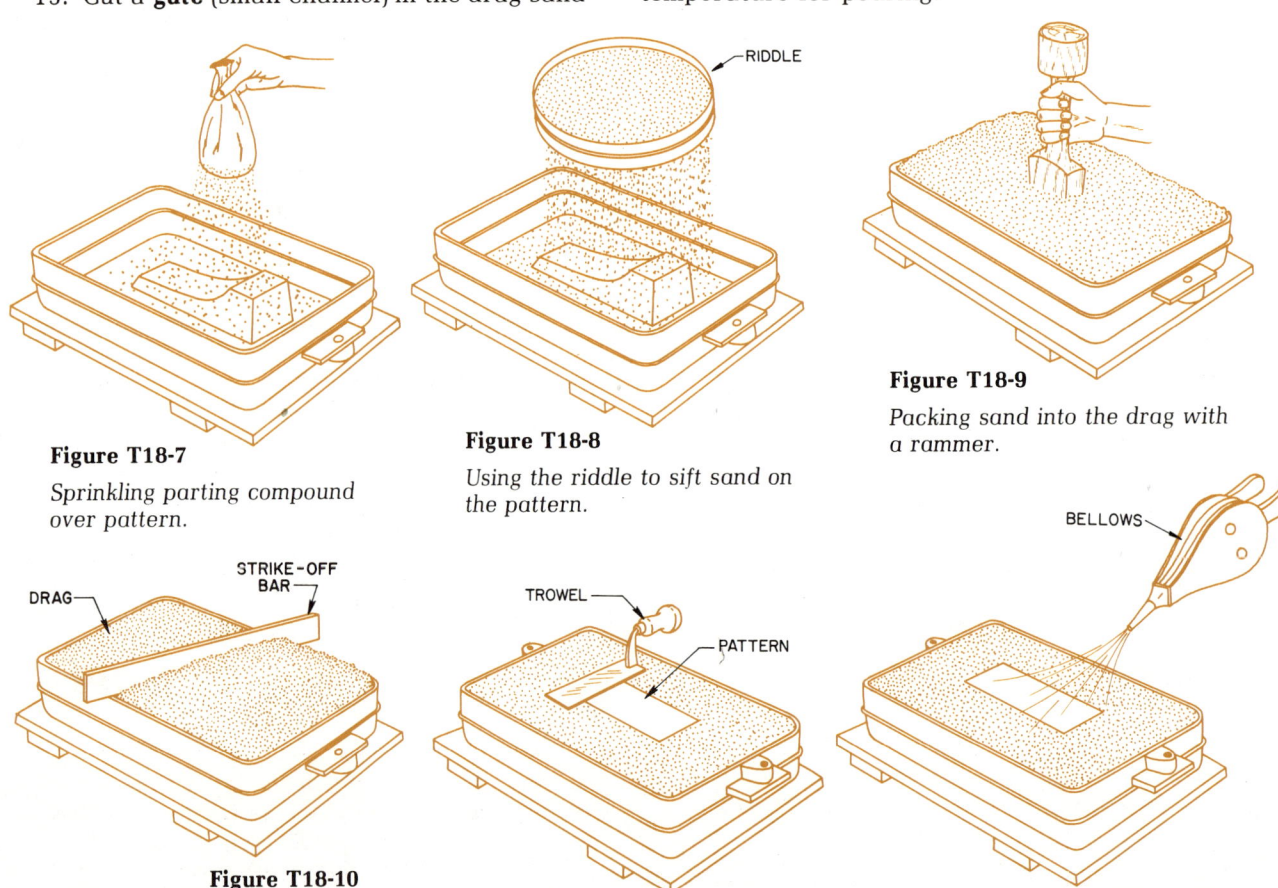

Figure T18-7
Sprinkling parting compound over pattern.

Figure T18-8
Using the riddle to sift sand on the pattern.

Figure T18-9
Packing sand into the drag with a rammer.

Figure T18-10
Leveling the sand with a strike-off bar.

Figure T18-11
Smoothing the surface with a trowel.

Figure T18-12
Removing excess sand from the pattern by blowing with the bellows.

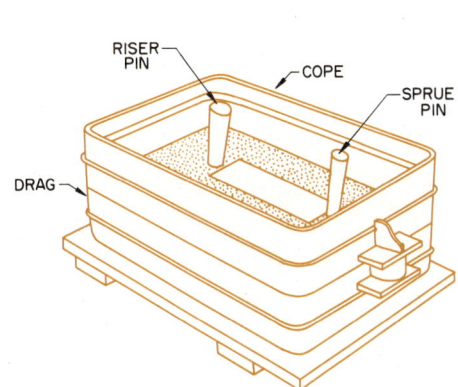

Figure T18-13

Correct placement of the sprue and riser pins.

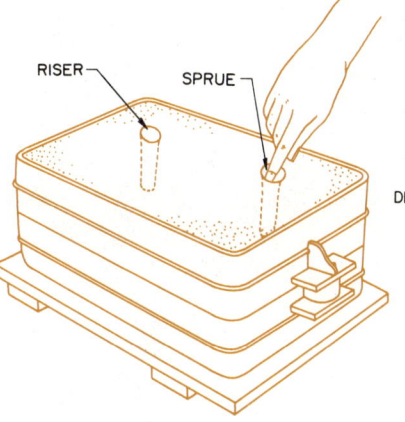

Figure T18-14

Rounding the top of the sprue hole into the shape of a funnel.

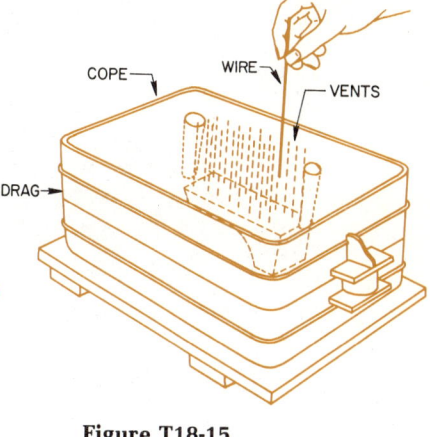

Figure T18-15

Using wire to make vents in sand.

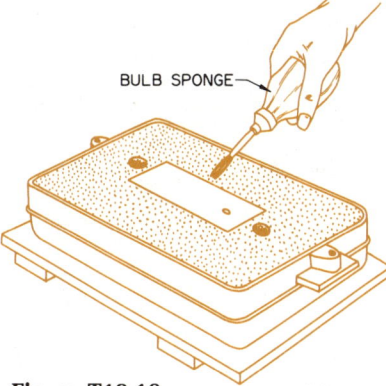

Figure T18-16

Using bulb sponge to lightly dampen sand around the edges of the pattern.

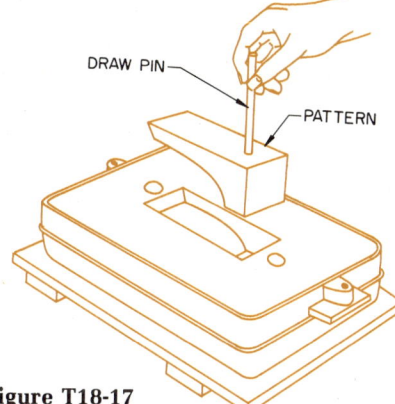

Figure T18-17

Using a draw pin to lift pattern from the sand.

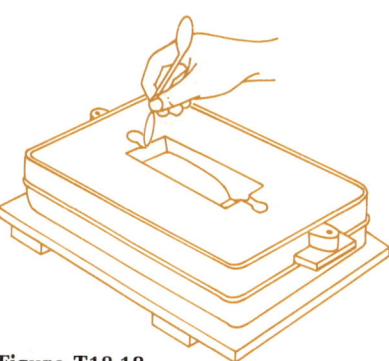

Figure T18-18

Cutting gates using the spoon and gate cutter.

Figure T18-19

A crucible furnace for melting aluminum.

Johnson Gas Appliance Company

Figure T18-20

Using crucible tongs to remove crucible from furnace.

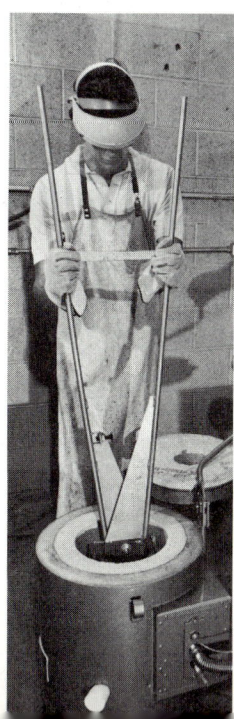

Johnson Gas Appliance Company

McEnglevan Heat Treating and Manufacturing Co.

Figure T18-21

Pouring molten aluminum into a mold. Note that the crucible is held with a crucible shank.

Figure T18-22

A finished metal casting made by using the pattern in Figure T18-1.

The correct pouring temperature is determined by using a **pyrometer.** Aluminum should be poured at a temperature 100°-200° hotter than its melting point. Pure aluminum melts at 1218° F (659° C). Alloys, however, melt at lower temperatures. The temperature of the metal can be checked at the time it becomes melted to determine the pouring temperature. If the metal is too hot when it is poured, the casting will be defective (have flaws). If the temperature is too low, the metal will stop flowing, and the casting will not be completed.

When the aluminum has nearly reached the correct pouring temperature, the furnace is turned off. Flux (metal-cleaning substance) is added to it to raise the slag (impure metal) to the surface.

After the slag has been skimmed off, **crucible tongs** are used to carefully remove the crucible from the furnace. See Figure T18-20. It is placed into a **crucible shank** for pouring. Stand to one side of the mold while the metal is being poured into the sprue. Molten metal may splash during pouring. Pouring is continued until the mold can hold no more or until metal begins coming up the riser. Once pouring has been started, it must not be stopped until it is completed. See Figure T18-21. Any metal that is left is poured into an **ingot mold.** (See Topic 27, **Metals**.)

When the metal has cooled, the sand and casting are shaken onto the molding bench. The sprue, riser, and gates are removed from the casting. The casting generally must be machined or finished. See Figure T18-22.

Casting

1. Wear protective clothing, gloves, shoes, and eye protection while working with molten metals.
2. Do not touch castings until you are **sure** that they are cool.
3. Always stand to one side of the mold when metal is being poured. **Never** stand directly over it or in front of it.

Topic 19 Metal Forging and Bending

Forging

Forging is the process of hammering metal into shape. Forging increases the strength and toughness of a metal.

In the school shop, most metal forging is done using a hammer and an **anvil**. See Figure T19-1. Ordinarily, hot metal is used. See Figure T19-2. It is held with a gripping tool, such as tongs. The metal workpiece is generally heated in a special furnace called a **forge**. See Figure T19-3. A welding torch could be used. (See Topic 13, **Oxyacetylene, Brazing, Welding, and Cutting**.)

The metal is heated to a cherry red and then forged (hammered) into shape on an anvil. Hammering is done with a ball peen hammer or a blacksmith's hammer. Do not work metal that is cooler or hotter than cherry red. Cooler metal may crack from hammering. Too much heat may burn the metal and cause sparks to fly. Tools, such as cold chisels, screwdrivers, and punches may be forged in a school shop.

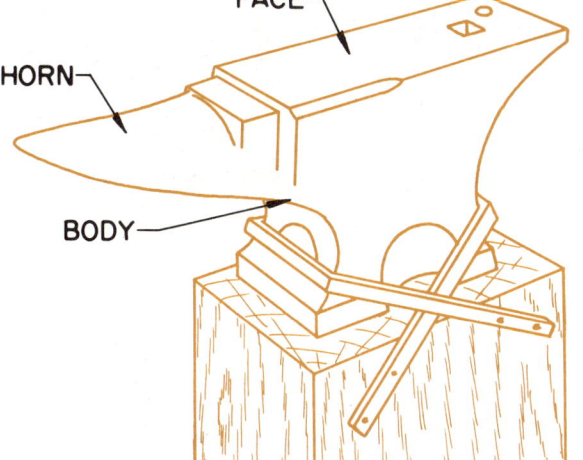

Figure T19-1
Hot metal is forged on an anvil.

Figure T19-3
Forging furnace. Note wide hearth to allow use by several students at a time.
Johnson Gas Appliance Company

Figure T19-2
Forging a piece of hot metal.

Wherry Machine and Welding

376   Tools, Materials, Processes

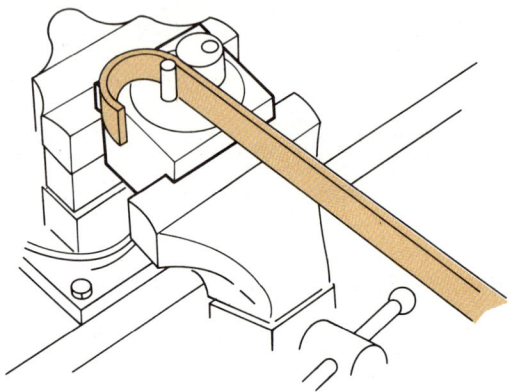

Figure T19-5
Using a bending jig to bend metal.

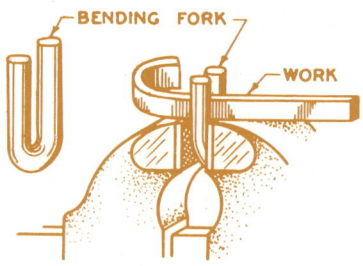

Figure T19-6
A bending fork can be easily made in the shop for use in bending metal.

Figure T19-4
Twisting bar stock.

Bending

Bending is the process of forming metal into different shapes by such methods as twisting, hammering, and clamping. Metals thicker than 1/4" can be heated to make bending easier. Metals less than 1/4" thick can be bent without being heated.

Bending Heavy Metal

A metal bar may be held in a vise and hammered to the correct angle. Bending may also be done with a variety of tools, jigs, and machines. (See Topic 3, **Bench Tools,** and Topic 25, **Production Tooling.**)

A **twisted** form can be made by turning bar stock with an adjustable **wrench** while one end is held in a **vise.** See Figure T19-4. Flat metal can be made into **irregular** shapes by clamping the metal in a vise between **forms** or blocks.

Bending Jigs

Bending jigs are used when bending bar stock into different shapes. Several kinds of bending jigs are available, but they all work in basically the same way. See Figure T19-5.

A **bending fork** is a nonadjustable bending jig. You could easily make one. It is clamped in a vise and used in the same way as other bending jigs. See Figure T19-6.

Using a Bending Jig

1. Clamp the bending jig in a vise.
2. Insert a strip of metal between the posts. Adjustable bending jigs have a series of holes. Two pins or posts are positioned in holes so that the metal fits snugly between them.
3. Pull on the metal strip until it is the desired shape. For larger-diameter bends, you will have to bend the metal slightly, move the strip between the posts, then bend a little more. Repeat these steps until the metal is the proper shape.
4. Remove the metal from the jig.

Metal Forging and Bending

Bending Machines

Special machines used to bend heavy metal are the **Di-Acro®** bender and the **Universal® bender.** These bending machines are commonly used to bend metal bar, rod, or pipe. The long handles on these machines provide leverage (increases the force you apply). Heavy stock can be bent quickly by using these machines. See Figure T19-7.

Using Bending Machines

Many different shapes can be made on bending machines. A few of the shapes you might make are circles, scrolls, loops, coils, and angles. The **pins** on the Di-Acro® bender and the special interchangeable **blocks** on the Universal® bender determine the shape that will be produced. As the handle of the bender is moved, the metal workpiece is pressed against the pins or block. This causes the metal to be bent to the correct shape.

Bending Sheet Metal

Many different shapes can be produced by using a **wooden mallet** to pound sheet metal against a metal form called a **stake.** Stakes are made in a variety of shapes, such as cones, tubes, and angles. Other tools used for bending sheet metal are hand seamers, bar folders, and box and pan brakes.

Hand Seamers

Hand seamers are plier-like tools used for making straight-line bends in thin sheet metal. See Figure T19-8. They are used along the edges of flat metal. Because of their small size, hand seamers can only be used to make small bends (5/16″ to 7/8″ wide). Large bends are made with a bar folder or a box and pan brake.

Using a Hand Seamer

1. Clamp the metal in the hand seamer.
2. Make sure that the bend is even with the front of the jaws.
3. Bend the metal.

Figure T19-7

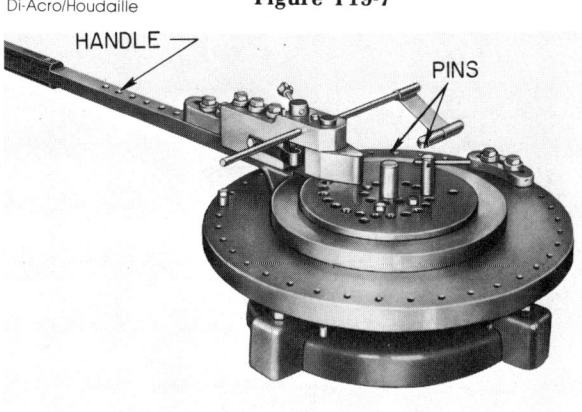

The parts of a Di-Acro® bender.

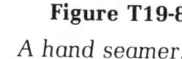

Figure T19-8
A hand seamer.

A Universal® bender.

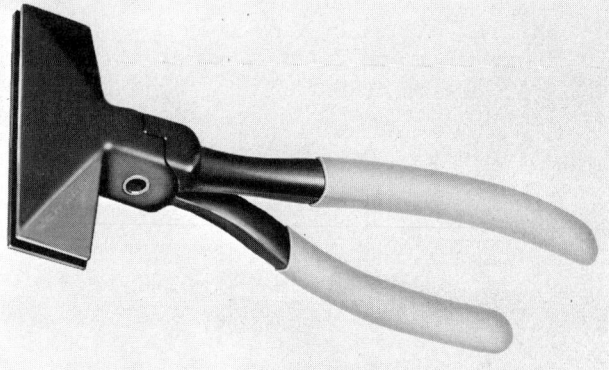

Bar Folders

Bar folders are used to bend seams, hems, folds, and edges on sheet metal. See Figure T19-9. These machines are also useful for joining pieces of sheet metal. Two bent edges can be joined together to form a **seam** (two overlapped or interlocked edges). This seam is then soldered or riveted in place or secured with a **hand groover.** See Figure T19-10. A **hem** (folded edge) strengthens the edge of a sheet-metal object. See Figure T19-11.

Using a Bar Folder

To make a hem:
1. Adjust the **depth gage** to the desired hem width by turning the **adjusting screw.**
2. Insert the metal between the **blade** and the **wing.**
3. Pull the **handle** up and toward you.
4. Remove the metal from between the blade and the wing of the machine.
5. To close the hem, place the metal fold (bend up) on top of the wing and bend the fold down with the handle.

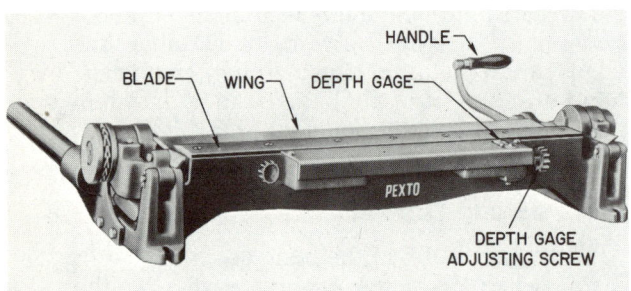

Figure T19-9

The parts of a bar folder.

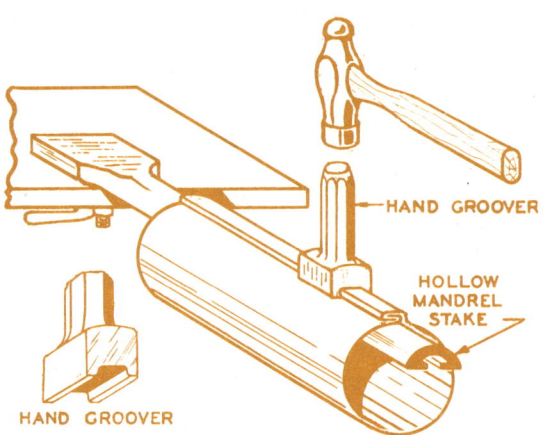

Figure T19-10

A hand groover is used to secure a seam.

Figure T19-11

Steps in making a hem in sheet metal using a bar folder.

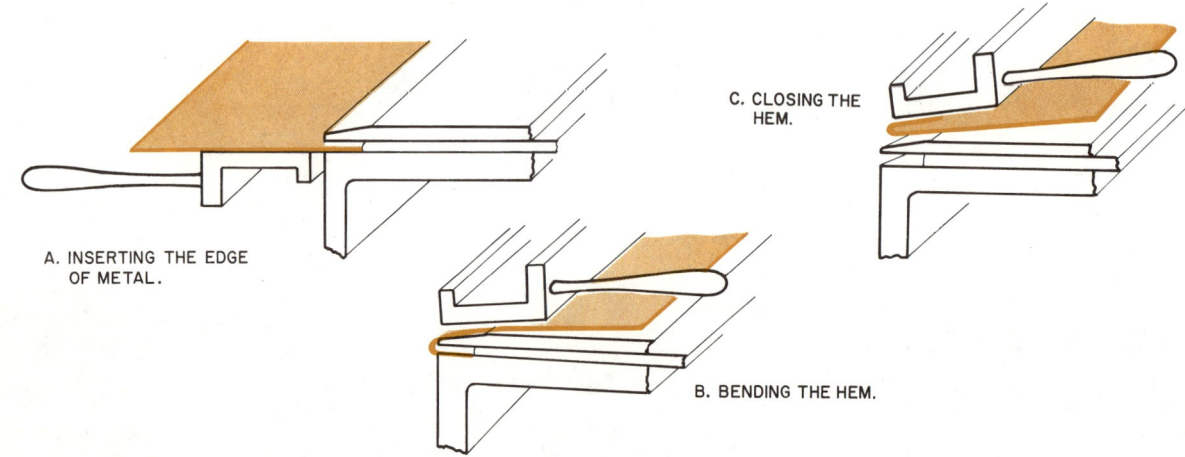

Box and Pan Brakes

Box and pan brakes are used to bend angles on sheet metal. See Figure T19-12. A box and pan brake can be adjusted to bend the sides of sheet-metal boxes. A series of metal blocks (called **fingers**) form the **upper clamp bar.** These fingers can be removed in sections to leave a bending edge of the desired length. The **bending leaf** is located under the upper clamp bar. It provides support for the sheet metal during bending.

Using a Box and Pan Brake

1. Lay out all bend lines on the sheet metal.
2. Set up the machine so the fingers make up the proper width for the bend.
3. Raise the upper clamp bar, and insert the metal.
4. Move the metal until the bend lines are even with the tips and sides of the fingers.
5. Lower the clamp bar, and clamp the sheet metal in place.
6. Move the bending leaf until the metal is bent to the correct angle.
7. Remove the metal from the box and pan brake.
8. Set up the proper width of fingers for each length to be bent, and repeat the process.

Slip-Roll Formers

Curved shapes can be made in sheet metal and wire using a slip-roll former. Items such as cans and funnels can be made. Three **rolls** form the metal into curved shapes. See Figure T19-13.

Using a Slip-Roll Former

1. Lock the **top** (or upper) **roll.**
2. Move the **bottom** (or lower) **roll** up or down until the opening is the right width for the thickness of the metal.
3. Move the **back** (or rear) **roll toward** the top roll to make a smaller circle. Move the back roll **away from** the top roll for a larger circle.
4. Set the position of the rolls by tightening the **adjusting screws.**
5. Place the sheet metal between the top roll and the bottom roll.
6. Turn the handle to move the metal through the rolls.
7. When you have finished making the circular form, press the **roller release** to raise the top roll.
8. Remove the metal.

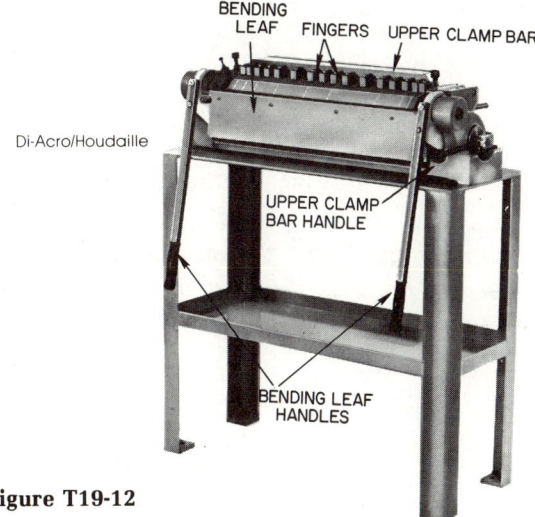

Figure T19-12
The parts of a box and pan brake.

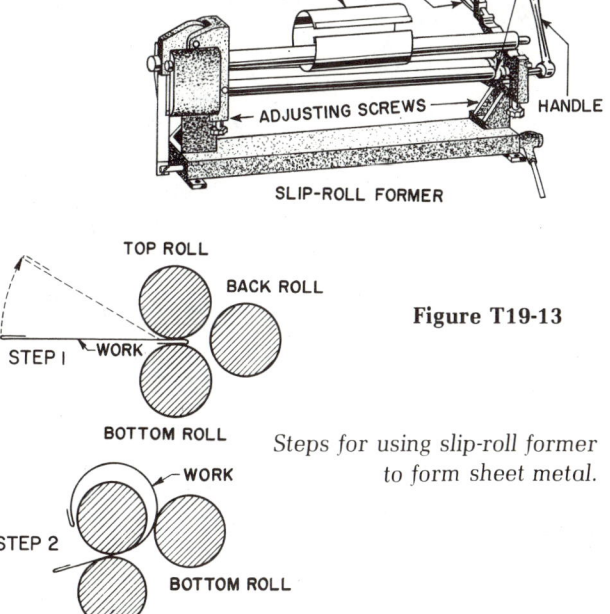

The parts of a slip-roll former.

Figure T19-13

Steps for using slip-roll former to form sheet metal.

Slip-Roll Former

Keep your fingers out of the way of the rolls.

Topic 20 Plastic Casting, Molding, and Forming

Plastic casting and forming change the shape and form of plastic materials. Polyester-resin casting, injection molding, thermoforming, and vacuum forming are common methods of shaping plastics. You should read Topic 28, **Plastics**, along with this topic to understand which types of plastics can be formed by these methods.

Casting Polyester Resin

Plastic casting is the process of pouring plastic resin into a mold and allowing it to harden. Clear **polyester resins** are generally used for this process. These casting resins are sometimes tinted to produce a variety of colors and effects. Polyester resins become hard when mixed with a **catalyst** (a substance that makes a chemical reaction). The resin and catalyst must be mixed together very well or parts of the casting may not cure (harden). See Figure T20-1.

Casting Procedure

1. Mix the resin and catalyst together in a paper cup.
2. Pour the mixture into the mold. See Figure T20-2.
3. Items such as dried flowers, coins, or sea shells can be embedded in the casting plastic for display. First pour half of the casting mixture into the mold. Carefully place the embedments on this layer. Pour the rest of the plastic over the embedments. Use a toothpick or needle to force trapped air bubbles from the plastic.
4. Allow the resin to cure for the length of time recommended in the directions.
5. If more than one layer of resin is to be cast, repeat the above process while the top of the resin is still tacky (sticky).

Casting Polyester Resin

1. Wear eye protection when working with plastic resins.
2. Do all resin casting in a well-ventilated area.
3. Keep the resin off your skin and clothing.
4. Immediately wash your eyes if you get catalyst into them. Use lots of water quickly. Permanent damage can result in a matter of seconds.

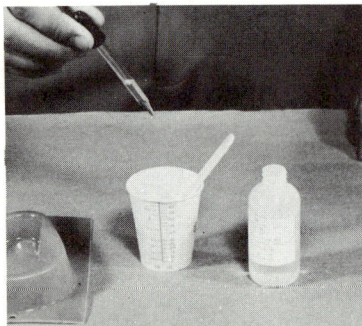

Figure T20-1

Mixing polyester casting resin and hardener (catalyst).

The Castolite Company

Figure T20-2

Pouring mixed resin into a mold. The medal will be embedded in the resin.

Plastic Casting, Molding, and Forming   381

Figure T20-3

Fiber glass materials used for laminating.

Fiber glass cloth

Fiber glass mat

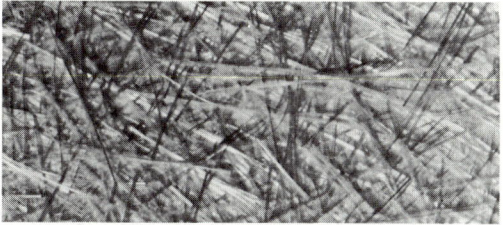

Figure T20-4

Forcing laminating resin through the cloth.

Laminating Fiber Glass

Plastic resins can be reinforced (made stronger) with **fiber glass.** The fiber glass is laminated (layered) in the plastic material. Laminated fiber glass parts are strong, lightweight, and will not rust or decay.

The fiber glass used for laminating is available in **cloth** or **mat** form. See Figure T20-3. Fiber glass cloth is woven, while fiber glass mat is made by pressing loose fibers together. A plastic resin mixed with a catalyst (hardener) is used to coat the cloth. When the resin cures, the cloth becomes stiff. Usually, cloth and mat are laminated together for strength and appearance.

A **mold** is used to shape the plastic. The mold must be prepared by waxing or applying **mold release.** Mold release is a substance that is applied to the mold so that the resin will not stick. It can be a liquid, powder, paste, or polyester film (such as Mylar®). Mold release is most often used in liquid form. It is usually applied by **spraying.**

Laminating Procedure

1. Apply wax or mold release to the mold.
2. Mix the right amount of resin and hardener. Mix only as much as you will need for one layer. You must mix a new batch for each layer.
3. Brush a thick coat of resin on the mold. Make sure there are plenty of newspapers under the mold to protect the working table from resin.
4. Place the fiber glass cloth or mat onto the wet resin. The first layer is usually made with fiber glass cloth. This forms a smoother surface than mat. Dab with a brush to push down on the cloth or mat. This will push the resin up through the cloth and coat all fibers. See Figure T20-4.
5. Repeat steps 2-4 for each layer you are laminating. Allow each layer to dry until tacky before applying another layer.
6. Brush the last coat of resin until it is as smooth as you can make it.
7. Let the product cure for 24 hours.
8. Remove the laminated product from the mold. Generally, you will need to trim, sand, and polish the product.

Laminating Fiber Glass

1. Wear eye protection.
2. Make sure that the work area is well-ventilated.
3. Wear plastic or rubber gloves.
4. Remember fiber glass is glass. It can cut you. If you have been handling fiber glass, do not rub your skin or eyes.

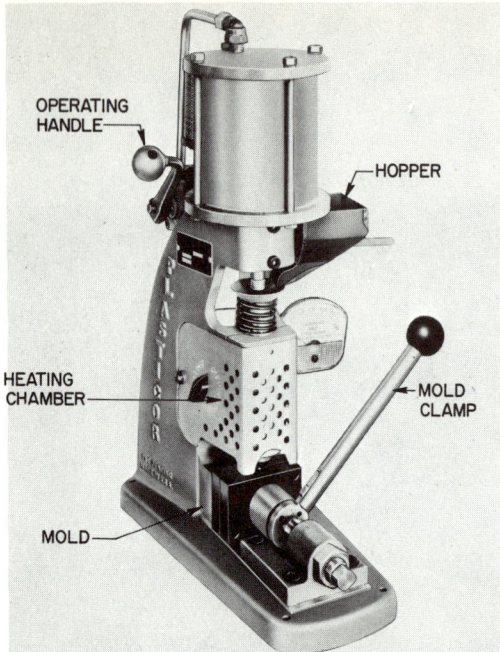

Figure T20-5 Simplomatic Manufacturing Company
An injection molder.

Injection Molding

Injection molding is a process of forcing hot, soft plastic into a mold. Injection molding is used to make products or product parts. It is a common way of making many small parts quickly and inexpensively.

Two types of plastics that are frequently used are **polystyrene** and **polyethylene**. Plastic used for injection molding is in the form of small pellets. These pellets are placed in the **hopper** (storage chamber) of the injection molder. See Figure T20-5. The plastic falls into a **heating chamber,** where it becomes hot and soft. The plastic is then injected (pushed) into the mold by a plunger-like part called a **ram.** Once the plastic is in the mold, it quickly cools and hardens.

Injection Molding Procedure

1. Turn on the injection molder, and allow the heating chamber to warm up.
2. Fill the hopper with plastic pellets. Do not overfill the hopper.
3. Purge (clean) the heating chamber of old material by pulling down on the operating handle.
4. Clean the mold, if necessary, and spray it with **mold release.**
5. Clamp the mold into the machine, and allow it to become warm from the heat of the heating chamber.
6. Pull down on the operating handle. This will cause the plastic to flow into the mold. Continue to hold the handle down for a few seconds to allow the plastic to fill the mold.
7. When the mold is filled, unclamp it and remove the product. Be careful because the plastic may still be hot.
8. Trim off all unwanted material. Keep the scrap plastic you have trimmed. It can be remelted and used again.

Injection Molding

Wear gloves when working with hot molds and plastic parts.

Thermoforming Sheet Plastics

Thermoforming is the process of forming heated plastics, usually over or into a mold. Sheets of acrylic or styrene plastic are normally used for thermoforming. The plastic is heated in a special **oven** or with a **strip heater.** See Figure T20-6. Strip heaters are used to heat narrow strips of plastic. Materials 1/4″ or thicker must be heated on both sides. The plastic should be hot enough to be easily formed, but not so hot that it blisters and bubbles. Acrylics are heated to a temperature of between 240° and 360° F (116° and 182° C).

The plastic is ready for forming when it can be bent easily. If the plastic does not take the correct shape the first time it is formed, it may be reheated and formed again. Plastic materials that can be reheated and shaped are called **thermoplastic.** If thermoplastic materials are reheated too many times, however, they become brittle and lose their surface luster (shine).

There are many types of thermoforming. Some thermoforming is done without a mold. This is called **free forming.** Free forming can be done in many ways. One method of free forming involves twisting heated acrylic plastic that is clamped in a vise. See Figure T20-7.

Another type of thermoforming is called **mechanical forming.** This method uses some type of simple setup, such as clamps or blocks. See Figure T20-8.

The most common thermoforming process is called **vacuum forming.** Vacuum forming uses a mold and vacuum pressure.

Plastic Casting, Molding, and Forming   **383**

Quincy Lab Incorporated

A plastic-heating oven.

Figure T20-6

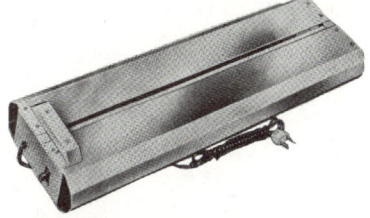

A strip heater is used to soften plastic. Only the plastic above the slot of the heater will be softened.

Hydor Therme Corporation

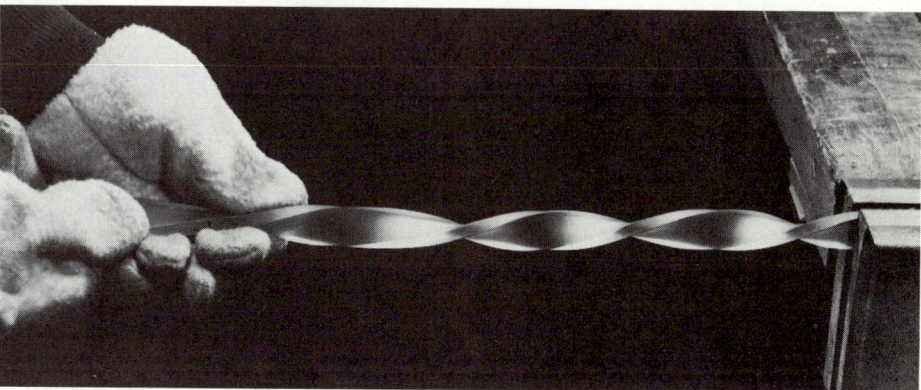

Figure T20-7
Twisting a heated plastic strip.

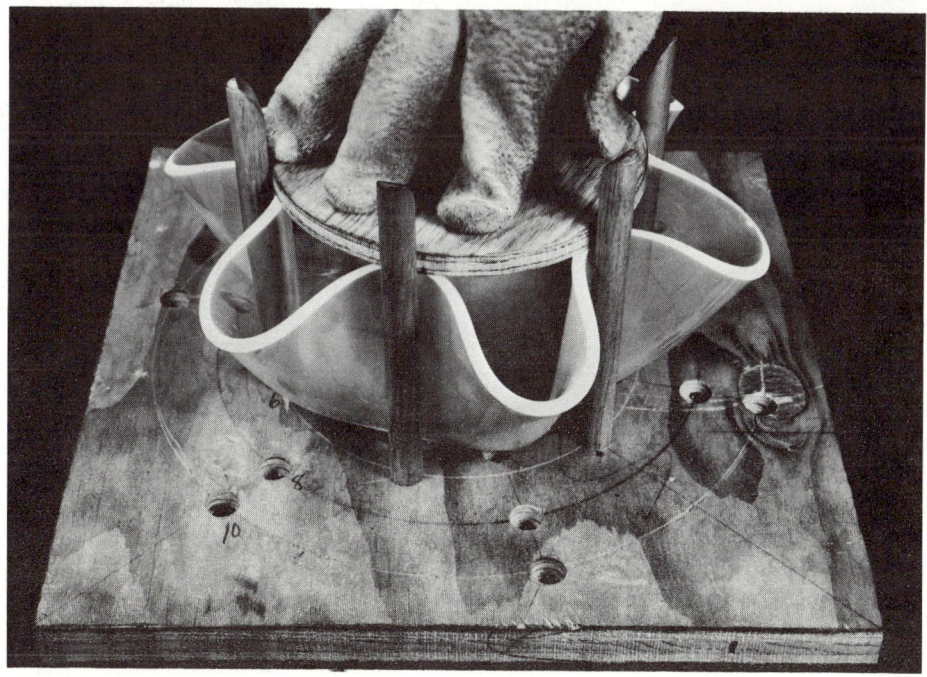

Figure T20-8
Forming heated plastic in a shop-made mold.

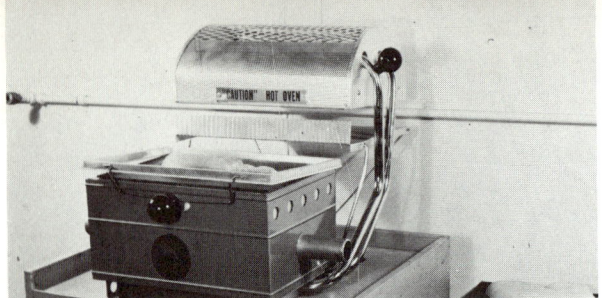

Figure T20-9
A vacuum thermoformer.

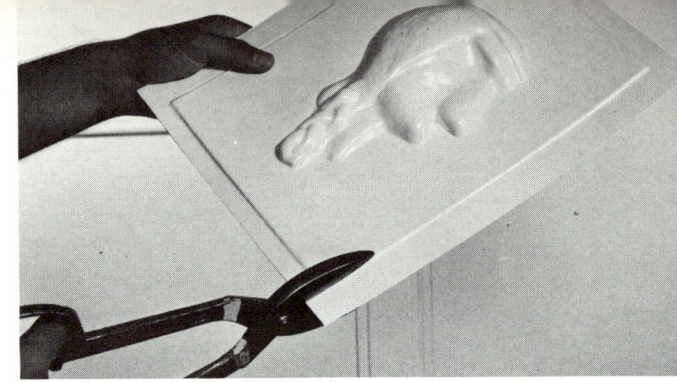

Figure T20-10
After the plastic has completely cooled, trim away unwanted edge.

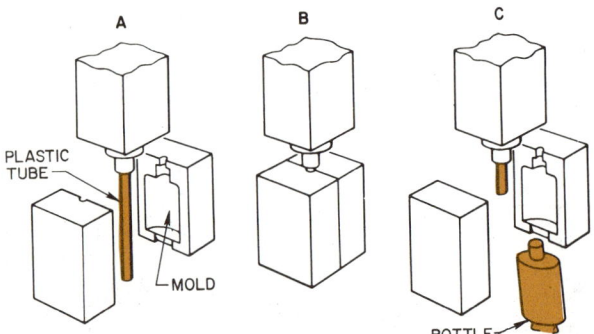

A. Plastic tube in position.
B. Mold closed and air blown in.
C. Cooled plastic bottle removed from mold.

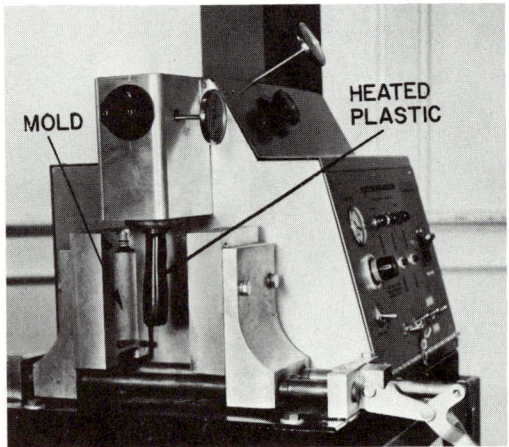

A blow molding machine.

Figure T20-11
Blow molding process.

Vacuum Forming

Vacuum formers shape sheets of plastic by drawing (pulling) the heated plastic into a mold. See Figure T20-9. Sheets of acrylic, polystyrene, and polyethylene are most often used for vacuum forming.

In vacuum forming, the plastic is held tightly in a frame and pressed against a mold. The plastic is then heated until it is soft. The air is sucked out on one side of the material, causing the plastic to form around the mold and take its shape.

Vacuum Forming Procedure

1. Clamp the sheet of plastic in a holding frame.
2. Apply heat to the plastic, but do not melt it.
3. Bring the plastic and mold together until they form a good seal. Allow extra plastic material around the edges so that there will be enough to make the seal.
4. Apply the vacuum.
5. Wait three to five minutes for the plastic to cool and harden. During this time, the vacuum former should remain on, but the heat of the vacuum former should be turned off.
6. When the plastic has hardened, remove the product.
7. Trim off excess (extra) plastic. See Figure T20-10.

SAFETY DIRECTIONS

Vacuum Forming

Wear gloves when working with hot plastics.

Blow Molding

Blow molding is similar to vacuum forming, but is less commonly used in shops. Plastic is heated and then formed into a mold. In blow molding, however, heated plastic is placed into a special mold and air is pumped in. This causes the plastic to expand much like a balloon. It takes the shape of the mold as it cools. Blow molding is used to make hollow plastic items, such as bottles. See Figure T20-11.

Topic 21 Stamping Tools

Stamping means making marks on a material. Stamping can be done for identification or decoration. Metal and leather are commonly stamped with metal stamps. (See Topic 27, **Metals,** and Topic 29, **Other Materials.**)

Die (Metal) Stamping

Stamping is a quick and inexpensive way to label metal parts with permanent lettering and numbering. **Die stamps** are made of steel and may be purchased in sets. Letters and figures usually come in separate sets. See Figure T21-1.

Most stamping done in the shop is done by hand. Guidelines are needed to ensure a neatly spaced job.

Using a Die Stamp

1. Carefully position the stamp on the material.

2. Strike the stamp firmly and carefully with a ball peen hammer. See Figure T21-2.

3. If the impression left is deeper on one side than the other, tilt the stamp over the lighter impression and strike again. Be sure to place the stamp directly on top of the first mark. If you do not do this, you may produce a double image.

4. Repeat this stamping operation for each letter or numeral. Make sure that each impression is evenly spaced.

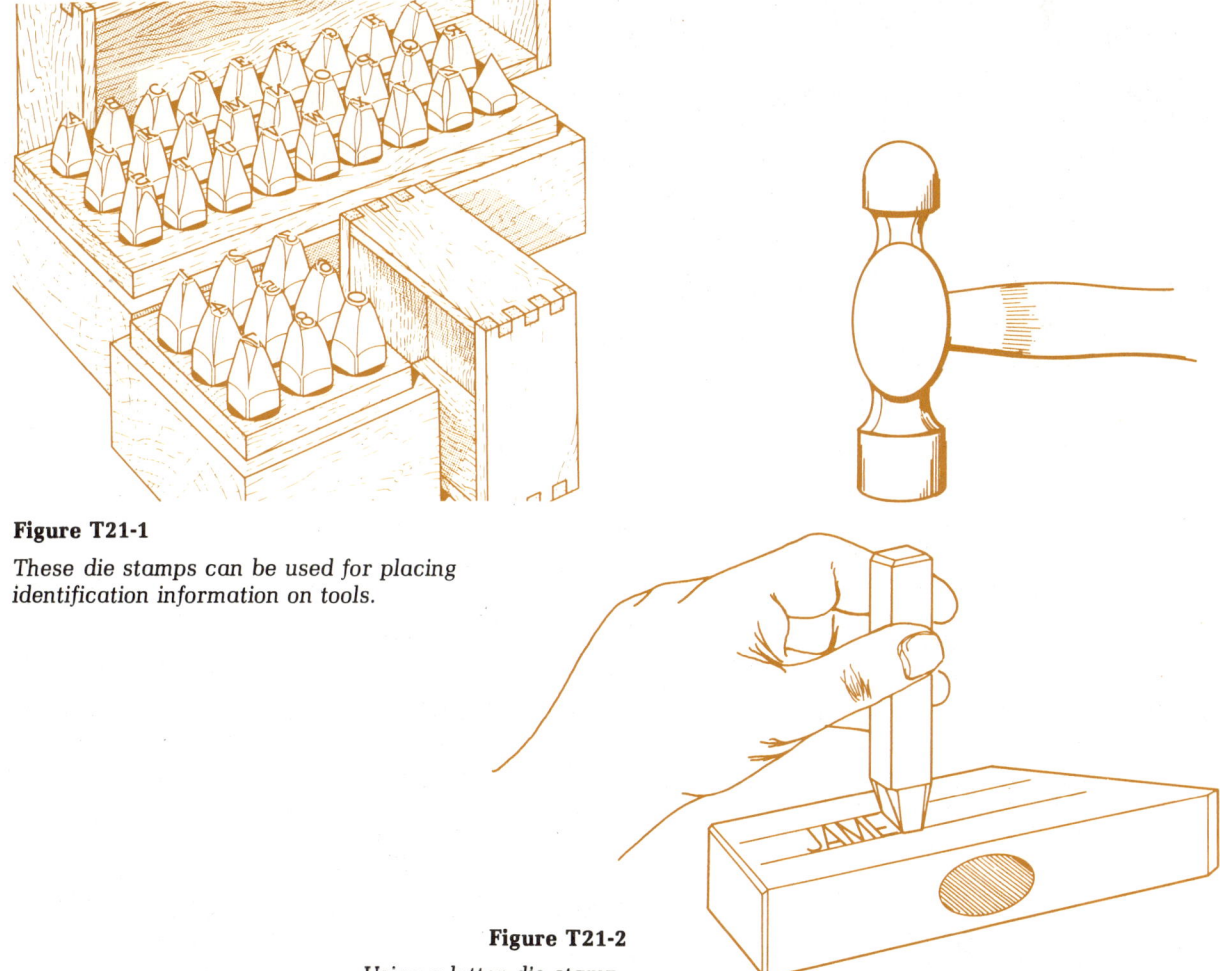

Figure T21-1

These die stamps can be used for placing identification information on tools.

Figure T21-2

Using a letter die stamp.

Stamping Leather

Preparing the Leather

Before stamping can be done successfully, the leather must first be prepared. This is done by applying a small amount of cool water to the back or rough side of the leather. The leather should become damp but not soggy.

Be careful when handling wet leather. It is very easy to mark. Even a fingernail can leave a permanent impression.

Leather Stamps

Leather stamps are used to decorate leather products. Impressions are left on the leather surface when the stamp is struck with a mallet. See Figure T21-3. Stamping should be done over a hard surface.

Stamps are available with a wide variety of designs, such as flowers, and shells. See Figure T21-4. Inexpensive stamps may also be made by filing the head of a large nail.

A special background stamp is used to shade and texture areas around a design. Letter stamps are used to spell out words, such as names, on belts or similar items. See Figure T21-5. Shading stamps and beveling stamps are used to make smooth impressions in leather. These are used to highlight and add depth to the design. See Figure T21-6.

Figure T21-3
Leather stamping tools can be purchased in sets.

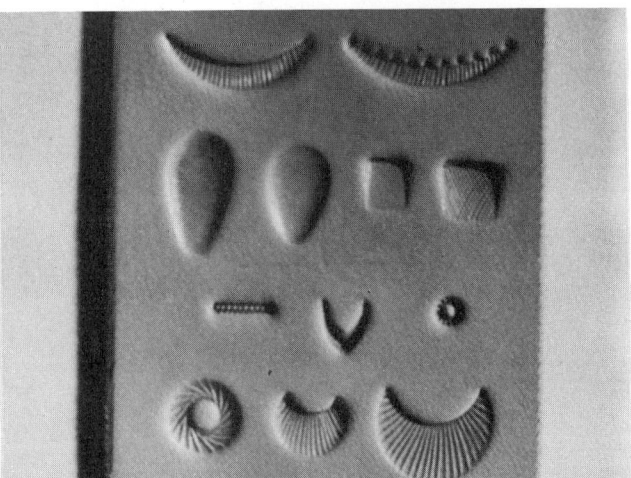

Figure T21-4
Patterns produced by leather stamping tools.

Figure T21-5
Alphabet stamps for leather have been used to stamp a name on this dog collar.

Figure T21-6

A. Leather must be lightly moistened to prepare it for stamping.

B. A pattern is made on the leather when the stamping tool is struck with a mallet.

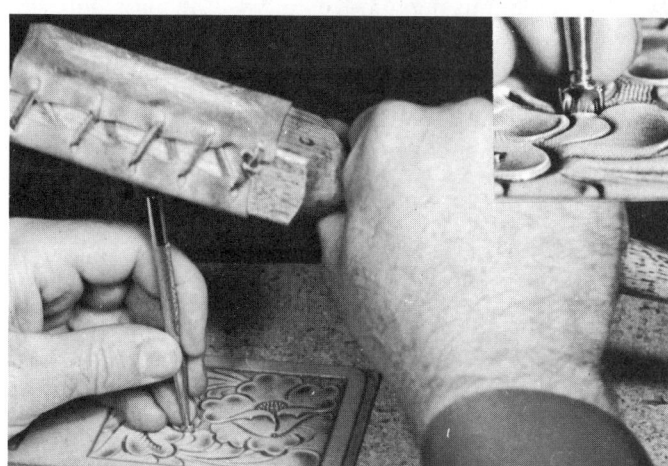

C. This backgrounding stamp is used to shade around a design.

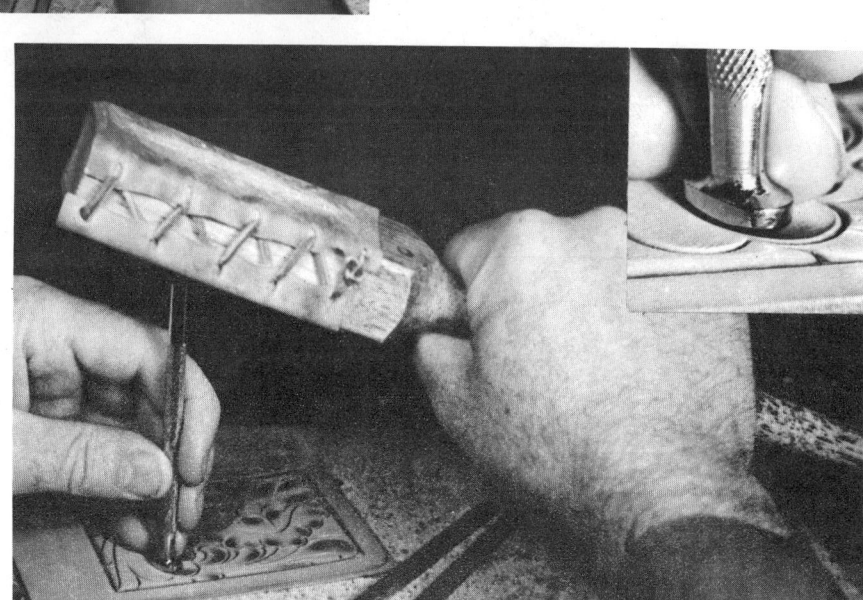

D. Using a shading tool.

Topic 22 Heat Treating

Heat treating is the process of heating and cooling metal to change and improve it. Four common methods of heat treating are **annealing**, **hardening**, **tempering**, and **case hardening**.

Annealing

Metals are annealed to make them easier to work or machine. High-carbon steel, for instance, can be very hard. When it is hard, it is difficult to cut or machine. Annealing softens the metal to improve its **workability** (make it easier to work).

To anneal steel, heat it in a heat-treating furnace until the metal is bright red. See Figure T22-1. Then, allow the steel to cool **slowly**. This may be done by turning off the furnace and leaving the metal inside until it is cool. Another method is to put the red-hot metal in an insulating material such as sand or ash. Wait until the metal is cool before removing it.

Nonferrous (non-iron) metals, such as aluminum, copper, and silver, can also be annealed. As these metals are shaped they can become hard from being worked. Work-hardened metals are brittle and difficult to shape. Annealing can improve their workability. The metals are heated and then **quickly quenched** (cooled) in water. See Figure T22-2.

Hardening

Tools made of carbon or alloy (mixed) steel are hardened to improve their strength and to make them last longer. Hardening is done after final shaping and before tempering.

To harden steel, slowly heat it in a heat-treating furnace until it is red-hot. Different types of steels require different heating temperatures. After the steel has reached the proper temperature, remove it from the furnace and quench it rapidly. Carbon steels are quenched in water. Some alloys of steel use oil for quenching. When quenching, move the tool around in the water or oil to provide even cooling. See Figure 22-3.

Johnson Gas Appliance Company

Figure T22-1

Steel can be heated in a heat treating furnace.

Wherry Machine and Welding

Figure T22-2

Hot metal can be quenched by dipping it into either water, as shown here, or into oil.

Heat Treating 389

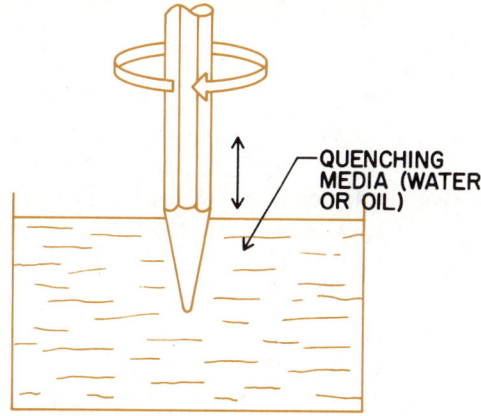

Figure T22-3

Quenching a cold chisel. Note that you move the chisel up and down and in a circular motion.

3. Reheat the metal to a bright red. As the metal is heating, the carbon will sink into the metal, hardening its surface.
4. Quench the metal in cold water.
5. If deeper hardening is needed, repeat the process.

Tempering

Fully hardened steel is brittle (breaks easily). A cold chisel, for example, would crack and break if it were hit hard with a hammer. Tempering makes steel less brittle.

Tempering a Tool

1. Harden the tool.
2. Polish the tool with abrasive cloth. This is done so that **tempering colors** can be seen as the tool is heated. See Figure T22-4.
3. Determine the desired color on a special color chart. Your teacher will help you.
4. Use an oxyacetylene torch to heat the tool. (See Topic 13, **Oxyacetylene Brazing, Welding, and Cutting**.) Direct the flame at the end farthest from the point of the tool. As the metal is heated, it will change colors.
5. Watch the colors "move" along the tool.
6. When the desired color reaches the tip, quench the tool.

Case Hardening

Low-carbon or **mild** steel cannot be hardened by simple heating and cooling. Instead, carbon must be added to the outside of the steel. This is called **case hardening**. Case hardening hardens only the **surface** of the steel. See Figure T22-5.

Case Hardening Low-Carbon Steel

1. Heat low-carbon steel to a bright red color in a heat-treating furnace.
2. Remove the steel, and roll it in a **carburizing compound**, such as Kasenit®, until it is covered with a **case** (shell) of carbon.

Tempering Colors

Color of Heated Metal	Degrees °C	Degrees °F	Tool
Pale Yellow	221-232	430-450	Scribers and Hammer Faces
Full Yellow	243	470	Center Punches
Brown	254-266	490-510	Cold Chisels
Purple	275	530	Screwdrivers

Figure T22-4

Tempering temperatures for various common tools. The heated metal turns different colors for each temperature level reached.

Heat Treating

1. Wear eye protection.
2. Wear protective gloves.
3. Work in a well-ventilated area. Fumes from case-hardening process are poisonous.
4. Handle hot metals carefully.

Figure T22-5

Cross section of case hardened steel. The outer, lighter-colored area is the hardened steel. The inner portion of the steel remains soft.

American Gas Furnace Company

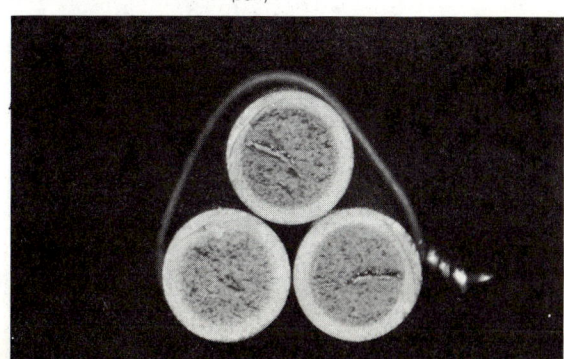

Topic 23 Finishing

Finishes

A **finish** is a coating that is applied to a material for protection or for decoration. See Figure T23-1.

Finishes for **wood** include stains, fillers, sealers, and top coatings such as varnish, lacquer, paint, oil finish, wax, and many special finishes. Wood must be sanded smooth before finish is applied. Flaws show through most finish coatings.

Finishes for **metal** are usually a primer (first or primary coating) and a top coating of paint.

Applying Finishes

Finishes are usually brushed or sprayed onto surfaces. See Figure T23-2. Small parts or parts with surfaces that are difficult to paint with a brush or spray may be finished by dip coating.

For the **dip coating**, the finish is poured into a pan or other container, and the part to be coated is then dipped into the finish. Dip coating usually gives a good, even finish. You must have an area for hanging the object to dry. Wear gloves if you are dipping parts by hand.

Finishing Wood

Finishes are applied to wood to improve or change its appearance. Finishes also make the wood more durable (last longer) and usable.

Stains

Stains are used to **change the color** of wood without hiding the grain pattern. Generally stains darken wood. Common types of stain are oil stains, vinyl stains, and water stains. The most common stain used in the school shop is oil stain.

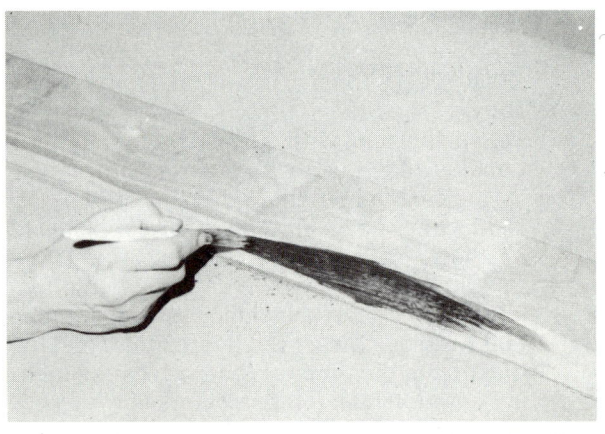

Figure T23-3
Applying stain to a wood surface.

Applying Oil Stain

Oil stain is easy to apply. It can be brushed or wiped on with a clean cloth. See Figure T23-3.
1. To test for color, apply stain first to a scrap piece of the wood that you are using. If you like the natural color of the wood, you may not want to use a stain.
2. Using a brush or a rag, apply the stain over all surfaces of the wood.
3. Allow the stain to soak into the wood. Use a clean cloth to wipe off excess (extra) stain before it dries.
4. Allow the stained wood to dry.

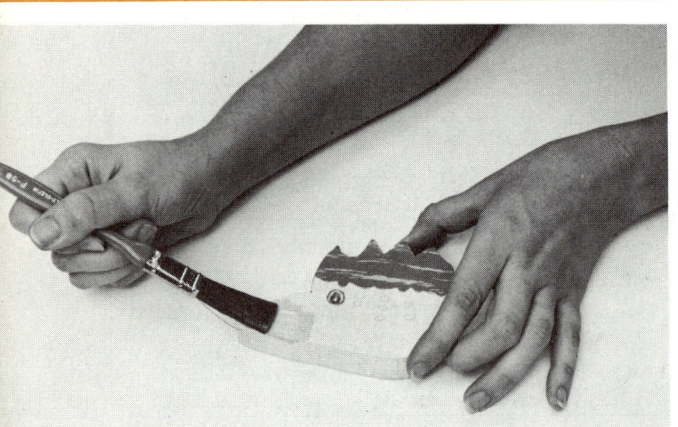

Using a brush to apply paint.

Figure T23-2

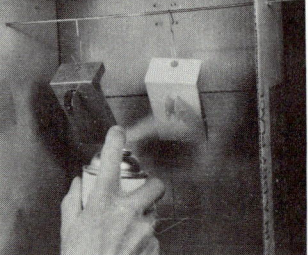

Spray-painting metal.

Figure T23-1
Select the finish that best suits your needs.

Common Finishes	Materials Used On	Solvent*	Drying Time Between Coats	Characteristics Good	Poor
STAIN					
Water	Wood	Water	8 hrs.	Easy to use / Good penetration / Inexpensive	Raises the wood grain**
Vinyl	Wood	Water	1 hr.	Water cleanup / Choice of colors / Brush or wipe on	
Oil	Wood	Mineral Spirits	8 hrs.	Easy to use / Rich color / Brush on and wipe off when desired shade is reached	
Clear Wood Finishes					
Clear Wood Finish (Deft®)	Wood	Lacquer Thinner	20 min.	Can be brushed or sprayed / Shows no brush marks / Easy to apply / Doesn't darken with age / Easy touch-up / Resists water, heat, and alcohol	
Acrylic (Wood Armor®)	Wood	Water	1 hr.	Easy to use / Water cleanup / Dries clear	Two or more coats should be applied
Lacquer	Wood	Lacquer Thinner	15 min.	Usually sprayed, but can be brushed on or applied by dipping	Requires two or more coats / Toxic (poisonous) fumes
Varnish, Polyurethane	Wood Metal	Mineral Spirits	24 hrs.	Clear, tough, hard / Resists oil, water, and alcohol	Hard to touch-up
Shellac	Wood	Alcohol	2 hrs.	Easy to apply / Good penetration / Good sealer	Poor resistance to heat / Not waterproof
PENETRATING OILS					
Mineral Oil	Wood	Mineral Spirits	4 hrs.	Easy to apply / Non-toxic / Good to use on cutting boards	Not permanent
Danish Oil (Watco®)	Wood	Mineral Spirits	4 hrs.	Easy to apply / Toughens wood surface	May discolor finish coat if not sealed
PAINT					
Enamel	Wood Metal	Mineral Spirits	12 hrs.	Waterproof / Can be brushed or sprayed / Tough, hard	
Latex	Wood	Water	4 hrs.	Odorless / Water cleanup	
Lacquer	Metal	Lacquer Thinner	15 min.	Usually sprayed, but can be brushed on or applied by dipping / Resists water, heat, and alcohol	Requires two or more coats / Toxic fumes
WAX	Wood Metal Plastic	—	10 min.	Protects surface / Makes the surface shine / Can be used alone or applied over other finishes	

*Material used to thin the finish and for cleanup.
**Causes the wood fibers to rise up. This makes the surface rough.

Wood Fillers

Wood filler is used to give wood a **smooth, finished surface.** Woods that are **close-grained,** such as pine, cherry, maple, and poplar, have small surface pores (openings) and do not need to be filled. Woods that are **open-grained,** such as oak, mahogany, and walnut, have large surface pores. These pores must be filled.

Wood should be filled after it has been stained and before a finish coat is applied. The filler may be tinted (colored) to match the color of the wood.

Applying Wood Filler

1. Brush the filler into the wood pores by brushing across the grain of the wood.
2. To pack the filler into the pores, rub the surface across the grain with a piece of burlap. See Figure T23-4.
3. Remove filler from the surface of the wood by rubbing it with a cotton cloth. Rub with the grain. Be sure not to remove any filler from the pores.
4. Let the filler dry.

Sealers

Sealers are used to **seal the stain and filler** from the finish coat. Stain can sometimes discolor the finish coat. A sealer also provides a smooth **base** for the finish coat. Without the sealer, more finish coats would probably be needed.

Shellac is the most common material used as a sealer. To be used as a sealer, one part shellac should be mixed with two parts alcohol. It is used most often when the finish coat is varnish.

Finish Coatings

Unmixed **shellac** may be used as a finish. It is applied with a brush. Shellac dries quickly and is very attractive. It is not durable however, and is used mainly for small objects that are not used or handled often.

Varnish, such as polyurethane, is a clear finish. It produces a tough finish that is resistant to heat, oil, alcohol, and water. Varnish is available in three gloss types. **Gloss** varnish dries very shiny, **Semigloss** varnish dries to a medium shine. **Satin** varnish dries to a **flat** (non-shiny) finish.

Applying Varnish

1. Prepare the surface for varnishing. It should be smooth, clean, and dry. The wood should be stained, filled, and sealed if needed.
2. Thin the first coat of varnish with turpentine. Read the label on the can of varnish for proper thinning instructions.

Figure T23-4

Rubbing tinted wood filler into a stained wood surface.

Finishing **393**

Figure T23-5
Applying varnish as a finish coating.

3. Apply a coat of varnish with a soft brush. Spread the varnish by brushing across the grain. Smooth the varnish by brushing with the grain. See Figure T23-5.
4. When the varnish is dry, lightly sand the surface with very fine abrasive paper (See Topic 15, **Filing and Sanding**.)
5. Apply another coat of varnish. Varnish should not be thinned.
6. Repeat steps 4 and 5. After the third coat of varnish, allow the surface to dry thoroughly.
7. Wax the surface to protect the finish.

Lacquers

Lacquers are very fast-drying and provide an attractive finish. Lacquer finishes are thin, however. The best way to apply lacquer is to use a spray gun. They are clear and have moderate (medium) durability. Lacquers are available as gloss or flat finishes and are thinned with lacquer thinner.

Spraying Lacquer

For best results, lacquer should be applied with spray equipment. See Figure T23-6. This generally includes a spray gun and an air compressor. Special lacquers are available, however, that may be applied with a brush.

1. To spray lacquer, thin it with lacquer thinner. Read the label on the can for proper thinning procedures.
2. Set up the spray equipment. This will require (1) filling the **spray gun cup** with the lacquer, (2) attaching the cup to the **gun,** (3) adjusting the air pressure from the **air compressor** to the gun, and (4) adjusting the **spreader adjustment valve** and **fluid adjustment valve** on the gun. Ask your teacher to help you make the adjustments.
3. Using back and forth strokes, spray the product. Keep the nozzle of the gun about 10" from the surface of the product. Keep the gun moving. Do not spray too much finish on one spot.

   The gun is controlled by pulling the **trigger** as you move the gun in a straight line in front of the object. You release the trigger as you move the gun past the object. Begin each pass about halfway back into the previous pass. Continue this overlapping method until the entire surface is covered with a thin coat.

   The technique for controlling a spray gun will take a little practice. Once the method is learned, it is very easy to apply smooth, even coats of finish.
4. Apply several coats of lacquer. Allow each coat to dry, and lightly sand between coats.
5. When finished with the spray gun, clean it with lacquer thinner.
6. Allow the final coat to dry thoroughly.
7. Wax the surface to protect the finish.

Figure T23-6

Portable spray equipment.
Binks Manufacturing Company

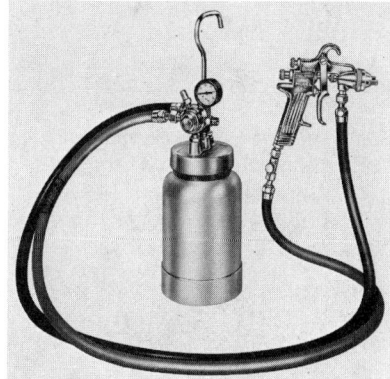

A spray gun setup.

A compressor.

Binks Manufacturing Company

Penetrating Oil Finishes

Penetrating oil finishes include mineral oil, Danish oil, tung oil, and many linseed-oil or tung-oil mixtures. Oil finishes do not lay on the top of the wood as do varnish, lacquer, or paint. Instead, the oil soaks into the wood to provide a durable, natural finish. Mineral oil is not used on most wood products. However, it is used on special items such as salad bowls or cutting boards. All oil finishes are easy to apply. See Figure T23-7.

Figure T23-7

Applying a penetrating oil finish using a soft cloth.

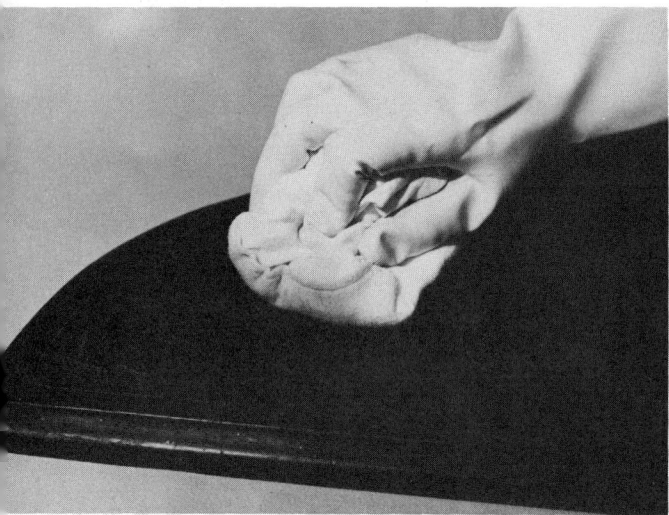

Applying Penetrating Oil Finish

1. Wipe or brush large amounts of the oil finish on all surfaces of the product.
2. Allow the oil to soak in. Then, add more oil.
3. After a few minutes (read the label on the can), wipe off the excess (extra) oil. This will leave the finished surface.
4. For more of a build-up and to produce a deep finish, more coats of oil can be applied.
5. After the oil is completely dry, the surface may be waxed.

Paints for Wood

Most paints are applied by spraying, brushing, or dipping. Many bright and shiny paints can be used to create very colorful products.

Paints for wood include Latex paints and enamels. **Latex paints** can be thinned and cleaned up with water. **Enamels** are thinned with mineral spirits. Enamels are generally more durable than latex paints.

See Figure T23-8.

Wax

Wax is a coating that is applied over a finish. It protects the surface from wear and also improves the shine of the finish. It is applied by simply wiping it on. See Figure T23-9. After it is dry, it is buffed to bring out the shine.

Figure T23-8

Using solvent to clean a paintbrush. Select the proper solvent for the type of paint you are using.

Figure T23-9

Applying a protective coating of paste wax to a finished surface.

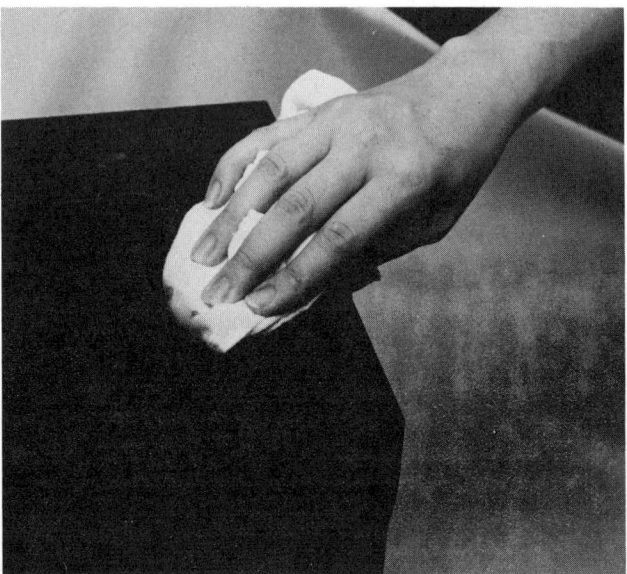

Figure T23-10
Painting metal.

Applying a coat of primer before painting.

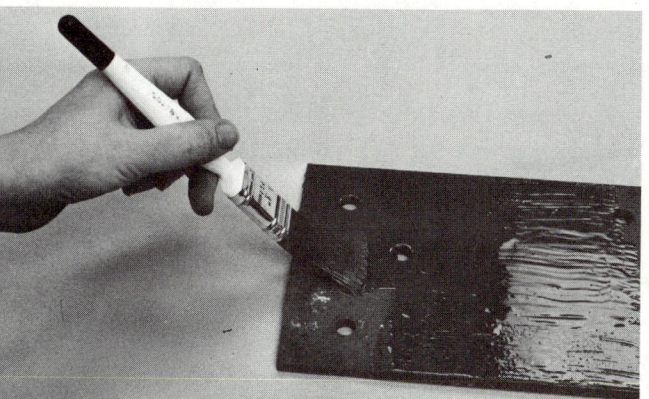

Applying paint after the primer coat is completely dry.

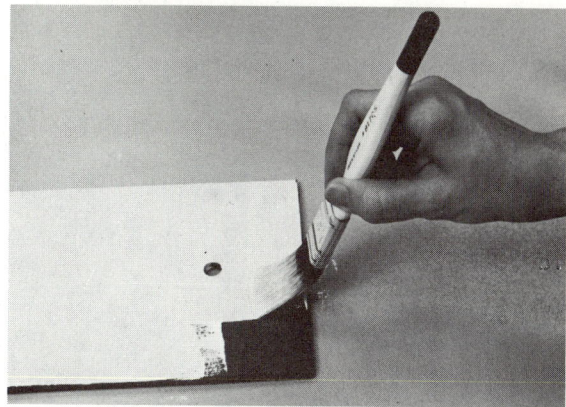

Finishing Metal

Metal is finished to keep it from oxidizing (rusting) and to make it more attractive. Some metals do not have to be finished. Gold and silver, for example, do not oxidize and are attractive without a finish. Steel, however, rusts very easily and is not very attractive without some kind of coating.

Paints for Metal

Paint is the major finishing material used on metal. Paint used on metal is generally lacquer or enamel. These are available in spray cans and in liquid form. Liquid paint is usually applied with a brush or with spray equipment. See Figure T23-10.

Paints come in a wide variety of colors. Clear paint is also available. It is applied to attractive metals, such as copper, aluminum, and brass. It protects them without hiding their surfaces.

Applying Paint to Metal

1. Prepare the surface. It must be clean and dry. Remove any oxidation. Lightly sanding the surface will ensure a good bond between the surface (metal) and the paint.
2. Apply a primer coat. A primer reduces the chances of oxidation and provides a good base for the paint.
3. After the primer is dry, apply the paint. Apply two thin coats rather than one heavy coat.
4. Lacquers dry rapidly. Another coat may be applied within minutes. Enamels dry much more slowly. Overnight drying is recommended between coats.

Applying Finishes

1. Store rags with finishes on them in closed metal containers.
2. Make sure fire extinguishers are near the area where finishes are being used.
3. Work in a well-ventilated area.
4. Keep paints away from open flames.
5. Wear eye protection.
6. Wear a respirator (face mask) when spray painting.

Other Finishing Processes

Silk-Screen Printing

Silk-screen printing can be used to print designs on a product or a package. A stencil (cut-out pattern) is adhered (made to stick) to a silk or polyester screen held in a frame. Ink or paint is forced through the stencil onto a material. Silk-screen printing can be done on wood, metal, plastics, cloth, and paper.

Generally lacquer film stencils are used in the shop. The film has two layers: a gelatin layer and a backing.

Tools, Materials, Processes

Figure T23-11 Kiss Screen Printing
A silk-screen and stencil in position for printing a design on a T-shirt.

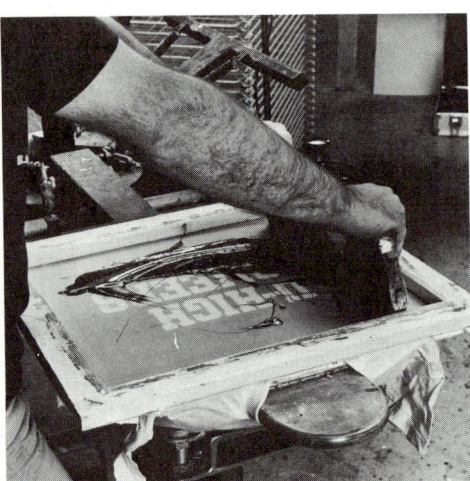

Figure T23-12 Kiss Screen Printing
Using a squeegee to force ink through the stencil and onto the T-shirt.

Figure T23-13 Kiss Screen Printing
After the ink is completely dry, the T-shirt is ready to wear.

Silk Screen Printing Procedure

1. Cut the design into the gelatin layer of the stencil using a sharp knife.
2. Carefully peel the gelatin from the backing. Remove only the areas that are to be printed.
3. After making sure that the screen is clean, place the gelatin side of the stencil against the underside of the screen.
4. Adhere it to the screen by rubbing it gently with a cloth dipped in lacquer thinner. Rub only a small area of the stencil at a time.
5. Allow the stencil to dry. Drying generally takes about 20 minutes.
6. Peel off the backing.
7. Mask off (tape over) all parts of the screen where unwanted ink could come through.
8. Place the paper, cloth, or other material to be printed underneath the screen, and lower the screen. See Figure T23-11. You should practice on some scrap paper before you start printing on your material.
9. Apply a small amount of ink to the top of the screen.
10. Use the **squeegee** to pull the ink toward you. See Figure T23-12. Press down as you pull. This forces the ink through the stencil and onto the material. Try to use only one stroke.
11. Slowly lift the frame off the material.
12. Remove the printed material, and allow the ink to dry. See Figure T23-13.
13. When you have finished printing, use lacquer thinner to clean the screen and squeegee. If the screen is cleaned thoroughly, it can be used over and over again.
14. Once you start printing, do not stop for more than a few minutes or the ink will dry on the screen. If a delay is necessary, clean the screen before leaving it.

Ceramic Enameling

Ceramic enameling is the process of coating metal with glass. Powdered glass is applied to a metal surface and then melted to form a coating. See Figure T23-14. Products such as stoves, refrigerators, and bathtubs have ceramic coatings.

Copper enameling is a type of ceramic enameling done on copper products, such as belt buckles. The powdered glass is available in many colors. The metal must be heated to a temperature of about 1,500° F (815° C) in order to melt the glass. An enameling **kiln** (oven), shown in Figure T23-15, or welding **torch** can be used to provide the necessary heat.

Applying Ceramic Enamel

To apply ceramic enamel to copper:
1. Cut and form the copper to the shape that you want.
2. Clean it with fine steel wool until the surface is bright and shiny.
3. Apply a light coat of mineral oil to the surface of the copper.
4. Sprinkle on the powdered glass.
5. Place the copper on a piece of corrugated sheet metal, and place it in the kiln (oven). Or if you are using a torch as the heat source, use a #1 or #3 tip with a neutral flame (See Topic 13, **Oxyacetylene Brazing, Welding, and Cutting**.) Begin to heat the bottom of the copper with the end of the flame. Keep the tip about 8″ away from the surface of the copper.
6. Watch the powdered glass. It will start to flame as the oil is burned away. Then it will begin to turn to liquid on top and will darken in color. Keep heating the metal until it has turned to liquid all the way to the edges. No color will be visible at this time. The color will return as the product cools.
7. Remove the metal from the kiln, or stop heating it with the torch. Allow it to cool thoroughly before touching it. Do not cool it in water. The glass will crack if cooled too quickly. See Figure T23-16.

Figure T23-14

Glass powders for ceramic enameling.

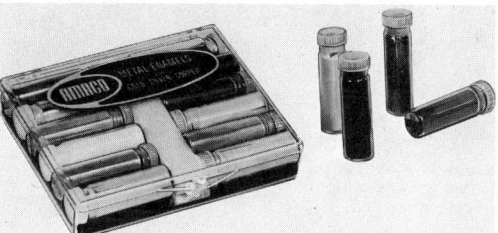

American Art Clay Co.

B. Jadow and Sons, Inc.

Figure T23-15

An enameling kiln.

Figure T23-16

An enameled belt buckle.

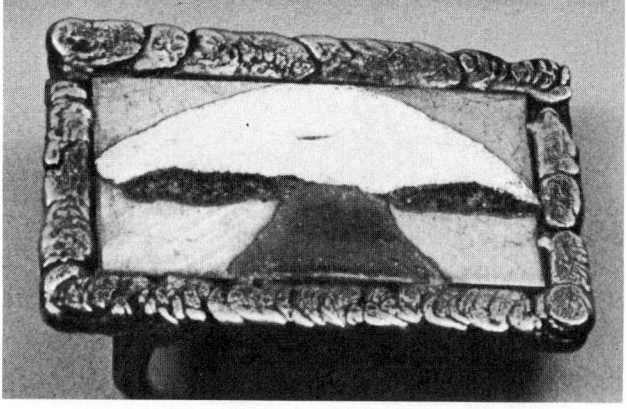

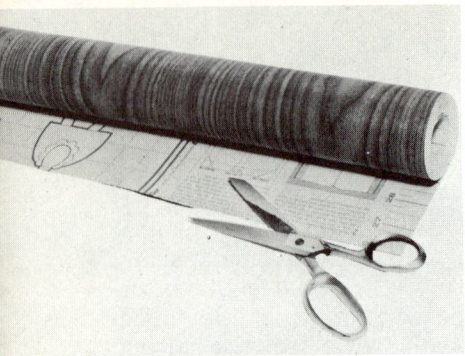

CON-TACT® Brand Self-Adhesive Plastic

Figure T23-17
Self-stick sheet vinyl plastic.

Figure T23-18
Applying sheet vinyl.

A. *Remove the paper backing and smooth the plastic sheet onto the workpiece.*

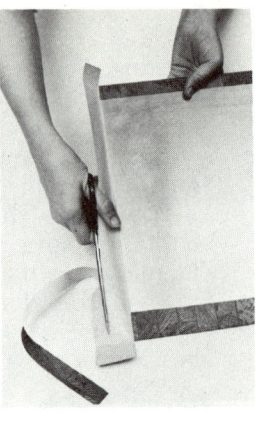

B. *Carefully trim excess plastic and finish the edges.*

Sheet Vinyl

Sheet vinyl (self-stick) has an adhesive coating on one side. (See Topic 28, **Plastics**.) It sticks easily to most surfaces. A protective backing paper must be removed before the vinyl can be applied to a surface. See Figure T23-17.

Sheet vinyl is useful as a coating for materials that do not have an attractive natural appearance. A metal lamp shade, for example, looks good with a sheet-vinyl covering. Apply vinyl carefully. You may need to practice on some scrap material first.

Applying Sheet Vinyl

1. Cut the material slightly larger than the finished dimensions.
2. If you are coating a product, place the product on the vinyl for a "dry run" in folding the vinyl over the product.
3. Remove the paper backing, and lay the vinyl carefully on the surface to avoid folds and air bubbles. See Figure T23-18.
4. Start adhering the vinyl in the center and work toward the edges. Stretch it slightly as you stick it down. If air bubbles develop, release the air by puncturing the bubble with a needle.
5. Use scissors or a sharp knife to trim the excess sheet vinyl from the edges.
6. Smooth the vinyl making certain the edges are stuck tightly.

Topic 24 Clamping

Materials are clamped to hold them solidly in position during processing. For example, wood is often clamped while it is being sawed or sanded. Clamps are also used to hold materials together when they are being assembled or glued. Clamping devices include a variety of clamps and vises.

Clamps

Spring Clamps

Spring clamps are used to grip objects that do not require a great amount of holding power. See Figure T24-1. Some spring clamps have jaws that are coated with vinyl to protect work surfaces from being dented.

Using a Spring Clamp

A spring clamp can be applied quickly and easily. You do not have to adjust it. See Figure T24-2.

If you plan to use a spring clamp to hold glued parts together, make sure that the parts fit closely. A spring clamp will not apply enough pressure to force materials together.

C-Clamps

C-clamps can apply a great amount of pressure on a **small area** of wood, metal, or plastic material. C-clamps come in a variety of shapes and sizes. See Figure T24-3. The size of a clamp is measured by the maximum opening between the **jaws.** The lower jaw is moved by turning the **screw handle.** It swivels to provide secure clamping on irregular or slanted surfaces.

Using a C-Clamp

1. Open the jaws of the clamps slightly wider than the size of the work.
2. Put together the pieces that are to be clamped.
3. To protect the work surfaces from being dented, place protective pads of hardboard or cardboard between the work and the clamp jaws. See Figure T24-4.
4. Tighten the clamp just enough to hold the work in place. When using glue, clamp tightly enough to hold the glued materials together without forcing the glue out of the joint.

Figure T24-1
A spring clamp.

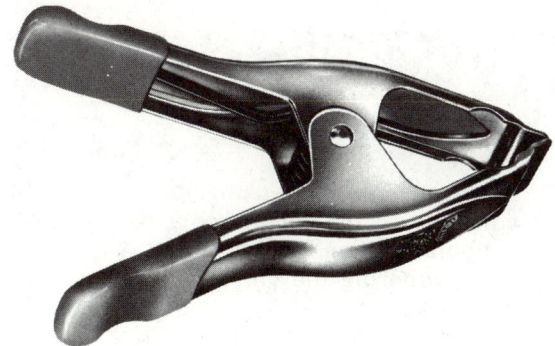

Adjustable Clamp Company

Figure T24-2
Using spring clamps to secure pieces in a rounded assembly.

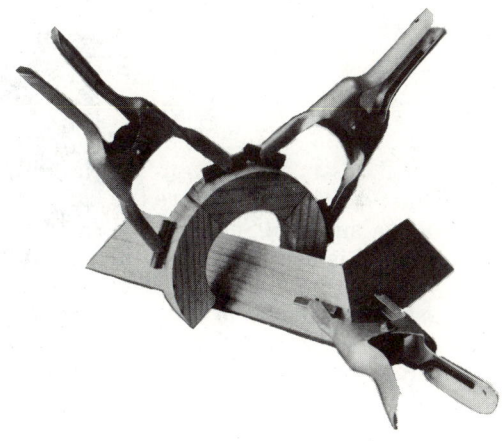

Arvids Iraids

Figure T24-4
Using a C-clamp to secure a workpiece to a drill press table.

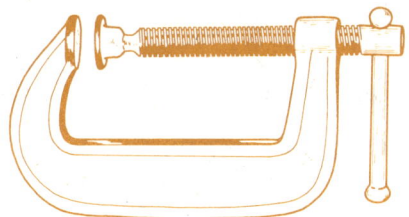

A regular-throat clamp.

 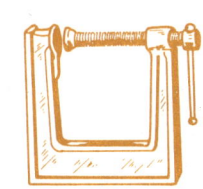

A deep-throat clamp.   A square-throat clamp.

Figure T24-3
Types of C-clamps.

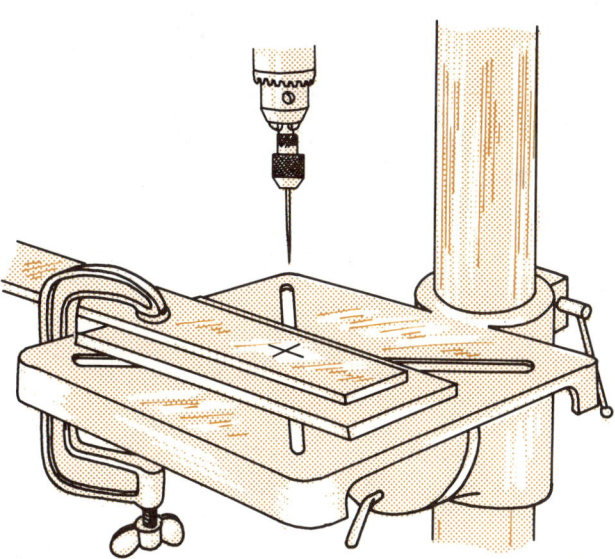

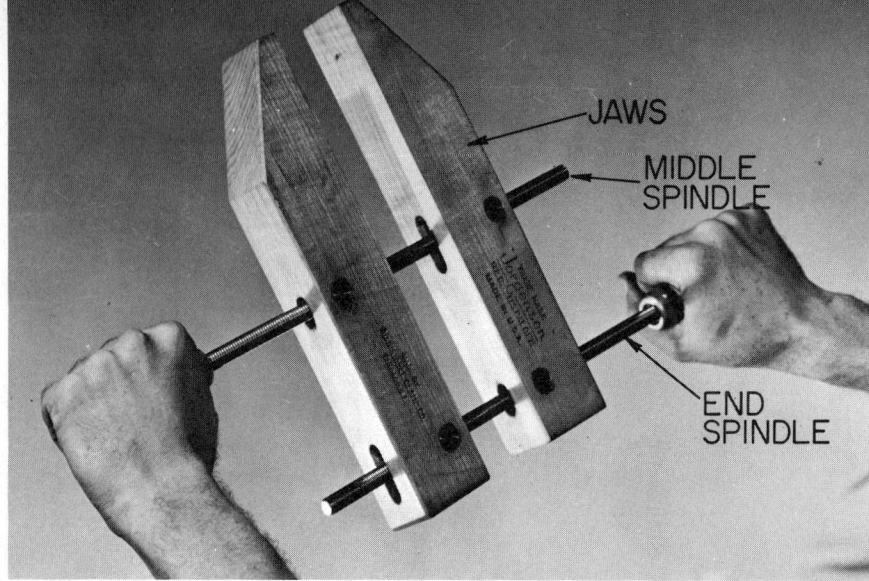

Figure T24-5
The parts of a hand screw clamp.

Adjustable Clamp Company

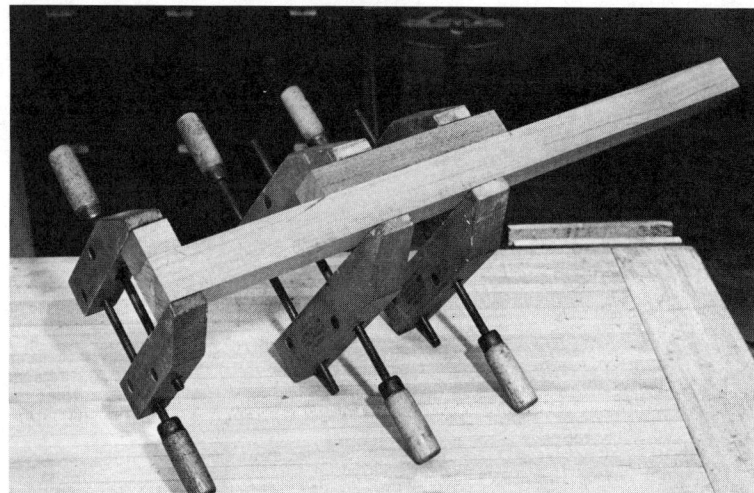

Figure T24-6
Using hand screws to secure assembled pieces.

Hand Screws

Hand screws are also called hand-screw clamps. They are wooden clamps. Hand screws are used chiefly for holding **wooden parts** together as they are being glued or assembled. Hand screws have movable wooden **jaws.** Each jaw has two steel **spindles.** The spindles are used to open and close the jaws. The spindles have **handles** to give you a better grip. See Figure T24-5.

Using a Hand Screw

When preparing to clamp stock together for gluing:
1. Assemble the parts without glue.
2. Adjust all hand screws to slightly wider than the size of the work to be clamped.
3. Check to be sure that the clamp jaws are parallel (same distance apart).
4. Waxed paper may be placed between the jaws of the clamp and the work surface. This protects the work surface from glue and damage.
5. Clamp the hand screw onto the wood. See Figure T24-6.
6. Wipe all excess glue from the work and clamps.

Bar Clamps

Bar clamps are used to glue **wide boards** together. See Figures T24-7 and T24-8. For example, when the edges of cutting boards and tabletops are glued, bar clamps are often used. Bar clamps can be adjusted to fit both **small and large objects.** The same clamp may be used for products of many sizes when they are glued or assembled.

A bar clamp is made up of a long metal bar (2′ to 8′ long) with a **clamp jaw** at each end. On one end, the jaw has a **screw.** Pressure is applied to pieces being glued by turning a **handle** on the screw part of the clamp. The jaw on the other end is movable. It is clamped close to the work before pressure is applied.

Clamping **401**

Figure T24-7

A bar clamp.

Adjustable Clamp Company

Figure T24-8

Using bar clamps to hold glued boards together.

Adjustable Clamp Company

Using a Bar Clamp

1. Assemble the pieces without glue.
2. Set the clamp jaws to the correct opening.
3. Slip a scrap of wood between the board and the clamp jaws to prevent damage to the board.
4. Apply glue to the boards.
5. Clamp the boards together with the bar clamp.
6. Alternate the jaws from one side to the other to help prevent warping. See again Figure T24-8.

Band Clamps

Band clamps are designed to clamp wooden objects that have **irregular shapes**. This clamp is made up of a flexible **band,** usually canvas, and looks much like a belt. See Figure T24-9. It is laced through a steel **base** and is tightened by a screw.

Using a Band Clamp

When gluing objects together, adjust the band to the correct size before applying glue to the wood. Waxed paper may be placed between the wood and the band to keep the band from being glued to the wood. Avoid applying too much pressure with the clamp. This will force the glue out of the joints.

Miter Clamps

Miter clamps are designed to clamp two pieces of **wood** glued to form a **miter joint** (square corner). Picture frames are a good example of stock that can be held together in a miter clamp. See Figure T24-10.

Using a Miter Clamp

1. Position the wood in the clamp before applying glue.
2. Make certain that the corner is square (90°).
3. Clamp one piece of wood in position.
4. Apply glue to both surfaces that are to be joined.
5. Clamp the second piece of wood in position.

Figure T24-9

Using band clamps to hold the parts of a cabinet together after gluing.

Figure T24-10

A miter clamp is used to secure corner joints. Here the glued corner of a picture frame is being held by a miter clamp.

Vises

Using a Woodworking Vise

A woodworking vise is made to hold **wide, flat stock,** such as standard-size lumber. It is mounted on a workbench. See Figure T24-11. A woodworking vise may have "quick action." This means that it can be opened and closed by simply pushing or pulling the vise jaw. After the jaw is moved close to the object to be clamped, the handle is used to tighten the vise.

A vise may have a metal dog on top. This is a device that can be pulled up to clamp materials on top of the bench. A bench stop is placed in a hole in the bench. Material is clamped between the vise dog and the bench stop.

Using a Bench Vise

A bench vise is also called a **machinist's vise.** It is used to hold **metal** as it is being worked. See Figure T24-12. A bench vise may have a swivel base (base that turns).

One jaw of a bench vise is stationary (does not move). The other jaw is adjustable. It can be moved in or out by turning a handle. The jaws of the vise will damage soft materials. To prevent this, place **vise-jaw caps** (protective caps) on the vise jaws. These are usually made out of soft aluminum.

The bench vise can apply great pressure. It is often used for straightening small pieces of band iron.

Using a Drill Press Vise

A drill press vise is used to securely hold **work that is being drilled** on the drill press. See Figure T24-13. Place a wood block under the work. This will keep the drill from hitting the vise. Many drill press vises have a "V" groove in one of the jaws to hold round stock. A vise should be clamped to the drill press table if accurate or heavy machining is to be done.

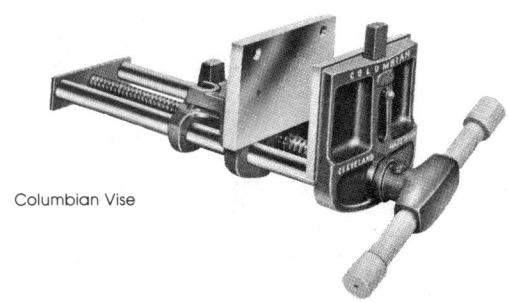

Figure T24-11
A woodworking vise.

Figure T24-12
A bench vise.

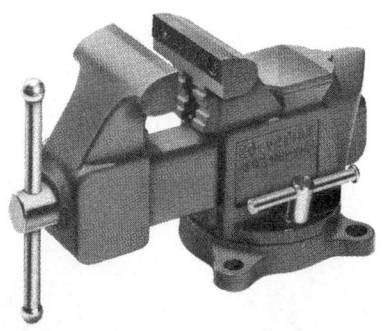

Figure T24-13
A drill press vise.

Topic 25 Production Tooling

Many times, special tools are used in manufacturing. Making special tools for use in production is called **production tooling**. Some of these special tools are **fixtures, jigs, templates,** and **gages**. These special tools help to hold, guide, and lay out work so that it can be done quickly and accurately. These tools are useful when mass-producing a product and when making only one product.

The Special Tools

Using Fixtures

Fixtures are **devices** used to **hold** materials in place during drilling, shearing, sawing, and assembling operations. Production fixtures are specially designed for certain jobs. Often they are designed for only one type of product.

Using Jigs

Jigs are **devices** that serve two purposes. Not only are they used to **hold** materials in place, they are also used to **guide** cutting tools into materials at the right locations. A jig is frequently used with a portable power tool, such as a portable electric drill.

Using Templates

Templates are **patterns** generally used during layout operations. (See Topic 2, **Layout**.) They are used for **marking** desired shapes and hole locations on work materials. Templates are very common in production jobs. They are usually made of sheet metal or hardboard.

Using Gages

Gages are special **measuring** tools. (See Topic 2, **Layout**.) They are used whenever the same dimension must be measured over and over again. Gages make it possible to **check** a dimension or size without having to make measurements each time.

A **plug gage** is used to measure the **diameter** of a drilled hole. A **depth gage** is used to measure the **depth** of a drilled hole. These two gages, are useful for checking how accurately a product is made. They can even be combined into one as shown in Figure T25-1.

Production Tooling for Accuracy

Fixtures, jigs, templates, and gages must be made carefully. If a template, for example, is the wrong size or shape, all parts made from that template will also be the wrong size or shape. See Figure T25-2. Accuracy is very important in any kind of production tooling.

A few of the more common types of fixtures, jigs, templates, and gages used in the school shop are explained and shown here. You can design and make others to meet your own special needs for tooling.

Figure T25-1
A combination plug and depth gage.

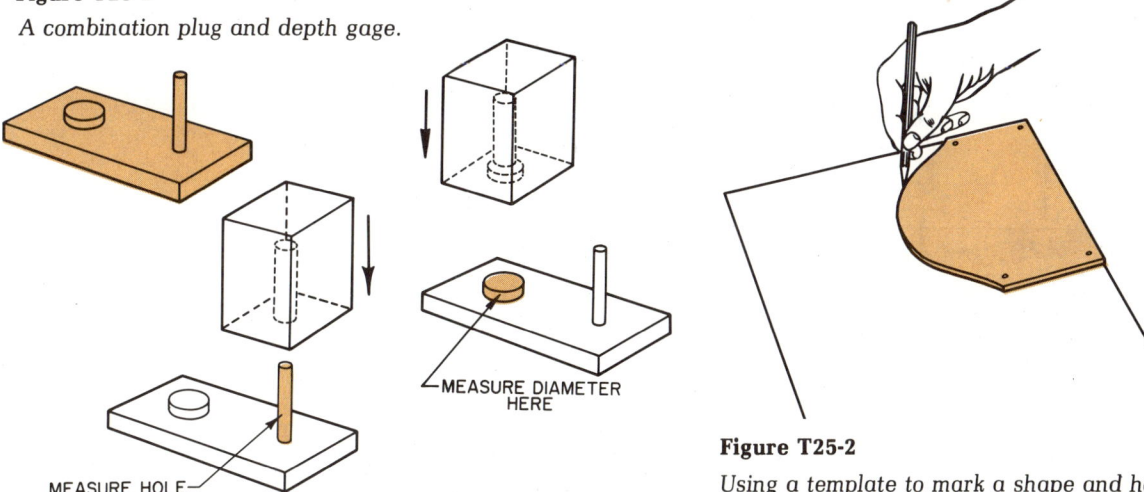

Figure T25-2
Using a template to mark a shape and hole locations.

Jigs and Fixtures for Sawing and Sanding

Sawing Equal Lengths

Wood for many different products can be positioned and cut correctly with the use of a **stop block.** See Figure T25-3. A stop block is a block of wood that is used to hold a board at a certain distance from the saw blade.

With a stop block, you can **cut many pieces of wood the same length.** Set the stop block so that the distance between it and the saw blade is equal to the board length you wish to make. Make sure the measurement between the stop block and the blade is **correct.** Move one end of the board against the stop block, and make the cut.

You can obtain this same board length over and over simply by moving the end of the board against the stop block and repeating the saw cut. This fixture can be used on a miter saw, a radial arm saw, a band saw, or a table saw.

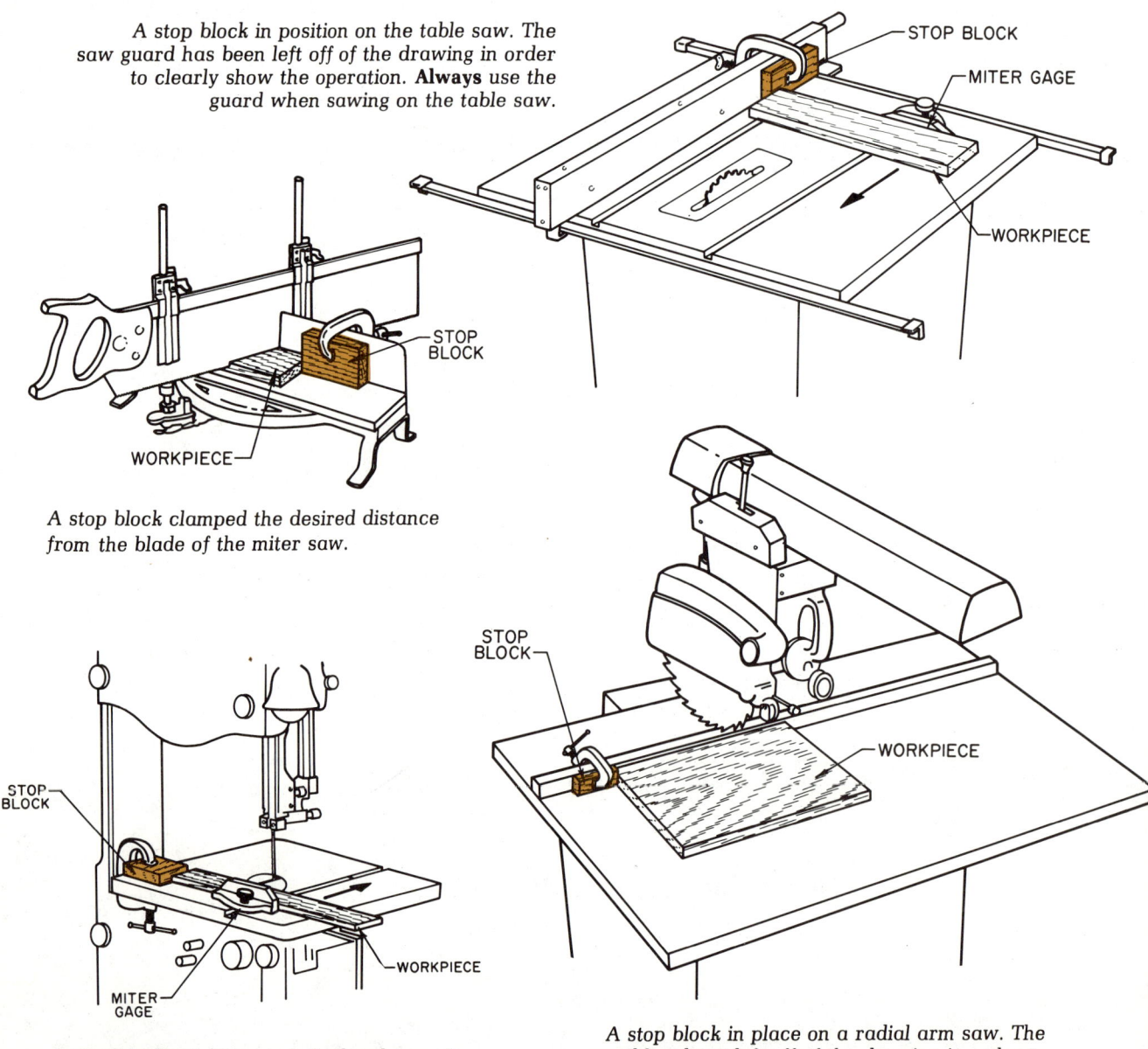

A stop block in position on the table saw. The saw guard has been left off of the drawing in order to clearly show the operation. **Always** *use the guard when sawing on the table saw.*

A stop block clamped the desired distance from the blade of the miter saw.

A stop block set for use on the band saw. A miter gage is used to guide the workpiece into the saw blade.

A stop block in place on a radial arm saw. The guard has been left off of the drawing in order to clearly show how the wood is cut. **Always** *use the guard when sawing on the radial arm saw.*

Figure T25-3

A stop block is used as a guide when sawing boards into equal lengths.

If it is necessary to set up a stop block more than once for the same distance, you might want to make a **setup gage.** See Figure T25-4. The setup gage should be cut or marked at the desired length of the cut pieces. You simply butt (set squarely) one end of the setup gage against the saw blade and position the stop block at the other end. The saw is now set for the proper length of cut. This saves you from having to measure this distance with a rule each time.

Sawing a Taper

A piece of wood, called a **taper jig,** is shaped to the desired taper. See Figure T25-5. A board is placed against the tapered edge. The board and the taper jig are moved against the fence as a unit. The saw cuts the part of the board that extends past the edge of the taper jig. This produces a taper on the board. Saw carefully, keeping the stock firmly in the jig and against the fence. Be **sure** the guard is in place.

Sawing a Slot

Sometimes a sawed slot is used for decoration on a product. This slot is a **saw kerf.** (See Topic 5, **Sawing**.) For example, salt and pepper shakers made from blocks of wood may be decorated in this manner. A fixture can be fastened to the band saw table to **locate** the block. See Figure T25-6. The fixture holds the block as each side is placed against the saw blade and sawed. Using the fixture will ensure that the saw kerf is cut evenly on all sides. The **depth** of the saw kerf will also be the same.

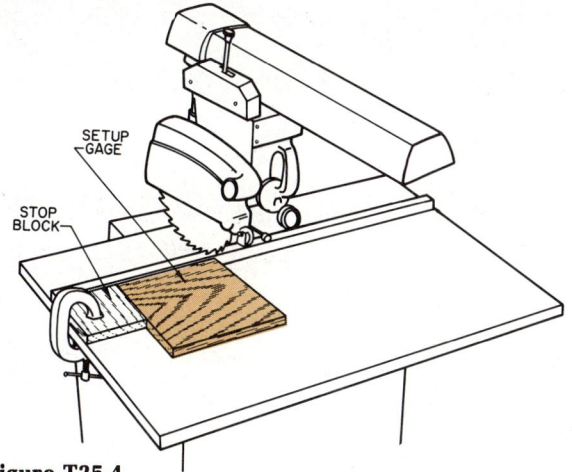

Figure T25-4

A setup gage being used to correctly set a stop block on a radial arm saw. The guard has been left off of the drawing in order to clearly show the positioning of the setup gage. **Always** *use the guard when sawing on the radial arm saw.*

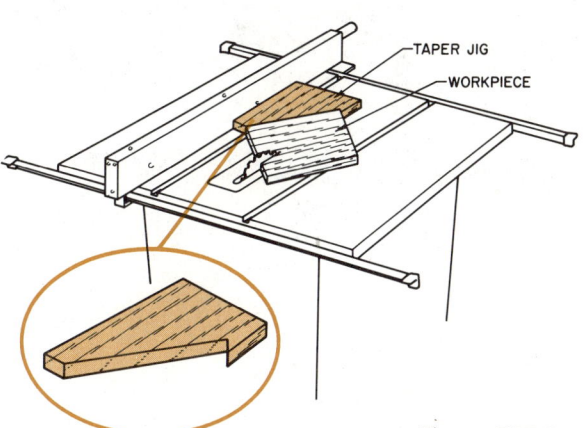

Figure T25-5

A taper jig used to position the workpiece and to guide it into the saw blade. The guard has been left off of the drawing in order to clearly show how to cut the wood. **Always** *use the guard when sawing on the table saw.*

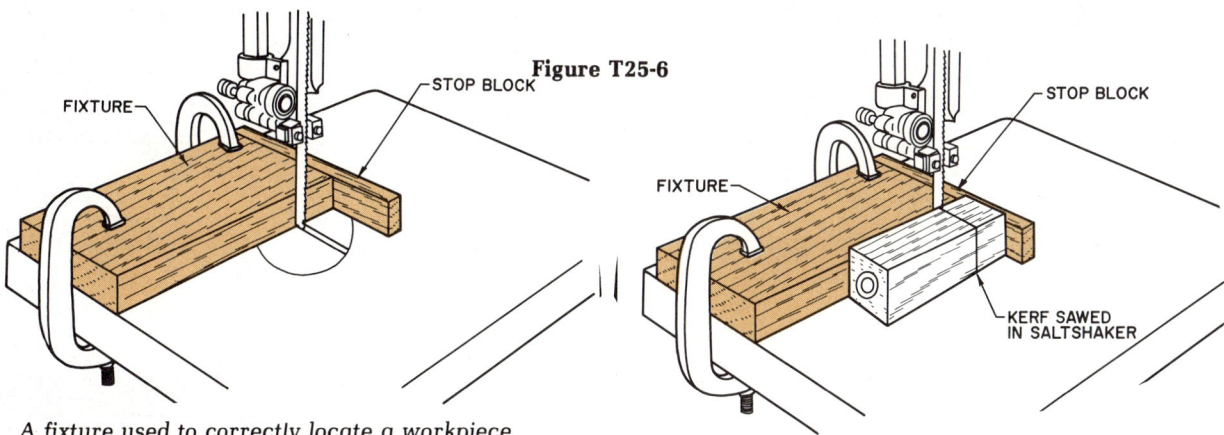

Figure T25-6

A fixture used to correctly locate a workpiece on a band saw. Slots can be cut evenly into all four sides of a wood block.

A workpiece in position against the fixture.

Sawing and Sanding Wooden Circles

Cutting a piece of wood into a perfect circle on the band saw can be difficult. To make this job easier and to make a better circle, a **circle-cutting jig** can be made. The jig consists of a plywood base with a **pivot point** mounted on the surface of the base. This may be a nail or dowel. A pivot point may be made to be adjustable for cutting circles of various sizes. See Figure T25-7.

The jig is clamped on the saw table. When the wood is mounted on the pivot point and turned, the **saw** cuts a circle. Cut the circle a little larger than finished size.

To sand the edges of a newly cut circle, a jig much like this one can be made and used with the **disk sander.** See Figure T25-8. Sand the circle to its finished size.

Figure T25-7

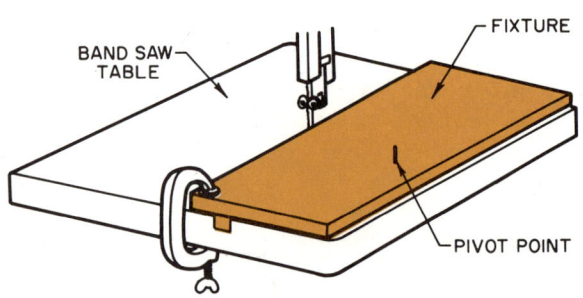

A circle-cutting jig in position on a band saw table. The center of the workpiece will be placed on the pivot point (nail). Note that the pivot point is lined up with the cutting edge of the band-saw blade.

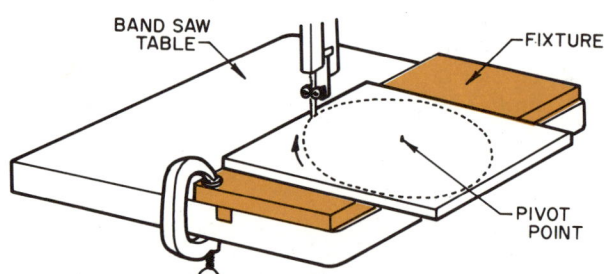

Workpiece in position on the circle-cutting jig. Slowly rotate the workpiece on the pivot point, guiding it into the saw blade.

Figure T25-8

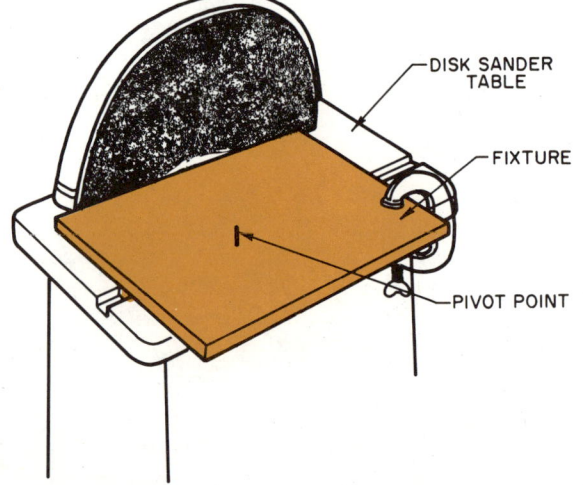

A circle-sanding jig clamped to the disk-sander table. The workpiece is rotated on a pivot point.

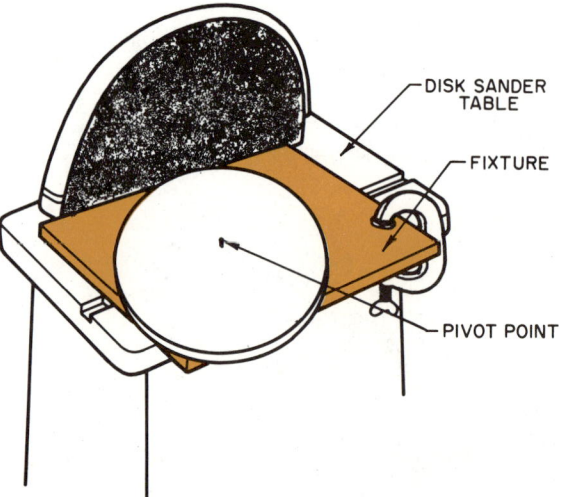

Workpiece in position on circle-sanding jig. Turn the work against the side of the sanding disk that is moving downward.

Production Tooling 407

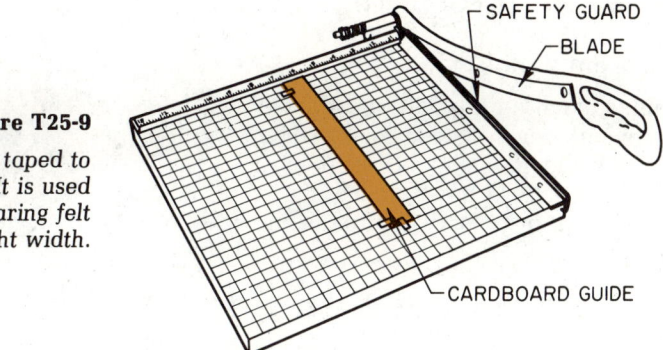

Figure T25-9

A piece of cardboard taped to a papercutter table. It is used as a guide when shearing felt to the right width.

Figure T25-10

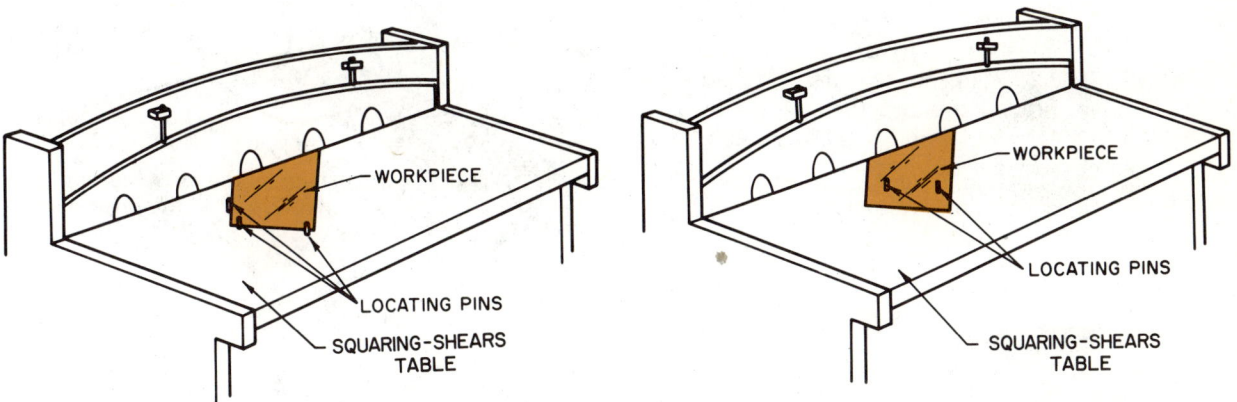

Locating pins placed in the squaring-shears table. A workpiece can be positioned against these pins.

Locating pins placed up through the squaring-shears table. The workpiece can be positioned over the pins. Properly spaced holes must first be made in the workpiece.

Jigs and Fixtures For Shearing

Shearing Felt

Cutting large amounts of felt to size is difficult using a scissors. A paper cutter with a **locating fixture** (cardboard guide) can be used to do the work rapidly. See Figure T25-9.

Tape a cardboard guide on a paper-cutter table. Place the edge of the felt against the cardboard. Lower the blade, shearing the felt to the correct width.

Shearing Sheet Metal

Locating pins (guides) in a squaring-shears table are useful in some mass-production jobs.

Paper Cutter
1. Use the safety guard if the paper cutter has one.
2. **Never** place your hands or fingers under the blade.
3. Keep the cutting blade locked to the base when the cutter is not in use.

(Machine screws can serve as pins.) If there are locating holes in the sheet metal, these can be positioned **over** the pins. Material can also be located **against** the pins. See Figure T25-10.

Jigs and Fixtures For Drilling

Drilling Several Small Holes

Some products, such as the Salt and Pepper Shakers, need to have several small holes drilled. A drilling jig may be made to be used as a guide. See in Figure T25-11 how the drilling jig is used to guide the bit or twist drill into the end of a block of wood.

Drilling Holes Equal Distances From an Edge

Suppose that you wish to drill axle holes in several pieces of wood for race cars. All of the axle holes need to be the same distance from the bottoms of the cars. A drill-press fixture is needed.

The fixture for locating the wood for drilling the axle holes could look like that shown in Figure T25-12. The bottom of the wood to be drilled is placed against the guide. By doing this, every piece of wood will be drilled exactly the same distance from the edge of the wood.

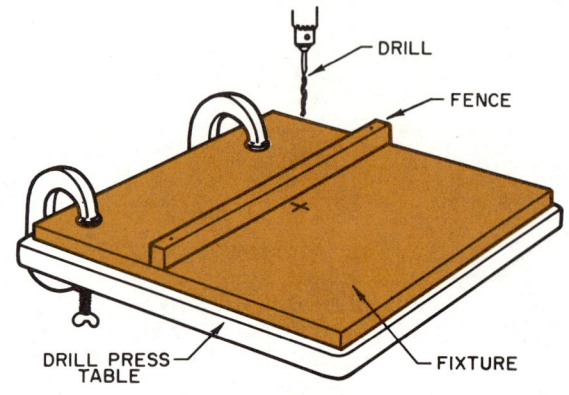

A fixture clamped to a drill press table to position a workpiece for drilling.

Figure T25-12

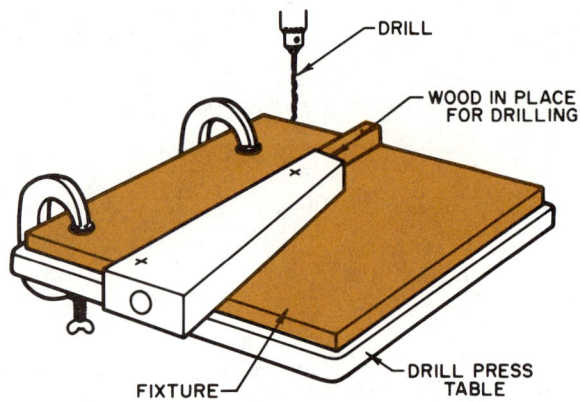

Workpiece in position against the drill-press fixture. Now the drill can be used to make holes the same distance from the edge of the wood.

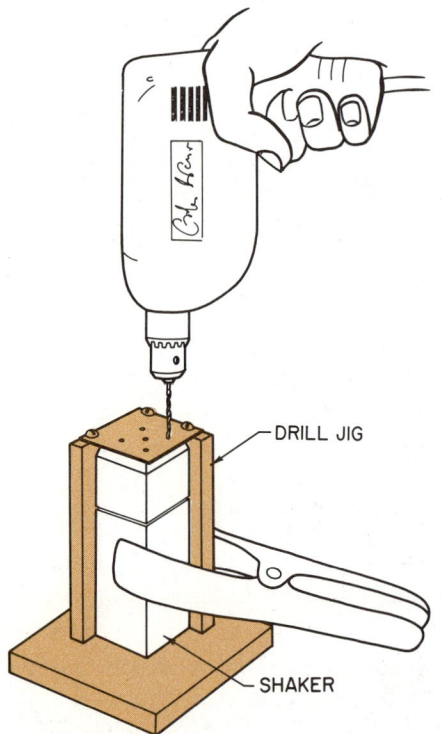

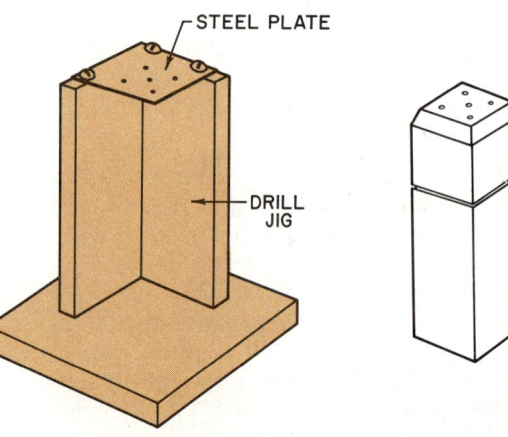

Figure T25-11

A drilling jig used when drilling holes in the top of a salt or pepper shaker. The workpiece should be clamped to the jig while being drilled.

Drilling Holes of Different Depths

Another product, the Whistle, requires that two holes be drilled in the end of a piece of wood. One hole is drilled to be deeper than the other. The fixture shown in Figure T25-13 will help you place the wood in the right position. The first hole is drilled at point **A**.

After this hole is drilled, the spacer under the wood is removed. The material is then positioned for drilling a hole at point **X**. Because the spacer has been removed, the hole drilled at point **X** will not be as deep as the hole drilled at point **A**.

Drilling or Boring Evenly Spaced Holes

A Marble Shoot game needs six evenly spaced holes bored partway through the board. A simple fixture makes repeated drilling (or boring) easy. See Figure T25-14.

Using Toggle Clamps

Toggle clamps are handy, quick-release devices for holding materials being worked on with either power or hand tools. They are often used with fixtures. Clamps are necessary for drilling holes in sheet metal. See Figure T25-15. A toggle clamp saves much time in clamping materials.

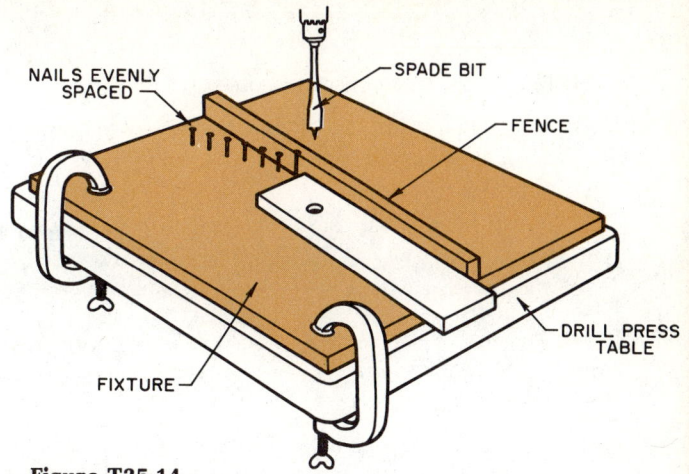

Figure T25-14

Fixture used when drilling (boring) several evenly spaced holes in a board.

1. Place the workpiece against the fence.
2. Move it forward until it barely touches the first nail.
3. Drill the first hole.
4. Remove the first nail, and slide the board forward until it touches the second nail.
5. Drill the second hole.
6. Continue removing the nails and moving the board forward until all holes are drilled.

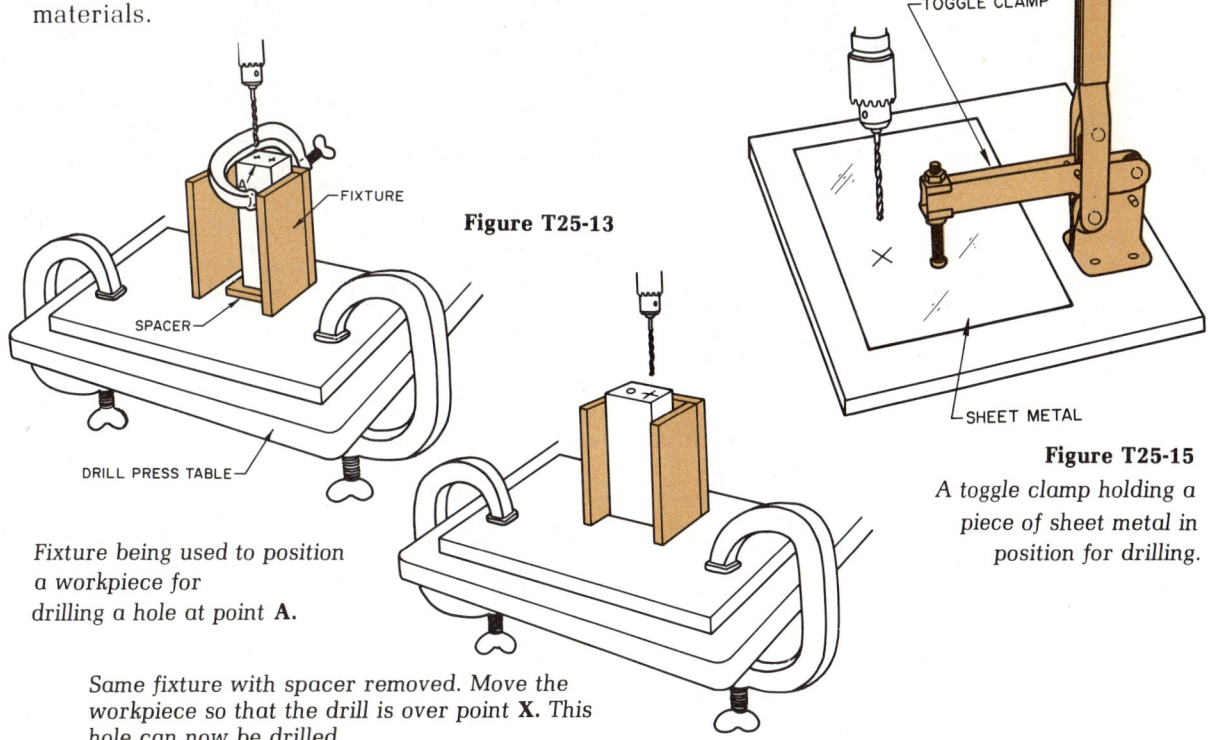

Figure T25-13

Fixture being used to position a workpiece for drilling a hole at point **A**.

Same fixture with spacer removed. Move the workpiece so that the drill is over point **X**. This hole can now be drilled.

Figure T25-15

A toggle clamp holding a piece of sheet metal in position for drilling.

Figure T26-3

Dimensions of Lumber
Commonly Used in the School Shop*

Lumber Surface	Thickness		Width		Length
ROUGH-SAWED	1" 1 1/4" (5/4)** 1 1/2" (6/4)** 2" 3" 4"		Sawed to random widths to obtain greatest number of boards from log.		**Hardwoods:** Sawed to random lengths to obtain greatest number of boards from log. Softwoods: 6' 8' 10' 12' 14' 16'
	Note: Order by number of **quarters (fourths) in a mixed number. There are **five quarters** in 1 1/4" and **six quarters** in 1 1/2". Therefore, you would order a board 1 1/4" thick as 5/4 and 1 1/2" as 6/4.				
SURFACED (planed)	Sales Unit	Actual Thickness	Sales Unit	Actual Width	
	1" 2" 3" 4"	3/4" 1 1/2" 2 1/2" 3 1/2"	4" 6" 8" 10" 12"	3 1/2" 5 1/2" 7 1/4" 9 1/4" 11 1/4"	8' 10' 12' 14' 16'
	Boards are often purchased in units of 2" × 4" and 4" × 4". These are called "two-by-fours" (2 × 4's) and "four-by-fours" (4 × 4's). The first number in the dimensions 2 × 4 and 4 × 4 is the board's **thickness**. The second number refers to the board's **width**. From the information shown above, you can see that **sales units** (standard amounts and sizes sold by a dealer) are different from the **actual** widths and thicknesses. Refer to the numbers shown under the headings "Actual Thickness" and "Actual Width" in the columns above. You can see that: a 2 × 4 is actually 1 1/2" thick and 3 1/2" wide and a 4 × 4 is actually 3 1/2" thick and 3 1/2" wide. Order lumber by the sales unit, but remember what the actual sizes will be when you use it to make a product.				

* Only customary measurements are given here. Metric standard sizes are not expected to be **exactly** the same as customary standard sizes. (See Topic 1, **Metrics**.)

Topic 26 Woods

Because of its many special qualities, wood is the material used most often by students in school shops. Wood is attractive, easy to work with, requires common tools, and can be made into a variety of useful and interesting objects. See Figure T26-1.

Wood is also easy to obtain. It may be purchased, **rough-sawed** or **surfaced**. See Figure T26-2. Surfaced lumber (wood) is that which was planed on both sides at the mill. Schools often buy rough-sawed lumber. It is then planed in the shop to the desired thickness. Refer to Figure T26-3 to learn more about lumber dimensions (sizes).

Wood Classifications

Woods are classified as hardwoods or softwoods. See Figure T26-4. These are **general** classifications, however. Although hardwoods **are** generally harder than softwoods, this is not always true. Balsa wood, for example, is a hardwood but is very soft and light.

Woods are also classified as open-grained woods and close-grained woods. Open-grained woods have large surface pores (openings). Close-grained woods have very small surface pores. See Figure T26-5. All softwoods are close-grained woods. Hardwoods are generally considered to be open-grained. However, some hardwoods have such small pores that they are classified as being close-grained.

Hardwood

Hardwood lumber comes from **deciduous** trees. Deciduous trees are those whose leaves fall off in autumn. Oak, maple, and cherry are examples of hardwoods.

Softwood

Softwood lumber comes from **coniferous** (evergreen) trees. These trees have needle-like leaves that remain green year-round. Pine, redwood, and fir are examples of softwoods.

Grades of Lumber

One factor that determines the price of lumber is its **grade**. This grade is determined by the type and number of **defects** (flaws such as knots or cracks) in the lumber. You should not use a board with many defects to make a product that should have an attractive appearance. However, a board free of defects can be very expensive. A board with a **few** defects costs less and may suit your purpose just as well. Become familiar with the appearance of the different grades of lumber. Then you can choose the best and most economical (lowest priced) for your product.

Figure T26-1

Wood can be made into many decorative and useful items, such as this coffee table made of oak.

Mark Levin

Figure T26-2

Rough-sawed lumber has a rough surface. It has not been planed. Surfaced lumber was planed at the mill. It has a smooth surface.

Figure T26-4

Characteristics and Uses of Common Woods

Name	Color	Workability	Cost	Holding Power Nails & Screws	Holding Power Glue	Grain	Examples of Uses
HARDWOODS							
Ash	Creamy white to light brown	Difficult	Expensive	Medium	Medium	Open	Tool handles, baseball bats, skis
Birch	Light tan to white	Difficult	Expensive	Good	Good	Close	Furniture, cabinets, dowels, toothpicks
Basswood	White	Easy	Average	Medium	Good	Close	Drawing boards, picture frames, piano keys, veneers
Cherry	Reddish brown	Medium	Expensive	Very good	Good	Close	Furniture, paneling, toys, novelties
Elm, American	Light tan	Medium	Average to low	Medium	Medium	Open	Bent handles, furniture, framing
Hickory	Brown	Difficult	Average	Poor	Medium	Open	Handles, ladder rungs, skis, dowels
Mahogany, African	Reddish brown	Medium	Expensive	Good	Good	Open	Musical instruments, veneer, furniture
Mahogany, Philippine	Reddish brown	Medium	Average	Good	Good	Open	Furniture, paneling, trim, boats
Maple	White to reddish	Medium	Average	Medium	Good	Close	Furniture, bowling pins, boxes, toys
Oak, Red	Tan, reddish white	Difficult	Average	Good	Medium	Open	Furniture, floors, handles, carving
Oak, White	Creamy white to tan	Difficult	Average	Good	Medium	Open	Tanning materials (bark), furniture, floors, handles
Pecan	Dark brown	Medium	Average	Poor	Poor	Open	Furniture, flooring, novelty items
Poplar	Pale yellow to brown, some greenish streaks	Medium	Average	Good	Good	Close	Drawer bottoms, boats, musical instruments, toys
Walnut	Chocolate brown	Medium	Very expensive	Good	Good	Open	Furniture, paneling, TV cabinets
Willow	Light green to brown	Medium	Average to low	Good	Good	Open	Toys, paneling, furniture, turnings
SOFTWOODS							
Cedar, Eastern Red	Red and white	Medium	Average	Medium	Medium	Close	Chests, closet linings, pencils, shingles
Pine, White	Creamy white	Easy	Average to low	Good	Good	Close	All parts of house construction, furniture, boxes, match sticks
Redwood	Deep red	Medium	Expensive	Good	Good	Close	Siding, shingles, fences, lawn furniture

Grades of Hardwood Lumber

Three grades of hardwoods may be used in the school shop.

- **Firsts and Seconds (FAS)** lumber — The top grades of hardwood lumber.
- **Selects** — the next lower grade of hardwood.
- **Number One Common** — The lowest grade used for school shop products.

Grades of Softwood Lumber

Six grades of softwoods may be used in the school shop.

- **Numbers One and Two Clear** — The best grades of softwood lumber. These grades are free of knots and knotholes.
- **C Select** — The next best grade. It may have a few slight defects, but is good for products that will be given a clear finish.
- **D Select** — The next lower grade. One side of D-Select lumber may have many flaws. The other side should only have minor flaws.
- **Numbers One and Two Common** — Not suitable for finish work. This lumber may have knots and other defects, but can still be used for many products made in the school shop.

Drying Lumber

Wood that is not **evenly** dried may warp or twist. A board that has not been **thoroughly** dried will shrink. Lumber companies dry lumber to help prevent boards from shrinking and warping. You will need kiln-dried or air-dried lumber for your products.

Special precaution: Use only wood that is **evenly** dried. See Figure T26-6.

Kiln-Drying

The **kiln-drying** process reduces moisture in lumber by drying it in a big kiln (oven). Kiln-dried lumber should be used for products which are intended for **inside use.** Even after wood has been dried, the wood will take in or give off moisture as the humidity in the air changes. You can control this by putting a finish on your product.

Air-Drying

In the air-drying process, lumber is stacked outside for a long period of time. It is arranged so that air circulates around each board. The air carries off excess (extra) moisture from the wood. This process is inexpensive, but it is slow. Air-dried lumber should be used for products which are to be used **outdoors.**

Manufactured Wood Materials

Wood materials suitable for making products are made from pieces or thin sheets of wood. See Figure T26-7. Natural wood sizes are limited by the sizes of trees. Manufactured wood materials can be made larger than natural lumber and to **exact** sizes. Some manufactured wood materials are stronger than natural wood.

Figure T26-5

An open-grained wood, such as oak, has large surface pores (openings). Forest Products Laboratory

A close-grained wood, such as poplar, has small, tightly knit pores. Forest Products Laboratory

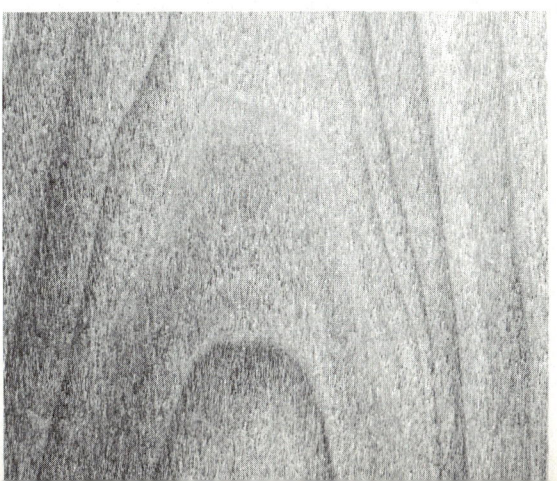

Plywood

Plywood is made of thin layers (plies) of wood glued together. The grain of each ply is placed at right angles to the plies below and above it. This adds strength. The inner layers are usually made of less expensive wood. The surface layer may be an expensive veneer (thin layer of high-quality wood). Plywood is usually sold in 4′ × 8′ sheets in thicknesses of 1/4″, 1/2″, or 3/4″. Plywood can be used to provide broad, strong surfaces.

Hardboard

Hardboard is a dark brown material made entirely of wood fibers. The fibers are pressed into a solid board. Lignin, a natural adhesive present in the wood, holds the fibers together. Most hardboard is either 1/8″ or 1/4″ thick. It is commonly sold in 4′ × 8′ sheets. You may wish to use this material for such products as game boards.

Particleboard

Particleboard is made by compressing wood shavings and chips together in a hot press. Resin is used to bind the particles. Particleboard has a smooth surface but lacks the strength of plywood. It is often used for making shelves and other products with wide, flat surfaces.

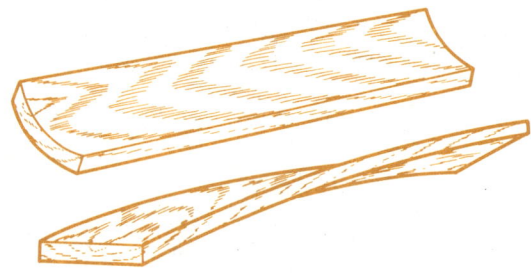

Figure T26-6
Wood that has not dried evenly may warp.

Figure T26-7
Manufactured wood products.

Topic 27 Metals

Metal is a durable and useful material. Products such as nails, tools, cars, and most machine parts are made of metal. See Figure T27-1. Metal is strong and resists wear. It can be made into many alloys (combined with other metals or materials) to produce metals with special characteristics. An alloy of aluminum and copper, for example, is much stronger than either metal separately.

Figure T27-1
Many products and parts for products are made of metal.

Textron Incorporated

Figure T27-2

Characteristics and Uses of Common Metals

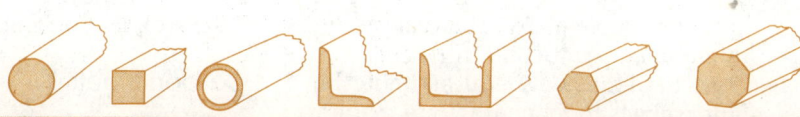

SHEET · BAR · ROD · SQUARE · TUBE · ANGLE · CHANNEL · HEXAGONAL · OCTAGONAL

	Type	Common Shapes	Appearance	Workability	Sample Products
FERROUS METALS	HOT-ROLLED STEEL	Sheet, Bar, Rod, Angle	Black scale on surface	Easy to bend, weld, braze, drill, or saw	Tables, shelf brackets, frames, supports
	COLD-ROLLED STEEL	Sheet, Bar, Rod, Square	Shiny, smooth surface	Easy to weld. Thinner metals easy to bend	**Sheet:** Air-conditioning ducts. **Bar:** Hammerheads. **Rod:** Hammers
	DRILL ROD	Rod	Shiny, smooth surface	Must be heated and hammered or ground to shape	Screwdrivers
	TOOL STEEL	Octagonal, Square	Shiny, smooth surface	Must be heated and hammered to shape, then hardened	Chisels, punches
NONFERROUS METALS	ALUMINUM	Sheet, Angle, Channel, Tube, Rod, Bar	Shiny silver-white	Machines quickly. Good conductor of electricity and heat. Can be brazed, welded. Sheet aluminum bends well	**Sheet:** Boxes, trays, caps, spun bowls, cans. **Angle:** Box-corner reinforcement. **Channel:** Supports, table panels. **Tube:** Lamp stems. **Rod:** Hammer handles, pulleys. **Bar:** Lamp bases, machine parts
	COPPER	Sheet, Rod, Bar	Shiny reddish-brown. **Note:** The surface of the finished product must be coated with lacquer to keep a bright appearance	Easily worked. Very good conductor of electricity. Hardens when hammered but can be softened by annealing	Bowls, costume jewelry, wiring, boxes, ornaments, springs, pipes
	BRASS	Sheet, Rod, Square, Hexagonal	Shiny yellow	Easily shaped. Hardens when hammered but can be softened by annealing	Decorative products, door hinges, musical instruments, screws, locks, small gears and parts for clocks and watches
	LEAD	Bar	Bluish-gray. Bright when freshly cut, but dulls with exposure to air	Soft, easily scratched. Resists corrosion by water and air	Pipes, storage batteries. Most often an alloy. Can be combined with tin to make solder

Metals are classified as ferrous or nonferrous. See Figure T27-2. **Ferrous** metals contain iron as the main element. **Nonferrous** metals, such as aluminum, do not contain iron.

A good test to identify a metal as being ferrous or nonferrous is to touch it with a magnet. Metal containing iron will attract the magnet. Nonferrous metals are not magnetic.

Ferrous Metals

Iron is the most common metal used by industry. It is obtained from iron ore. The iron is melted from the iron ore in huge furnaces. It is then mixed with different elements to make cast iron, wrought iron, carbon steels, and alloy steels. See Figure T27-3.

Cast Iron

Cast iron is made when iron is mixed with carbon and other elements, such as silicon. It is used to make strong, durable castings. Many tools, machine parts, and engine parts are made of cast iron.

Wrought Iron

Wrought iron is almost pure iron. It can be bent and worked easily. This makes it excellent for ornamental work. See Figure T27-4. It is expensive, however. Mild steel (low-carbon steel) is less expensive and can be used instead.

Carbon Steels

Carbon steels are those made of iron and **small** amounts of carbon. It contains less carbon and is more easily worked than cast iron. The more carbon that the steel contains, the harder it is. Carbon steels are classified by the amounts of carbon that they contain. See again Figure T27-3.

- **Low-carbon steel** is often referred to as **mild steel**. It contains very little carbon. This steel is easily worked and bent. It can not be hardened.
- **Medium-carbon steel** contains enough carbon to make it moderately hard and strong.
- **High-carbon steel** contains a larger amount of carbon. This type of steel is often called **tool steel**. It is used for tools, such as cold chisels and center punches. This type of steel can be hardened and can withstand hard use.

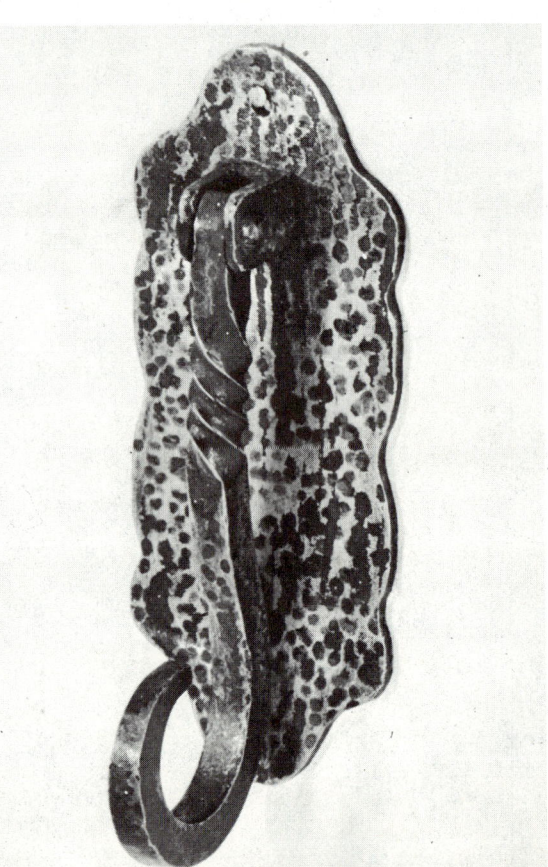

Figure T27-4
A wrought-iron knocker for a door.

Mrs. Morris J. Ruley

Figure T27-3

Common Types of Ferrous Metals

Metal	Characteristics	Uses
CAST IRON	Melts easily, 2-5 percent carbon, some types are very brittle	Machines and machine tool parts, other castings
WROUGHT IRON	Almost pure iron, contains little or no carbon, can be worked easily	Ornamental iron work
LOW-CARBON STEEL	Also called mild or machine steel, 0.05 to 0.30 percent carbon, easily welded, machined, and formed	Wire, nails, rivets, chains, machine parts, school shop products
MEDIUM-CARBON STEEL	Harder and stronger than low-carbon steel, 0.30 to 0.60 percent carbon, can be heat-treated	Machine parts, nuts and bolts, axles
HIGH-CARBON STEEL	Also called tool steel, 0.60 to 1.50 percent carbon, can be heat-treated and hardened, difficult to cut and bend	Tools — drills, taps, dies, cold chisels, center punches
ALLOY STEEL	Metals or other materials are added to iron to produce steels with special characteristics: **chromium** — adds toughness, makes it more resistant to wear **nickel** — adds strength and toughness, makes it more resistant to corrosion **molybdenum** — make it more resistant to heat	Gears, bearings, cutting tools, wire cables, rails, springs
SHEET STEEL	Hot- or cold-rolled sheets, usually coated with zinc to protect from moisture	Sheet metal products, heating and air-conditioning ducts, shelf brackets

Alloy Steels

When iron is combined with certain other elements, alloy steels are formed. Alloy steels are stronger, harder, and tougher than iron. They are more resistant to wear and corrosion (rust and other chemical reactions). Common alloying elements include nickel, chromium, manganese, vanadium, silicon, tungsten, and molybdenum.

Shaping Steel

Steel may also be classified according to the method by which it was shaped. **Hot-rolled** steel is steel that was formed into shape while the steel was still hot. This type of steel is least expensive and usually has a black or scaly surface crust.

Cold-rolled steel refers to steel that was worked into shape without being heated. Cold-rolling is done by running the steel through rollers or dies until it is the proper shape. Cold-rolled steel has a smooth surface finish. It is more expensive than hot-rolled steel. See Figure T27-5.

Figure T27-5

Two forms of sheet steel. The darker metal was cold-rolled. The silver-colored metal is galvanized sheet steel which has been coated with zinc.

Nonferrous Metals

Aluminum, copper, brass, and lead are nonferrous metals. They contain no iron. See Figure T27-6.

Aluminum

Aluminum is made from an ore called **bauxite**. The bauxite is crushed and processed first into aluminum oxide and then into aluminum. The aluminum is cast into shapes called **pigs**. These are later melted and alloyed with other metals.

Aluminum Alloys

The alloys of aluminum have a wide range of characteristics. Pure aluminum is very soft. Alloys that are nearly pure aluminum can be hammered, spun on a lathe, or stretched into many shapes. However, some aluminum alloys are very tough. One aluminum alloy, for instance, is so tough that it is used for bullet deflectors.

Aluminum alloys are light and strong. They resist weather damage. They are used to make outdoor products such as lawn chairs. Aluminum also conducts heat rapidly. This characteristic makes this metal useful in solar heat collectors.

Figure T27-6

Nonferrous metal products. Shown here are items made from aluminum, copper, and brass.

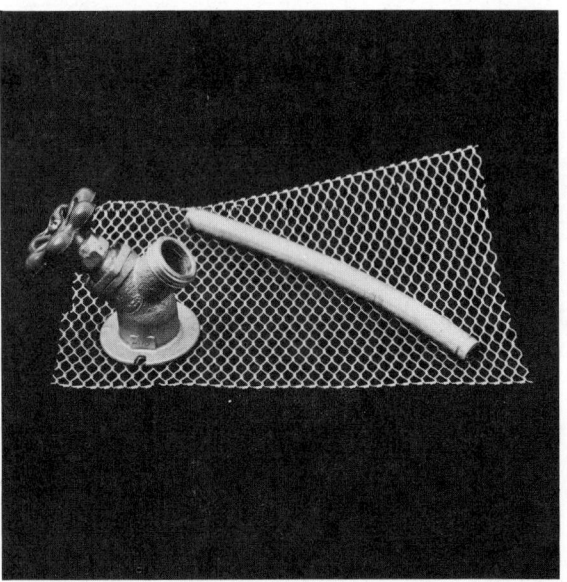

Aluminum alloys (casting aluminum) are good materials to use for **casting** in the school shop. (See Topic 18, **Metal Casting**.) Aluminum has a lower melting temperature than many other metals. See Figure T27-7. It also has a shiny appearance. Products made from aluminum are lightweight and attractive.

Figure T27-7

Melting Points of Common Metals

Metal	°C	°F
Solder, 50-50	204	400
Pewter	216	420
Tin	232	449
Lead	327	621
Zinc	420	787
Aluminum	659	1218
Bronze	912	1675
Brass	926	1700
Silver	961	1761
Gold	1062	1945
Copper	1081	1981
Iron, Cast	1204	2200
Steel	1427	2500
Nickel	1452	2646
Iron, Wrought	1482	2700

Casting aluminum is purchased in **ingots**. An ingot is a mass of metal shaped for easy storage and casting.

Aluminum cannot be soldered or welded easily with school shop equipment. Because of this, products made of aluminum are assembled with mechanical fasteners, such as rivets or bolts.

Aluminum flashing (for houses) is thin but strong. It can be used in making many products in the school shop. Products made from this aluminum are usually inexpensive. Building supply companies sell the aluminum by the lineal foot or roll. It generally comes in 50-foot rolls in widths of 8", 10", 12", 14", and so on up to 28".

Copper

Copper is a soft metal. It can easily be hammered into different shapes. One of the major uses of copper in industry is for electrical **wiring**. Copper is an excellent conductor of **electricity**. Copper is also used in the school shop for jewelry, metal spinning, decorative boxes, and ornaments. It can be combined with tin to form **bronze**.

Brass

Brass is a bright yellow metal made of copper and zinc. Brass can be made harder by adding tin to the alloy. Brass sheet is often used for decorative products. The hardness of the metal varies. Soft brass is easy to hammer or stretch into various shapes. Both brass and copper are easy to solder.

Brass sheet is sold in thicknesses of .025", .032", .040", and so on up to .064". See Figure T27-8. Sheets are 24 1/2" wide and 96" long. Brass is also sold in round rods, square rods, and hex rods in diameters of 1/8" to 3" for various school products. The rod may be cut to the length that you need up to 12' long.

Figure T27-8

Gages Given in Decimal Measurements (Inch)

U.S. Standard Gage Number	Measurement in Decimals for Ferrous Metals (Steel and Iron)	Measurement in Decimals for Nonferrous Metals
16	0.0625	0.0508
18	0.0500	0.0403
20	0.0375	0.0320
22	0.0313	0.0253
24	0.0250	0.0201
26	0.0188	0.0159
28	0.0156	0.0126
30	0.0125	0.0100

Lead

Lead is a heavy, nonferrous metal which is extracted (removed) from various ores. This is done by melting the ore and separating the lead. The primary use of lead is in storage (electric) **batteries**. Lead is also used in large quantities for **solder**. See Figure T27-9.

Solder is made of an alloy of lead and tin. The tin improves the strength and hardness of the lead. Lead also flows more easily when combined with tin. (See Topic 12, **Soldering**.)

Lead may be used as a casting metal to make small, heavy objects. The melting temperature of lead is only 621° F (327° C). Lead melts easily both as solder or as a casting metal.

Figure T27-9
Solder is made from lead and tin.

Bow Solder Products, Inc.

Topic 28 Plastics

Plastics are **synthetic** materials. They do not occur in nature. People make plastics from such materials as petroleum, coal, natural gas, and wood. These materials are chemically changed to make plastics.

Plastics can be formed or molded using heat and pressure. Some plastics are **thermoplastic**. These can be reheated and reshaped. Other plastics are **thermosetting** and cannot be reshaped. They are changed **chemically,** and the chemical change cannot be reversed.

Each plastic material whether thermoplastic or thermosetting has unique characteristics. They are chosen for use according to these characteristics.

Many kinds of plastics in many forms can be used in school shops. The kinds most often used are acrylics, polyesters, styrenes, cellulosics, vinyls, and ethylenes. Standard stock for plastics may be sheets, beads, pellets, foams, liquids, or tubing. See Figure T28-1.

Acrylics

Acrylic plastics (such as Plexiglas®) are commonly used in school shops. This type of plastic is a **thermoplastic** material. The method most often used to shape it is bending. (See Topic 20, **Plastic Casting, Molding, and Forming**.) Acrylic plastic may also be sawed, drilled, sanded, and polished. Separate parts made of acrylic may be bonded together by using methylene chloride. (See Topic 11, **Glues, Cements, and Other Adhesives**.)

Acrylic plastic sheets have protective coverings on each side. The coverings protect the plastic from scratches and make layout and cutting easier. Do not remove the covering until the plastic is cut to the proper size and shape.

You may choose from many colors of acrylic plastic. Many acrylics are **transparent.** You can see through them. Many are **translucent.** Light passes through them, but you cannot see through them. **Opaque** acrylics do not allow light to pass through. Acrylic plastic is available in rods and tubes as well as sheets and other shapes. See Figure T28-2.

Acrylic products that you can make in school include calculator stands, letter openers, and bracelets.

Figure T28-1

Common Plastics

Type of Plastic	Thermoplastic or Thermosetting	Form	Characteristics	Products
ABS	TP	Sheets	Excellent stretching qualities, strong, resists most stains and chemicals	Funnels, suitcases
ACRYLIC	TP	Rods Tubes Sheets	Weather and heat resistant, strong	Airplane windshields, taillight lenses
ACETATE	TP	Sheets	Clear, glossy, slow-burning, easy to process	Packaging
POLYESTER	TP & TS	Liquid	Clear, used with fiber glass, rigid, heat resistant, strong	Car bodies, skateboards, trivets
POLYETHYLENE	TP	Pellets Powder Film	Lightweight, easily processed, resistant to most chemicals	Bread wrappers, toys, squeezable bottles
POLYSTYRENE (HI-IMPACT)	TP	Beads Foam Sheets	Easily processed variety of colors, poor chemical resistance. **Caution:** Burns easily	Housewares, insulation, toys, model cars and airplanes
POLYVINYL CHLORIDE	TP	Liquid Pipe	Scratch resistant, weather resistant, tough. **Caution:** Gives off poisonous gas if overheated	Raincoats, shoes, furniture covering

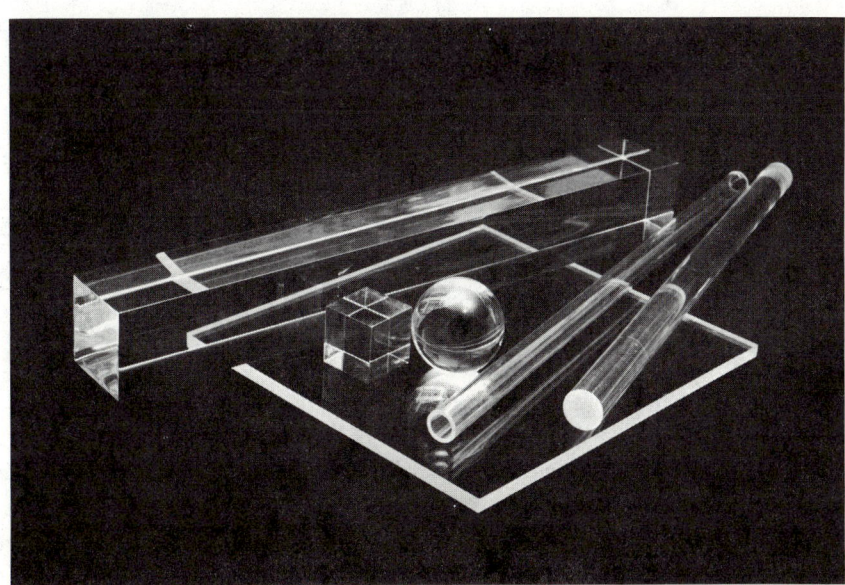

Figure T28-2
Acrylic plastic is available in many shapes and sizes.

Polyesters

Polyester resin is a clear material that is used for embedding (enclosing) small objects in castings. See Figure T28-3. Polyester resin is also used to **laminate** fiber glass in molds to make skateboards, boats, and other products. (See Topic 20, **Plastic Casting, Molding, and Forming**.)

Polyester resin comes in liquid form. It may be purchased in quart or gallon cans.

Styrenes

Styrenes are very versatile (can be used in many ways). Most styrenes are easily worked. Some kinds are good for insulating electrical wires. Others may be used to insulate your home. Styrenes are available in many forms. See Figure T28-1. Be careful when working with styrenes. Many styrenes burn easily.

Beads

Expandable polystyrene beads are used to make plastic foam products such as ice buckets. See Figure T28-4. These beads are first heated using boiling water. Then they are placed into molds and heated again. The beads expand (swell) in the water and are fused in the mold.

Pellets

Polystyrene pellets are used for injection molding. The pellets are heated in an injection molding machine. See Figure T28-5. They are then forced into molds to form products such as golf tees.

Figure T28-3

Polyester resin was used to cast this paperweight. Notice how clear the plastic is.

Figure T28-5

An injection molder and the plastic pellets used in it.

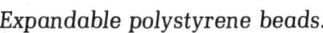

Expandable polystyrene beads.

Figure T28-4

An ice chest made from expandable polystyrene beads.

Figure T28-6
Some plastics are tough enough to be used as handles for tools.

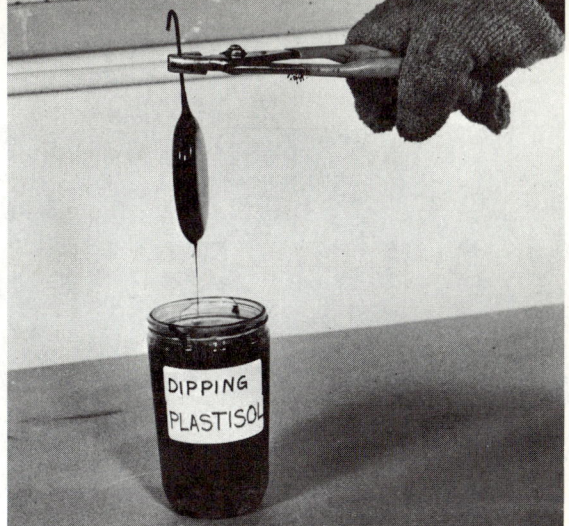

Figure T28-7
Making a small plastic purse by dip casting a metal mold into plastisol.

Sheets

Styrene sheets used for vacuum forming include the following:

- **Hi-impact polystyrene,** 0.010″ to 0.060″ thick, is available in a variety of colors.
- **ABS plastic,** 0.025″ thick, is strong and resists damage.

See Figure T28-1.

Foams

Polystyrene foam (such as Styrofoam®) is often used for making mock-ups of products. This material is commonly used for insulation. It is sold in sheets (boards) 3/4″ to 3″ thick, 24″ wide, and 8′ long.

Cellulosics

Cellulosics are thermoplastic materials. They are easy to process. Generally they are clear and glossy. They can be combined with special chemicals to give them a wide variety of characteristics.

Sheets

Cellulosic sheets used for vacuum forming include the following:

- **Acetate,** 0.0075″ thick, is easy to process.
- **Tenite**® is a cellulosic plastic that is often made into handles for tools such as screwdrivers. See Figure T28-6. Most Tenite® is yellow in color, but a variety of colors and color combinations are available. Tenite® is a very tough plastic. It is formed into rods and sold by the foot in lengths up to 4′. Diameters range from 1/2″ to 1-1/2″.

Vinyls

A wide variety of vinyl plastics are available. They may be flexible or rigid. Generally, they are tough and weather-resistant. Vinyls are not always easy to process, however. They generally react poorly to heat. Some kinds give off a poisonous gas when overheated.

Tubing

Clear vinyl plastic tubing is occasionally used for school shop products. Outside diameters of the tubing range from 1/4″ to 1″. The material may be purchased by the foot or in rolls of 100′.

Plastisols

Plastisols are plastics in a very thick liquid form. They are often used in **dip coating** or **casting.** Polyvinyl chloride plastisol resin is used to make products such as flexible coin purses. A heated mold is dipped into the plastisol. The plastisol coats the mold. See Figure T28-7. Additional heating cures (sets) the plastic. Then the product is removed from the mold. There are also air-curing plastisols. These are used to coat tool handles and various other products. Plastisols are available in different colors.

Figure T28-8
Self-adhesive vinyl sheet can be used to make a product more attractive.

Rainville Company **Figure T28-9**
Polyethylene is used to make plastic containers.

Sheet Coatings

Vinyl coating materials (self-stick shelf paper) may be used to improve the appearance of a product. (See Topic 23, **Finishing**.) These vinyls are made in different patterns and colors and are sold by the foot. See Figure T28-8.

Ethylenes

Polyethylene is known as the **squeeze-bottle plastic**. It is often melted and used in blow molding bottles and other hollow plastic products. It is also used to make a mold for casting plastic resin. The most common polyethylene plastics are lightweight. See Figure T28-9.

Tubing

Flexible polyethylene plastic tubing (milky translucent) is not as flexible as vinyl plastic tubing. However, it costs about half as much. You may wish to use polyethylene tubing instead of the more expensive vinyl tubing.

Pellets

Polyethylene pellets are used for injection molding. The processing is the same as for polystyrene pellets.

For Further Information

A few of the more common plastics have been described in this topic. More can be learned about these and many other plastics in books about plastics or in such publications as plastics catalogs.

Topic 29 Other Materials

Materials other than wood and metal are often used to make products in the school shop. Leather, fabric, ceramic material, enameling material, and glass are some of these materials.

Leather

Leather is made from animal hides. In tanneries, the hides are processed (tanned) into usable leather. See Figure T29-1.

Tanning Leather

The tanning of leather involves soaking the hides in large vats of chemical solutions. Different types of solutions are used for different types of tanning. **Chrome tanning** is done with a chromium compound solution. Hides tanned in this manner are resistant to water. They are used for such items as shoes and coats.

Vegetable tanning is done with a solution of water and tree bark. This process takes much longer than chrome tanning. Leather that is to be tooled and carved is tanned by this method.

Thicknesses of Leather

The thickness of the leather is given in ounces. One ounce is equal to 1/64" thick. Thus, 3-ounce leather would be 3/64" thick. See Figure T29-2.

The various thicknesses are obtained by splitting apart the layers of tanned leather. The lower layers of leather are often made into **suede.** See Figure T29-3. The surface of suede leather is brushed by special machines. It has a very soft texture.

Only one layer will have a top-grain side. The **top grain** is the surface from which the animal hair was removed. This surface is smooth. It is the only layer suitable for tooling.

Figure T29-2

Thicknesses of leather and common leather products.

Leather Weight	Thickness	Common Products
3-4 oz.	3/64"-4/64"	Billfolds, bookmarks, leather jewelry
6-7 oz.	6/64"-7/64"	Purses, key cases, pictures, handbags
7-8 oz.	7/64"-8/64"	Ornate (highly decorated) belts
8-9 oz.	8/64"-9/64"	Wider belts
9-10 oz.	9/64"-10/64"	Leather products that are tooled or carved, heavy belts, tool bags, other items that require strength and thickness

Figure T29-1

Jewelry and belt buckles made of leather.

Figure T29-3

A belt made from suede.

Figure T29-4
Top grain leather that has been carved and made into a product.

Tooling Leather

Tooling leather is a natural, unfinished, vegetable-tanned, top-grain leather. This type of leather is often sold by the hide or half-hide. The cost of the hide depends on the thickness, quality, and surface area (square footage) of the leather.

Leather work requires only a few basic tools. (See Topic 4, **Shearing**.) When tooling leather is moistened, it is easy to press or cut designs into its surface. See Figure T29-4. (See Topic 21, **Stamping Tools**.)

Fabrics

Fabric is another name for **cloth**. Fabrics are made of **fibers** (threads). See Figure T29-5. Natural fibers, such as cotton and wool, come from plants and animals. Synthetic fibers are made from chemicals. Different fibers can be combined in the same piece of fabric. You might be wearing a shirt made of 50% cotton and 50% polyester. This gives the fabric the good characteristics (qualities) of both kinds of fibers.

Generally, fibers are made into fabric in one of three ways: weaving, knitting, or felting. **Woven** fabrics have threads or fibers that cross over and under each other at right angles to form the cloth. See Figure T29-6. This kind of fabric is strong, but stiffer than knitted fabrics.

In **knitted** fabrics, the fibers are knitted or looped together. See Figure T29-7. This kind of fabric is very flexible and wrinkle-resistant.

Felted fabrics are made by pressing fibers together into cloth. See Figure T29-8. Because of the way it is made, felt is softer and thicker than other fabrics, but it is not as strong. Felt is usually made from wool. Felt pads may be placed on the bottom of a desk lamp to protect the desk top from being scratched.

See the chart in Figure T29-9 for characteristics and examples of common fabrics.

Figure T29-5

Fibers that are made into fabrics.

Wool	Cotton	Polyester	Nylon
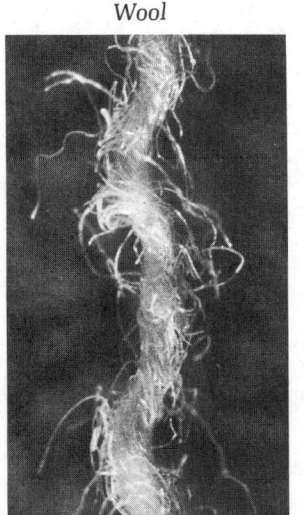	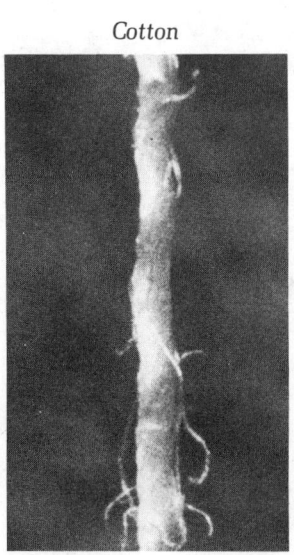	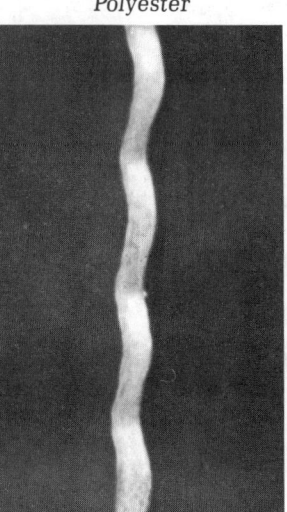	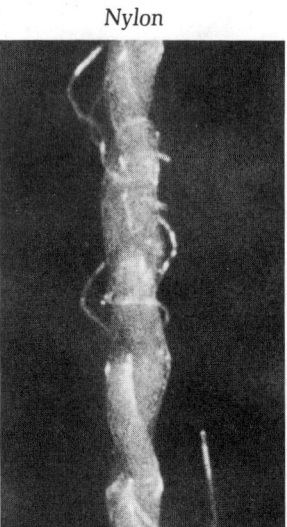

Other Materials   **427**

Figure T29-6
Woven fabric.

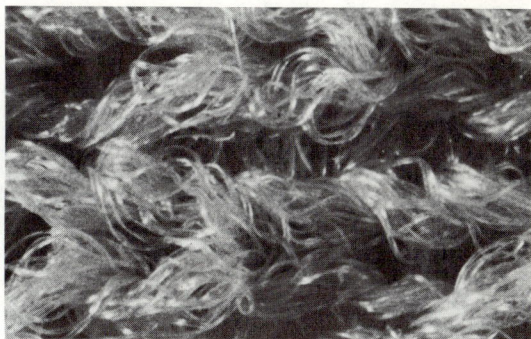

Figure T29-7
Knitted fabric.

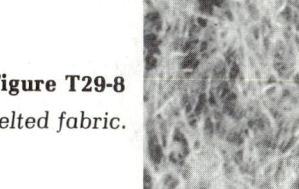

Figure T29-8
Felted fabric.

Figure T29-9

Common Fabrics

Fabric (Fiber)	Source	Characteristics	Common Uses
COTTON	Natural Cotton plant	Strong, wears well, inexpensive, absorbs moisture	Blue jeans, T-shirts, shop aprons, tote bags
WOOL	Natural Sheep	Resists soiling, wears well, retains shape, absorbs moisture	Clothing, coats, sweaters, carpets, felt protective pads or feet
NYLON	Synthetic Chemicals	Lightweight, very strong, does not absorb moisture, wrinkle resistant	Clothing, Backpacks, jackets, carpets, tote bags
POLYESTER	Synthetic Chemicals	Very wrinkle resistant, very strong, washes easily	Wash-and-wear clothing, and household items such as sheets and curtains

Ceramic Materials

Ceramics are clay items which have been fired (baked) at a high temperature. The firing is done in a special oven called a **kiln**. Before firing, clay is easily shaped when it is soft and moist. After firing, however, it is hard. The shape is permanent. See Figure T29-10. Ceramic products are resistant to moisture and acids.

A clay object is usually fired twice. The first firing is called **bisque firing**. It hardens the clay and removes the moisture from it. The second firing is **glazing**. This is done to decorate and seal the clay. A special glass-like finish (glaze) is melted and bonded to the surface of the clay.

Glazes are made of powdered glass and metal oxides. You can buy these ready-to-use. Many colors are available.

Ozark Ceramics

Figure T29-10
A glazed ceramic figurine.

Ceramics

CAUTION! Some glazes contain **poisonous** lead. Avoid using these if possible. If you must use a glaze that contains poisonous lead:
1. Keep your hands away from your mouth.
2. Wash your hands immediately after glazing is complete.
3. Avoid breathing the fumes.
4. Do not use lead glaze on a product to be used for food or drink.

Remember: **Lead** glazes are poisonous!

Ceramic-Enameling Materials

Enameling materials are glass and glass-like substances used to decorate copper. See Figure T29-11. (See Topic 23, **Finishing**.)

The process used to make enameling materials takes a long time. Ground oxides of metals are fused (melted together) with a special sand mixture to make colored glass. The colored glass is then ground into a powder. Generally, this powder is purchased at a store that sells craft supplies.

During the enameling process, the powder is applied to the surface of a metal. These are heated in a kiln to complete the enameling.

Figure T29-11
Copper enameling can be used to make products attractive in appearance.

Mary R. Walker

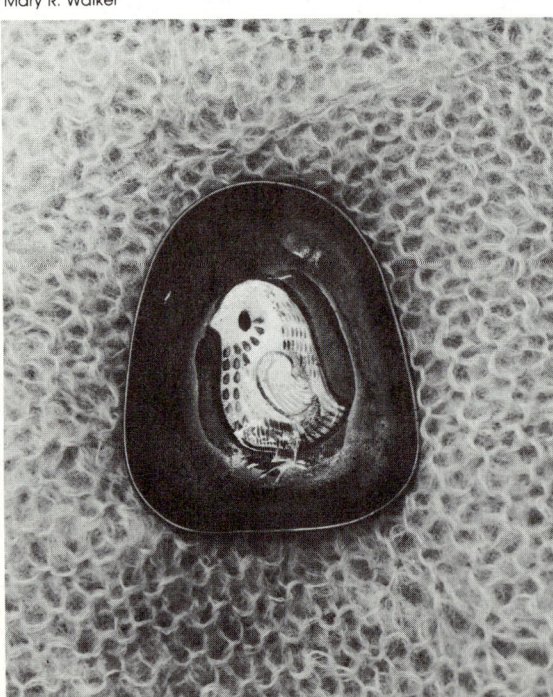

Glass

Glass is made from silica, an element which is found in sand. **Window** glass is brittle (easily broken). **Tempered** glass is stronger. It is sometimes used for tabletops.

Glass may be used in products made in the school shop. A pane of glass (any size) is referred to in the glass industry as a **lite.** A 30-lite box of glass contains 30 pieces of glass. When you buy glass, check for the size on the box or ask a salesperson to help you.

Single-strength window glass is 1/16" or 3/32" thick. **Double-strength** glass is 1/8" thick. You can also buy glass 3/16" thick and 1/4" thick for shop products.

Glass is cut to size by scoring (scratching) a line along its surface with a glass cutter and fracturing it along this line. (See Topic 17, **Induced Fracture and Etching.**)

Colored glass pieces are used to make decorative lampshades and leaded windows. See Figure T29-12. Copper foil or lead strips can be formed around the edges of the glass. This framework is then soldered to hold the pieces of glass in place. (See Topic 12, **Soldering.**)

Figure T29-12
"Sun-catchers" are decorative items that are made of colored glass.

A Touch of Glass

Photo Notes

Union Carbide Corporation

The Bettmann Archive, Inc.

PPG Industries, Inc.

Ford Archives/Henry Ford Museum, Dearborn, Michigan

Norton Company

Borg-Warner Corporation

Brown Brothers

Aluminum Company of America

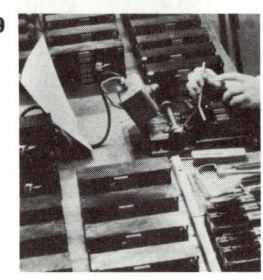
Westinghouse Electric Corporation

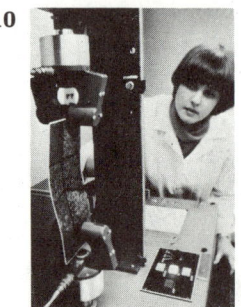
GAF Corporation

1. This smelting plant turns iron ore into a chromium alloy used in stainless steel.

2. A barrel serves as a workbench in this makeshift factory during World War I.

3. A worker buffs a tuba before final lacquer is applied.

4. All industries depend on products for safety. Hard hats, goggles, hearing protectors, and gloves are necessary equipment for many jobs.

5. This person worked four months to develop a special seal for pipeline pumps. Because of his idea coal and other minerals may be moved by pipeline rather than by conveyor or railcar.

6. Their dreams are our realities. Yesterday's American researchers paved the way for things we take for granted, such as medicines, machines, materials. They gave us new ones and found new uses for others.

7. This new smelter improves the production of aluminum.

8. This is one of Ford Motor Company's first assembly lines in the Highland Park, Michigan, plant.

9. With the operation sheet in view, a worker assembles a circuit breaker.

10. The strength of this vinyl flooring is being tested to make sure it will stand years of use. Tests such as this help to ensure high-quality products.

11. These students in 1881 developed these products in a class at Yale.

11

Brown Brothers

12

PPG Industries, Inc.

13

Brown Brothers

14

Fairchild Camera and Instrument Corporation

15

IBM

16

Masonite Corporation

17

Industry and Tourism, Ontario, Canada

18

The Gillette Company

19

GAF Corporation

20

Western Electric

12. Researchers test the strength of a special plastic which may replace heavy steel parts in a car. This plastic box was molded on a 500-ton press.

13. The basics of manufacturing are the same now as they were for this worker in the 1900's — tools, materials, processes.

14. A single-crystal silicon ingot is grown in a materials plant. The ingot is sliced into wafers and made into memory chips.

15. This production line appears to belong in the future, but, today, silicon wafers cut from an ingot and imprinted with circuits are moved along by air jets. These wafers are later separated into individual memory chips that will be used in computers.

16. A researcher studies wood fiber chemistry in order to find new ways to use wood.

17. A geodesic dome structure designed by R. Buckminster (Bucky) Fuller. An amazing fact about this design is that the bigger it is, the stronger it is. That is, the number of parts that support the structure increases faster than the weight that needs to be supported as the structure is built larger and larger.

18. Designers and engineers work together to make sure the product works properly.

19. A marketing program is developed to ensure that the product sells once it is manufactured.

20. Telephones are assembled in an Indianapolis, Indiana, plant.

On the Cover —

A night view of a chemical plant in Plaquemine, Louisiana. Chemicals produced at this plant are used in paper products.

The photo on pages ii and iii is a daytime view of the same plant.

INDEX

Abrasive belts, 361
Abrasive compound, used in buffing, 364
Abrasive paper, *illustrated*, 359
ABS plastic, 423
Accidents, 55, 57
Acetate, 423
Acid, used in etching, 368, 369
Acrylic plastic, *illustrated*, 420
    fracturing, 368
Addis, William, 19
Adhesion, *defined*, 50
Adhesives, 343-346
Administration, people, 111-112
Advertising, 159, 161-164
    illustrated, 160
    packages used for, 157
Aluminum, 418
    alloys, 418-419
    etching, 368
    melting and pouring, 372-374
America, manufacturing in, 29-33
American system, 30-31, 169-171
Analysis, of new product ideas, 124
Angles, cutting in wood, 312-313
Annealing, 388
Anvil, *illustrated*, 375
Appearance mock-up, 89
Applied research, 72
Arc welding, 356-357
Arc-welding machine, *illustrated*, 355
Articles of Incorporation, 109
Askins, Barbara, 75
Asphaltum, used in etching, 368
Assemblies, 145
Assembling, 46
Assembly, 151, 153-154
Assembly drawings, 91
Assembly line, 31-32, 128-129, 153
Assets, *defined*, 109
Attitudes, importance of good, 52-54
Automation, *defined*, 34

Band saw, *illustrated*, 315
Band saw table, *illustrated*, 406
Bar folder, using, 378
Bar stock, twisting, *illustrated*, 376
Bartering, *defined*, 26
Basic machines, 39-40
Basic research, 72
Beads, polystyrene, 422
Belt —
    installing on portable belt sander, *illustrated*, 361
    on stationary belt sander, 362-363
Belt sanders, 360-361, 362-363
Bench punches, 310
Bench tools, 305-308
Bending —
    defined, 48
    and forging, 375-379

Bending fork, *illustrated*, 376
Bending jig, *illustrated*, 376
Bending machines, using, 377
Beveling —
    on jointer, 325
    with Uniplane®, 326
Beveling tool, *illustrated*, 331
Biddle, Nannie, 76
Bit brace, *illustrated*, 318
Bits, 320-322
Black, Samuel, 40
Blades, of saws, 312-317
Blimps, 164
Block plane, 39-40, 323
Blodgett, Katherine, 75-76
Blow molding, 384
Board —
    boring holes in, *illustrated*, 409
    crosscutting, 311, 312, 314, 316-317
    jointing edge of, *illustrated*, 325
Board of directors, 109
Bolt cutter, 310
Bolts, 307, 334
Bonding, 50-51, 343
Boring and drilling, 318-322
Box and pan brake, using, 379
Bracelet, optional product, 258-259
Brainstorming, 82
Brass, 45, 419
Brazing, oxyacetylene, 353
Breaking even, *defined*, 168
Briggs, Steve, 107
Bronze, 419
    used in brazing, 353
Brush —
    cleaning, *illustrated*, 394
    using to apply paint, *illustrated*, 390
Buffers, 364

Calculator case, optional product, 278-281
Calipers, using, 301
Capital, *defined*, 161
Capitalism, *defined*, 169
Carburizing compound, 389
Cardboard guide, on paper cutter, *illustrated*, 407
Careers, in research and development, 74
Carver, George Washington, 72
Case hardening, 389
Casting, *defined*, 47
    of metal, 369-374
    of plastic, 380
Casting aluminum, 419
Catalyst, *defined*, 48-49, 380
Cellulosics, 423
Cementing, 345-346
Center drill, 322

Center-drilling, 332
Center punch, 303-304
Centers, turning stock between, 328-329, 332-333
Center square, 302
Ceramic enameling, 396-397
    materials, 428
Ceramics, 46
Chamfer, planing on jointer, *illustrated*, 325
Chamfering, on Uniplane®, 326
Chauncey, Jerome, 30-31
Chemical etching, 368-369
Chemicals —
    and safety, 58
    used for separating, 49
Chip removing —
    defined, 49
    illustrated, 48
Chisels, 322-323
    used for shearing, 309
Chlorinated solvents, using, 346
Chuck, of brace, *illustrated*, 318
Circle-cutting jig, *illustrated*, 406
Circles —
    dimensioning, *illustrated*, 100
    drawing, *illustrated*, 99
    marking, *illustrated*, 304
    sawing and sanding, 406
Clamping, 398-402
Classifications, of woods, 411
Cloth, 426
    bonding, 345, 346
    fasteners for, 338-341
Clothing —
    and safety, 57
    illustrated, 55
Coating, *defined*, 50
Coatings, finish, 392
Cochran, Michael, 169
Cold chisel, 309
    sharpening, *illustrated*, 366
Colors, tempering (table), 389
Combination square, 302
    illustrated, 303
Combining —
    defined, 46
    forming process, 49-51
Commission, in sales, *defined*, 165
Companies, starting and organizing, 104-116
Company —
    defined, 104
    departments of, 110-114
    healthy, 169
    selecting name, 117-118
    selecting trademark, 118
Compass, using, *illustrated*, 98, 99
Competitors, 146
Components, 22, 23, 38, 145, 150, 151, 153
Compressing, forming process, 48
Compression molding, *defined*, 48

Computer —
   as drafter, 91
   used in assembly lines, 153
   used to control production, 153, 154
   used in manufacturing, 34-35
   used in production, 133
   used to store information, 11
Conditioning, forming process, 48-49
Consumer demand, 19, 80, 83-84
Consumers, *defined*, 15
Consumer surveys, 93-94
   use of mock-ups in, 89
Contact cement, 345
Contract, role in partnership, 107
Control —
   of inventory, 154-155
   of production, 153
   of quality, 146
Conveyor system, 18
   and trial run, 150
Cooperation —
   during production process, 154
   importance of, 114, 116
   as quality of good worker, 52-53
Copper, 419
   etching, 368
Copper enameling, 397, 428
Coppers, for soldering, 347, 348
Corner joints, securing, *illustrated*, 401
Corporations, 107-109
Corrugated fastener, *illustrated*, 342
Cost estimator, 134
Costs —
   and design decision, 90
   estimating, 134-137
   figuring total, 136
   fixed, 137
   importance of, 123, 124, 126-127
   labor, 133, 135-136
   material, 134-135
   setup, 126
   shown on records, 168
Counterbored, 320
Countersink, *illustrated*, 321
Crosscutting, 314, 316-317
Crucible furnace, *illustrated*, 373
Curing methods, when gluing, 343
Customary measurement, 296
   (table) 299
Custom production, 16, 17, 119
   defined, 16
   illustrated, 121
Cut, of files, *defined*, 358
Cutoff tool, *illustrated*, 331
Cuts —
   made by router bits, *illustrated*, 324
   making curved, 312, 314, 315
   making in metal, 330, 332, 333
   making in wood, 32-33, 322, 323, 324, 328, 329, 358
Cutting —
   of glass, 49
   oxyacetylene, 352
   wood to length, 404
Cutting torch, 354

Debts, *defined*, 109
Debugging, of production system, 149
Decker, Alonzo, 40
Defects, in lumber, 411
Delays, shown on flow chart, 124
Demand, 19, 80, 83-84, 169, 170
Departments, of manufacturing company, 110-114, 140
Depth gage, 302, 403
   on bar folder, 378
Designers, role of, 81, 82, 87, 88, 89
Design ideas, 89, 94
Designing, of products, 19, 87-90
Design problem, stating, 72, 87
Detail drawings, 91
Development —
   of manufacturing industry, 26-37
   research and, 70-77, 79
Die, 338
   illustrated, 138
   punch and, 22
   used in drawing, 48
Die punching, 139
Die stamping, 385
Diestock, *illustrated*, 338
Dimensioning —
   hole, *illustrated*, 100
   on working drawings, *illustrated*, 100, 101
Dimensions, of lumber (table), 410
Dimensions list, used in product planning, 60-61
Dip coating, 423
Disk sanders, 362
Distribution, 165
   illustrated, 160
   preparing for, 23
Dividend, paid to stockholders, 169
Dividers, using, 304
Division of labor, 29
Downtime, *defined*, 114
Down-up method, of arc welding, 357
Draft, on casting pattern, *defined*, 370
Drafting, *illustrated*, 98-100
   of plans, 91
Drawing, forming process, 48
Drawings —
   dimensioning, *illustrated*, 100, 101
   working, 60-61, 76, 91, 98-100
   illustrated, 72
Drill —
   centering for drilling hole, 303-304
   early, 40
Drilling, 408-409
Drilling and boring, 318-322
Drill press, *illustrated*, 319
Drills, 318-319, 321
   illustrated, 43
Drive punch, 310
Drying, of lumber, 413
Dual dimensioning, 137
Dymaxion, 76-77

Economic system, American, 30-31, 169-171

Economy, *defined*, 169
Edison, Thomas, 19
Efficiency experts, 33
Electric welding, 355-357
Emery cloth, 359
Enameling, ceramic, 396-397
Enameling materials, 428
Enamel paint, 394
End cutting nipper, 309
Engineer, job of, 119, 121, 124, 126, 127
Engineering, of products, 19, 87, 91, 93
Engineers, manufacturing, 22, 87, 91, 93
Equipment, *defined*, 41
   industrial machines, and, 39, 42
   power machines and, 39, 41-42
Etching, 368-369
   defined, 49
Ethylenes, 424
Extension lines, on working drawings, 100

Fabrics, 426
Faceplate turning, 329-330
Facing —
   on metalworking lathe, 330, 332
   illustrated, 331
Factories, conditions in early, 28-29
Fagan, Grace M., 156
Fasteners, 334-343
   for aluminum, 419
Fastening, mechanical, 51
Felt, shearing, 407
Felted fabrics, 426
Ferrous metals, 416, 418
   (table) 417
Fiberboard, 44
Fiber glass, laminating, 381
File card, *illustrated*, 359
Filing, 358-359
Fillers, for wood, 392
Financial backing, 71
Finish coatings, 392
Finished goods, 154-155
Finishes —
   applying, 390
   defined, 390
   (table), 391
Finishing, 390-398
Finishing sanders, 360
Finishing tool, broad-nose, *illustrated*, 331
Fire extinguishers, labels on, 57
Fire safety, 57
Fixture, for belt buckle, 237
Fixtures, 137, 138, 139
   defined, 403
   illustrated, 138
   see also Jigs and fixtures
Flames, in oxyacetylene welding, *illustrated*, 351
Flatness, checking for, 302

Flow chart, 60, 61, 124
  illustrated, 125, 131
  and plant layout, 127-128
Flux —
  used in brazing, 353
  used when pouring aluminum, 374
  used in soldering, 348
Fluxcored solder, 348
Flycutter, 322
Foam, polystyrene, 423
Foamed polystyrene plastic pattern, 370
Football kicking tee, 60, 62, 63
Ford, Henry, 31-32, 128
Forging —
  and bending, 375-379
  defined, 48, 375
Forklifts, 129
Forming —
  defined, 47
  of plastic, 382, 384
  illustrated, 383
Forming processes, 46-51
Foundry, and sand casting, 370
Foundry tools, 371
Fracturing, 367-368
Framing square, illustrated, 302
Free enterprise, 170-171
Free forming, defined, 382
Fuels, types of, 36
Fuller, R. Buckminster (Bucky), 76-77
Function, and design decision, 90
Fusible tape, 343, 346
Fusion, defined, 51
Fusion bonding, 343

Gages —
  automatic, illustrated, 147
  back, of squaring shears, 309
  defined, 403
  to measure tolerance, 148
  miter, used in chamfering, 326 (table) 419
  used in layout, 302, 304
Gilbreth, Lillian and Frank, 32-33
Glass, 429
  coating metal with, 396-397
  non-glare, 74-75
  used in enameling, 428
  window, 429
Glass cutter, illustrated, 367
Glass cutting, defined, 49
Glazing, 428
Glues, 47, 50, 343-345
Gluing, of boards, 400-401
Goggles, 56, 350
Gouge —
  illustrated, 328
  for shearing leather, 310-311
Grades, of lumber, 411, 413
Grinding, 365-366
Grit, of abrasives, defined, 359
Grommets, 341
Gross profit, defined, 169
Guards, safety, 55, 58

Hall, Charles, 29
Hall, John, 30
Hammers, illustrated, 305
Hand drill, using, 318
Hand groover, illustrated, 378
Hand plane, using, 323
Hand punches, 310
Hand sanding, 359
Handsaws, 311-314
Hand seamer, using, 377
Hand tools, 39, 40
Hand-tool safety, 58
Hardboard, 414
Hardening, 388
Hard mock-up, 89
Hardwood, 411, 413
Heat, used for separating, 49
Heat treating, of metals, 388-389
Hiring, of workers, 139-142
History, of manufacturing industry, 26-37
Hole, dimensioning, illustrated, 100
Holes —
  boring in metal, 318, 319, 321
  boring in plastic, 318, 321
  boring in wood, 318-319, 320-322
  drilling in wood, 408-409
  drills and bits for making, 320-322
Honda, Soichiro, 132
Howland, Esther, 18
Hyatt, John Wesley, 45

Ideas —
  as key to new companies, 104-105, 106
  gathering for products, 117
  for new products, 80-83, 88-89, 160
  role of, 70-79
Inboard faceplate turning, 329
Inclined plane, illustrated, 39
Income, defined, 168
Induced fracture, 367-368
Industrial machines and equipment, 39, 42
Industrial materials, 38, 42-46
  defined, 42
Industrial Revolution, 26-28
Industry —
  agricultural, 5, 6-7
  communications, 5, 8-9, 11
  construction, 5, 11-12
  lumbering, 5, 7-8
  manufacturing, see Manufacturing industry
  role of, 2, 4
  textile, 27
  transportation, 5, 11
Ingot, metal, 374, 419
Injection molding, 382
Input-process-output system, 104
Inspection, 146-149
Interchangeable parts, 30
Interview, for job, 140, 141

Invention —
  of airplane, 77, 79
  of automatic switching system, 81
  of automatic valve, 29
  of flat-bottomed bag, 70
  of "invisible" glass, 75
  of toothbrush, 19
  of type of vacuum cleaner, 81-82
Inventions, as key to new companies, 104-105
Inventors, role of, 81-82
Inventory, 34, 133
Inventory control, 154-155
Inventory records, 155
Iron, 416
  (table), 417

Jack plane, 323
Jig, 22, 137, 138, 139
  bending, 376
  circle cutting, illustrated, 406
  defined, 403
Jigs and Fixtures —
  for drilling, 408-409
  for sawing and sanding, 404-406
  for shearing, 407
Jigsaw, using, 315
  see also scroll saw
Job interview, 140, 141
Jot-lot production, 16, 17-18
  defined, 16
Job openings, 140
Job planning, development of, 32-33
Jobs, applying for, 140-141
Job shop, defined, 17-18
Job specialization, 32
Johnson, Samuel C., 106
Jointer, 324-326
Jointer plane, 323
Joints, securing, illustrated, 401

Kerf, 311, 405
Kettering, Charles, 113
Kickback, defined, 326
Kiln, 428
  illustrated, 397
Knife —
  swivel, 310
  utility, illustrated, 308
Knight, Margaret E., 55, 70, 171
Knitted fabrics, 426
Knurling, 333
  defined, 330
Krieble, Vernon, 47
Kroc, Ray, 105-106

Labor, division of, 29
Labor cost, 133, 135-136
  and choice of machines, 126-127
Labor force, defined, 15
Lacing, 341
Lacquers, 394

Index

Laminating, of fiber glass, 381, 422
Laser beam, used in inspection, 148
Latex paint, 394
Lathe, metalworking, 330, 332-333
Lathe dog, *illustrated*, 332
Lathe tools, 328-330
Laws, and free enterprise, 171
Layout, 300-304
   defined, 61, 64
   making for school shop, 130-131
   using flow chart to plan, *illustrated*, 127
Lead, 419-420
Lead time, *defined*, 133
Leather, 46, 424-426
   bonding, 345
   cutting, 308, 310
   fasteners for, 338-341
   stamping, 386
   stitches, 341
Leather, tools used with, *illustrated*, 310-311
Length —
   cutting wood to, 317
   measuring, 300
Lengths, sawing equal, 404-405
Lettering, on drawings, 99
Lever, *illustrated*, 39
Lewis, *Tillie*, 80-81
Licensing, in free enterprise, 170
Lifting, correct way of, 57
Limitations, determine product selection, 84-86
Line production, 16-17, 18, 31-32, 119, 121
Lines —
   drawing on working drawings, 98, 99, 100
   scribing, 303-304
Lite, in glass industry, 429
Locating fixture, 407
Locating pins, 407
Loss, figuring, 165, 168
Lumber —
   dimensions of (table), 410
   grades of, 411, 413

Machine guards, 58
Machines —
   basic, 39-40
   controlling speed of, 34-35
   industrial, 39, 42
   and labor cost, 126
   limit product choice, 83-84
   power, 39, 41-42
   safety when using, 58, 59
   used for bending, 377
Machine screws, 336
Machinist's vise, 402
Mallet, 305
Management —
   leaders of company, 110-111
   role in design decisions, 89, 90, 94
Management personnel, and product selection, 82-83

Manufacture, preparing to, 132-142
Manufactured wood materials, 413-414
Manufacturing —
   in America, 29-33
   and American Industries, 1-12
   computers used in, 34-35
   defined, 38
   development of, 26-37
   future of, 35-37
   introduction to, 14-25
   modern, 33-35
   people in, 52-56
   primary, 145
   safety in, 55
   steps in (chart), 95, 123, 167
Manufacturing company —
   activities for starting, 117-118
   starting and organizing, 104-116
Manufacturing engineer, 22, 91, 93, 119, 121, 124, 126, 127
Manufacturing industry, 2, 5, 14-15
Manufacturing system, planning, 18-19, 22
   illustrated, 20, 21
Manufacturing technologists, 121
Marketing, 159-171
   activities for, 172
Marketing channels, 165
Marketing department, 110, 114
Market research, 159, 161
Marking tools, 303-304
Marking up, *defined*, 165
Mass production, 30
Material cost, 134-135
Material inspection, 146-147
Materials, 15
   choosing for shop product, 60
   cost, 134-135
   industrial, 38, 42-46
   defined, 42
   limit product choice, 85
   manufactured wood, 413-414
   natural, 43-44
   ordering, 132-133
   raw, 5, 22, 145, 146, 154
   removing from machine, 58
   synthetic, *defined*, 44
   used to make shop products, 424-426, 428-429
Materials-handling system, 128-129
Materials list, used in product planning, 60, 61
McDonald's Corporation, 105-106
Measurement —
   conversion tables, 299
   metric and customary (table), 296
   systems of, 296-298, 300
Measuring, diameter, 301
Measuring devices, 148
Measuring tools, 300-303
Mechanical fastening, 51
Mechanical forming, 382
Melting temperatures, of metals (table), 419

Metal, 43, 44
   bending, 376, 379
   boring holes in, 318, 319, 321
   clamping, 398-399
   cutting, 308, 309, 313-314
   etching, 368-369
   finishing, 390, 395, 396-398
   forging and bending, 375-379
   holding when working, 402
   making cuts in, 330, 332-333
   painting, *illustrated*, 395
   polishing, 364
   processed in outer space, 35
   punching holes in, 310
   sanding, 359, 362-363
   shaping, 305
   shearing, 407
   stamping, 385
Metal casting, 369-374
Metals, 414, 416, 418-420
   characteristics and uses (table), 415
   heat treating, 388-389
   joining, 350-357
   and safety, 58
   soldering, 347-348
Metal turning, 330, 332-333
Metric measurement, 137
Metric system, 61, 296-298, 300
   conversion tables, 299
Microcomputer, 71
   illustrated, 169
Micrometer, using, 301
Miter box, 312-313
Miter gage, used in chamfering, 326
Mixing, *defined*, 50
Mock-ups, 70-71, 89, 97, 124
   illustrated, 89, 90
Model, used to plan manufacturing system, 122
Mold, 139
   for forming funnel, 263
   illustrated, 138
   preparing, 370-372
   for skateboard, 270
   for trivet, 267
Molding —
   compression, 48
   defined, 47
   injection, 382
Mold release, *defined*, 381
Money, limits product choice, 85
Money system, 26

Nails, 342
Name, selecting for company, 117-118
Natural materials, 43-44
Natural resources, saving, 36
Net profit, *defined*, 169
Noise —
   control of, 55
   and safety, 56
Nonferrous metals, 418-420
   annealing, 388
   defined, 416
Non-glare glass, 74-75

436 Index

Non-threaded fasteners, 51, 338-342
North, Simeon, 30
Notcher, *illustrated,* 308
Nuts and bolts, 307, 334
Nylon, 44

Oil finishes, 394
Operation sheet, 127, 141
   illustrated, 126
Ordering, of materials and supplies, 132-133
Ore, *defined,* 44
OSHA, 55
Outboard faceplate turning, *illustrated,* 329
Outlets —
   for selling products, 157
   factory, 165
Output, *defined,* 104
Overhead, *defined,* 136
Ownership, types of, 106-108
Oxyacetylene brazing, welding, and cutting, 350-354

Packaging, 155-157, 172
Paints, 394, 395
Paper —
   abrasive, see Abrasive paper
   bonding, 343, 345
   placing on drafting board, *illustrated,* 98
Paper cutter, 407
Parker Brothers Company, 108-109
Particleboard, 44, 414
Parting compound, used in sand casting, 370
Partnership, *defined,* 107
Parts —
   interchangeable, 30
   made by casting, 369
   making, 150-151
   purchased, 154
Parts list, 124
   sample, 93
   used in product planning, 60, 61
Paste-up mock-up, 89
Patents, in free enterprise, 170-171
Pattern, for trivet, 267
Patterns —
   making with stamping tools, *illustrated,* 387
   used in sand casting, 369-370
Pay scale, 140
Pedestals, for machines, 41
Pellets, plastic, 422, 424
Penetrating oil finishes, 394
Penny system, of nail numbering, 342
People, in manufacturing, 52-56
Performance test, taking, 143-144
Personnel department, 110, 113, 140
Petersen, William, 40
Pilot, on router bit, 324
Pins, in squaring-shears table, 407
Pivot points, on jigs, 406

Plane, inclined, 39
Planer, thickness, 327
Planes, *illustrated,* 323
Planning —
   job, 32-33
   of manufacturing system, 18-19, 22
   of production, 19, 22, 119-129
Plans —
   drafting, 91, 98-100
   in manufacturing, 60-64
Plastic —
   bonding, 343, 346
   boring holes in, 318, 321
   clamping, 398, 399
   fracturing, 368
   sanding, 359, 362-363
   sheet vinyl, 398, 424
Plastic casting, molding, and forming, 380-384
Plastics, 43, 44, 420-424
   characteristics and uses (table), 421
   cutting, 314, 315
Plastisols, 423
Pliers, 306-307
   locking, 40
Plug gage, 403
Plywood, 44, 414
Polishing, of wood and metal, 364
Polyester resin, casting, 380
Polyesters, 422
Polyethylene, 382, 424
Polystyrene, 382
   forms, 422-423
Polyvinyl acetate, 343
Portable belt sander, 360-361
Portable finishing sander, 360
Portable power tools, 39, 40
Portable spot welder, *illustrated,* 355
Potter, Humphrey, 29
Power machines and equipment, 39, 41-42
Power sanding, 360
Power saws, 314-317
Power tools, portable, 39, 40
Price, determining selling, 159
Primary manufacturing, *defined,* 145
Primer, *defined,* 390
Printing, silk-screen, 395-396
Problem, stating, step in design, 72, 87
Procedure chart, 61
   making, 101
Procedures, used in production, 22-23
Processes, 15
   used in manufacturing, 46-51
Processing, *defined,* 38
Producibility, and design decision, 90
Producing products, 145-157
Product —
   final, selection of, 86
   selecting for class, 117
Product analysis, 124
Product choice, limitations in, 84-86
Product ideas, 80-83
   analysis of, 124

Production —
   activities for, 158
   control of, 153
   planning of, 19, 22
   preparing for, 143-144
   types of, 15-17, 30, 31-32, 119, 121
Production department 110, 114
Production flow chart, 60, 61, 124
   illustrated, 125
Production planning, 119-129
   activities for, 130-131
Production procedure, 22-23
   illustrated, 25
Production run, 150-151, 153
Production system —
   choosing, 60
   planning, 119-129
   and trial run, 149-150
Production time, estimating, 133-134
Production tooling, 403-409
Productivity, *defined,* 145
Product plans, understanding, 60-64
Product selection, 80-86, 94
Products —
   choosing for school shop, 60
   and company limitations, 84-86
   designing and engineering, 19, 87-94
   production of, 145-157
Profit —
   defined, 104
   figuring, 165, 168-169
Proprietorship, *defined,* 106
Prototype, 124
   in engineering process, 93-94
   making, 101
Punch, used in layout, 303-304
Punch and die, 139
   defined, 22
Punches —
   see *also* name of punch
   used for shearing, 310
Push shoe, on jointer, 325
Push stick, on jointer, 325
Pyrometer, 374

Qualities of successful workers, 52-54
Quality control, 22, 146
Quenching, 388, 389

Rasp, 358
Raw materials, 5, 145, 146, 154
   defined, 22
Records, importance of keeping, 111-112, 165
Recycling, 36
   defined, 35
References, and job interview, 141
Relief cuts, *illustrated,* 315
Rendering, *illustrated,* 88, 96
Research, 72
   illustrated, 160
   marketing, 159, 161
Research and development, 70-77, 79
   activities for, 96-118
   careers in, 74

Research and development
  department, 110, 113
  (chart,)95
Resources, saving natural, 36
Resin —
  casting, 380
  polyester, 422
  used to laminate fiber glass, 381
Resist, applying, *illustrated*, 368
Resistance welding, *see* spot welding
Retail outlets, 157
Retooling, 137
Retraining, *defined*, 84
Ripping, of wood, 314, 316, 317
Riveter, *illustrated*, 42
Rivets, 338-340
Rods —
  brass, 419
  grinding to angle, *illustrated*, 338
  used in welding, 352, 356
Rolling, *defined*, 48
Rotary punch, 310
Rough sketches, 88, 96
Router, 323-324
Rule, using, 300

Safety, 54-59
  when bending metal, 379
  when casting, 374, 380
  with ceramic materials, 428
  and design decision, 90
  in drilling, 319, 322
  when etching, 368, 369
  when filing and sanding, 359, 360, 361, 362, 363
  when finishing, 395, 397
  when grinding, 366
  when heat treating metal, 388, 389
  with lathes, 330
  when laminating, 381
  with metalworking lathe, 333
  when molding, 382
  with paper cutter, 407
  with plastics, 422, 423
  and saws, 311, 314, 316, 317, 405
  when soldering, 348
  when thermoforming, 384
  when using adhesives, 344, 346
  when using chisels, 323
  when using jointer, 326
  when using nails, 342
  when using planers, 327
  when using router, 324
  when using Uniplane®, 327
  when welding, 350, 351, 352, 353, 355, 356, 357
Safety zone, *defined*, 58
Sales, *see also* Marketing
Sales, *illustrated*, 160
Sales department, 110, 114
Sales forecast, 159, 161
Sales manager, 164-165
Sales representatives, 165
Samples, product, 84

Sampling, in inspection process, 149
Sand casting, 369-374
Sanding, 359-365
  jigs and fixtures for, 404-406
Sanding block, 359
Sanding drum, 364
Sand mold, preparing, 370-372
Sandpaper, *see* abrasive paper
Sawing,
  jigs and fixtures for, 404-406
Saw kerf, 405
Saws, 311-317
Schedule, for assembly line, 153
Schematic drawings, 91
Scissors, 308
Scoring, *illustrated*, 367
Scratch awl, used in layout, 303
Screwdriver, 246-247, 306
Screws, 336
  hand, 400
  illustrated, 39
  (table),335
Scroll saw, using, 315
Sealers, *defined*, 392
Seam, in sheet metal, *defined*, 378
Seamer, hand, 377
Separating —
  forming process, 49
  by oxyacetylene cutting, *illustrated*, 354
Separating processes, *illustrated*, 48
Setting up, 139
Setup cost, importance of, 126
Setup gage, *illustrated*, 405
Shaping, of wood, 322-330
Shaping steel, 418
Shares, of ownership, 108, 109
Shearing, 308-311, 407
  defined, 49
  illustrated, 48
Sheet coatings, 424
Sheet metal —
  bending, 377-379
  cutting, 308, 309
  punching holes in, 310
  shaping, 305
  shearing, 407
Sheet plastics, thermoforming, 382-384
Sheets, plastic, 398, 423
Shellac, 392
Shielded metal-arc welding, *see* arc welding
Shipping —
  packages used for, 156
  of products, 165
Shop —
  choosing products for, 60
  safety in, 55-56
Shop layout —
  illustrated, 130, 131
  making, 130-131
Shortage, 133
Silk-screen printing, 395-396
Sketches, 76, 88
  making, 96
  of new product ideas, 124

Slag, 357, 374
Slater, Samuel, 29
Slip-roll former, *illustrated*, 379
Slot, sawing, *illustrated*, 405
Smith Brothers, 155
Smooth plane, 323
Snap fasteners, 340
Snips —
  aviation, *illustrated*, 308
  tin, *illustrated*, 308
Socket, of snap fastener, 340
Softwood, 411, 413
Solar collectors, 36
Solder, 419-420
Soldering, 347-348
Soldering copper, 343
Solvent, *defined*, 343
Space —
  limits product choice, 85
  outer, manufacturing, in, 35
Spacing, of drilled holes, *illustrated*, 408-409
Spain, Jayne Baker, 140
Spangler, Murray, 81-82
Spark lighter, *illustrated*, 351
Specialization, of workers, 32
Spot welding, 355
Spray equipment, *illustrated*, 393
Spray painting, *illustrated*, 390
Square cut, making, 32-33
Squareness, checking for, 302
Squares, used in layout, 302
Squaring shears, 41-42, 309
Squeegee, in silk screening, 396
Squeeze-bottle plastic, *see* Polyethylene
Stains, finish, 390
Stake, *defined*, 377
Stamping, forming process, 48
Stamping leather, *illustrated*, 387
Stamping tools, 385-386
Standard stock, 132-133, 145
  defined, 22
Staples, 342
Stationary belt sander, 362-363
Stationary disk sander, 362
Steam engine, 27, 28
Steel, 44
  alloy, 418
  annealing, 388
  case hardening, 387
  hardening, 388
  shaping, 418
  tempering, 389
Steels, carbon, 416
Steel wool, 364
Stencils, used in screen printing, 395, 396
Stitches, in leather, 341
Stock —
  in ownership of company, 108-109
  standard, 22, 132-133, 145
  turning between centers, 328-329, 332-333
Stockholders, 109
Stop block, *illustrated*, 404, 405
Storage, of products, 154, 156, 157

Straight turning, 332-333
Stratton, Harry, 107
Stretching, forming process, 48
Strip heater, for thermoforming, 382
Strowger, Almon, 81
Styrenes, 422-423
Subassemblies, 23, 145
Suede, 425
Surfacing, of board, *defined*, 327
Surveys, to find consumer demand, 84, 89, 93-94
Symbols, used on flow chart, 124
Synthetic, *defined*, 420
Synthetic materials, *defined*, 44

Table saw, using, 316-317
Tacks, 342
Tang, of bit, *defined*, 318
Tanning, of leather, 424-425
Tape, fusible, 346
Taper, using jig when sawing, 405
Tap drill, sizes of (table), 337
Taps, using, 336
Tap wrench, *illustrated*, 337
Taylor, Frederick, 32
T-Bevel, used in layout, 303
Technologists, manufacturing, 121
Tempering, 389
Templates —
    defined, 403
    used in layout, 303
Tenite®, 423
Terry, Eli, 30
Testing, of finished products, 149
Textiles, 46
    bonding, 345, 346
Thermoforming, of sheet plastics, 382-384
Thermoplastic, 420
    defined, 382
Thermoplastics, 423
Thermosetting plastics, 420
Thickness —
    of leather (table), 425
    measuring, 301
Thickness planer, 327
Thonging chisel, *illustrated*, 341
Threaded fasteners, 51, 334-338
Threading die, *illustrated*, 338
Threading tool, *illustrated*, 331
Three-jaw chuck, 330
    illustrated, 331
Throatless shear, 309
Thumbnail sketches, 88, 96
Time —
    estimating for production, 133-134
    limits product choice, 85
Time-and-motion studies, 33
Timing —
    assembly line, 129, 153
    and trial run, 150

Tips, torch, used in welding (table), 351
Toggle clamps, 409
Tolerance, in measurement, 148
Tooling, production, 403-409
Tooling leather, 426
Tooling-up, 22, 137-138, 139
Tools, 15, 32, 38-42
    bench, 305-308
    for cutting and knurling, 333
    for drilling and boring, 318-322
    for layout, 300-304
    limit product choice, 84-85
    safety when using, 58
Torch —
    in oxyacetylene cutting, 354
    in oxyacetylene welding, 350-353
    for welding, 397
Trademark —
    defined, 155
    selecting for class company, 118
Training, during trial run, 150
Transportation, of goods, *illustrated*, 166
Trial run, 22, 149-150, 158
Triangles, use in drafting, *illustrated*, 98
Try square, *illustrated*, 302
Tubing cutter, 309
Tubing, plastic, 423, 424
Turning —
    of metal on lathe, 330, 332-333
    of wood on lathe, 328-330
Turning chisel, 328-330
Turning tools, *illustrated*, 331
Twist drill, 319
    illustrated, 318, 321

Unions, 29, 141-142
Uniplane®, using, 326-327

Vacuum forming, 382, 384
Varnish, 392
    applying, 392-393
V-block, used in drilling, *illustrated*, 319
Veneer, applying, *illustrated*, 345
Views, showing on working drawings, 99, 100
Vinyls, 423-424
Vise, 58, 402
    for drill press, 319
    used in bending, 376
    used when filing, 358
Vise-Grip®, 40
    see also pliers, locking
Visual inspection, 148

Warehouse, *defined*, 110
Warp, lumber defect, 413
Washers, 336

Water —
    used in etching, 368, 369
    used when grinding, 366
    used for quenching, 389
Watt, James, 27
Wax, 394
Wedge, 39-40
    illustrated, 39
Welding, 51
    arc, 356-357
    electric, 355-357
    oxyacetylene, 352
    spot, 355
Welding equipment, *illustrated*, 350
Welding torch, 350, 351, 352, 353
    used in enameling, 397
Welds, oxyacetylene, *illustrated*, 353
Wheel, *illustrated*, 39
Wholesale outlets, 157
Wholesalers, 165
Width, cutting board to, 311
Wire, cutting, 306, 307, 309, 310
Wire strippers, 307
Wood, 43, 44
    bonding, 343, 344, 345
    boring holes in, 318, 319, 320-322
    clamping, 398-402
    classifications of, 411
    cutting, 311-317
    finishing, 390, 392-394
    gluing, 400-401
    planing, shaping, and turning, 322-330
    polishing, 364
    rough-sawed, 411
    sanding, 359, 360-361, 362-363
    sawing and sanding, 404-406
    sizes, 413
    surfaced, 411
    used for mock-ups, 97
Wood chisels, using, 322-323
Wood fasteners, 342
Wood fillers, 392
Wood lathe, 328-330
Woods, 411, 413-414
    characteristics and uses (table), 412
Wood screws, 336
Workers —
    attitudes of, 52-54
    hiring and training, 139-142
    qualities of successful, 142
Working conditions, in early factories, 28-29
Working drawings, 60-61, 76, 91, 98-100
    illustrated, 72
Work measurement, *defined*, 134
Woven fabrics, 426
Wrench, used in bending, 376
Wrenches, *illustrated*, 307
Wright, Wilbur and Orville, 78
Wrought iron, 416